16th International Conference on Strangeness in Quark Matter (SQM2016)

Journal of Physics: Conference Series Volume 779

Berkeley, California, USA
27 June – 1 July 2016

Editors:

Huan Zhong Huang
Richard Seto
Jochen Thader
Nu Xu

ISBN: 978-1-5108-3683-9
ISSN: 1742-6588

Printed from e-media with permission by:

Curran Associates, Inc.
57 Morehouse Lane
Red Hook, NY 12571

Some format issues inherent in the e-media version may also appear in this print version.

Printed by Curran Associates, Inc. (2017)

For permission requests, please contact the Institute of Physics
at the address below.

Institute of Physics
Dirac House, Temple Back
Bristol BS1 6BE UK

Phone: 44 1 17 929 7481
Fax: 44 1 17 920 0979

techtracking@iop.org

Additional copies of this publication are available from:

Curran Associates, Inc.
57 Morehouse Lane
Red Hook, NY 12571 USA
Phone: 845-758-0400
Fax: 845-758-2633
Email: curran@proceedings.com
Web: www.proceedings.com

Table of contents

Volume 779

16th International Conference on Strangeness in Quark Matter (SQM2016)
27 June to 1 July 2016, Berkeley, USA

Accepted papers received: 30 November 2016
Published online: 8 February 2017

Preface

011001
OPEN ACCESS
16th International Conference on Strangeness in Quark Matter (SQM2016)

011002
OPEN ACCESS
Peer review statement

Papers

012001
OPEN ACCESS
Experimental Overview of Open Heavy Flavor

Kai Schweda.....1

012002
OPEN ACCESS
Heavy flavor Production and Interactions in Relativistic Heavy-Ion Collisions in CMS Experiment

Wei Xie for the CMS collaboration.....7

012003
OPEN ACCESS
Latest Results of Open Heavy Flavor and Quarkonia from the PHENIX Experiment at RHIC

Rachid Nouicer (for the PHENIX Collaboration).....13

012004
OPEN ACCESS
Heavy-Flavor Production at RHIC with STAR

Zhenyu Ye (for the STAR Collaboration).....19

012005
OPEN ACCESS
Highlights from the LHCb ion physics program

Michael Schmelling on behalf of the LHCb collaboration.....25

012006
OPEN ACCESS
Recent development on heavy quark theory in heavy-ion collisions

Shanshan Cao.....31

012007
OPEN ACCESS
Strangeness in ALICE at the LHC

Francesca Bellini for the ALICE Collaboration.....37

012008
OPEN ACCESS
Strangeness in STAR experiment at RHIC

Shusu Shi (for the STAR collaboration).....43

012009
OPEN ACCESS
Collectivity and thermalization in *p*+*p*, *p*+A, *d*+Au and ^{3}He+Au collisions

Shengli Huang.....49

012010
OPEN ACCESS
Λ production in p+p collisions at 40 and 158 GeV/c

Herbert Ströbele for the NA61 collaboration.....55

012011
OPEN ACCESS
Collectivity in small systems: Initial correlations or final state flow?

Björn Schenke.....60

012012
OPEN ACCESS
Hadron yields, the chemical freeze-out and the QCD phase diagram

A Andronic, P Braun-Munzinger, K Redlich and J Stachel.....64

012013
OPEN ACCESS
Experimental Overview of the Search for Chiral Effects at RHIC

Gang Wang.....70

012014
OPEN ACCESS
From gluon topology to chiral anomaly: Emergent phenomena in quark-gluon plasma

Jinfeng Liao.....76

012015
OPEN ACCESS
Lattice QCD results on cumulant ratios at freeze-out

Frithjof Karsch.....82

012016
OPEN ACCESS
Bulk properties of QCD matter from lattice simulations

Claudia Ratti.....88

012017
OPEN ACCESS
Sub-threshold strangeness and charm production in UrQMD

J. Steinheimer and M. Bleicher.....94

012018
OPEN ACCESS
Experimental overview on hadronic resonance production in high-energy nuclear collisions

Yosuke Watanabe.....100

012019
OPEN ACCESS
A new detector at RHIC, sPHENIX goals and status

Rosi Reed for the sPHENIX Collaboration.....106

012020
OPEN ACCESS
Experimental summary for heavy flavor production

Rongrong Ma.....112

012021
OPEN ACCESS
Summary of recent experimental results on strangeness production

Alexander Kalweit.....118

012022
OPEN ACCESS
SQM2016: Theory Summary

Vincenzo Greco.....124

012023
OPEN ACCESS
Open heavy-flavour measurements in p-Pb and Pb-Pb collisions with ALICE at the LHC

Cristina Terrevoli for the ALICE Collaboration.....130

012024
OPEN ACCESS
Heavy-flavour production in pp collisions and correlations in pp and p-Pb collisions measured with ALICE at the LHC

Fabio Colamaria on behalf of the ALICE Collaboration.....134

012025
OPEN ACCESS
Heavy-flavour multiplicity dependence in p-Pb collisions with ALICE at the LHC

Jan Wagner for the ALICE Collaboration.....138

012026
OPEN ACCESS
D^0 Meson Production in Heavy Ion Collisions in CMS experiment

Jing Wang for the CMS Collaboration.....142

012027
OPEN ACCESS
Recent Heavy-Flavor Results Utilizing the FVTX in PHENIX at RHIC

Xuan Li for the PHENIX Collaboration.....146

012028
OPEN ACCESS
$D_s^{\pm}$ meson production in Au+Au collisions at $\sqrt{s_{NN}} = 200$ GeV in STAR

Long Zhou (for the STAR collaboration).....150

012029
OPEN ACCESS
Heavy flavor production in heavy-ion collisions from soft collinear effective theory

Zhong-Bo Kang, Felix Ringer and Ivan Vitev.....154

012030
OPEN ACCESS
Heavy-quark dynamics in heavy-ion collisions

T Song, H Berrehrah, E L Bratkovskaya, D Cabrera, W Cassing, T Tolos and J M Torres-Rincon.....158

012031
OPEN ACCESS
Heavy quark dynamics in QCD matter

S. K. Das, F. Scardina, S. Plumari and V. Greco.....162

012032
OPEN ACCESS
Constraining in-medium heavy-quark energy-loss mechanisms via angular correlations between heavy and light mesons

M Rohrmoser, P-B Gossiaux, T Gousset and J Aichelin.....166

012033
OPEN ACCESS
D-meson observables in Pb-Pb and p-Pb collisions at LHC with EPOSHQ model

V Ozvenchuk, J Aichelin, P B Gossiaux, B Guiot, M Nahrgang and K Werner.....170

012034
OPEN ACCESS
T-Matrix Approach to Strongly Coupled QGP

Shuai Y.F. Liu and Ralf Rapp.....174

012035
OPEN ACCESS
Heavy flavor R_{AA} and v_n in event-by-event viscous relativistic hydrodynamics

Caio A. G. Prado, Jacquelyn Noronha-Hostler, Mauro R. Cosentino, Marcelo G. Munhoz, Jorge Noronha and Alexandre A. P. Suaide.....178

012036
OPEN ACCESS
Charmonium production in pp collisions with ALICE at the LHC

Hugo Pereira Da Costa for the ALICE Collaboration.....182

012037
OPEN ACCESS
Charmonium production in Pb-Pb collisions at $\sqrt{s_{NN}} = 2.76$ and 5.02 TeV with ALICE

Biswarup Paul on behalf of the ALICE Collaboration.....186

012038
OPEN ACCESS
Charmonium production in pp, pPb and PbPb with CMS

Songkyo Lee on behalf of the CMS collaboration.....190

012039
OPEN ACCESS
Excess of J/ψ yield at very low transverse momenta in Au+Au collisions at $\sqrt{s_{NN}} = 200$ GeV and U+U at $\sqrt{s_{NN}} = 193$ GeV with STAR

Wangmei Zha (for the STAR collaboration).....194

012040
OPEN ACCESS
Quarkonium measurements via the di-muon decay channel in p+p and Au+Au collisions with the STAR experiment

Takahito Todoroki (for the STAR collaboration).....198

012041
OPEN ACCESS
Dynamical bottomonium-suppression in a realistic AA background

P-B Gossiaux and R Katz.....202

012042
OPEN ACCESS
$\psi(2S)$ Production at the LHC

Xiaojian Du and Ralf Rapp.....206

012043
OPEN ACCESS
Strangeness production in p–Pb and Pb–Pb collisions with ALICE at LHC

Domenico Colella for the ALICE Collaboration.....210

012044
OPEN ACCESS
Studies of final state interactions via femtoscopy in ALICE

Lukasz Kamil Graczykowski (for the ALICE Collaboration).....214

012045
OPEN ACCESS
Multi-strange hadrons and the precision extraction of QGP properties in the RHIC-BES domain

Jussi Auvinen, Krzysztof Redlich and Steffen A. Bass.....218

012046
OPEN ACCESS
Signatures of Flux Tube Fragmentation and Strangeness Correlations in *pp* Collisions

Cheuk-Yin Wong.....222

012047
OPEN ACCESS
Flavour Oscillations in Dense Baryonic Matter

Peter Filip.....226

012048
OPEN ACCESS
An Application of Functional Renormalization Group Method for Superdense Nuclear Matter

G. G. Barnaföldi, A. Jakovác and P. Pósfay.....230

012049
OPEN ACCESS
Phenomenological Implications of the p_T Spectra of ϕ and Ω produced at LHC and RHIC

Rudolph C. Hwa and Lilin Zhu.....234

012050
OPEN ACCESS
Strangeness at finite temperature from Lattice QCD

J. Noronha-Hostler, R. Bellwied, J. Günther, P. Parotto, A. Pasztor, I. Portillo Vazquez and C. Ratti.....238

012051
OPEN ACCESS
Production of π^0, $K^\pm$ and η mesons in Pb-Pb and pp collisions at $\sqrt{s_{NN}}$ = 2.76 TeV measured with the ALICE detector at the LHC

Astrid Morreale on behalf of the ALICE collaboration.....242

012052
OPEN ACCESS
Measurement of dielectrons in pp, p–Pb and Pb–Pb collisions with ALICE at the LHC

Alberto Calivà on behalf of the ALICE Collaboration.....246

012053
OPEN ACCESS
Measurement of bottom contribution to the non-photonic electron production in p+p collisions at STAR

Wei Li for the STAR Collaboration.....250

012054
OPEN ACCESS
Transverse-momentum spectra of strange particles produced in Pb+Pb collisions at $\sqrt{s_{NN}}$ = 2.76 TeV in the chemical non-equilibrium model

Wojciech Florkowski.....254

012055
OPEN ACCESS
Anisotropic flow measurements in Pb–Pb collisions at $\sqrt{s_{NN}}$ = 5.02 TeV with ALICE

You Zhou for the ALICE Collaboration.....258

012056
OPEN ACCESS
Measurement of higher harmonic flow coefficients of identified hadrons in Pb-Pb collisions at $\sqrt{s_{NN}}$ = 2.76 TeV

Naghmeh Mohammadi for the ALICE Collaboration.....262

012057
OPEN ACCESS
Study of the long-range azimuthal correlations in *pp* and *p*+Pb collisions with the ATLAS detector at the LHC

Krzysztof W. Wozniak on behalf of the ATLAS Collaboration.....266

012058
OPEN ACCESS
Collectivity of (non-)strange hadrons in high-multiplicity pp with CMS

George S. F. Stephans For the CMS collaboration.....270

012059
OPEN ACCESS
Measurement of D^0 elliptic and triangular flow in Au+Au collisions at $\sqrt{s_{NN}}$ = 200 GeV at RHIC

Michael R. Lomnitz and STAR collaboration.....272

012060
OPEN ACCESS
Viscous Damping of Anisotropic Flow in 7.7 – 200 GeV Au+Au Collisions

Niseem Magdy (For the STAR Collaboration).....276

012061
OPEN ACCESS
Insight from elliptic flow of open charm mesons using quark coalescence model at RHIC and LHC energies

Roli Esha, Md. Nasim and Huan Zhong Huang.....280

012062
OPEN ACCESS
Investigating the correlations of flow harmonics in 2.76A TeV Pb–Pb collisions

Xiangrong Zhu, You Zhou, Haojie Xu and Huichao Song.....284

012063
OPEN ACCESS
The physics mechanisms of light and heavy flavor v_2 and mass ordering in AMPT

Hanlin Li, Zi-Wei Lin and Fuqiang Wang.....288

012064
OPEN ACCESS
The fluctuations of quadrangular flow

Giuliano Giacalone, Li Yan, Jacquelyn Noronha-Hostler and Jean-Yves Ollitrault.....292

012065
OPEN ACCESS
A viscous blast-wave model for heavy-ion collisions

Amaresh Jaiswal and Volker Koch.....296

012066
OPEN ACCESS
Directed flow of Λ, $\bar{\Lambda}$, $K^{\pm}$, K^0_S and ϕ mesons from Beam Energy Scan Au+Au collisions using the STAR experiment

Subhash Singha (for the STAR collaboration).....300

012067
OPEN ACCESS
Systematic Searches for the Chiral Magnetic Effect and the Chiral Vortical Effect Using Identi ed Particles at RHIC/STAR

Liwen Wen for the STAR Collaboration.....304

012068
OPEN ACCESS
Vorticity in the QGP liquid and hyperon polarization at the RHIC BES energies

Iurii Karpenko and Francesco Becattini.....308

012069
OPEN ACCESS
Vorticity and Λ polarization in event-by-event (3+1)D viscous hydrodynamics

Long-Gang Pang, Ren-Hong Fang, Hannah Petersen, Qun Wang and Xin-Nian Wang.....312

012070
OPEN ACCESS
Event-by-event generation of vorticity in heavy-ion collisions

Wei-Tian Deng and Xu-Guang Huang.....316

012071
OPEN ACCESS
Identified particle production in pp collisions at $\sqrt{s}$ = 7 and 13 TeV measured with ALICE

R. Derradi de Souza (for the ALICE Collaboration).....320

012072
OPEN ACCESS
Recent hadronic resonance measurements at ALICE

A G Knospe (for the ALICE Collaboration).....324

012073
OPEN ACCESS
Higher Moments of Net-Kaon Multiplicity Distributions at STAR

Ji Xu (for the STAR Collaboration).....328

012074
OPEN ACCESS
Behavior of universal critical parameters in the QCD phase diagram

Marcus Bluhm, Marlene Nahrgang, Steffen A. Bass and Thomas Schäfer.....332

012075
OPEN ACCESS
A particlization procedure for the NχFD model and the evolution of the net-proton kurtosis

C Herold, M Nahrgang, A Limphirat, C Kobdaj and Y Yan.....336

012076
OPEN ACCESS
Fast Dynamical Evolution of Hadron Resonance Gas via Hagedorn States

M Beitel, K Gallmeister and C Greiner.....340

012077
OPEN ACCESS
High baryon densities achievable in the fragmentation regions at RHIC and LHC

Joseph Kapusta and Ming Li.....344

012078
OPEN ACCESS
Analysis of hadron yield data within hadron resonance gas model with multi-component eigenvolume corrections

Volodymyr Vovchenko and Horst Stoecker.....348

012079
OPEN ACCESS
On the Use of the Tsallis Distribution at LHC Energies.

J. Cleymans.....352

012080
OPEN ACCESS
Updates to the p+p and A+A chemical freeze-out lines from the new experimental data

V. V. Begun, V. Vovchenko and M. I. Gorenstein.....356

012081
OPEN ACCESS
Near and Far from Equilibrium Power-Law Statistics

T.S. Biró, G.G. Barnaföldi, G. Bíró and K.M. Shen.....360

012082
OPEN ACCESS
Prototype sector production for the STAR inner TPC upgrade

Chi Yang (for the STAR Collaboration).....364

012083
OPEN ACCESS
A fixed-target programme at the LHC (AFTER@LHC)

Barbara Trzeciak on behalf of the AFTER@LHC study group[1].....368

Journal of Physics: Conference Series **779** (2016) 011001 doi:10.1088/1742-6596/779/1/011001

Preface

The Strangeness and Heavy Flavor in Quark Matter in 2016 (SQM2016) conference was held on the beautiful Clark Kerr Campus of the University of California, Berkeley. SQM2016 conference follows a series of conferences with a long tradition of focusing on special topics of strangeness and heavy flavor physics in nuclear collisions. The most recent two conferences in the series were held at Dubna, Russia (SQM2015) and Birmingham, UK (SQM2014).

The SQM2016 conference has taken place while heavy-ion research programs at both RHIC and the LHC continue to generate exciting results. The completion of the Heavy Flavor Tracker (HFT) for STAR and the VTX for PHENIX as well as the campaign for heavy quark physics at RHIC have ushered in an exciting era for heavy quark physics at RHIC. The experiments at the LHC - ALICE, ALTAS, CMS and LHCb – continue to advance the energy frontier. We are very pleased and grateful that many experimental and theoretical results were presented at the conference, mostly by young physicists who represent the future of our growing field.

We have also used the occasion to dedicate one session for the celebration of scientific achievements made by a pioneer physicist in our community, namely Dr. Grazyna Odyniec. She has made remarkable contributions to measurements of strangeness production and collective motion in nuclear collisions. Moreover, she is also a founding member of the SQM conference series.

We would like to thank the International Advisory Committee members for their efforts in shaping the scientific program of the SQM2016 conference. The Local Organizing Committee includes X. Dong (LBNL), H.Z. Huang (UCLA), V. Koch (LBNL), A. Schmah (LBNL), R. Seto (UCR), J. Thäder (LBNL), J. Thomas (LBNL) and N. Xu (LBNL). The conference was hosted by the Nuclear Science Division, Lawrence Berkeley National Laboratory. We gratefully acknowledge Ms. Lady Bonifacio for her service as a conference coordinator. We thank BNL, CCNU, EMMI, HIC for FAIR, LBNL, NWO, ORNL, UCLA and UCR for their sponsorship of the conference.

The SQM2017 conference will be hosted by University of Utrecht, Netherlands in July 2017.

Huan Zhong Huang (UCLA), Richard Seto (UCR), Jochen Thäder (LBNL) and Nu Xu (LBNL)

Editors

Committees

International Advisory Committee

J. Aichelin (University of Nantes, France)

F. Antinori (INFN Padova, Italy)

F. Becattini (University and INFN Firenze, Italy)

T. Biró (Wigner Institute Budapest, Hungary)

M. Bleicher (University of Frankfurt, Germany)

P. Braun-Munzinger (GSI, Germany)

J. Cleymans (University of Cape Town, South Africa)

G. David (BNL, USA)

S. Esumi (Tsukuba University, Japan)

W. Florkowski (University of Krakow, Poland)

M. Gaździcki (University of Frankfurt, Germany)

H.Z. Huang (UCLA, USA)

V. Koch (LBNL, USA)

Y.G. Ma (SINAP. China)

A. Mischke (Utrecht University, Netherlands)

B. Mohanty (NISER, India)

B. Müller (Duke University, USA)

G. Odyniec (LBNL, USA)

J. Rafelski (University of Arizona, USA)

G. Roland (MIT, USA)

R. Seto (UC Riverside, USA)

A. Sorin (JINR, Russia)

H. Stöcker (University of Frankfurt, Germany)

J. Stroth (University of Frankfurt, Germany)

O. Villalobos Baillie (University of Birmingham, UK)

F. Wang (Purdue University, USA)

N. Xu (LBNL, USA)

P. Zhuang (Tsinghua University, China)

Local Organizing Committee

L. Bonifacio, LBNL (Conference Coordinator)

X. Dong, LBNL

H.Z. Huang, UCLA, Co-Chair

V. Koch, LBNL

A. Schmah, LBNL

R. Seto, UC Riverside, Co-Chair

J. Thäder, LBNL

J. Thomas, LBNL

N. Xu, LBNL, Co-Chair

Peer review statement

All papers published in this volume of *Journal of Physics: Conference Series* have been peer reviewed through processes administered by the proceedings Editors. Reviews were conducted by expert referees to the professional and scientific standards expected of a proceedings journal published by IOP Publishing.

IOP Conf. Series: Journal of Physics: Conf. Series **779** (2017) 012001 doi:10.1088/1742-6596/779/1/012001

Experimental Overview of Open Heavy Flavor

Kai Schweda

Research Division, GSI Helmholtzzentrum für Schwerionenforschung, Darmstadt, Germany

E-mail: kschweda@cern.ch

Abstract. These are the proceedings of the experimental overview of the production of open heavy flavor at the international conference *Strangeness in Quark Matter 2016*. Instead of a comprehensive overview, I focus on a few topics which the reader might find particularly interesting.

1. Introduction

Heavy-flavor particles containing a heavy quark (charm or beauty) are unique probes for studies into QGP bulk properties. Their heavy masses constitute a new scale that is much larger than the QCD scale, $m_c, m_b \gg \Lambda_{\rm QCD}$, making their production cross sections at least in principle accessible to calculations in perturbative QCD. Their masses are also much larger than the maximum initial QGP temperature. Heavy quarks acquire their mass almost entirely from the electroweak sector due to their coupling to the Higgs field. Therefore, they remain heavy even when chiral symmetry is at least partially restored in a QGP. This makes heavy quarks a calibrated probe. Heavy-flavor particles provide access to the degree of thermalization among quarks and gluons in the QGP. Due to their relatively long lifetime, heavy-flavor particles can be identified through their decay topology with the decay vertex being displaced from the collision vertex.

2. Results

The Relativistic Heavy Ion Collider (RHIC) had just completed run 16 on the very day when this presentation was given. Meanwhile, an impressive data set has been collected ranging from polarized protons at up to 510 GeV and systems as large as uranium-uranium collisions. As part of the beam energy scan program, the center-of-mass energy was taken as low as 7 GeV. The PHENIX Collaboration has upgraded its detector with vertexing capabilities at mid- and forward rapidity, enabling the separation of electrons stemming from the semi-leptonic decays of beauty and charm quarks [1]. A ten-fold increased data set is presently being analyzed, which will extend the momentum reach substantially and decrease uncertainties in order to look for potential differences in suppression of beauty and charm quarks in the electron decay channel. The PHENIX Collaboration has also separated prompt J/ψ production from J/ψ production stemming from the decay of beauty hadrons, with the latter being displaced from the primary vertex. The somewhat increased fraction of B $\rightarrow$ J/ψ decays in the overall J/ψ production in Au–Cu collisions when compared to Au–Au collisions is attributed to a smaller level of suppression of beauty hadrons in Au–Au collisions than for J/ψ mesons.

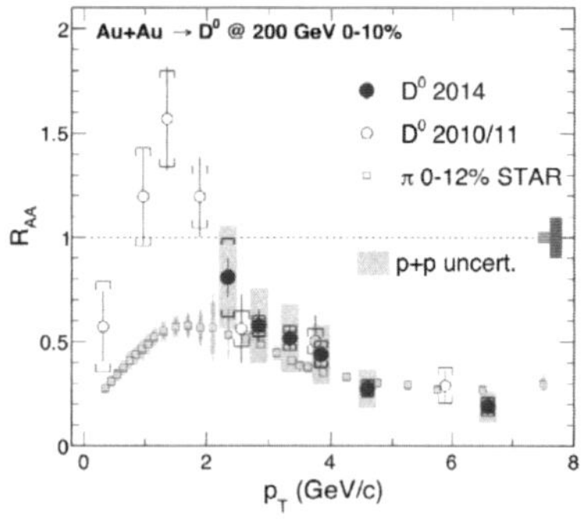

Figure 1. Nuclear modification factor R_{AA} of D^0 mesons in Au–Au collisions at RHIC.

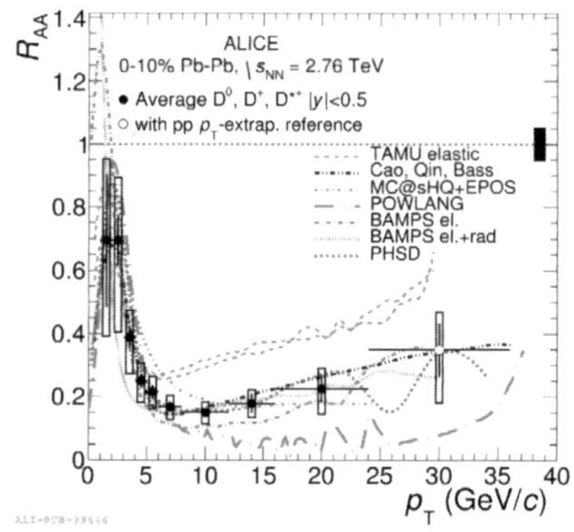

Figure 2. Nuclear modification factor R_{AA} of D mesons in Pb–Pb collisions at the LHC.

The STAR Collaboration is taking advantage of its enhanced vertexing capabilities by means of the Heavy-Flavor Tracker, which is based on monolithic active pixel sensor technology [2]. STAR has recorded some 3 billion Au–Au collisions at 200 GeV, and as of this conference a subset (1/4) of the data has been analyzed. The nuclear modification factor in Au–Au collisions at 200 GeV shows increasingly large suppression of neutral D mesons above 2 GeV/c, while around 1.5 GeV/c, a characteristic bump arises with values above unity, see Fig. 1. Since charm is conserved throughout the evolution of the system, the depletion of the spectrum at large momentum should correspond to an enhancement at lower momentum. In addition, if charm participates in the collective expansion of the medium, charm flow would lead to a depopulation at very low momentum and further enhance the spectrum at intermediate momentum. Both effects would lead to a nuclear modification factor above unity, with a maximum of R_{AA} = 1.2–1.5 around 1.5 GeV/c, as experimentally observed. In order to simultaneously describe the nuclear modification factor and elliptic flow measurement of D mesons at top RHIC energies, model calculations have to take into account charm diffusion with values of $(2\pi T)D$ between 2–11. These values coincide with results from the lattice.

All four large-scale LHC experiments participate in the open heavy-flavor program. This offers the unique opportunity to measure heavy quark production over the entire transverse momentum range from 0 up to 400 GeV/c over a large rapidity range of more than seven units. At low momentum, where most of the production cross sections is located, particle identification of decay daughters is key since topological selection is limited. This is especially true for detecting the short-lived charmed baryon Λ_c^+.

At LHC energies, the ALICE Collaboration has pushed the measurement of D meson production to zero momentum [3] in pp and p–Pb collisions. At mid-rapidity, these D mesons are virtually at rest and topological selection by means of a displaced vertex measurement is no longer applicable. Rather than using vertexing at these ultra-low momenta, ALICE extracts the D meson yield by simply subtracting the large combinatorial background in the invariant mass spectrum of identified pions and kaons. This provides a precise experimental reference of the charm production cross section at mid-rapidity without any model dependence, which serves as an essential input to model calculations describing charmonium production at the phase boundary as a signal for deconfinement. Further measurements are also made in Pb–Pb collisions.

In contrast to RHIC, at LHC energies the nuclear modification factor of neutral D mesons remains substantially below unity at all measured momenta, see Fig. 2. Assuming conservation of charm in the system, i.e. charm quarks only shifting in momentum due to interactions in the

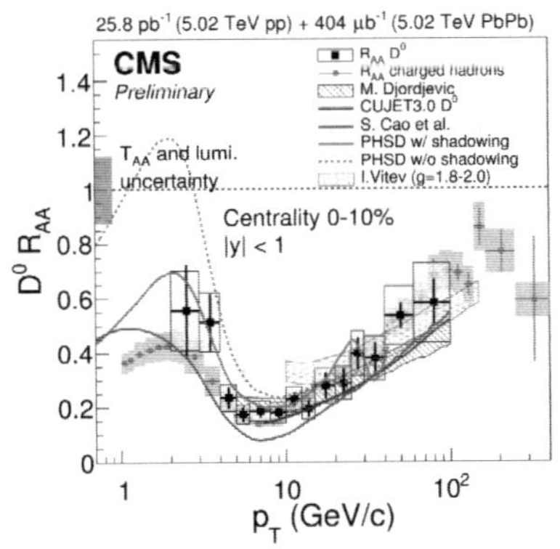

Figure 3. Nuclear modification factor R_{AA} of D^0 mesons in Pb–Pb collisions at the LHC.

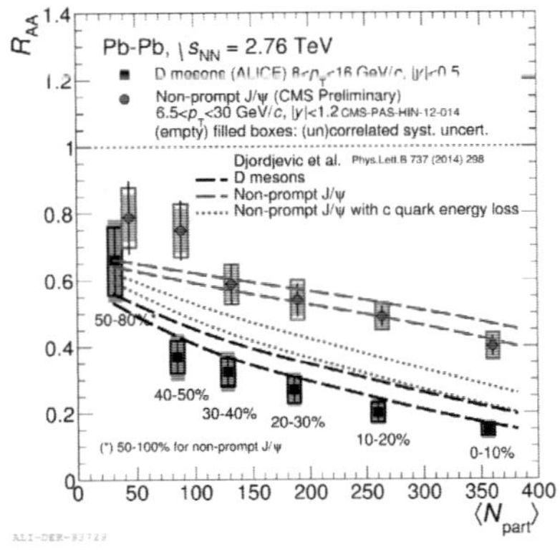

Figure 4. R_{AA} of D mesons and non-prompt J/ψ mesons versus centrality.

medium such as energy loss processes and participating in collective flow, the initial production of charm in Pb–Pb collisions at LHC must be smaller than in pp collisions at the same energy. Transport models assume a reduction in the gluon density in each lead nucleus of up to 20%, resulting in an initial suppression factor of 0.64 in Pb–Pb collisions. Various microscopic models are more or less able to describe the observed modification factor. While the R_{AA} values vary drastically between different models, none of them is able to consistently describe the nuclear modification factor for light- and heavy-flavor hadrons.

The nuclear modification factor of neutral D mesons has been measured up to 100 GeV/c by the CMS Collaboration [4], see Fig. 3. At low momentum, this measurement is restricted to above 2 GeV/c. In the absence of particle identification, all pairs of opposite charge have to be considered as D meson decay candidates, resulting in an overwhelming combinatorial background at low momentum. Additionally, decay daughters from a D meson decay are identified as an anti-D meson when a kaon is mistaken for a pion and vice versa, leading to an irreducible correlated background. Note that the pp reference was recorded over 3 days of data taking, with 10^{12} minimum bias pp collisions being inspected. This large statistical base may allow for other experiments with lower rate capabilities in pp collisions to use reference data from other detectors.

Despite the CMS measurements showing similar values of the R_{AA} for the 10% most central collisions when compared to minimum bias collisions (0-100%), the centrality dependence of the R_{AA} cannot be excluded. If binary collision scaling holds for charm production, the 10% most central collisions contain 40% of all D mesons in the minimum bias sample, and thus both measurements are largely correlated. Indeed, the ALICE Collaboration has already published results showing a distinct centrality dependence. The apparent rise in R_{AA} with transverse momentum is solely due to the shape of the reference pp spectrum [5], and not a particular prediction of any model.

The long predicted mass dependence of beauty and charm suppression has been observed by comparing reconstructed D mesons from ALICE and displaced J/ψ mesons from B mesons in a comparable kinematic range as measured by CMS, see Fig. 4. As has been pointed out at this conference [6], elastic collisions naturally lead to a mass hierarchy in suppression for simple kinematic reasons. It would be interesting to see whether these mass differences to disappear starting at $p/m > 4$, i.e. in a momentum range above 20 GeV/c, where rest masses no longer matter. In order to address the collective flow of beauty quarks, these measurements should also be extended to low momenta.

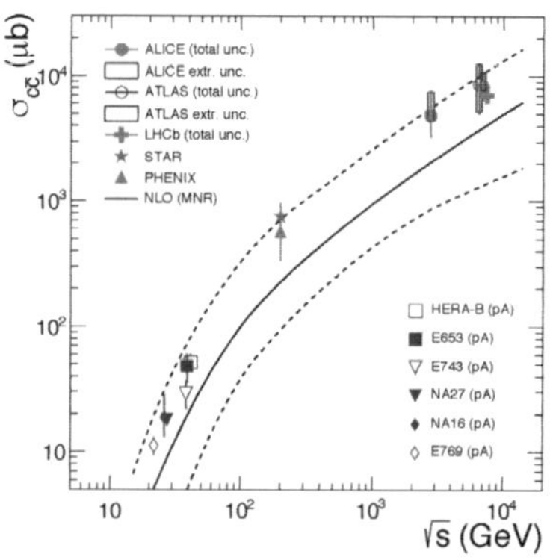

Figure 5. Charm production cross section versus collision energy.

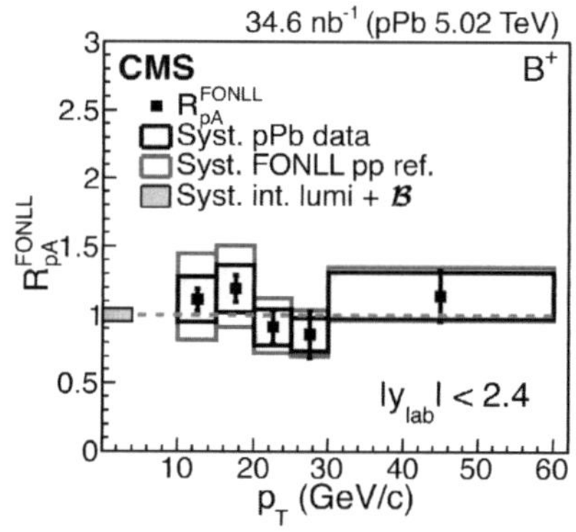

Figure 6. Nuclear modification factor R_{pA} of B^{+} mesons in p–Pb collisions at the LHC.

While ALICE has improved its precision in the charm production cross section by a factor two by measuring down to $p_T = 0$ GeV/c, STAR has superseded its measurement on the total charm production cross section in pp collisions at 200 GeV by using run 12 data with a factor of 10 larger statistics sample, see Fig. 5. The latter value has come down by a factor of two when compared to the first publication by STAR [7]. Interestingly, now all experimental data sit at the upper end of results in perturbative QCD at next-to-leading order within the NMR framework. This might hint to a smaller actual charm quark mass than the central value of $m_c = 1.5$ GeV/c^2 assumed in the calculations. A recent combined next-to-leading order QCD analysis of charm production in deep inelastic electron-proton scattering at HERA using the $\overline{\mathrm{MS}}$ running mass scheme determined a lower charm quark mass of $m_c \approx 1.26$ GeV/c^2. A reduced charm mass would implicitly enhance the cross section of gluons splitting into pairs of charm and anti-charm quarks. A deficit in gluon splitting processes to the production of charm has been reported at Tevatron in the associated production of Z bosons with charm quark jets and photon-tagged heavy-quark jets as well as at the LHC in the measurement of associated charm production in W final states [8].

Reference data from p–Pb collisions serve as a control measurement, allowing the modification of parton distribution functions in the nuclear environment to be disentangled from final-state effects in Pb–Pb collisions. The CMS Collaboration has fully reconstructed B mesons in the hadronic channel with a J/ψ meson in the final state, see Fig. 6. Note, that at present these measurements are kinematically restricted above 10 GeV/c largely due to the kinematical constraint of the muons from the J/ψ decay.

Charmed jets are also now available in p–Pb collisions at the LHC, with comparisons to pQCD calculations showing no deviation within rather large systematical uncertainties, see Fig. 7. Overall in p–Pb collisions, the measurement of D mesons, electrons and muons from the decay of heavy quarks, and heavy-flavor jets stemming from the hadronization of charm and beauty quarks as detected by the CMS Collaboration over a large kinematical range from 50 GeV/c up to 400 GeV/c, all still leave room for nuclear modification of the initial gluon distribution of up to 20%.

ALICE has measured the self-normalized yields of D mesons and electrons from semi-leptonic heavy-flavour hadron decays as a function of the charged-particle multiplicity [9]. The data exhibits a faster-than-linear increase with multiplicity at mid-rapidity for both the D mesons and the electrons from semi-leptonic heavy-flavour hadron decays. This might hint to the development of collectivity in high-multiplicity collisions of small systems. The seed for this

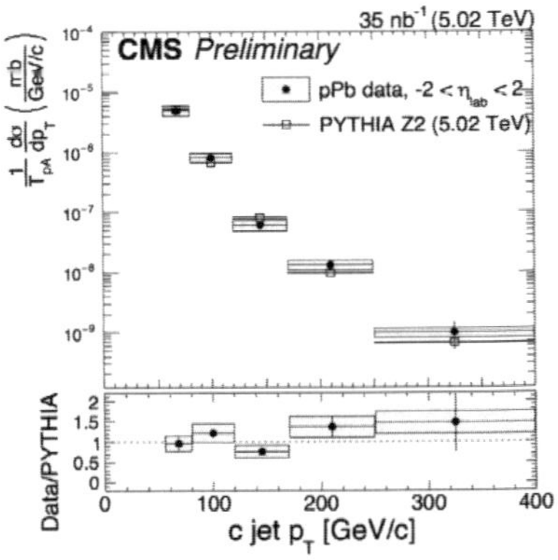

Figure 7. Charmed-jet spectrum in p–Pb collisions at LHC.

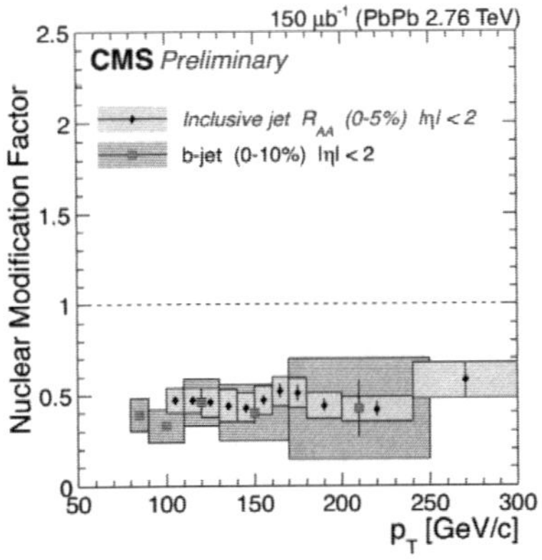

Figure 8. Nuclear modification factor $R_{\rm AA}$ of beauty-jets in Pb–Pb collisions at the LHC.

collectivity seems to be related to multi-parton interactions which become more important with increasing collision energy and are yet to be fully understood.

Beauty jets have been measured in Pb–Pb collision by the CMS Collaboration in the jet momentum range from 80 to 250 GeV/c, showing suppression by a factor of two, see Fig. 8. Beauty jets are identified by a secondary vertex of at least 3 charged particles which is significantly displaced from the collision vertex. An estimate of the secondary vertex mass from reconstructed charged particles, taking missing mass into account, is used to extract the relative contributions of jet flavors, showing that a clean sample of beauty jets is identified at rather high masses. The suppression of beauty jets is comparable with that of inclusive jets, which should be dominated by gluons. With higher statistics in run 2 at the LHC, the color factor of quarks and gluons might become visible as differences in the suppression of those jets.

The LHCb Collaboration operated in fixed target mode by using their SMOG detector, a gas-filled device for precise luminosity measurements, as a gas target which can contain the noble gases ^{4}He, ^{20}Ne or ^{40}Ar [10]. Depending on the initial proton energy, the center of mass energy covers the SPS and RHIC energy range and converts LHCb into a mid-rapidity experiment. Due to the large boost along the beam direction, LHCb can apply precise vertexing even at vanishing transverse momentum and is thus complementary to the RHIC experiments. The feed-down contribution to D mesons stemming from the decay of beauty hadrons is experimentally disentangled by a simultaneous two-dimensional fit of the impact parameter and invariant mass, giving access to prompt open charm production without any model dependence, see Fig. 9. While LHCb has participated in data taken already in run 1 of LHC, it has also recently recorded peripheral to mid-central Pb–Pb collisions.

3. Outlook

Open heavy flavor measurements in ultra-relativistic heavy-ion collisions have come a long way. From the first sighting of single electrons from semi-leptonic decays and reconstructed D mesons at the Quark Matter 2004 conference at Oakland, we now have high-precision vertex capabilities, full reconstruction of D and B mesons and displaced J/ψ mesons, flavor-tagged jets, correlations of heavy flavor with other particles, multiplicity dependence in pp and p–Pb collisions, etc. over an enormous range of collision energy and phase space.

The RHIC experiments are still analyzing their high-statistics data sets and all four large-scale LHC experiments are participating in the heavy-flavor program. The near future should see detection of the elusive Λ_c^+ baryon at both RHIC and the LHC in heavy-ion collisions, to see

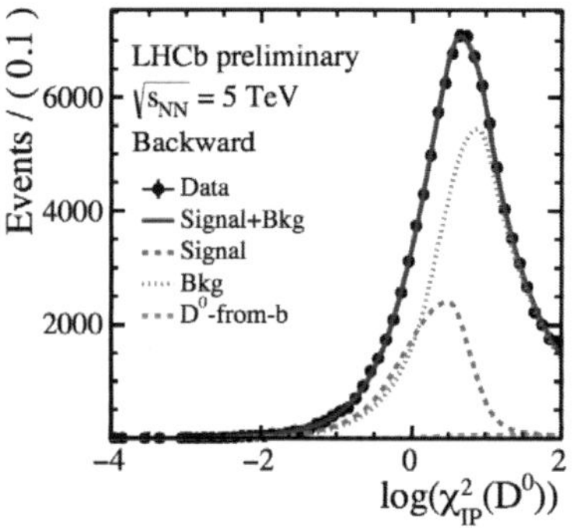

Figure 9. Impact parameter fit of D meson candidates with LHCb.

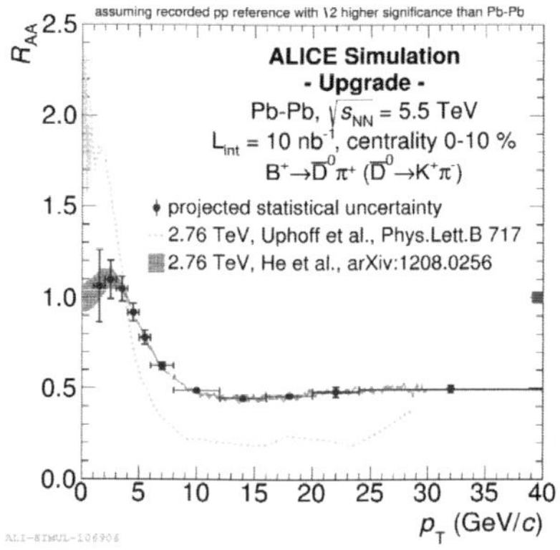

Figure 10. Nuclear modification factor R_{AA} of B$^+$ mesons in Pb–Pb collisions at the LHC.

whether the baryon-over-meson enhancement observed in the light flavor sector is also present in the heavy flavor sector. With the factor ten increase in statistics at the ongoing run 2 at the LHC, initial state effects such as potential modifications of the gluon distribution in the environment of a lead nucleus should be constrained to significantly better than 10%. A serious attempt must be made in order to experimentally constrain the production of charm in central lead-lead collisions to better than 10% in order to provide essential input to all model calculations on J/ψ production. Since charm apparently does not fully equilibrate [11], it gives access to determining transport coefficients as a further characterization of the medium. This necessitates precise measurements of the centrality dependence of heavy-quark energy loss and elliptic flow, at RHIC and LHC. The observed mass hierarchy of heavy-quark suppression between charm and beauty should disappear above 20 GeV/c, if our present understanding is correct. It remains to be seen whether the color factor appears in the different amount of suppression of gluon versus heavy-quark jets.

The ALICE Collaboration will extend full kinematic reconstruction of B mesons down to lowest momenta of 2 GeV/c in the decay channel $B^+ \to D^0 + \pi^+$, see Fig. 10, after the upgrade during the second long shutdown at LHC in 2019/2020. The issue of beauty quark collectivity will be decisively addressed [12] in run 3 and 4, which is presently scheduled to start in 2021.

3.1. Acknowledgments
The author wishes to thank the organizers for their kind invitation and the opportunity to give a presentation at this conference and all speakers for providing their results in advance.

References
[1] Noucier, R. (PHENIX Collaboration), *these proceedings*
[2] Ye, Z. (STAR Collaboration), *these proceedings*
[3] Castillo, J. (ALICE Collaboration), *these proceedings*
[4] Wei, X. (CMS Collaboration), *these proceedings*
[5] Zapp, K., Krauss, F., and Wiedemann, U., JHEP 1303 (2013) 080
[6] Gossiaux, P., *these proceedings*
[7] STAR Collaboration, Phys. Rev. Lett. 94 (2005) 062301
[8] for references, see e.g. Schweda, K., arXiv:1402.1370
[9] Wagner, J. (ALICE Collaboration), *these proceedings*
[10] Schmelling, M. (LHCb Collaboration), *these proceedings*
[11] Greco, V., *these proceedings*
[12] Silvermyer, D. (ALICE Collaboration), *these proceedings*

IOP Conf. Series: Journal of Physics: Conf. Series **779** (2017) 012002 doi:10.1088/1742-6596/779/1/012002

Heavy flavor Production and Interactions in Relativistic Heavy-Ion Collisions in CMS Experiment

Wei Xie for the CMS collaboration
Department of Physics and Astronomy, PURDUE University, West Lafayette, IN
E-mail: wxie@purdue.edu

Abstract. This paper presents the CMS measurements of quarkonia and open heavy flavor production in *p*-*p* , *p*-Pb , and Pb-Pb collisions at $\sqrt{s_{NN}} = 2.76$ and 5.02 TeV. A brief outlook of the near-future CMS heavy flavor physics analyses is provided at the end.

1. Introduction

Measurements of heavy quarkonia and open heavy flavor production may provide further insight in understanding the properties of the strongly coupled Quark-Gluon plasma (sQGP). The quarkonium production is sensitive to the medium's temperature and the state of deconfinement due to different binding energies for different quarkonia. Besides the large suppression of production rate caused by mechanisms like the Debye screening [1], the coalescence process was found to be important for quarkonium production in heavy ion collisions at RHIC and LHC because of the large number of heavy quarks produced [2]. Open heavy flavor measurements not only serve as a good reference in studying quarkonia, but also provide unique inputs in understanding the parton energy loss in the sQGP medium. The interaction between heavy quarks and the medium involves different mechanisms [3–6] from those of light quarks due to the much larger masses and can be studied through measuring open heavy flavor production. Furthermore, cold nuclear matter effects (CNM) [7] can also lead to enhancement or suppression of production yield in heavy-ion collisions. To disentangle all these effects, we need to study a wide variety of species of heavy flavor probes in broad p_T ranges in *p*-*p* , *p*-A , and A-A collisions.

The CMS experiment is a large acceptance, high performance, and multipurpose experiment. In CMS, heavy quarkonium production is measured through dilepton decay channels. The open heavy flavor measurements are done either indirectly through decay daughters or directly through full reconstruction of D, B hadrons, and heavy flavor jets. In this article, Sec. 2 describes the CMS heavy quarkonia measurements including charmonia and bottomonia production in Pb-Pb , *p*-Pb , and *p*-*p* collisions. Sec. 3 presents the measurements of open heavy flavor production, including D^0 , non-prompt J/ψ and heavy flavor jets in in Pb-Pb , *p*-Pb , and *p*-*p* collisions. At the end of this article, a summary and a brief outlook of the CMS heavy flavor physics analyses in the near future is provided.

2. Heavy Quarkonium Measurements

The left panel of Figure 1 shows the prompt J/ψ R_{AA} as a function of N_{part} at high p_T ($6.5 < p_T < 30$ GeV/c) measured within three rapidity regions in 2.76 TeV Pb-Pb collisions [8].

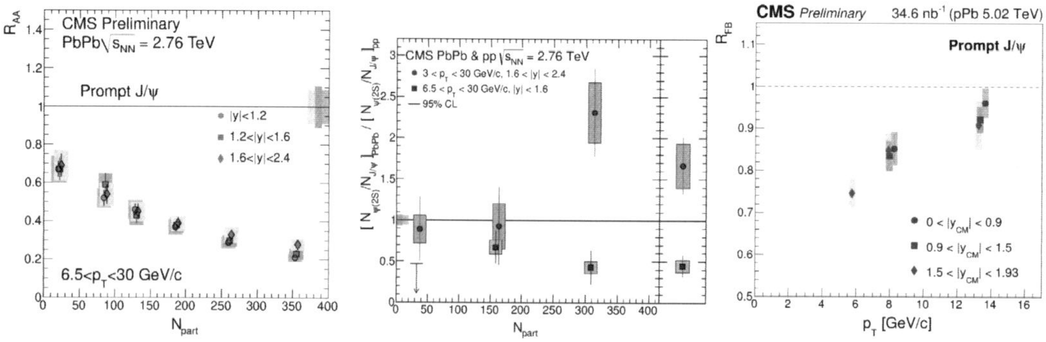

Figure 1. (Left) Prompt J/ψ R_{AA} as a function of N_{part} for $6.5 < p_T < 30$ GeV/c in different rapidity regions. (Middle) Double ratio $[N_{\psi(2S)}/N_{J/\psi}]_{PbPb}/[N_{\psi(2S)}/N_{J/\psi}]_{pp}$ as a function of N_{part} for $3<p_T<30$ GeV/c in $1.6 < |y| < 2.4$ and for $6.5<p_T<30$ GeV/c in $|y| < 1.6$. (Right) Prompt J/ψ R_{FB} as a function of p_T in different y_{CM} regions.

Compared to the low p_T measurements at the same energy [9], high p_T J/ψ is more suppressed. This p_T dependence is opposite to what is observed at RHIC in Au-Au collisions at $\sqrt{s_{NN}} =$ 200 GeV [10] and may suggest interplay among suppression, regeneration from coalescence in sQGP medium and CNM effects in different collision energies. The middle panel shows the double ratio $[N_{\psi(2S)}/N_{J/\psi}]_{PbPb}/[N_{\psi(2S)}/N_{J/\psi}]_{pp}$ as a function N_{part} for two different p_T regions. The results indicate $\psi(2S)$ is less suppressed than J/ψ for low p_T ($3<p_T<30$ GeV/c) and more suppressed for higher p_T region ($6.5<p_T<30$ GeV/c), which is possibly caused by the coalescence production at different time scale during the medium evolution due to different binding energies [11]. The right panel shows the prompt J/ψ $R_{FB} = \frac{d^2\sigma(y_{CM}>0)/dp_T dy_{CM}}{d^2\sigma(y_{CM}<0)/dp_T dy_{CM}}$ as a function of p_T in 5.02 TeV p-Pb collisions, where positive y_{CM} is defined as the proton-going direction. The prompt J/ψ production in $y_{CM} > 0$ region is found to be less than that in $y_{CM} < 0$ region, presumbly due to CNM effects. The R_{pA} of prompt J/ψ is needed and is expected to be ready soon to tell how CNM effects play a role in R_{AA} measurements in Pb-Pb collisions. CMS also measured the coherent J/ψ photoproduction in ultra-peripheral Pb-Pb collisions at $\sqrt{s_{NN}} = 2.76$ TeV [12]. The results are consistent with the leading twist approximation which includes nuclear gluon shadowing.

In Figure 2, the left and middle panels show the $\Upsilon(nS)$ R_{AA} as a function of N_{part} and p_T in 2.76 TeV Pb-Pb collisions, respectively [13]. The R_{AA} of $\Upsilon(nS)$ are found to be ordered with the binding energy of each state for all centralities and for all p_T regions. The suppression in Pb-Pb collisions is mostly from the sQGP medium effect since the double ratio $\frac{\Upsilon(nS)/\Upsilon(1S)]_{PbPb}}{\Upsilon(nS)/\Upsilon(1S)]_{pp}}$ observed in p-Pb collisions is much higher than that in Pb-Pb collisions as shown in the right panel. Contrary to what is observed from the $\psi(2S)$ and J/ψ double ratio measurement at low p_T , $\Upsilon(2S)$ is more suppressed than $\Upsilon(1S)$ for all p_T . These different trends may come from the different roles the regeneration contribution plays in Υ and J/ψ production due to their large difference in mass.

The left panel of Figure 3 shows the ratio of $\Upsilon(nS)$ cross section to $\Upsilon(1S)$ cross section, i.e. $\Upsilon(nS)/\Upsilon(1S)$, in p-p and p-Pb collisions as a function of the charged tracks multiplicity in $|\eta| < 2.4$, where the Υ signals are selected in $|y_{CM}| < 1.93$ [14]. The $\Upsilon(nS)/\Upsilon(1S)$ is found to decrease toward higher multiplicity. Similar trend is observed at RHIC for $\psi(2S)$ and J/ψ production in d-Au collisions [15]. The more suppression of excited states with weaker binding energy indicate final state effects in quarkonia production in high multiplicity p-p and p-Pb or

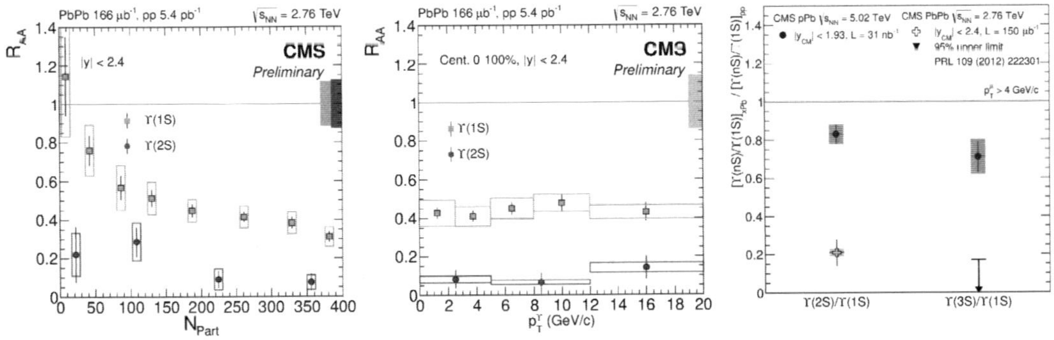

Figure 2. (Left) R_{AA} as a function of N_{part} for $\Upsilon(1S)$ and $\Upsilon(2S)$ in Pb-Pb collisions. (Middle) R_{AA} as a function of p_T for $\Upsilon(1S)$ and $\Upsilon(2S)$ in Pb-Pb collisions. (Right) Double ratio $[\Upsilon(nS)/\Upsilon(1S)]_{PbPb}/[\Upsilon(nS)/\Upsilon(1S)]_{pp}$ for $\Upsilon(2S)$ and $\Upsilon(3S)$ in p-Pb and Pb-Pb collisions.

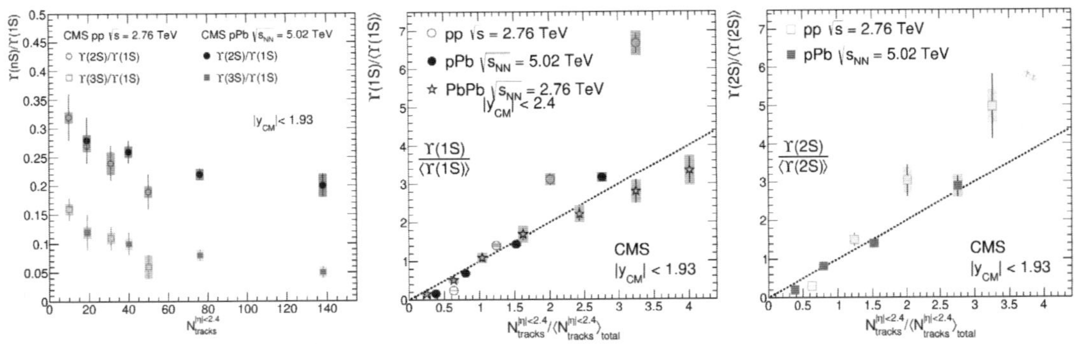

Figure 3. (Left) Ratio $\Upsilon(nS)/\Upsilon(1S)$ as a function of $N_{tracks}^{|\eta|<2.4}$ in 2.76 TeV p-p and 5.02 TeV p-Pb collisions for $\Upsilon(2S)$ and $\Upsilon(3S)$ in $|y_{CM}|$ <1.93. (Middle) Normalized $\Upsilon(nS)$ cross section as a function of the normalized charged track multiplicity for $\Upsilon(1S)$ in 2.76 TeV p-p , 5.02 TeV p-Pb , and 2.76 TeV Pb-Pb collisions for $|y_{CM}| < 1.93$. (Right) Same as middle panel but for $\Upsilon(2S)$.

d-Au collisions.

Since a $Q\bar{Q}$ interact with the surrounding QCD medium before forming a quarkonium state, it is essential to understand how this interaction may affect quarkonia formation mechanism and thus the production rate in high multiplicity p-A collisions. A measurement of quarkonia polarization as a function of charged particle multiplicity can provide the needed clues and CMS took the first step toward that direction. Figure 4 shows measrements of the frame-invariant parameter $\tilde{\lambda}$, in 7 TeV p-p collisions, as a function of charged particle multiplicity in the center-of-mass helicity frame as well as in Collins-Soper frame for the three Υ states in $10 < p_T < 15$ GeV/c and $15 < p_T < 35$ GeV/c [16]. No significant variation of Υ polarization as a function of charged particle multiplicity is observed for the two high p_T bins.

3. Open Heavy Flavor Measurements

Figure 5 shows the prompt D^0 R_{AA} as a function of p_T within $|y|$ <1.0 in 0-100% and 0-10% centrality selection of Pb-Pb collisions at $\sqrt{s_{NN}} = 5.02$ TeV [17]. One significant improvement over the previous measurements at 2.76 TeV [19] is that the reference of the new R_{AA} results

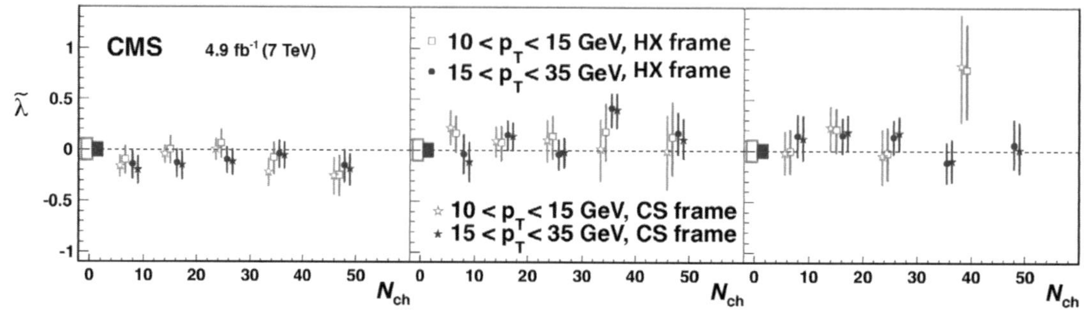

Figure 4. The $\tilde{\lambda}$ parameter for the $\Upsilon(1S)$ (left), $\Upsilon(2S)$ (middle), and $\Upsilon(3S)$ (right) states in the center-of-mass helicity frame (HX) and in Collins-Soper frame (CS) as a function of N_{ch} for p_T = 10-15 GeV/c (open symbols) and 15-35 GeV/c (closed symbols). The boxes at the zero horizontal line represent the global uncertainties.

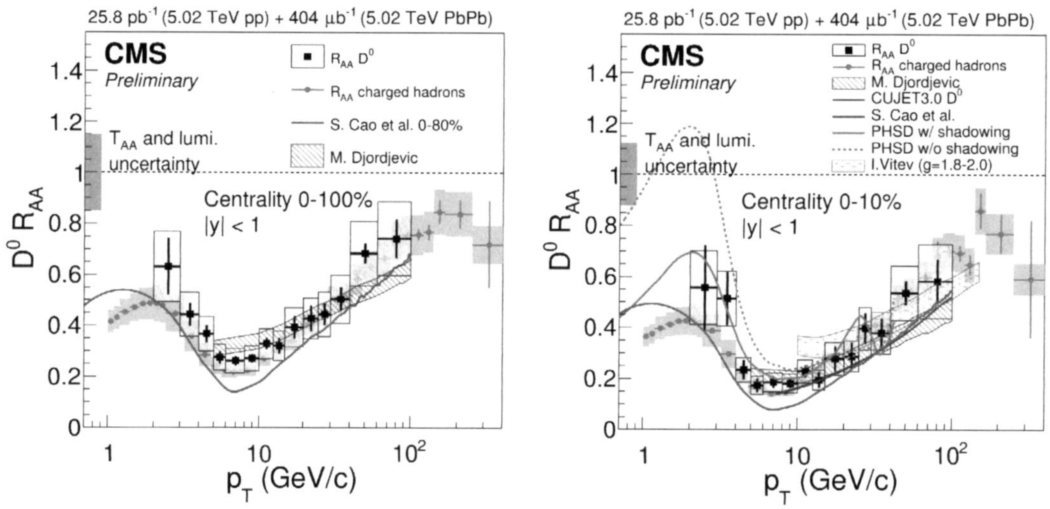

Figure 5. D^0 R_{AA} as a function of p_T in 0-100% (left) and 0-10% (right) centrality selection of Pb-Pb collision at $\sqrt{s_{NN}}$ = 5.02 TeV. The charged particle R_{AA} are shown for comparison. The bands and curves are model predictions.

is from the *p-p* measurement at the same collisions energy. The prompt D^0 R_{AA} is found to decrease with p_T for $p_T < 10$ GeV/c and increase toward higher p_T , which can be qualitatively described by various model calculations [18]. The charged particle R_{AA} is found to be consistent with D^0 R_{AA} within uncertainties for p_T = 2-100 GeV/c despite the much larger charm quark mass.

To further study the mass dependence of energy loss in sQGP medium, the non-prompt J/ψ from B-hadron decays is measured. The left panel of Figure 6 shows non-prompt J/ψ R_{AA} as a function of p_T in 2.76 TeV Pb-Pb collisions in 0-100% centrality region for 6.5 GeV/c $< p_T <$ 30 GeV/c [8]. For non-prompt J/ψ with $p_T > 6.5$ GeV/c , the parent B-hadron p_T is roughly above 8.0 GeV/c . The D^0 R_{AA} measurement with $p_T > 8.0$ GeV/c can thus be compared to the non-prompt J/ψ results. With large uncertainties, the comparison indicates larger non-prompt

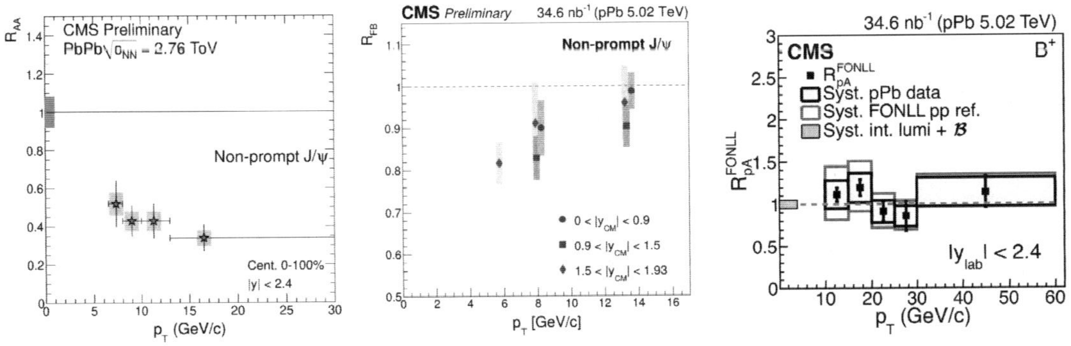

Figure 6. (Left) Non-prompt J/ψ R_{AA} as a function of p_T in 0-100% centrality selection of Pb-Pb collisions at $\sqrt{s_{NN}}$ = 2.76 TeV. (Middle) Non-prompt J/ψ R_{FB} as a function of p_T in three different y_{CM} regions in 5.02 TeV p-Pb collisions. (Right) B^+ meson R_{pA}^{FONLL} in $|y_{lab}| <$ 2.4 as a function of p_T in 5.02 TeV p-Pb collisions.

J/ψ R_{AA} than that of D^0 at $p_T \sim 10$ GeV/c . Further information can be provided when the non-prompt J/ψ at 5.02 TeV becomes available to compare with the more precise D^0 result at 5.02 TeV. The right panel of Figure 6 shows the B^+ R_{pA}^{FONLL} as a function of p_T in 5.02 TeV p-Pb collisions, where B^+ signal is obtained via full reconstruction and the reference is from FONLL calculation [20]. No significant CNM effects are observed within uncertainties in B hadron production. However, the more differential R_{FB} measurements from non-prompt J/ψ in p-Pb shows that the production in the $y_{CM} > 0$ region is less than that in the $y_{CM} < 0$ region at $p_T \sim 6$ GeV/c , as observed in the prompt J/ψ measurements.

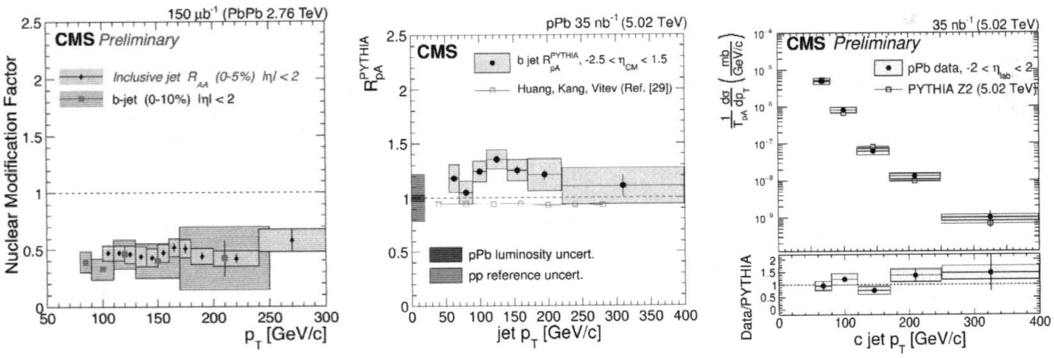

Figure 7. (Left) b-jet R_{AA} as a function of p_T within $|\eta| < 2$ in 0-10% Pb-Pb collisions at $\sqrt{s_{NN}}$ = 2.76 TeV. The measurement of inclusive jet R_{AA} is also shown for comparison. (Middle) b-jet R_{pA}^{PYTHIA} as a function of p_T in 5.02 TeV p-Pb collisions. (Right) c-jet invariant yield scaled by T_{pA} as a function of p_T 5.02 TeV p-Pb collisions. PYTHIA calculation is shown for comparison.

Figure 7 shows the CMS heavy flavor jets measurements. The b-jet reconstructed in $|\eta| < 2.0$ in Pb-Pb collision is more suppressed in central collisions than in peripheral collisions [21]. In the left panel of the figure the R_{AA} of b-jet and inclusive jet are compared and found to be consistent within uncertainties for $p_T > 80$ GeV/c . The middle panel shows the R_{pA} measurements for b-jet as a function of p_T in 5.02 TeV p-Pb collisions [22]. The result indicates that the b jet suppression measured in Pb-Pb collisions, is not due to CNM effects. In the right panel of the figure, the invariant yield of the first charm-jet measurement normalized by T_{pA} as a function of p_T is compared with PYTHIA [23]. As in the case of b-jet production, the data/PYTHIA ratio indicates no significant CNM effect in charm-jet production.

4. Summary and Outlook

The suppression of the three Υ states is found to be ordered, i.e. $R_{AA}\,(\Upsilon(1S)) > R_{AA}\,(\Upsilon(2S)) > R_{AA}\,(\Upsilon(3S)$, for all centrality and p_T regions. This ordering is consistent with the expectation that a quarkonium state with weaker bounding energy is less likely to survive in the sQGP medium. However, the $\psi(2S)$ is found less suppressed than J/ψ in $3 < p_T < 30$ GeV/c , which may be caused by the coalescence production at different time scale during the medium evolution. The dependence of Υ production on charged particle multiplicity indicates final state effects in p-Pb collisions.

The prompt D^0 R_{AA} is consistent with charged particle R_{AA} in 5.02 TeV Pb-Pb collisions, with the trend that R_{AA} decreases with p_T for $p_T < 10$ GeV/c but increases toward higher p_T , which can be qualitatively described by many models. There are some indications that non-prompt J/ψ is less suppressed than prompt D^0 in intermediate p_T region, while at high p_T the b-jet R_{AA} is consistent with the inclusive jet R_{AA} within uncertainties. The R_{pPb} measurements suggest that CNM effects do not have a large role in open heavy flavor R_{AA} measurements in Pb-Pb . However, R_{FB} measurements show significant CNM effects in non-prompt J/ψ production in p-Pb collisions.

At the time of writing this paper, many new physics results in CMS are expected to be publicly available soon, part of which include R_{AA} of fully reconstructed B meson and non-prompt D^0 meson, di-b-jet correlation, prompt D^0 v_n. These new measurements should provide further inputs in understanding the heavy quark and sQGP medium interaction.

References

[1] T. Matsui, and H. Satz, Phys. Lett. **B 178**, 416 (1986).
[2] P. Braun-Munzinger and J. Stachel, Phys. Lett. **B 490**, 196 (2000); R.L. Thews *et al.*, Phys. Rev. **C 63**, 054905 (2001).
[3] M. Djordjevic *et al.*, Phys. Lett. **B 632**, 81 (2006).
[4] H Van Hees *et al.*, Phys. Rev. Lett. **100**, 192301 (2008); P.B. Gossiaux *et al.*, Nucl. Phys. **A 830**, 203 (2009); X. Zhao *et al.*, Phys. Rev. **C 82**, 064905 (2010); A. Buzzatti *et al.*, arXiv:1207.6020.
[5] R. Sharma *et al.*, Phys. Rev. **C 80**, 054902 (2009).
[6] W.A. Horowitz *et al.*, J. Phys. **G 35**, 104152 (2008).
[7] K. Tuchin, Nucl. Phys. **A 904-905**, 627c-630c (2013).
[8] CMS Collaboration, CMS PAS HIN-12-014 (2012).
[9] B. Abelev *et al.*, Phys. Lett. **B 734**, 314 (2014).
[10] A. Adare *et al.*, Phys. Rev. **C 84**, 054912 (2011); L. Adamczyk *et al.*, Phys. Lett. **B 722**, 55 (2013).
[11] X. Du, R. Rapp, Nucl. Phys. **A 943**, 147 (2015).
[12] CMS collaboration, arXiv:1605.06966 (2016).
[13] CMS collaboration, CMS PAS HIN-15-001 (2015).
[14] CMS collaboration, JHEP **04**, 103 (2014).
[15] A. Adare *et al.*, Phys. Rev. Lett. **107**, 142301 (2011).
[16] CMS collaboration, arXiv:1603.02913 (2016).
[17] CMS collaboration, CMS PAS HIN-16-001 (2016).
[18] J. Xu *et al.*, JHEP **02**, 169 (2016); M. Djordjevic *et al.*, Phys. Rev. **C 92**, 024918 (2015); Shanshan Cao *et al.*, arXiv:1605.06447v1; T. Song *et al.*, Phys. Rev. **C 93**, 034906 (2016); Z.-B. Kang *et al.*, Phys. Rev. Lett. **114**, 092002 (2015); Y.-T. Chien *et al.*, Phys. Rev. **D 93**, 074030 (2016).
[19] CMS collaboration, CMS PAS HIN-15-005 (2015).
[20] S. Chatrchyan *et al.* (CMS collaboration), Phys. Rev. Lett. **116**, 032301 (2016).
[21] S. Chatrchyan *et al.* (CMS collaboration), Phys. Rev. Lett. **113**, 132301 (2014).
[22] S. Chatrchyan *et al.* (CMS collaboration), Phys. Lett. **B 754**, 59 (2016).
[23] CMS collaboration, CMS PAS HIN-15-012 (2015).

Latest Results of Open Heavy Flavor and Quarkonia from the PHENIX Experiment at RHIC

Rachid Nouicer (for the PHENIX Collaboration)

Department of Physics, Brookhaven National Laboratory, Upton, New York 11973, USA

E-mail: rachid.nouicer@bnl.gov

Abstract. The PHENIX Collaboration carries out a comprehensive physics program which studies heavy flavor production in relativistic heavy ion collisions at RHIC. The discovery at RHIC of large high-p_T suppression and flow of electrons from heavy quarks flavors have altered our view of the hot and dense matter formed in central Au+Au collisions at $\sqrt{s_{NN}}$ = 200 GeV. These results suggest a large energy loss and flow of heavy quarks in the hot, dense matter. In recent years, the PHENIX has installed a silicon vertex tracker both in central rapidity (VTX) and in forward rapidity (FVTX) regions, and has collected large data samples. These two silicon trackers enhance the capability of heavy flavor measurements via precision tracking. This paper summarizes some of the latest PHENIX results concerning open heavy flavor and quarkonia production as a function of rapidity, energy and system size.

1. Physics Motivation

Understanding the properties of strongly interacting matter at high temperatures has been a central goal of many researchers' numerical simulations of lattice QCD. Recent results from lattice QCD calculations were presented in Refs. [1, 2, 3]. wherein the energy density is normalized by temperature (T^4) as a function of temperature for a system of n_f = 2 + 1 flavors of dynamical quarks. These calculations reveal a rapid increase in the number of degrees of freedom associated with this deconfinement of quarks and gluons from the hadronic chains. The transition point is at a temperature $T_c = 154 \pm 9$ MeV and energy density ϵ_c = 0.18 to 0.5 GeV/fm^3 [3]. Therefore, scientists have experimentally searched for the signatures both of QGP formation and the in-medium effects of hadron properties. For instance, the Relativistic Heavy Ion Collider (RHIC) was designed as a heavy ion machine, able to provide collisions over a large range of energies. We have attained collisions of gold-gold (Au + Au), copper-copper (Cu + Cu), uranium-uranium (U + U), copper-gold (Cu + Au), deuteron-gold (d + Au), helium-gold (^{3}He + Au), proton-gold (p + Au), proton-Aluminum (p + Al) and proton-proton $p + p$ at center of mass energies from $\sqrt{s_{NN}}$ = 7.7 to 200 GeV. Figure 1 summarizes the achievements of RHIC accelerator over the years [4]. Researchers at RHIC have made a major physics discovery, namely the creation of a new form of matter in high-energy central gold–gold collisions. This matter, dense and strongly interacting, is called the strongly coupled quark-gluon plasma, or sQGP. This finding of a new form of matter was published in four independent white papers [5] from the four RHIC experiments. Without a doubt, the physics program at RHIC has enabled remarkable advances in the study of hot strongly interacting matter [5, 6, 7]. The extended particle momentum range at RHIC allowed the use of hard probes to study the behavior of the created medium that was difficult to access at lower collision energies. The hard penetrating probes now such as open heavy flavor and quarkonia production are one of the major experimental tools successfully exploited by RHIC and LHC collaborations [5, 8, 9].

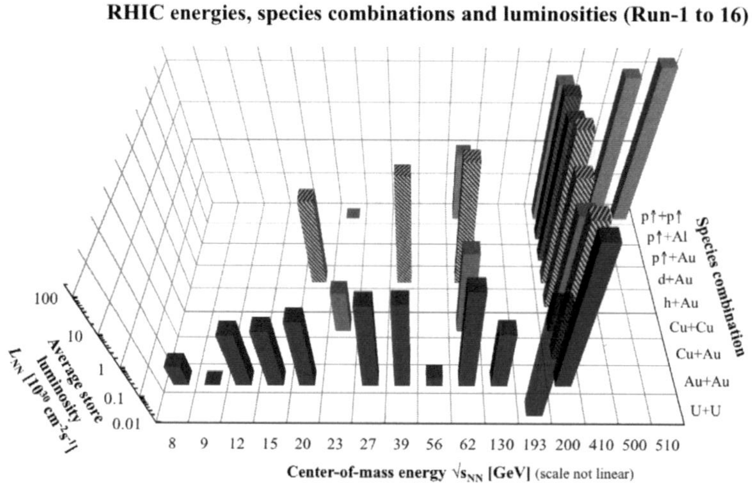

Figure 1. RHIC operating modes and total integrated luminosity delivered to all experiments for each run [4].

2. Recent Open Heavy Flavor and Quarkonia Results

The measurement of inclusive hadron yields in central Au+Au collisions at RHIC led to the discovery of the suppression of hadron production at large transverse momenta (p_T) compared to $p + p$ collisions [5]. This is generally attributed to the energy loss of light partons in the dense nuclear matter created at RHIC. Heavy quarks, i.e. charm and beauty, are believed to be mostly created in initial hard scattering processes of partons [10] and thus are excellent probes of the hot and dense matter formed in nucleus−nucleus collisions at high energy. Adding to this discovery, two very striking results were seen for open heavy flavor from the PHENIX experiment via the measurement of electrons from the semi-leptonic decays of hadrons carrying charm or bottom quarks [7, 11, 12, 13]. First, heavy mesons, despite their large mass, exhibit a suppression at high transverse momentum compared to that expected from $p + p$ collisions. This suppression is similar to that of light mesons, which implies a substantial energy loss of fast heavy quarks while traversing the medium. Second, an elliptic flow is observed for heavy mesons, electrons from semi-leptonic decays of hadrons carrying charm or bottom quarks, that is comparable to that of light mesons like pions. This implies that the heavy quarks in fact are sensitive to the pressure gradients driving hydrodynamic flow giving new insights into the strongly coupled nature of the QGP fluid at these temperatures. However, the interpretation of heavy ion collision data is complicated by the fact that heavy quark production is modified in a nuclear target by cold nuclear matter (CNM) effects. These include modification of the parton density functions in a nucleus, parton energy loss in CNM, and (for quarkonia states), breakup due to collisions with nucleons in the target. To disentangle such effects, multiple measurements were made, such as collisions of different ions and at different collision energies.

Heavy quark production has been studied by the PHENIX experiment via measurements of single leptons from semi-leptonic decays in both the electron channel at mid-rapidity and in the muon channel at forward-rapidity. In recent years, the PHENIX has installed a silicon vertex tracker both in central rapidity (VTX) and in forward rapidity (FVTX) regions, and has collected large data samples. These two silicon trackers enhance the capability of heavy flavor measurements via precision tracking.

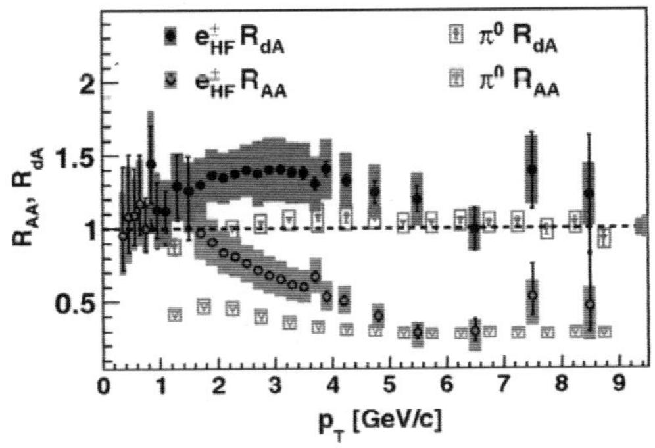

(a) $e^{\pm}_{HF}$ and π^0: Au + Au vs. d + Au at $\sqrt{s_{NN}}$ = 200 GeV

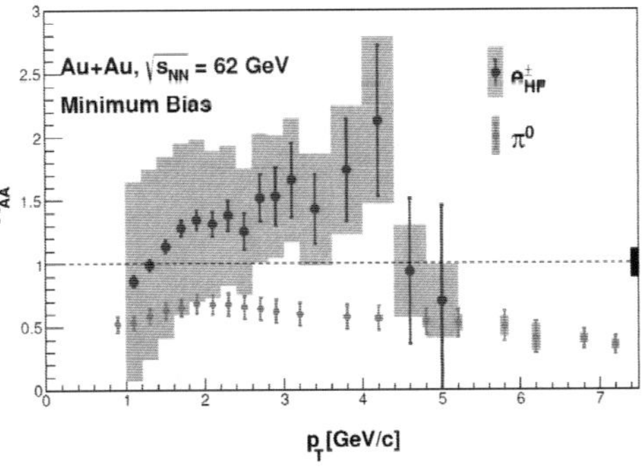

(b) $e^{\pm}_{HF}$ and π^0: Cu + Cu at $\sqrt{s_{NN}}$ = 62.4 GeV

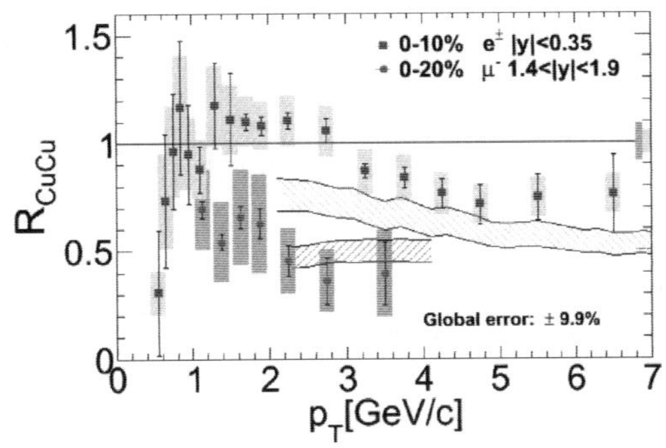

(c) $e^{\pm}_{HF}$ and μ^{-}_{HF}: Cu + Cu at $\sqrt{s_{NN}}$ = 62.4 GeV

Figure 2. The nuclear modification factors $R_{dAu}(p_T)$ and $R_{AA}(p_T)$ for the π^0, $e^{\pm}_{HF}$ and μ^{-}_{HF}: panels (a) and (b) for minimum-bias d + Au, Au + Au and panel (c) for central collisions Cu + Cu [12, 13].

2.1. Open Heavy Flavor - Single Electron/Muon Measurements

Open heavy flavor production is measured in PHENIX through the measurement of inclusive electrons or muons [14, 15]. These analyses use a cocktail method to remove fake and real electrons/muons from the data sample. i.e. for the electron measurement, the electrons that come from either meson decay or photon conversions are measured and subtracted from the inclusive spectrum and the remainder is attributed to electrons coming from the semi-leptonic decay of D and B mesons. The effects of CNM are expected to be present in the initial state of A + A collisions; however, this CNM enhancement is convolved with the suppressing effects of hot nuclear matter. Figure 2(a) shows $R_{dAu}(p_T)$ and $R_{AA}(p_T)$ for $e^{\pm}_{HF}$ and the neutral pion (π^0), for which only small CNM effects are observed between $e^{\pm}_{HF}$ and π^0 for p_T < 5 GeV/c. Above $p_T \approx$ 5 GeV/c, where the CNM effects on both species, $e^{\pm}_{HF}$ and π^0, are small, their $R_{AA}(p_T)$ values are consistent within uncertainties. However, in the range where CNM enhancement is large for $e^{\pm}_{HF}$ and small on π^0, the corresponding $e^{\pm}_{HF}$ $R_{AA}(p_T)$ values are consistently above the π^0 values. This could suggest that the difference in the initial state cold nuclear matter effects due to the mass-dependent Cronin enhancement is reflected in the final state spectra of these particles in Au + Au collisions, although alternate explanations involving mass-dependent partonic energy loss in the hot medium are not ruled out. To study the interplay between initial-state effects and final-states effects for heavy flavor ($e^{\pm}_{HF}$) productions, PHENIX measured the $R_{AA}(p_T)$ in Au + Au collisions at low beam energy, $\sqrt{s_{NN}}$ = 62.4 GeV. These $R_{AA}(p_T)$ values for $e^{\pm}_{HF}$ in Au + Au collisions at 62.4 GeV are compared to π^0 $R_{AA}(p_T)$ at the same energy as shown in Figure 2(b). At 62.4 GeV, the competition, if present, favors heavy flavor enhancement over suppression. This is consistent with previous results

IOP Conf. Series: Journal of Physics: Conf. Series **779** (2017) 012003 doi:10.1088/1742-6596/779/1/012003

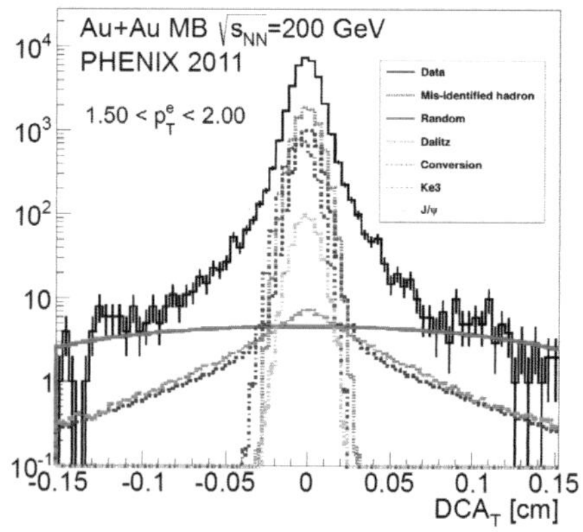

Figure 3. DCA$_T$ distribution for electrons with $1.5 < p_T^e < 2.0$ GeV/c in Au + Au minimum-bias collisions at $\sqrt{s_{NN}} = 200$ GeV. Also shown are the normalized contributions for the various background components detailed in Ref. [19].

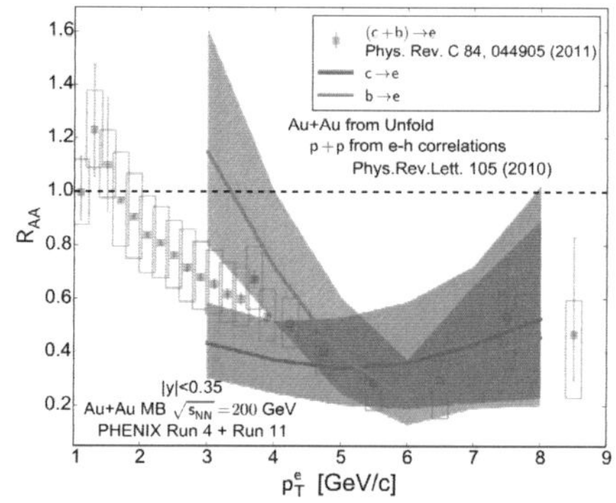

Figure 4. R_{AA} of electrons from charm ($c \to e$), bottom ($b \to e$) decays and combined heavy flavor as a function of p_T^e in Au + Au minimum-bias collisions at $\sqrt{s_{NN}} = 200$ GeV. For more details see Ref. [19].

with hadrons where the Cronin enhancement increases as the collision energy decreases [16]. PHENIX measurements of heavy flavor muons (μ^-_{HF}) at forward rapidity ($1.4 < y < 1.9$) in central Cu+Cu collisions at $\sqrt{s_{NN}} = 62.4$ GeV show a significant suppression [16]. The magnitude of this suppression at forward-rapidity in Cu + Cu (shown in Fig.2(c)) is compared to the suppression of $e^\pm_{HF}$ in central Au+Au collisions at mid-rapidity at the same energy. As open heavy flavor is significantly more suppressed at forward-rapidity than at mid-rapidity in Cu + Cu, additional nuclear effects, such as gluon shadowing at low-Björken-x or partonic energy loss in the nucleus, may be significant. The heavy flavor $e^\pm_{HF}$ and μ^-_{HF} are compared in Fig.2(c) to a theoretical prediction that combines the effects of partonic energy loss, energy loss from fragmentation and dissociation, and includes nuclear matter effects such as shadowing and Cronin enhancement due to parton scattering in the nucleus [17]. While consistent within uncertainties, the model predicts more suppression for heavy flavor electrons than seen in the data.

The large suppression R_{AA} of $e^\pm_{HF}$ at high-p_T shown in figure 2(a) in Au + Au collisions at $\sqrt{s_{NN}} = 200$ GeV, and similar to that of light mesons π^0, is the more striking because perturbative QCD calculations indicate a substantial contribution from bottom quark decays for $p_T > 5$ GeV/c [18]. For the specific purpose of separating the contributions of charm and bottom quarks at mid-rapidity, the PHENIX collaboration has added micro-vertexing capabilities in the form of a silicon vertex tracker (VTX). The different lifetimes and kinematics for charm and bottom hadrons decaying to electrons enables separation of their contributions with measurements of displaced tracks. Figure 3 shows the DCA$_T$ distribution (in transverse plane) for electrons ($e^\pm$) in $1.5 < p_T^e < 2.0$ GeV/c range obtained in Au + Au minimum-bias collisions at $\sqrt{s_{NN}} = 200$ GeV. In order to extract the heavy flavor contributions, all background components, as shown in figure 3, were fully accounted for and their DCA$_T$ shapes as a function of p_T incorporated. The nuclear modifications, $R_{AA}^{c \to e}$, and $R_{AA}^{b \to e}$, for charm and bottom hadron decays respectively are presented in figure 4 and detailed in Ref [19]. We find that electrons from bottom hadron decays are less suppressed than those from charm for the region $3 < p_T < 4$ GeV/c. We note that a significantly larger data set of Au + Au collisions at $\sqrt{s_{NN}} = 200$ GeV was collected in 2014 with an improved performance of the VTX detector. The 2014 Au + Au data coupled with the $p + p$ data taken in 2015 should yield both an important baseline measurement of the bottom electron fraction and a more precise measurement in Au + Au.

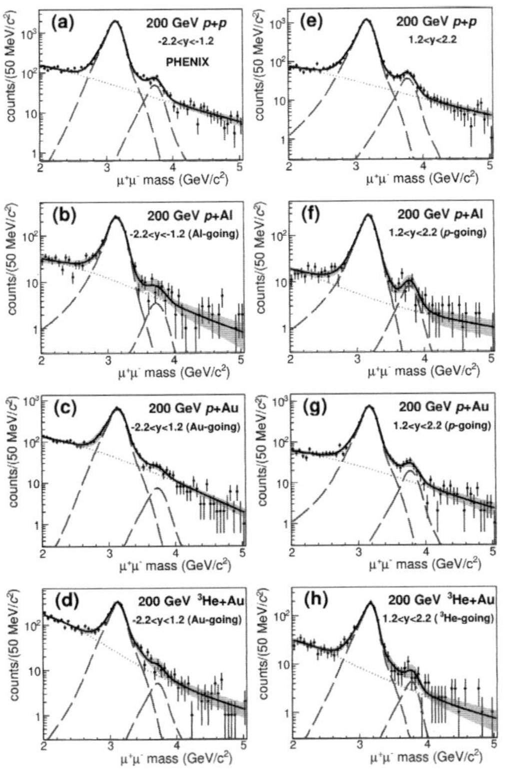

Figure 5. The measured dimuon mass spectra with fits from the (a–d) South and (e–h) North PHENIX muon arms, for $p+p$, $p+$Al, $p+$Au, and ^{3}He + Au collisions at $\sqrt{s_{NN}}$ = 200 GeV. The total fit is the solid [black] line with a shaded [gray] band representing the 90% confidence level. The dashed [blue] and dotted [red] lines represent the contributions from the resonances and background, respectively. For more details see Ref. [21].

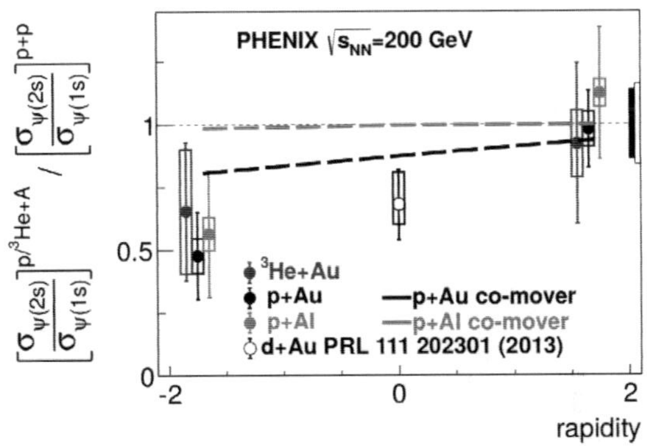

Figure 6. The double ratio of $\psi(2S)/\psi(1S)$ mesons measured in in $p+$ Al, $p+$ Au, and ^{3}He + Au collisions to that same ratio in $p+p$ collisions at $\sqrt{s_{NN}}$ = 200 GeV, with a calculation based on breakup by comoving particles. The bars (boxes) on the data points represent the statistical (systematic) uncertainties, and the shaded (open) box around unity represents the global uncertainty on the forward/backward (mid) rapidity data. For more details see Ref. [21].

2.2. *Quarkonia Measurements - $\psi(1S)$ and $\psi(2S)$ mesons production versus system size*

One way to isolate final-state effects is through studies of states with the same quark content but different binding energies, such as the charmonium states $\psi(1S)$ and $\psi(2S)$, with binding energies of 640 MeV and 50 MeV, respectively [20, 21]. While significant initial-state effects on open charm have been found in $d+$ Au collisions at RHIC, these should equally affect all charm pairs before projection onto a final state. Therefore any differences in the modification of $\psi(2S)$ and $\psi(1S)$ production are likely due to late time effects which are sensitive to differences in the fully-formed meson radius and binding energy. The measured dimuon mass spectra from $p+p$, $p+$ Al, $p+$ Au, and ^{3}He + Au collisions at $\sqrt{s_{NN}}$ = 200 GeV are shown in figure 5. These distributions comprise peaks at the $\psi(1S)$ and $\psi(2S)$ masses (3.1 and 3.7 GeV/c^2, respectively) on top of correlated background from charm and bottom hadron decays and Drell-Yan pairs, plus combinatorial background from light meson decays and hadrons which are not stopped in the absorbers. From figure 5 it is apparent that the $\psi(2S)$ peaks are suppressed relative to the $\psi(1S)$ peak in the columns on the left (in the A-going direction). Quantitative comparisons are accomplished by calculating the double ratio of $\psi(2S)/\psi(1S)$ production in p/He+A collisions to the ratio found in $p+p$ collisions, as shown in figure 6. We see that at forward rapidity, the double ratio is consistent with unity in all three collision systems, indicating that any possible nuclear effects on the two charmonium states are comparable. Because these states are not fully formed until after they exit the nucleus, the fact that any nuclear effects have an equal magnitude on both states suggests that there are no significant final state effects on the pair which occur in this rapidity region. At backward rapidity, the ratios in all collision systems are suppressed by a factor of 2. The mechanism for this preferential suppression of the $\psi(2S)$ relative to the $\psi(1S)$ is expected to occur after the $\psi(1S)$ formation time. A significant difference in the late stages of the collision between this region and forward rapidity is the presence of a larger number of comoving hadrons.

3. Conclusions

RHIC is an amazing heavy ion machine, able to provide collisions over a large range of energies and system size. PHENIX Measurements of heavy flavor made in heavy ion collisions at RHIC have led to a broad and deep understanding of the properties of hot QCD matter. The large suppression R_{AA} of $e^{\pm}_{HF}$ at high-p_T in Au + Au collisions at $\sqrt{s_{NN}} = 200$ GeV, and similar to that of light mesons π^0, is the more striking because perturbative QCD calculations indicate a substantial contribution from bottom quark decays for $p_T > 5$ GeV/c. For the specific purpose of separating the contributions of charm and bottom quarks at mid-rapidity, the PHENIX Collaboration has added micro-vertexing capabilities in the form of a silicon vertex tracker (VTX). We find that electrons from bottom hadron decays are less suppressed than those from charm for the region $3 < p_T < 4$ GeV/c. The PHENIX experiments also established measurements of $\psi(1S)$ (or $\psi(1S)$) and $\psi(2S)$ production as a function of system size, $p+p$, p + Al, p + Au, and ^{3}He + Au collisions at $\sqrt{s_{NN}} = 200$ GeV. We have found that in p/^{3}He + A collisions at forward rapidity, we observe no difference in the $\psi(2S)/\psi(1S)$ ratio relative to $p+p$ collisions. At backward rapidity, where the comoving particle density is higher, we find that the $\psi(2S)$ is preferentially suppressed by a factor of two.

References

[1] S. Borsanyi, G. *et al.*, Journal of High Energy Physics, **11**, 077 (2010).
[2] S. Borsanyi *et al.*, Phys. Lett. B **370**, 99 (2014).
[3] A. Bazavov *et al.*, Phys. Rev. D **90**, 094503 (2014).
[4] `http://www.agsrhichome.bnl.gov/RHIC/Runs/`
[5] I. Arsene *et al.*, Nucl. Phys. A **757**, 1 (2005); B. B. Back *et al.*, Nucl. Phys. A **757**, 28 (2005); J. Adams *et al.*, Nucl. Phys. A **757**, 102 (2005); K. Adcox *et al.*, Nucl. Phys. A, **757**, 184 (2005).
[6] M. Gyulassy and L. McLerran, Nucl. Phys. A **750**, 30 (2005).
[7] R. Nouicer, European Physical Journal Plus **131**, 70 (2016).
[8] G. Aad *et al.* (ATLAS Collaboration), Phys. Rev. Lett. **114**, 072302 (2015).
[9] K. Aamodt *et al.* (ALICE Collaboration), Phys. Lett. B **734**, 31 (2014).
[10] Z. Lin and M. Gyulassy, Phys Rev. C **51**, 2177 (1995).
[11] S.S. Alder *et al.* (PHENIX Collaboration), Phys. Rev. Lett. **96**, 032301 (2006).
[12] R. Nouicer *et al.* (PHENIX Collaboration), Nuclear Physics A **862**, 64 (2011).
[13] R. Nouicer *et al.* (PHENIX Collaboration), Journal of Physics : Conference Series **420**, 012021 (2013).
[14] A. Adare *et al.* (PHENIX Collaboration), Phys. Rev. Lett. **109**, 242301 (2012).
[15] A. Adare *et al.* (PHENIX Collaboration), Phys. Rev. Lett. **112**, 252301 (2014).
[16] A. Adare *et al.* (PHENIX Collaboration), Phys. Rev. C **86** 024909 (2012).
[17] R. Sharma *et al.*, Phys. Rev. C **80**, 054902 (2009).
[18] M. Cacciari *et al.*, Phys. Rev. Lett. **95**, 122001 (2005).
[19] A. Adare *et al.* (PHENIX Collaboration), Phys. Rev. Lett. **93**, 034904 (2016).
[20] H. Satz, Journal of Physics G **32**, R25 (2006).
[21] A. Adare *et al.* (PHENIX Collaboration), arXiv:1609.06550 [nucl-ex].

Heavy-Flavor Production at RHIC with STAR

Zhenyu Ye (for the STAR Collaboration)
University of Illinois at Chicago, Chicago, Illinois 60607, USA
Central China Normal University, Wuhan 430079, China

Abstract. Heavy-flavor particles have been suggested as an excellent probe to study properties of the Quark-Gluon Plasma, a new state of matter discovered in high-energy heavy-ion collisions. In these proceedings, we report a few highlighted results on heavy-flavor production from the STAR experiment, including the first physics results obtained with the newly installed Heavy Flavor Tracker for open heavy flavor and the Muon Telescope Detector for quarkonium studies. These results are compared to results from other experiments and theoretical calculations, and the physics implications are also discussed.

1. Introduction

Through measurements of heavy-flavor production, we can study parton interactions with the Quark-Gluon Plasma (QGP) and thus the properties of the QGP. Measurements of open heavy flavor hadrons and quarkonia at the Relativistic Heavy Ion Collider (RHIC) have revealed very interesting features, such as enhancement (suppression) of open charm hadron production at intermediate (high) transverse momentum (p_T) in Au+Au collisions compared to p+p collisions at $\sqrt{s_{NN}} = 200$ GeV [1], small beam energy dependence of J/ψ suppression integrated over p_T in Au+Au collisions from $\sqrt{s_{NN}} = 39$ to 200 GeV [2]. The former can be described by theoretical calculations with strong charm-QGP interactions as well as charm-light quark coalescence, while the latter can be understood by taking into account both the color screen effect and regeneration of J/ψ from uncorrelated $c\bar{c}$ pairs.

Many progresses have been made on this front, but there are still important questions not fully addressed, such as how parton-QGP interactions depend on the flavor and mass of the partons, and how the properties of the QGP depend on its temperature. More precise measurements at RHIC are needed to answer these questions. The STAR experiment has recently completed data taking with two new detectors for heavy-flavor studies, the Heavy Flavor Tracker (HFT) [3] and Muon Telescope Detector (MTD) [4]. In these proceedings, we report recent results from STAR on heavy-flavor production, including the first physics results from the HFT and MTD.

2. STAR Experiment

STAR is a large acceptance, multi-purpose detector at RHIC designed to study the QGP. Enclosed in a 0.5 Tesla solenoidal magnet are a Time Projection Chamber (TPC), a Time-Of-Flight detector (TOF), and a Barrel ElectroMagnetic Calorimeter (BEMC). The TPC is the main tracking detector for STAR. It also provides particle identification capability through measurement of the specific energy loss (dE/dx). The TOF greatly improves particle identification at low p_T. The BEMC is used to trigger on and identify high-p_T electrons. These detectors cover full azimuth over a large pseudo-rapidity range ($|\eta| < 1$).

Since the beginning of 2014, two new detectors, the HFT and MTD, have been fully integrated into STAR. The HFT consists of four layers of silicon detectors, located inside the TPC. With thinned Monolithic Active Pixel Sensors (MAPS), the HFT provides excellent track impact parameter resolution even at p_T down to 0.2 GeV/c for open heavy flavor measurements. The MTD installed outside of the solenoidal magnet is based on the Multi-gap Resistive Plate Chamber (MRPC) technology. With precise time-of-flight information, the MTD can trigger on and identify muons with $p_T > 1.2$ GeV/c and $|y| < 0.5$. The HFT and MTD have recorded significant amount of data in p+p, p+Au, d+Au, and Au+Au collisions at $\sqrt{s_{NN}} = 200$ GeV in 2014-2016. Results presented below are obtained from around 1/4 of the Minimum-Bias (MB) data with the HFT and 1/2 of the dimuon trigger data with the MTD in Au+Au collisions.

3. Open Heavy Flavor Production

The total charm production cross-sections at mid-rapidity ($|y| < 1$) have been measured by STAR in p+p and Au+Au collisions at $\sqrt{s_{NN}} = 200$ GeV, and are found to scale with the number of binary nucleon collisions (N_{bin}) [1]. This observation supports that heavy-flavor quarks are predominantly produced in initial hard parton scatterings at RHIC. Moreover, because of their relatively long lifetime, heavy-flavor quarks can experience the whole evolution of the system. Therefore, they can serve as calibrated probes to study the QGP. Below, we present results on open heavy flavor production through the hadronic and semi-leptonic decay channels.

3.1. D^0-Meson R_{AA} and v_2

The HFT can significantly improve the signal-to-background (S/B) ratio for open charm hadron reconstruction, leading to greatly reduced uncertainties for the measurements of the nuclear modification factor R_{AA}, which is defined as the ratio between the yield in Au+Au collisions and that in p+p collisions scaled by N_{bin}. Shown in the left panel of Fig. 1 are the D^0-meson R_{AA} in 0-10% Au+Au collisions at $\sqrt{s_{NN}} = 200$ GeV extracted from about 780 million MB events with the HFT (black solid circles). Also shown are the STAR published results from 1.1 billion MB events recorded in 2010 and 2011 without the HFT (red open circles) [1]. The vertical bars (brackets) represent statistical (systematic) uncertainties, while the grey shaded areas p+p reference systematic uncertainties for the R_{AA} obtained from the HFT data. The latter were not present for the published results due to cancellation of efficiency corrections between the p+p and 2010/2011 Au+Au data, both of which did not have the HFT. The sizes of the boxes around $R_{AA} = 1$ in the left panel correspond to global normalization uncertainties for 0-10% Au+Au (left) and p+p (right) collisions. Despite using a smaller data sample, the HFT results have significantly better precision than the published results, confirming the strong suppression of the D^0-meson R_{AA} at high p_T. The HFT has also enabled the first v_2 measurements for open charm hadrons at RHIC. Shown in the middle panel of Fig. 1 are the D^0 v_2 in 0-80% Au+Au collisions extracted from the abovementioned data. The vertical bars (brackets) represent statistical (systematic) uncertainties, while the grey shaded areas estimate of non-flow effects. When plotted as a function of $(\sqrt{m_0^2 + p_T^2} - m_0)/n_q$, where m_0 is the hadron mass and n_q the number of constituent quarks, the D^0 v_2/n_q is below those of light hadrons, but still consistent within uncertainties, suggesting non-zero charm quark flow [5].

The measured R_{AA} and v_2 are compared to theoretical calculations, which use different approaches to describe transportation or interaction of charm quarks in the medium (for details see [6] and the references therein). The TAMU model uses a non-perturbative T-matrix approach with potentials obtained from Lattice QCD calculations to describe two-body interactions. The SUBATECH model uses perturbative QCD to evaluate hard interactions and resummation with Hard Thermal Loop approximation for soft collisions. The DUKE model uses a generalized Langevin approach with transport coefficients tuned to LHC data. As shown in Fig. 1, the

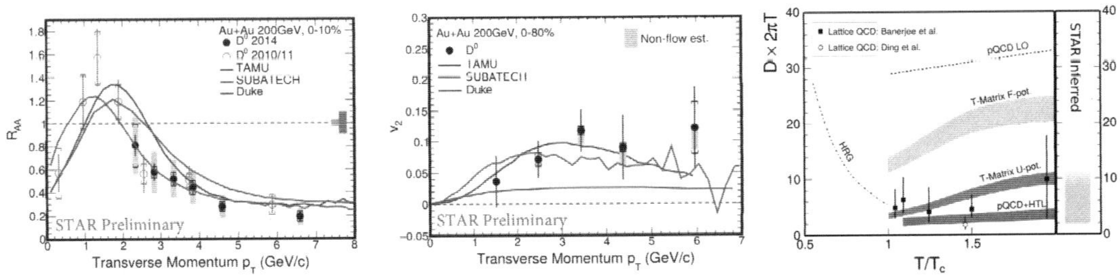

Figure 1. Left (middle): D^0 nuclear modification factor $R_{\rm AA}$ (elliptic flow v_2) in Au+Au collisions at $\sqrt{s_{\rm NN}} = 200$ GeV. Right: charm quark diffusion coefficient. See text for details.

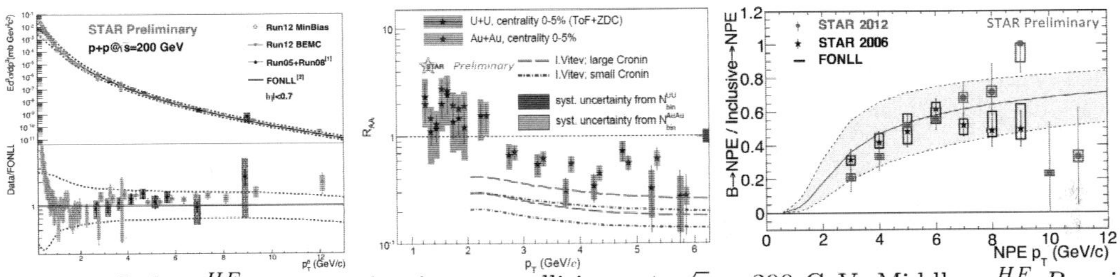

Figure 2. Left: e^{HF} cross-section in p+p collisions at $\sqrt{s} = 200$ GeV; Middle: e^{HF} $R_{\rm AA}$ in 0-5% Au+Au (U+U) collisions at $\sqrt{s} = 200$ (193) GeV; Right: relative contribution of bottom-decayed electrons in p+p collisions at $\sqrt{s} = 200$ GeV. See text for details.

TAMU and SUBATECH models describe both the $R_{\rm AA}$ and v_2, while the DUKE model describes the $R_{\rm AA}$ but underestimates the v_2. The charm quark diffusion coefficient $D \times 2\pi T$ extracted from the TAMU (SUBATECH) model ranges between 3-11 (2-4) for T/T_c = 1-2, shown as the green (blue) band in the right panel of Fig. 1, and agrees with Lattice QCD calculations [7].

3.2. Electrons from Semi-leptonic Decays of Open Heavy Flavor Hadrons

Heavy flavor production can also be studied through semi-leptonic decay channels. Using the data taken in 2010 and 2012, STAR has studied electrons from open heavy flavor hadron decays (e^{HF}) in p+p and Au+Au collisions at $\sqrt{s_{\rm NN}} = 200$ GeV and U+U collisions at $\sqrt{s_{\rm NN}} = 193$ GeV. The results are shown in Fig. 2, where the vertical bars (shaded areas or boxes) represent statistical (systematic) uncertainties. The 2012 p+p results (red circles) have much better precision and wider $p_{\rm T}$ coverage than the STAR published results from data taken in 2005 and 2006 (black circles) [8]. They agree with Fixed-Order-Next-to-Leading-Log (FONLL) predictions [9], with small tension at low $p_{\rm T}$. Using this updated p+p reference, the e^{HF} $R_{\rm AA}$ in 0-5% Au+Au and U+U collisions are found to be consistent with each other. The LANL model using perturbative QCD to calculate parton-QGP interactions underestimates the measured $R_{\rm AA}$.

In order to fully understand the parton-QGP interactions, we need to measure charm and bottom production separately. Shown in the right panel of Fig. 2 are the relative contributions from bottom hadron decays to e^{HF} in p+p collisions at $\sqrt{s} = 200$ GeV. They are determined by fitting the azimuthal correlation between e^{HF} and hadrons in data to the template distributions of charm- and bottom-decayed electrons extracted from PYTHIA, which have different shapes in the near-side peak due to different decay kinematics. Results obtained from the 2012 p+p data are consistent with the STAR published results from the 2006 data [10] and FONLL calculations [9]. Studies are underway to separate charm- and bottom-decayed electrons through their track impact parameters measured with the HFT in heavy-ion collisions.

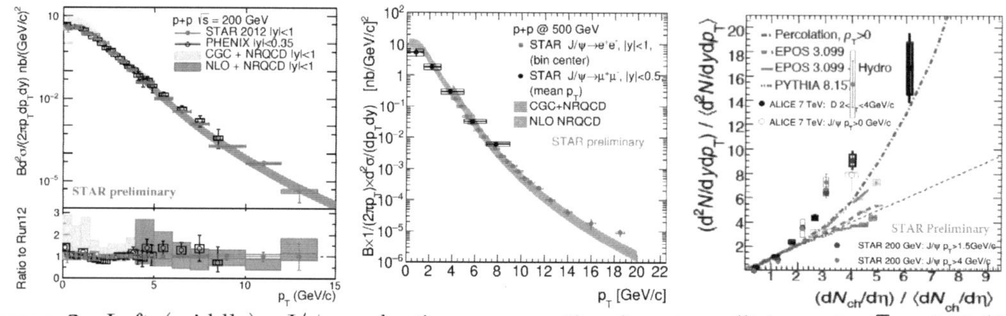

Figure 3. Left (middle): J/ψ production cross-section in p+p collisions at $\sqrt{s}$ = 200 (500) GeV; Right: relative J/ψ and D^0 yield versus charged-particle multiplicity in p+p collisions. The vertical bars (boxes) represent statistical (systematic) uncertainties. See text for details.

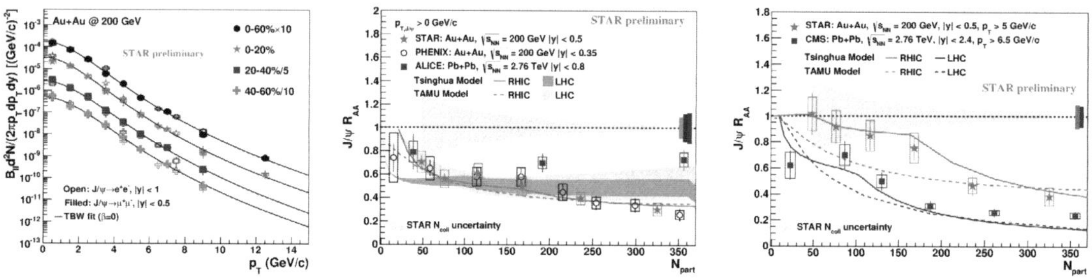

Figure 4. Left: J/ψ invariant yields in Au+Au collisions at $\sqrt{s_{\mathrm{NN}}}$ = 200 GeV; Middle: J/ψ R_{AA} for $p_{\mathrm{T}} > 0$ GeV/c in Au+Au (Pb+Pb) collisions at $\sqrt{s_{\mathrm{NN}}}$ = 200 GeV (7 TeV); Right: J/ψ R_{AA} for $p_{\mathrm{T}} > 5$ (6.5) GeV/c in Au+Au (Pb+Pb) collisions at $\sqrt{s_{\mathrm{NN}}}$ = 200 GeV (2.76 TeV). The vertical bars (boxes) represent statistical (systematic) uncertainties. See text for details.

4. Quarkonium Production

Quarkonia are made of bound heavy quark-antiquark pairs. While the heavy quark production can be calculated by pQCD, how they evolve into quarkonia are not fully known. Precise measurements of quarkonium production in p+p collisions can help understand the quarkonium production mechanisms. In heavy-ion collisions, quarkonia may dissociate due to the color screening of the heavy quark-antiquark potential in the QGP. Such dissociation depends on the temperature of the QGP and varies for different quarkonium states with different binding energies. Therefore, quarkonium production can be used to determine the QGP temperature once the color screening, regeneration, and cold nuclear matter effects are understood. Below, we present results on J/ψ and Υ productions in p+p, Au+Au and U+U collisions.

4.1. J/ψ Production in p+p Collisions

Differential cross-section of J/ψ production as a function of p_{T} has been measured in p+p collisions. Shown in the left panel of Fig. 3 are the STAR results (red circles) obtained from the 2012 data and PHENIX published results [11] (black circles) at $\sqrt{s}$ = 200 GeV. They are consistent with each other, but the STAR results have much better precision at high p_{T}. Shown in the middle panel are the results at $\sqrt{s}$ = 500 GeV from the dimuon (black circles) and dielectron channels (red circles). Both the 200 and 500 GeV results can be described by CGC+NRQCD calculations [12] at low p_{T} and NLO NRQCD calculations [13] at high p_{T}.

We have also studied how J/ψ production depends on charged-particle multiplicity (N_{ch}) in p+p collisions at $\sqrt{s}$ = 200 GeV. As shown in the right panel of Fig. 3, a faster-than-linear increase of relative J/ψ yield versus relative N_{ch} has been observed, suggesting not only correlation between soft and hard interactions, but also stronger suppression on soft interactions than hard interactions or stronger enhancement to hard interactions than soft interactions in high multiplicity p+p collisions. Such an increase is found to be independent of beam collisional energy when comparing these results with those at $\sqrt{s} = 7$ TeV [14]. The observed global trend independent of beam collisional energy is consistent with the Percolation model predication [15], but not with predications by the EPOS3 event generator with hydrodynamic evolution [16].

4.2. J/ψ $R_{\rm AA}$ in Au+Au Collisions

J/ψ invariant yields in Au+Au collisions at $\sqrt{s_{\rm NN}} = 200$ GeV have been determined using the data recorded with the MTD in 2014. The results (solid points) are shown as a function of $p_{\rm T}$ for different centrality bins in the left panel of Fig. 4, and are found to be consistent with the STAR published results from the dielectron channel (open points) [17]. The middle (right) panel shows the $R_{\rm AA}$ for $p_{\rm T} > 0$ (5) GeV/c determined from the MTD data (red stars). Also shown are the results from PHENIX in Au+Au collisions at $\sqrt{s_{\rm NN}} = 200$ GeV (black open circles), as well as the results from ALICE and CMS in Pb+Pb collisions at $\sqrt{s_{\rm NN}} = 2.76$ TeV (blue solid circles). For $p_{\rm T} > 0$ GeV/c, the STAR results are consistent with the PHENIX results, showing a descreasing trend towards more central collisions, while the ALICE results are almost independent of centrality and significantly higher than those at RHIC in the central collisions. Such a difference can be due to larger regeneration contributions to low $p_{\rm T}$ J/ψ at the LHC with much higher $c\bar{c}$ yields. At high $p_{\rm T}$, the $R_{\rm AA}$ at both RHIC and LHC have a descreasing trend towards more central collisions, and the LHC results are lower than those at RHIC. This is likely due to higher energy density and temperature reached at the LHC, leading to stronger J/ψ dissociation. Two different transport models [18, 19], both including dissociation and regeneration, can qualitatively describe the measured $R_{\rm AA}$.

4.3. Υ Production in Au+Au Collisions

Using the MTD data, STAR has studied Υ production in Au+Au collisions at $\sqrt{s_{\rm NN}} = 200$ GeV. Compared to the dielectron channel, the dimuon channel has the advantage of better invariant mass resolution, thanks to the significantly suppressed bremsstrahlung radiation. Shown in the left panel of Fig. 5 are the invariant mass spectra of dimuon pairs. The unlike-sign (red points) spectrum is fitted to the sum of $\Upsilon(1S)$, $\Upsilon(2S)$ and $\Upsilon(3S)$ signal distributions, and $B\bar{B}$, Drell-Yan and combinatorial background distributions. The shapes of the Υ signal distributions are Gaussian functions with means fixed to the PDG values, and widths determined from detector simulations. The shape of $B\bar{B}$ and Drell-Yan background is determined from PYTHIA6.416 [20]. Both the shape and normalization of the combinatorial background are obtained by fitting the like-sign (black points) spectrum. Shown in the right panel are the ratio of the excited Υ state yield with respect to the ground state yield, obtained from the abovementioned fit, the world-wide p+p data, and CMS measurements in Pb+Pb collisions at $\sqrt{s_{\rm NN}} = 2.76$ TeV. An indication of larger $\Upsilon(2S+3S)/\Upsilon(1S)$ ratio at RHIC than that at the LHC is observed.

4.4. Very Low $p_{\rm T}$ J/ψ Production in Au+Au and U+U Collisions

An enhancement of very low $p_{\rm T}$ J/ψ production has been reported in peripheral Pb+Pb collisions at $\sqrt{s_{\rm NN}} = 2.76$ TeV [21]. Such an enhancement has also been observed in peripheral Au+Au (U+U) collisions at $\sqrt{s_{\rm NN}} = 200$ (193) GeV by STAR. Shown in the left panel of Fig. 6 are the invariant mass spectra of dielectron pairs with $p_{\rm T} < 0.1$ GeV/c in Au+Au collisions. A clear J/ψ peak can be seen for the unlike-sign pairs. The middle panel shows the $t = p_{\rm T}^2$ spectrum of J/ψ candidates in the 40-80% centrality bin, together with an exponential function fit within $0.002 < t < 0.015$ (GeV/c)^2. The slope of the fitted exponential function, 199±31 (GeV/c)$^{-2}$, is consistent with the expected value for coherent photon-Au nucleus interactions. The measured

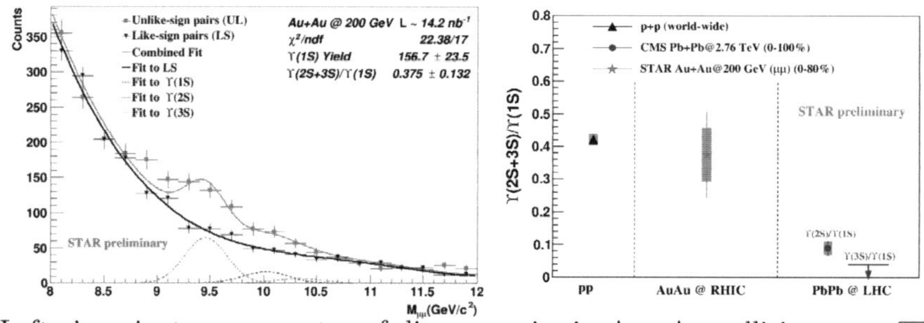

Figure 5. Left: invariant mass spectra of dimuon pairs in Au+Au collisions at $\sqrt{s_{\mathrm{NN}}} = 200$ GeV; Right: ratio of Υ excited state to the ground state measured in p+p collisions, Au+Au collisions at $\sqrt{s_{\mathrm{NN}}} = 200$ GeV and Pb+Pb collisions at $\sqrt{s_{\mathrm{NN}}} = 2.76$ TeV. See text for details.

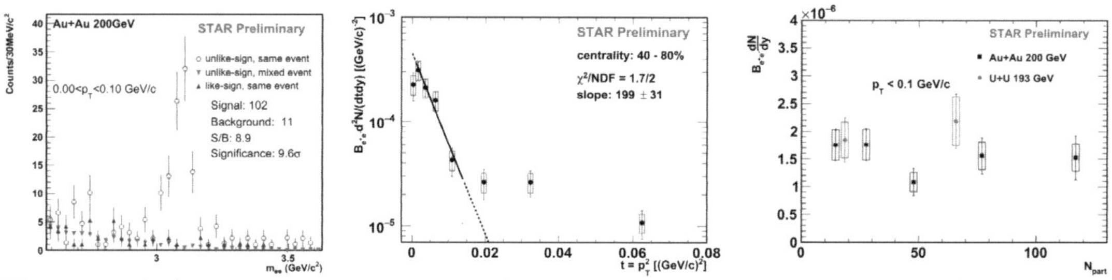

Figure 6. Left: invariant mass spectra of dielectron pairs with $p_{\mathrm{T}} < 0.1$ GeV/c in Au+Au collisions at $\sqrt{s_{\mathrm{NN}}} = 200$ GeV; Middle: p_{T}^2 spectrum of J/ψ candidates in Au+Au collisions at $\sqrt{s_{\mathrm{NN}}} = 200$ GeV; Right: invariant yields of J/ψ production with $p_{\mathrm{T}} < 0.1$ GeV/c in Au+Au (U+U) collisions at $\sqrt{s_{\mathrm{NN}}} = 200$ (193) GeV. See text for details.

J/ψ yield with $p_{\mathrm{T}} < 0.1$ GeV/c is independent of centrality between 30-80% and consistent between Au+Au and U+U collisions, as shown in the right panel. Such observations can not be explained by the current photoproduction models and warrents further investigation.

References

[1] L. Adamczyk *et al.* (STAR Collaboration), Phys. Rev. Lett. **113** (2014) 142301.
[2] L. Adamczyk *et al.* (STAR Collaboration), arXiv 1607.07517 (2016).
[3] Technical Design Report: The STAR Heavy Flavor Tracker (2011).
[4] L. Ruan *et al.*, J. Phys. G: Nucl. Part. Phys. **36** (2009) 095001.
[5] M. Lomnitz (for the STAR collaboration), these proceedings.
[6] A. Andronica *et al.*, arXiv 1506.03981 (2015).
[7] D. Banerjee *et al.*, Phys. Rev. D **85** (2012) 014510; H. Ding *et al.*, J. Phys. G **38** (2011) 124070.
[8] B. I. Abelev *et al.* (STAR collaboration), Phys. Rev. D **83** (2011) 052006.
[9] R. E. Nelson, R. Vogt, and A. D. Frawley, Phys. Rev. C **87** (2013) 014908.
[10] M. M. Aggarwal *et al.* (STAR collaboration), Phys. Rev. Lett. **105** (2010) 202301.
[11] A. Adare *et al.* (PHENIX collaboration), Phys. Rev. D **82** (2010) 012001.
[12] Y.-Q. Ma, and R. Venugopalan, Phys. Rev. Lett. **113** (2014) 192301.
[13] Y.-Q. Ma, K. Wang, and K.-T. Chao, Phys. Rev. Lett. **106** (2011) 042002.
[14] J. Adam *et al.* (ALICE collaboration), J. High Eenergy Physics **09** (2015) 148.
[15] E. G. Ferreiro and C. Pajares, Phys. Rev. C **86** (2012) 034903; arXiv:1501.03381 (2015).
[16] H. Drescher *et al.*, Phys. Rept. **350** (2001) 93; K. Werner *et al.*, Phys. Rev. C **89** (2014) 064903.
[17] L. Adamczyk *et. al.* (STAR Collaboration), Phys. Lett. B **722** (2013) 55; Phys. Rev. C **90** (2014) 024906.
[18] Y. Liu *et. al.* , Phys. Let. B **678** (2009) 72; K. Zhou *et. al.* , Phys. Rev. C **89** (2014) 054911.
[19] X. Zhao and R. Rapp, Phys. Rev. C 82 064905 (2010); Nucl. Phys. A **859** (2011) 114.
[20] T. Sjostrand, S. Mrenna, P. Skands, JHEP **05** (2006) 026.
[21] J. Adam *et al.* (ALICE collaboration), Phys. Rev. Lett. **116** (2016) 222301.

Highlights from the LHCb ion physics program

Michael Schmelling on behalf of the LHCb collaboration

Max Planck Institute for Nuclear Physics, Saupfercheckweg 1, 69117 Heidelberg, Germany

E-mail: Michael.Schmelling@mpi-hd.mpg.de

Abstract. Following the successful participation of LHCb in the 2013 proton-lead run of the LHC, in 2015 the collaboration decided to further extend its physics program to study also lead-lead collisions and fixed target interactions. These proceedings discuss the physics reach of the detector and the first results from the LHCb ion physics and fixed target program.

1. Introduction

Although the QCD Lagrangian is well understood, phenomenological models still dominate the description of strong interactions at small momentum transfers or in the regime of high pressure and density, *i.e.* generally in processes that cannot be handled in perturbation theory. At high energy density a so-called quark-gluon plasma (QGP) is expected to form. Properties of the QGP can *e.g.* be probed by heavy flavour particles [1], where heavy quarks are produced in hard parton-parton interactions and then evolve in the strongly interacting medium. Comparing the particle production properties in central Pb-Pb collisions, where a QGP is expected to form, with those in p-Pb collisions, allows to disentangle genuine QGP signatures from effects of the cold nuclear medium. Those so-called cold nuclear matter effects, such as modifications of the parton densities inside the bound nucleons [2] or energy-loss of heavy particles in the medium [3], can be studied by comparing production characteristics in pp and p-Pb collisions.

2. LHCb detector and physics reach

The LHCb detector [4, 5] is a single-arm forward spectrometer covering the pseudorapidity range $2 < \eta < 5$, designed for the study of particles containing b or c quarks. The detector includes a high-precision tracking system consisting of a silicon-strip vertex detector surrounding the pp interaction region, a large-area silicon-strip detector located upstream of a dipole magnet with a bending power of about 4 Tm, and three stations of silicon-strip detectors and straw drift tubes placed downstream of the magnet. The tracking system provides a measurement of momentum, p, of charged particles with a relative uncertainty that varies from 0.5% at low momentum to 1.0% at 200 GeV/c. The minimum distance of a track to a primary vertex, the impact parameter, is measured with a resolution of $(15 + 29/p_{\rm T})\,\mu{\rm m}$, where $p_{\rm T}$ is the component of the momentum transverse to the beam, in GeV/c. Different types of charged hadrons are distinguished using information from two ring-imaging Cherenkov detectors. Photons, electrons and hadrons are identified by a calorimeter system consisting of scintillating-pad and preshower detectors, an electromagnetic calorimeter and a hadronic calorimeter. Muons are identified by a system composed of alternating layers of iron and multiwire proportional chambers. The online event selection is performed by a trigger, which consists of a hardware stage,

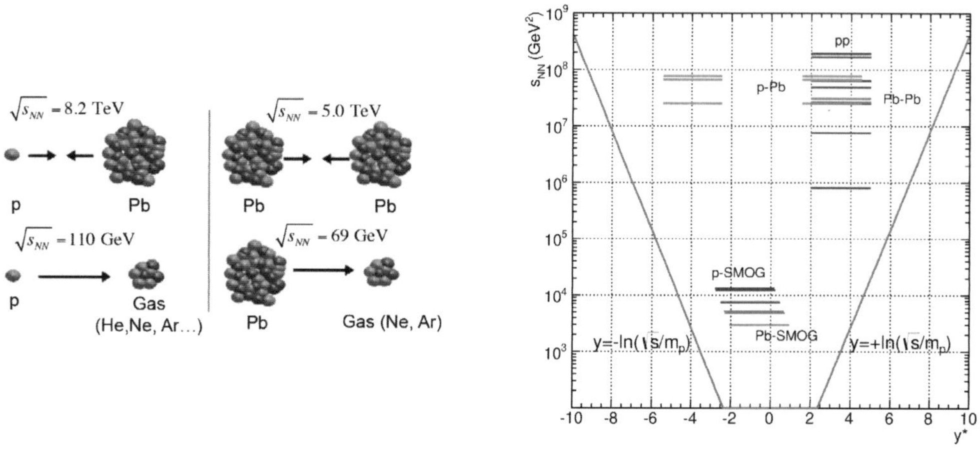

Figure 1. Ion physics running modes and kinematic coverage of LHCb

based on information from the calorimeter and muon systems, followed by a software stage, which applies a full event reconstruction.

Figure 1 (left) shows the heavy-ion operation modes of the LHCb experiment. With colliding beams, p-Pb and Pb-Pb collisions can be studied at nucleon-nucleon centre-of-mass energies in the multi-TeV regime. In addition, the LHCb SMOG system (*System for Measuring Overlap with Gas*) [6] allows to inject small amounts of gas into the beam vacuum. Initially designed for luminosity determination based on beam-profile measurements by means of beam-gas interactions [7], the SMOG system enables in addition the study of fixed target interactions with p-A and Pb-A collision at nucleon-nucleon centre-of-mass energies of $O(10^2)$ GeV. The possibility of doing fixed target physics was first demonstrated by recording p-Ne collisions in parallell to a test of the SMOG system during in the 2012 p-Pb pilot run [8], showing that LHCb is able to span the physics from SPS and RHIC energies to the LHC in a single experiment. The phase space where particle production measurements can be performed is sketched in Fig. 1 (right). For symmetric configurations the detector has forward coverage between 2 and 5 units in rapidity for all centre-of-mass energies. In fixed target mode, with the forward direction defined by the beam particle, the coverage is central to slightly backward, depending on the beam energy and type of projectile. For p-Pb collisions, depending on the orientations of the beams, both the forward and the backward hemisphere can be probed.

3. Proton-lead collisions

Analyses of p-Pb collisions performed to date by the LHCb collaboration cover quarkonium production [9–11], Z-boson production [12], D^0 production [13] and the study of long-range near-side correlations in high multiplicity events [14].

Because of space limitations, the following focuses on charm production measurements in p-Pb collisions, where J/ψ and $\psi(2S)$ mesons are reconstructed via decays into two muon final states. Here the excellent performance of the LHCb vertex detector allows to disentangle prompt from secondary production via b meson decays by means of the so-called pseudo proper-time $t_z = L_z m/p_z$, where L_z is the distance between primary vertex and secondary vertex along the beam direction z, p_z is the momentum component of the charmonium candidate along z and m the known mass of the charmonium state. Figure 2 illustrates how prompt and delayed components are disentangled [9]. Signal and background are separated by means of the invariant mass. The *sPlot* technique [19] is used to determine the lifetime distribution of the background, which is then fixed in the fit to the t_z distribution that determines the prompt and delayed components.

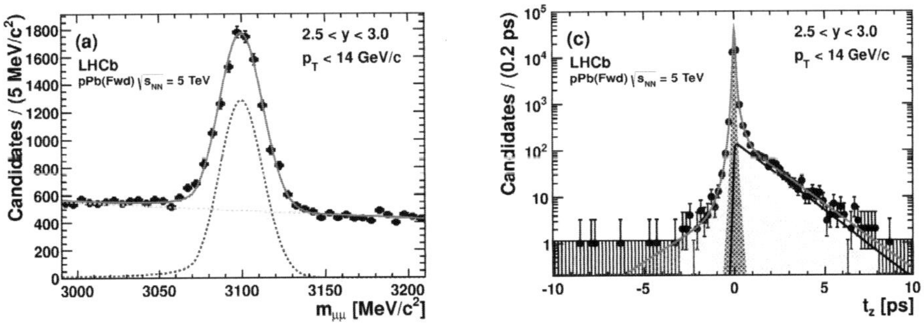

Figure 2. Example from the study of p-Pb collisions [9] for the use of (left) invariant mass and (right) pseudo proper-time to disentangle prompt and delayed production of charmonium states.

Variables which allow to quantify nuclear effects, such as modifications of the parton densities of the bound nucleons (shadowing) or energy loss effects of particles traversing the nuclear medium, are the nuclear modification factor R_{pPb} and the forward-backward asymmetry R_{FB},

$$R_{pPb} = \frac{1}{A_{Pb}} \frac{\sigma_{pPb}}{\sigma_{pp}} \quad \text{and} \quad R_{FB}(|y|) = \frac{\sigma_{pPb}(+|y|)}{\sigma_{pPb}(-|y|)} \,. \tag{1}$$

The nuclear modification factor measures by how much the p-Pb cross-section deviates from the trivial case that the Pb nucleus is a dilute system of independent free nucleons. Although conceptually simple, its determination requires knowledge of the pp cross-section at the nucleon-nucleon centre-of-mass energy of the proton-lead collision, which in absence of direct measurement needs to be estimated by inter- or extrapolation [13, 20, 21]. The forward-backward asymmetry probes nuclear effects by comparing particle production in the hemispheres corresponding to the direction of the proton and that of the nucleus. No reference cross-section is required, but the variable is not sensitive to overall factors in the production cross-sections.

Figure 3 shows results for the nuclear modification factors of J/ψ and $\psi(2S)$ production [11], both for prompt production and for secondary production from b-hadron decays. For prompt J/ψ mesons the data are in good agreement with predictions based on leading-order nuclear PDFs. The agreement degrades when using next-to-leading order predictions, but improves again if energy loss effects in the nuclear medium are accounted for. Although the theoretical uncertainties are large, it seems that an overall satisfactory description of the data is feasible when the known effects are taken into account.

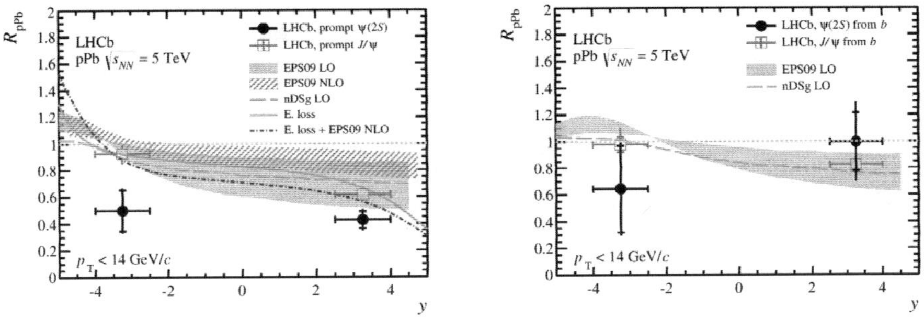

Figure 3. Nuclear modification factors for J/ψ and $\psi(2S)$ mesons measured in p-Pb collisions by LHCb, (left) for prompt production and (right) for charmonia from b-hadron decays, compared to theoretical predictions [15–18].

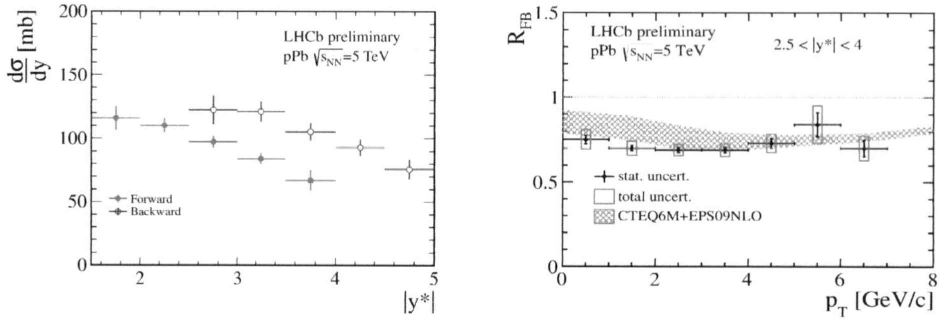

Figure 4. D^0 production in p-Pb collisions, (left) differential cross-section as a function of the absolute value of the rapidity in the nucleon-nucleon centre-of-mass system, and (right) forward-backward asymmetry, calculated from the same absolute rapidity ranges, as a function of the transverse momentum. Due to a boost of 0.47 units in rapidity between lab- and centre-of-mass system, more forward measurements are possible in the direction of the Pb nucleus.

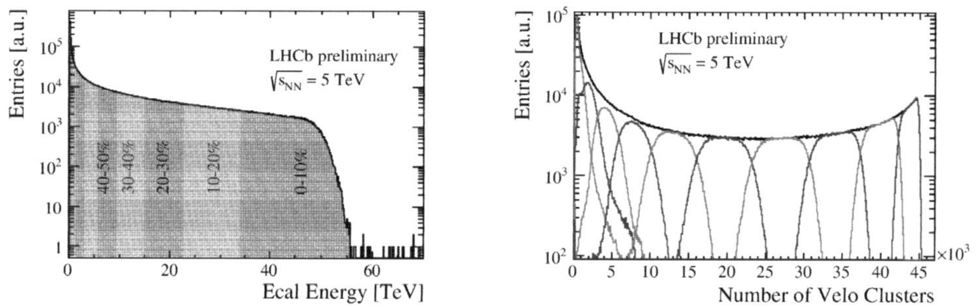

Figure 5. (left) ECAL energy deposit of minimum bias Pb-Pb collisions and (right) distributions of the number of VELO clusters for the 10%-quantiles of ECAL energy deposit. For the most central collisions (0-10% quantile) the number of clusters goes into saturation at 45×10^3.

The nuclear modification factors for prompt $\psi(2S)$ production, especially in the backward region, show larger deviations from unity than for the J/ψ, i.e. the data support the picture that the less strongly bound systems are stronger affected by cold nuclear matter effects. For charmonium states from b-hadron decays, the nuclear effects are only seen by the b-hadron. As a consequence of the tighter binding, nuclear modification factors closer to unity are expected. This is confirmed for the J/ψ from b; for the $\psi(2S)$ from b the experimental uncertainties are still to too large for any conclusions. The data are, however, consistent with the expectation that J/ψ and $\psi(2S)$ from b-hadron decays are affected in the same way.

Since the products of a breakup of a charmonium state most likely will recombine to open charm mesons, charmonia and open charm mesons are affected differently by the transport properties of the nuclear medium. Additional information is therefore gained from the study of D^0 production in p-Pb collisions [13], which because of the small charmonium to open charm ratio will not be measurably affected by contributions from charmonium break-up. Figure 4 shows a pronounced forward-backward asymmetry as a function of rapidity in the nucleon-nucleon centre-of-mass system. At the same absolute rapidity, the differential cross-section in the Pb-direction is significantly larger than in the proton direction. The forward-backward asymmetry as a function of transverse momentum is almost constant, in agreement with theoretical expectations. It is, however, worth noting that at small transverse momenta the theoretical uncertainties are much larger than the experimental ones.

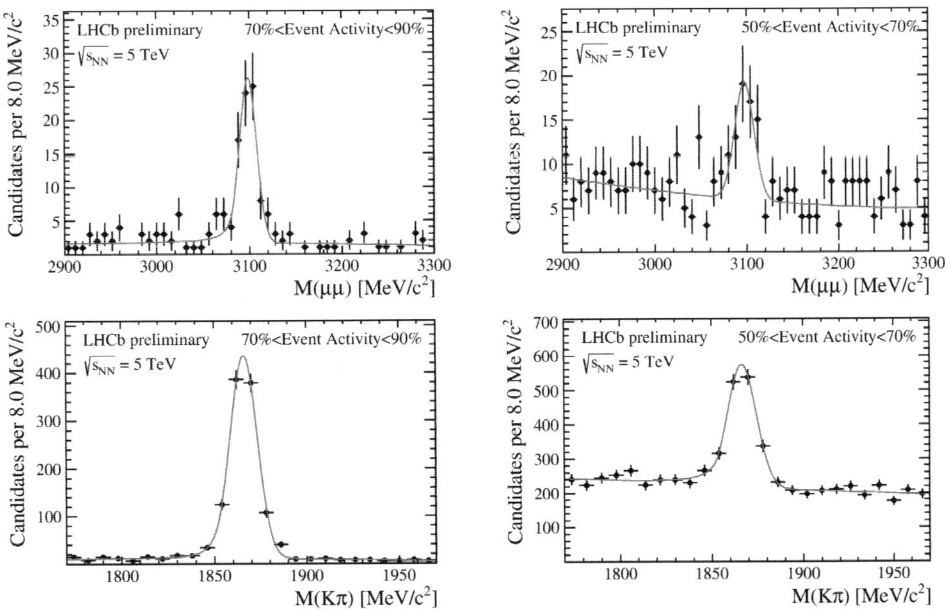

Figure 6. Production of (top) J/ψ and (bottom) D^0 mesons in (left) peripheral and (right) semi-central Pb-Pb collisions.

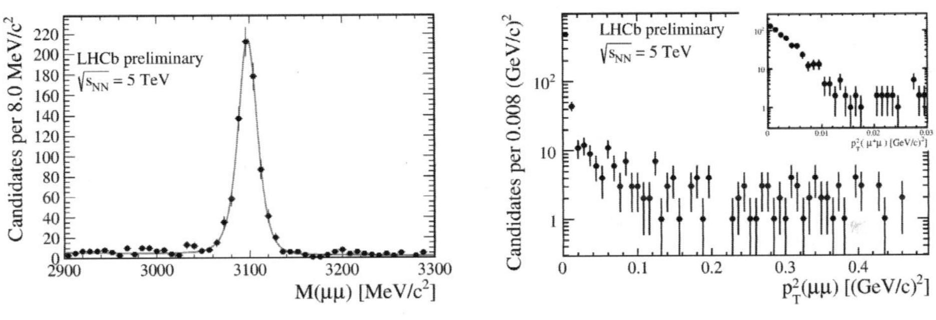

Figure 7. J/ψ photoproduction in Pb-Pb collisions, (left) di-muon invariant mass spectrum and (right) transverse momentum distribution. The inset is a zoom of the low-p_T part.

4. Lead-lead collisions

Depending on the centrality of the collision, Pb-Pb interactions at the LHC can produce extremely high multiplicity events with correspondingly large energy deposits in the detectors. Figure 5 shows the distribution of the ECAL energy deposit and the distribution of the number of clusters in the VELO. The number of VELO clusters saturates around 45×10^3. Since the standard pattern recognition program works up to $\sim$ 30% occupancy in the tracking detectors, physics studies appear feasible up to event activities in the range of the 60–50% quantile of the ECAL energy distribution, *i.e.* for peripheral to semi-central collisions.

Figure 6 illustrates the physics reach of LHCb for Pb-Pb collisions. Clear signals for J/ψ and D^0 production are observed in the event activity ranges from 90-70% and from 70-50%, but one also observes a sizeable increase on the background level with higher event activities.

A promising field of study for LHCb are also ultra-peripheral Pb-Pb collisions. An example is shown in Fig. 7 which shows a J/ψ signal in events where only two muons are recorded in the detector. A very clean signal is observed, with the very soft, exponentially falling p_T spectrum characteristic of coherent photoproduction processes.

5. Fixed target physics

Since 2015 fixed target data taking is an integral part of the LHCb ion physics program, which takes advantage of the SMOG system to inject noble gases into the interaction region. Typical gas pressures are of $O(10^{-7})$ mbar, which is sufficient to obtain sizeable samples of beam-gas interaction. Data samples have been recorded with proton or Pb beams on He, Ne and Ar gas. To illustrate the physics prospects, Fig. 8 shows first J/ψ and D^0 signals from the study of p-Ne collisions.

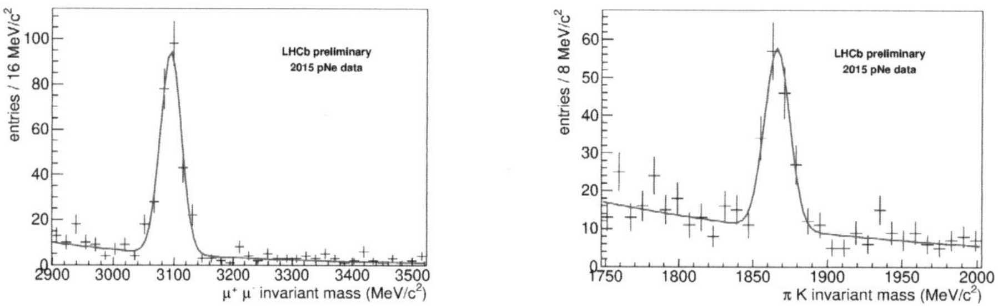

Figure 8. Production of (left) J/ψ and (right) D^0 mesons in fixed-target p-Ne collisions.

6. Summary

In 2013 the LHCb experiment successfully participated in the LHC proton-ion running. Two years later the ion physics program of the experiment was expanded to cover also Pb-Pb collisions and, taking advantage of the SMOG system that allows to inject gas into the LHCb interaction region, the study of fixed target interactions at the LHC. First studies of Pb-Pb collision show that LHCb is able to analyse peripheral to semi-central Pb-Pb collisions.

References

[1] Matsui T and Satz H 1986 *Phys.Lett.* **B178** 416
[2] Eskola K, Paukkunen H and Salgado C 2009 *JHEP* **04** 065
[3] Arleo F and Peigné S 2012 *Phys.Rev.Lett.* **109** 122301
[4] Alves Jr A A *et al.* (LHCb collaboration) 2008 *JINST* **3** S08005
[5] Aaij R *et al.* (LHCb collaboration) 2015 *Int. J. Mod. Phys.* **A30** 1530022
[6] Barschel C 2014 PhD thesis, RWTH Aachen
[7] Aaij R *et al.* (LHCb collaboration) 2014 *JINST* **9** P12005
[8] LHCb collaboration 2013 LHCb-CONF-2012-034
[9] Aaij R *et al.* (LHCb collaboration) 2014 *JHEP* **02** 072
[10] Aaij R *et al.* (LHCb collaboration) 2014 *JHEP* **07** 094
[11] Aaij R *et al.* (LHCb collaboration) 2016 *JHEP* **03** 133
[12] Aaij R *et al.* (LHCb collaboration) 2014 *JHEP* **09** 030
[13] LHCb collaboration 2016 LHCb-CONF-2016-003
[14] Aaij R *et al.* (LHCb collaboration) 2016 *Phys. Lett.* **B762** 473
[15] Ferreiro E G, Fleuret F, Lansberg J P and Rakotozafindrabe A 2013 *Phys.Rev.* **C88** 047901
[16] Arleo F and Peigné S 2013 *JHEP* **03** 122
[17] Albacete J, Armesto N, Baier R, Barnafoldi G G, Barrette J *et al.* 2013 *Int. J. Mod. Phys.* **E22** 1330007
[18] Conesa del Valle Z, Ferreiro E G, Fleuret F, Lansberg J P and Rakotozafindrabe A 2014 *Nucl.Phys.* **A926** 236
[19] Pivk M and Le Diberder F R 2005 *Nucl.Instrum.Meth.* **A555** 356–369
[20] ALICE and LHCb collaborations 2013 LHCb-CONF-2013-013 & ALICE-PUBLIC-2013-002
[21] ALICE and LHCb collaborations 2014 LHCb-CONF-2014-003 & ALICE-PUBLIC-2014-002

Recent development on heavy quark theory in heavy-ion collisions

Shanshan Cao
Nuclear Science Division, Lawrence Berkeley National Laboratory, Berkeley, CA 94720, USA
Department of Physics and Astronomy, Wayne State University, 666 W. Hancock St., Detroit, MI 48201, USA

E-mail: shanshan.cao@wayne.edu

Abstract. A brief overview is presented for the theory of open heavy flavor dynamics in ultra-relativistic nuclear collisions, including a summary of different transport models, recent development on heavy quark phenomenology, in particular possible solutions to the "heavy vs. light flavor" puzzle and the "R_{AA} vs. v_2" puzzle in the field.

1. Introduction

Heavy quarks serve as ideal probes of the color deconfined quark-gluon plasma (QGP) matter produced in relativistic nuclear collisions. Since their thermal production is suppressed by their large mass, they are primarily produced at the primordial stage of heavy-ion collisions via hard scatterings and then travel through and interact with the medium with their flavors conserved, and therefore probe the entire evolution history of the QGP fireballs. Over the past decade, experimental observations at both RHIC and the LHC have revealed a great many interesting data of heavy flavors, among which the most surprising ones are the small values of the nuclear modification factor R_{AA} and large values of the elliptic flow coefficient v_2 of heavy mesons which are almost comparable to those of light hadrons [1, 2, 3]. This seems contradictory to one's earlier expectation of the mass hierarchy of parton energy loss and is known as a "heavy flavor puzzle". Apart from this "heavy vs. light flavor puzzle", another puzzle related to not only heavy quarks but also to all hard partons is the difficulty in simultaneously describing their R_{AA} and v_2. Therefore, it still remains a great challenge to fully understand the heavy flavor dynamics in the same framework with light flavor partons.

Various transport models have been developed to study the heavy quark evolution inside the dense nuclear matter, such as the parton cascade model based on the Boltzmann equation [4, 5, 6], the linearized Boltzmann approach coupled to a hydrodynamic background [7, 8, 9] and the Langevin-based transport models [10, 11, 12, 13, 14, 15]. In this talk, I will first briefly summarize different theoretical treatments of heavy flavor dynamics in heavy-ion collisions and then discuss recent theoretical progress towards solving the heavy flavor puzzles in the field.

2. Transport Theory of Heavy Flavor in Heavy-Ion Collisions

In the most general form, the evolution of a heavy quark can be described using the Boltzmann equation

$$p \cdot \partial f_Q(x,p) = EC[f_Q], \tag{1}$$

where the right hand side represents the collision term which can be expressed as a gain term minus a loss term

$$C[f_Q] = \int d^3k \left[w(\vec{p}+\vec{k},\vec{k}) f_Q(\vec{p}+\vec{k}) - w(\vec{p},\vec{k}) f_Q(\vec{p}) \right], \tag{2}$$

in which $w(p,k)$ represents the transition rate of a heavy quark from momentum p to $p-k$ and can be directly evaluated from the microscopic cross sections.

For the quasi-elastic scattering process, one may assume the momentum change of heavy quark during each of its scattering with a light parton is small ($|\vec{k}| \ll |\vec{p}|$) and thus the collision term can be simplified as

$$C[f_Q] \approx \int d^3k \left(k_i \frac{\partial}{\partial p_i} + \frac{1}{2} k_i k_j \frac{\partial^2}{\partial p_i \partial p_j} \right) w(\vec{p},\vec{k}) f_Q(\vec{p}), \tag{3}$$

and the Boltzmann equation is reduced to the following Fokker-Planck equation:

$$\frac{\partial}{\partial t} f_Q = \frac{\partial}{\partial p_i} \left\{ A_i(\vec{p}) f_Q + \frac{\partial}{\partial p_j} \left[B_{ij}(\vec{p}) f_Q \right] \right\}. \tag{4}$$

Additionally, one may assume every heavy quark scatters multiple times with the medium background along its propagation, and thus the Fokker-Planck equation can be stochastically realized by the Langevin equation:

$$dx_i = \frac{p_i}{E_{\vec{p}}} dt, \tag{5}$$

$$dp_i = -\eta_D(\vec{p}) p_i dt + \xi_i dt. \tag{6}$$

The first term in Eq. (6) is known as drag and the second is related to the thermal random force. One may refer to Ref. [16] for calculations of the above mentioned transport coefficients – A_i, B_{ij}, η_D and ξ_i. Note that these simplifications from the Boltzmann equation to the Fokker-Planck equation and then to the Langevin equation are only valid for the collisional energy loss, but not for their radiative energy loss because the energy of each radiated gluon is not necessarily small and the number of emitted gluons from each heavy quark may not be large.

While collisional energy loss alone is successful in describing heavy flavor observables in the low p_{T} region where the phase space for the medium-induced gluon radiation is restricted by the large mass of heavy quarks [17, 18], it has been shown insufficient [14, 15] at high p_{T}. To incorporate gluon radiation into the Boltzmann transport model, one needs to evaluate the pQCD diagrams for the $2 \to 3$ processes for the collision term. Although a full evaluation is available [19], the result is tedious and hard to efficiently implement in numerical calculations. For this reason, the Gunion-Bertsch approximation is adopted by Refs. [7, 8, 20] that is derived at high energy limit and reproduce the exact calculation of the matrix elements over a wide rapidity range. The LO pQCD calculation does not include the LPM effect due to the coherent scatterings. To mimic this effect in the numerical simulation, Ref. [21] requires that the heavy quark mean free path is larger than the formation time of radiated gluons times an X factor.

An alternative approach to implement radiative energy loss is calculating the inelastic scattering probability based on the average number of medium-induced gluon, which has been successfully applied in the improved Langevin framework [14, 15] and the linear Boltzmann transport (LBT) model [9]. The average gluon number during each time interval Δt is obtained by integrating the radiated gluon spectrum:

$$\langle N_g \rangle (E, T, t, \Delta t) = \Delta t \int dx dk_\perp^2 \frac{dN_g}{dx dk_\perp^2 dt}, \tag{7}$$

and the gluon spectrum can be adopted from a higher-twist energy loss calculation [22, 23, 24]

$$\frac{dN_g}{dx dk_\perp^2 dt} = \frac{2\alpha_s C_A \hat{q} P(x)}{\pi k_\perp^4} \left(\frac{k_\perp^2}{k_\perp^2 + x^2 M^2} \right)^4 \sin^2 \left(\frac{t - t_i}{2\tau_f} \right), \tag{8}$$

in which x is the fractional energy of the radiated gluon from its parent hard parton, $k_\perp$ is the gluon transverse momentum. $\hat{q}$ is known as the quark or gluon transport coefficient. P is the splitting function and τ_f is the splitting time for gluon emission. The quartic term in Eq. (8) is the dead cone factor for heavy quarks, denoting the mass effect on parton energy loss. Multiple gluon emissions during each time step is allowed and the number of radiated gluons during Δt obeys a Poisson distribution with the mean value N_g. Thus the probability of inelastic scattering during Δt is $P_{\rm inel} = 1 - e^{-\langle N_g \rangle}$. If gluon radiation happens based on this probability, the energy and momentum of each radiated gluon is determined based on the differential spectrum Eq. (8). It has been shown in Ref. [15] that in 2.76 TeV central Pb-Pb collisions, quasi-elastic scattering dominates the energy loss of heavy quarks with low initial energy while gluon radiation dominates the high energy region. The crossing points are around 7 GeV for charm quark and 18 GeV for bottom quark. This indicates that including both energy loss mechanisms is necessary to study the heavy quark phenomenology at high $p_{\rm T}$ as observed at the LHC experiment.

One key ingredient of transport models is the transport coefficient. Apart from perturbative QCD calculations, evaluations based on quasi-particle models [25], non-perturbative T-matrix method [26, 27, 11, 28] and lattice QCD [29, 30, 31] have been investigated and shown successful. In addition to these first principle driven calculations of heavy quark transport coefficient, a data driven framework based on the Bayesian analysis [32] has also been developed for heavy-ion physics in which one is able to precisely extract the transport coefficient of hard probes from transport model to experimental data comparison and place systematic constraints on our understanding of the heavy quark – medium interaction.

After heavy quarks travel outside the QGP medium, they hadronize into color neutral bound states. Hybrid models of fragmentation plus heavy-light quark coalescence have been established to convert heavy quarks into heavy flavor hadrons. High $p_{\rm T}$ heavy quarks tend to fragment directly into hadrons. One may use either a proper fragmentation function to calculate the corresponding hadron spectra or use PYTHIA to simulate this process. On the other hand, it is more probable for lower $p_{\rm T}$ heavy quarks to combine with thermal partons from the medium to form new hadrons. This mechanism can be described using either an instantaneous coalescence model [33, 34, 15] or a resonance recombination model [26, 27, 11]. It has been shown in Refs. [11, 15] that while fragmentation dominates the high $p_{\rm T}$ region of heavy quark hadronization, coalescence significantly enhances the heavy flavor hadron production rate at medium $p_{\rm T}$. For this reason, the coalescence mechanism could generate the bump structure of the D meson $R_{\rm AA}$. In addition, coalescence also enhances the D meson v_2 since it adds the momentum space anisotropy of light partons onto heavy quarks when D mesons are formed.

3. Progress on Heavy Flavor Phenomenology

To obtain a thorough understanding of heavy flavor dynamics in heavy-ion collisions and investigate the mass effect on parton energy loss, one is expected to establish a unified theoretical framework for both heavy and light parton evolution inside the QGP. This has been realized in a Monte-Carlo based transport model for the first time within the LBT framework [9] in which both elastic and inelastic scatterings of heavy and light partons are treated on the same footing.

Figure 1 displays the flavor hierarchy of the nuclear modification factor at both parton and hadron levels. In Fig. 1(a), one observes that due to the mass effect of radiative energy loss, light quarks are slightly more suppressed than charm quarks in all the 3 colliding systems –

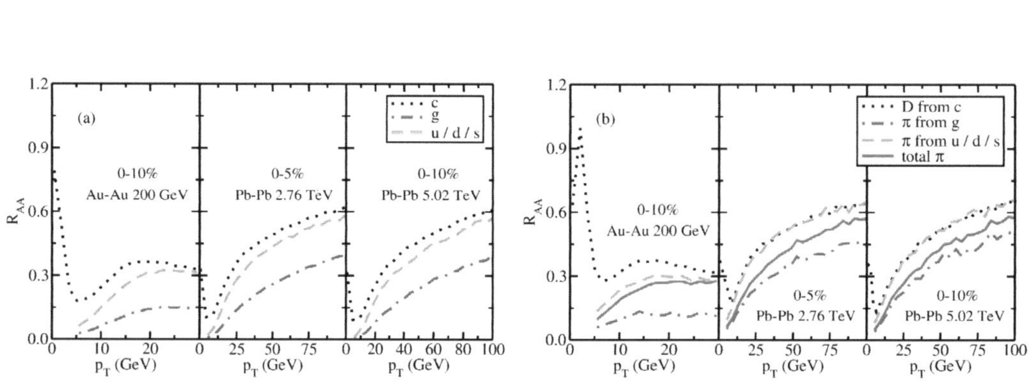

Figure 1. (Color online) Flavor dependence of R_{AA} from RHIC to the LHC energies at (a) parton and (b) hadron levels.

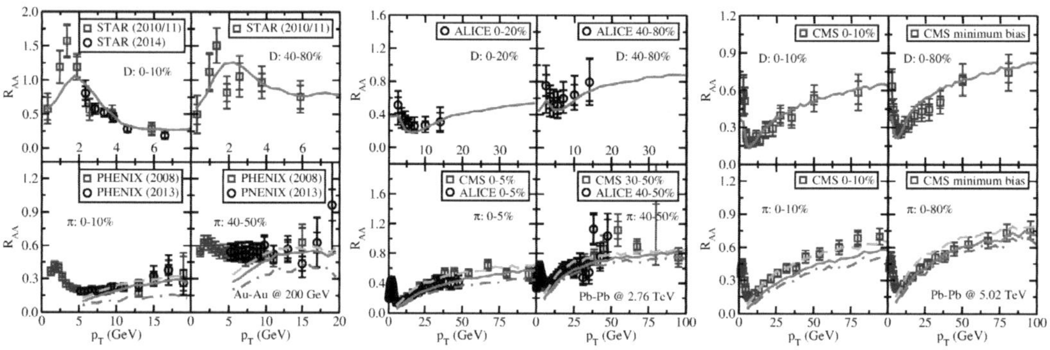

Figure 2. (Color online) R_{AA} of D and π from RHIC to the LHC energies. Data are taken from Refs. [2, 35, 36, 37, 38, 39, 40, 41].

200 GeV Au-Au, 2.76 TeV Pb-Pb and 5.02 TeV Pb-Pb collisions. Gluons are significantly more suppressed due to its larger color factor. However, this hierarchy can be slightly changed at the hadron level. As shown in Fig. 1(b), due to the harder fragmentation function of charm quark than that of light quark, together with the harder initial parton spectra that leads to the fast increasing parton R_{AA} with p_{T} at the LHC energy, D meson and π from quark jet have almost the same R_{AA}. On the other hand, the effect of different fragmentation functions on the flavor hierarchy of hadron R_{AA} is not so significant at the RHIC energy due to the soft initial spectra that leads to the relatively flat parton R_{AA} at high p_{T}. π from gluon jet is always significantly more suppressed in all the 3 systems.

After taking all effects into account, the LBT model provides a simultaneous description of heavy and light hadron suppression from RHIC to the LHC energies. In Fig. 2, upper panels show the R_{AA} of D mesons, and lower panels show the R_{AA} of π in which three curves are presented separately – upper for π from quark jet, lower for π from gluon jet and middle for the mixture. Note that since all heavy quarks are produced from initial hard scatterings, their full p_{T} spectra of R_{AA} can be obtained. On the other hand, reliable calculation for light hadron is only available at high p_{T} at this moment; contribution from the soft bulk matter at low p_{T} will be included in a future effort.

Another challenge for not only heavy flavor study but other hard probes as well is the simultaneous description of their R_{AA} and v_2. As presented in Ref. [42], although most model calculations provide reasonable descriptions of the D meson R_{AA} with proper tunings of heavy

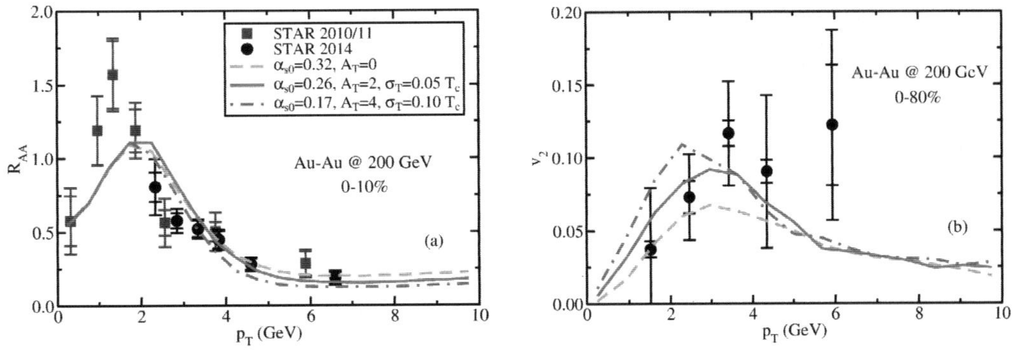

Figure 3. (Color online) Effects of the temperature dependence of the transport coefficient on (a) R_{AA} vs. (b) v_2 of D mesons.

quark transport coefficients, v_2 is often underestimated. Several recent studies target on solving this "R_{AA} vs. v_2" puzzle. For instant, the effect of different temperature dependence of quark transport coefficient on the D meson v_2 is investigated in Ref. [9]. The strong coupling constant is rescaled with a factor that peaks around the critical temperature T_c and returns to a constant α_{s0} at large T:

$$\alpha_s(T) = \alpha_{s0}\left[1 + A_T e^{-(T-T_c)^2/2\sigma_T^2}\right]. \tag{9}$$

As shown in Fig. 3(a), by adjusting α_s, different momentum dependences of the transport coefficient can provide similar R_{AA} of D mesons. However, as shown in Fig. 3(b), while their R_{AA} is fixed, the stronger α_s is around T_c, the larger v_2 one obtains. The physical picture is that if the interaction around T_c is stronger, larger part of heavy quark energy loss will be shifted towards the freeze-out hypersurface of the QGP where the anisotropic flow of the bulk matter is stronger, and therefore heavy quarks pick up a larger v_2 from the medium. This is consistent with the findings presented in Refs. [43, 44].

Other interesting topics on open heavy flavor presented at this conference include: effects of different path length dependence of heavy quark energy loss and event-by-event fluctuation on heavy flavor v_2 and v_3 [45]; heavy flavor production from soft collinear effective theory [46]; D-meson observables in p-Pb collisions; angular correlation between heavy and light mesons [47]; and effect of strong magnetic field on heavy quark diffusion [48].

4. Summary

To conclude, I have summarized different transport models and their implementations to heavy quark energy loss in heavy-ion collisions. Recent progress on open heavy flavor phenomenology has been discussed. In particular, it has been shown that with a delicate transport model that consistently incorporate elastic and inelastic scatterings of heavy and light partons inside the QGP, one can naturally obtain a simultaneous description of heavy and light hadron suppression from RHIC to the LHC energies and the "heavy vs. light flavor puzzle" no longer exists. A possible solution to the "R_{AA} vs. v_2 puzzle" has also been discussed: a more careful investigation of the temperature dependence of the interaction strength between hard probes and the QGP medium.

Acknowledgments

This work is supported by the Director, Office of Energy Research, Office of High Energy and Nuclear Physics, Division of Nuclear Physics, of the U.S. Department of Energy under

Contract No. DE-AC02-05CH11231 and DE-SC0013460, and within the framework of the JET Collaboration.

References

[1] Adare A *et al.* (PHENIX Collaboration) 2011 *Phys. Rev.* **C84** 044905 (*Preprint* 1005.1627)
[2] Adamczyk L *et al.* (STAR Collaboration) 2014 *Phys. Rev. Lett.* **113** 142301 (*Preprint* 1404.6185)
[3] Abelev B B *et al.* (ALICE) 2014 *Phys. Rev.* **C90** 034904 (*Preprint* 1405.2001)
[4] Molnar D 2007 *Eur. Phys. J.* **C49** 181–186 (*Preprint* nucl-th/0608069)
[5] Zhang B, Chen L W and Ko C M 2005 *Phys. Rev.* **C72** 024906 (*Preprint* nucl-th/0502056)
[6] Uphoff J, Fochler O, Xu Z and Greiner C 2011 *Phys. Rev.* **C84** 024908 (*Preprint* 1104.2295)
[7] Gossiaux P, Aichelin J, Gousset T and Guiho V 2010 *J. Phys.* **G37** 094019 (*Preprint* 1001.4166)
[8] Nahrgang M, Aichelin J, Gossiaux P B and Werner K 2014 *Phys. Rev.* **C90** 024907 (*Preprint* 1305.3823)
[9] Cao S, Luo T, Qin G Y and Wang X N 2016 *Phys. Rev.* **C94** 014909 (*Preprint* 1605.06447)
[10] Moore G D and Teaney D 2005 *Phys. Rev.* **C71** 064904 (*Preprint* hep-ph/0412346)
[11] He M, Fries R J and Rapp R 2012 *Phys. Rev.* **C86** 014903 (*Preprint* 1106.6006)
[12] Cao S and Bass S A 2011 *Phys. Rev.* **C84** 064902 (*Preprint* 1108.5101)
[13] Cao S, Qin G Y and Bass S A 2013 *J. Phys.* **G40** 085103 (*Preprint* 1205.2396)
[14] Cao S, Qin G Y and Bass S A 2013 *Phys. Rev.* **C88** 044907 (*Preprint* 1308.0617)
[15] Cao S, Qin G Y and Bass S A 2015 *Phys. Rev.* **C92** 024907 (*Preprint* 1505.01413)
[16] Rapp R and van Hees H *published in R. C. Hwa, X.-N. Wang (Eds.), Quark Gluon Plasma 4 (World Scientific, 2010)* 111–206 (*Preprint* 0903.1096)
[17] Dokshitzer Y L and Kharzeev D E 2001 *Phys. Lett.* **B519** 199–206 (*Preprint* hep-ph/0106202)
[18] Abir R, Jamil U, Mustafa M G and Srivastava D K 2012 *Phys. Lett.* **B715** 183–189 (*Preprint* 1203.5221)
[19] Kunszt Z, Pietarinen E and Reya E 1980 *Phys. Rev.* **D21** 733
[20] Fochler O, Uphoff J, Xu Z and Greiner C 2013 *Phys. Rev.* **D88** 014018 (*Preprint* 1302.5250)
[21] Uphoff J, Fochler O, Xu Z and Greiner C 2015 *J. Phys.* **G42** 115106 (*Preprint* 1408.2964)
[22] Guo X F and Wang X N 2000 *Phys. Rev. Lett.* **85** 3591–3594 (*Preprint* hep-ph/0005044)
[23] Majumder A 2012 *Phys. Rev.* **D85** 014023 (*Preprint* 0912.2987)
[24] Zhang B W, Wang E and Wang X N 2004 *Phys. Rev. Lett.* **93** 072301 (*Preprint* nucl-th/0309040)
[25] Song T, Berrehrah H, Cabrera D, Torres-Rincon J M, Tolos L, Cassing W and Bratkovskaya E 2015 *Phys. Rev.* **C92** 014910 (*Preprint* 1503.03039)
[26] van Hees H, Greco V and Rapp R 2006 *Phys. Rev.* **C73** 034913 (*Preprint* nucl-th/0508055)
[27] van Hees H, Mannarelli M, Greco V and Rapp R 2008 *Phys. Rev. Lett.* **100** 192301 (*Preprint* 0709.2884)
[28] Liu S Y F and Rapp R 2015 *Nucl. Phys.* **A941** 179–187 (*Preprint* 1501.07892)
[29] Ding H, Francis A, Kaczmarek O, Karsch F, Satz H *et al.* 2011 *J. Phys* **G38** 124070 (*Preprint* 1107.0311)
[30] Banerjee D, Datta S, Gavai R and Majumdar P 2012 *Phys. Rev.* **D85** 014510 (*Preprint* 1109.5738)
[31] Kaczmarek O 2014 *Nucl. Phys.* **A931** 633–637 (*Preprint* 1409.3724)
[32] Bernhard J E, Marcy P W, Coleman-Smith C E, Huzurbazar S, Wolpert R L and Bass S A 2015 *Phys. Rev.* **C91** 054910 (*Preprint* 1502.00339)
[33] Oh Y, Ko C M, Lee S H and Yasui S 2009 *Phys. Rev.* **C79** 044905 (*Preprint* 0901.1382)
[34] Gossiaux P, Bierkandt R and Aichelin J 2009 *Phys. Rev.* **C79** 044906 (*Preprint* 0901.0946)
[35] Xie G (STAR) 2016 *Nucl. Phys.* **A956** 473–476 (*Preprint* 1601.00695)
[36] Adare A *et al.* (PHENIX) 2013 *Phys. Rev.* **C87** 034911 (*Preprint* 1208.2254)
[37] Abelev B *et al.* (ALICE Collaboration) 2012 *JHEP* **1209** 112 (*Preprint* 1203.2160)
[38] Chatrchyan S *et al.* (CMS) 2012 *Eur. Phys. J.* **C72** 1945 (*Preprint* 1202.2554)
[39] Abelev B *et al.* (ALICE) 2013 *Phys. Lett.* **B720** 52–62 (*Preprint* 1208.2711)
[40] CMS 2016 (*Preprint* CMS-PAS-HIN-16-001)
[41] CMS 2016 (*Preprint* CMS-PAS-HIN-15-015)
[42] Andronic A *et al.* 2016 *Eur. Phys. J.* **C76** 107 (*Preprint* 1506.03981)
[43] Das S K, Scardina F, Plumari S and Greco V 2015 *Phys. Lett.* **B747** 260–264 (*Preprint* 1502.03757)
[44] Xu J, Liao J and Gyulassy M 2015 *Chin. Phys. Lett.* **32** 9 (*Preprint* 1411.3673)
[45] Prado C A G, Noronha-Hostler J, Cosentino M R, Munhoz M G, Noronha J and Suaide A A P 2016 (*Preprint* 1609.06093)
[46] Kang Z B, Ringer F and Vitev I 2016 (*Preprint* 1609.04908)
[47] Rohrmoser M, Gossiaux P B, Gousset T and Aichelin J 2016 (*Preprint* 1611.01854)
[48] Fukushima K, Hattori K, Yee H U and Yin Y 2016 *Phys. Rev.* **D93** 074028 (*Preprint* 1512.03689)

Strangeness in ALICE at the LHC

Francesca Bellini* for the ALICE Collaboration

*Dipartimento di Fisica dell'Universitá di Bologna, Via Irnerio 46 - 40126 Bologna (Italy)

E-mail: fbellini@cern.ch

Abstract. Strangeness production has been measured by the ALICE experiment in different collision systems at the unprecedented center-of-mass energies available at the CERN Large Hadron Collider. In Pb–Pb collisions at $\sqrt{s_{\rm NN}} = 2.76$ TeV the relative production of strange and multi-strange baryons relative to pions is observed to follow a saturating trend with increasing centrality, and reaching values that are consistent with those predicted by thermal model calculations in the Grand-Canonical ensemble. More recently, the multiplicity dependence of strangeness production in small systems such as pp and p–Pb has also been investigated. An overview of the most recent results on strangeness production is reported, including the first observation of strangeness enhancement with charged particle multiplicity in pp collisions.

1. Introduction

The study of strangeness production has had a central role in heavy-ion physics since the observation of its enhancement in heavy-ion collisions (A–A) with respect to proton-proton (pp) collisions. This observation was predicted as one of the signatures of the formation of the Quark Gluon Plasma (QGP) [1]. The enhancement, as quantified by comparing the particle yields per participant nucleon pair in A–A collisions to the yields in pp, was measured at SPS and RHIC and finally by ALICE at the LHC, where it was soon clear that a key ingredient for a proper interpretation of the results is the understanding of the evolution of strangeness production in the reference pp collisions with the collision energy. Since the average number of charged particles does not scale linearly with the average number of nucleons participating in the collision, the ratio of strange particle yield to pion yield is used to quantify the enhancement of strangeness production.

Recently, ALICE has performed measurements in pp and proton-lead (p–Pb) collisions to investigate strangeness production as a function of the charged-particle multiplicity. This effort is part of a growing interest of the heavy-ion community in the detailed study of the dynamics of small systems, after the observation by the LHC experiments of features in pp and p–Pb collisions traditionally associated with the presence of a QGP in Pb–Pb collisions and hinting at the presence of collective phenomena [2, 3].

This work is intended as a brief review of the new ALICE results on identified particle production presented at this conference, with a focus on two key topics that are presently subject of animated discussions in the community: the investigation of features associated to collectivity-driven in small systems and the multiplicity-differential measurements of strangeness production in pp.

2. Production of identified hadrons in pp, p–Pb and Pb–Pb

ALICE has measured the production of a large variety of identified particles[1], including pions, kaons and protons, strange and multi-strange hadrons (K^0_S, Λ, Ξ, Ω), short-lived hadronic resonances ($(\rho(770)^0$, $K^*(892)^0$, $\phi(1020)$, $\Sigma(1385)^{\pm}$, $\Xi(1530)^0$), and light nuclei (d, ^{3}H, ^{3}He, ^{4}He), in proton-proton (pp), proton-lead (p–Pb) and lead-lead (Pb–Pb) collisions. The yields of a selection of these particles relative to pions are reported in figure 1 as a function of the average charged particle multiplicity density at midrapidity ($|\eta| < 0.5$), $\langle dN_{ch}/d\eta \rangle$.

The reader can refer to [4] for details on the ALICE sub-detectors and their particle identification performance. For the measurements presented in this work, the forward-rapidity array scintillator counters V0A and V0C, and the T0 system have been used for triggering and background rejection. In order to avoid auto-correlation biases, events are classified in centrality in Pb–Pb and multiplicity in pp and p–Pb collisions, based on the total charge deposited in the forward V0A and V0C detectors. The values of $\langle dN_{ch}/d\eta \rangle$ for each event class are measured at midrapidity. $\langle dN_{ch}/d\eta \rangle$ spans over three orders of magnitude in Pb–Pb collisions (up to $O(10^3)$), while it ranges from few units to few tens in pp and p–Pb collisions, with a significant overlap between the two systems. The Inner Tracking System (ITS) and the Time Projection Chamber (TPC) are used for vertexing, tracking and particle identification (PID). The Time-Of-Flight (TOF) system and the High Momentum Particle Identification (HMPID) detector allow to extend the range of identification for charged particles and light nuclei to higher momentum. In addition, the topological reconstruction of weak decays is exploited to identify strange and multi-strange particles. Thanks to the combination of different PID techniques, ALICE is able to identify hadrons with momenta between 150 MeV/c and 20 GeV/c.

The centrality-dependent transverse momentum (p_T) spectra of pions, kaons and protons in Pb–Pb collisions at $\sqrt{s_{NN}} = 2.76$ TeV have been published recently in [5] together with their nuclear modification factors R_{AA}, defined as the ratio of the yield in Pb–Pb collisions to the pp yield scaled by the number of binary collisions. In figure 2 the preliminary measurements of the Ξ and Ω R_{AA} in central Pb–Pb collisions are compared to the modification of inclusive

[1] For brevity, the notation for particles is used in these proceedings to indicate both particles and antiparticles, and resonance masses will be omitted.

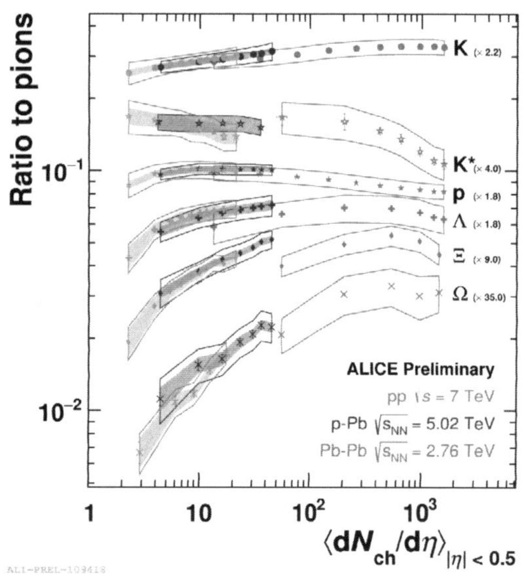

Figure 1. Yield of K, K^{*0}, p, ϕ, Λ, Ξ, Ω relative to the pion yield as a function of $\langle dN_{ch}/d\eta \rangle$ measured at midrapidity ($|\eta| < 0.5$) in pp, p–Pb, Pb–Pb collisions.

charged particles, π, K, p, ϕ and the average R_{AA} of D-mesons. A strong suppression at high p_T ($p_T > 8$ GeV/c) is observed for all particle species under study, including open-charm mesons, suggesting that the mechanism responsible for the suppression is flavour-independent. In the low and intermediate p_T range several effects can result in the difference observed in both the shapes and maxima of the distributions: the presence of radial flow in Pb–Pb collisions shifts the maxima of the spectra towards higher values according to the mass of the particle; the strangeness content might cause the enhancement of the R_{AA} of Ω above 1; the significantly different reference spectra of p and the ϕ-meson in pp collisions justifies the difference in the R_{AA} at $2 < p_T < 4$ GeV/c of these particles, whose ratio of spectra in Pb–Pb collisions is flat in the same p_T range [6].

The analysis of two small data samples of pp and Pb–Pb collisions at $\sqrt{s_{NN}} = 5.02$ TeV collected in late 2015 and corresponding to 3% and 25% of the available statistics, respectively, has led to a preliminary measurement of the R_{AA} of inclusive charged particles at the new energy. The measurement exploits an improved analysis technique that allows to reduce significantly the systematic uncertainty on the spectra with respect to the 2.76 TeV case [7]. The R_{AA} of charged particles at 5.02 TeV is reported in figure 3 for six different centrality intervals and compared to the previous measurement at 2.76 TeV. Particle production at high p_T is increasingly suppressed from peripheral to central events, where the suppression is consistent with the observations at 2.76 TeV. Since a hardening of the spectra is expected for increasing $\sqrt{s_{NN}}$, this observation may indicate a larger parton energy loss in a hotter and/or denser medium produced at the higher energy. A new analysis of the 2.76 TeV data is in progress with the same improved analysis technique used at 5.02 TeV in order to allow a more precise comparison of the two energies, especially for the 40-60% and 60-80% centrality classes, where some hints for differences at $p_T > 5$ GeV/c can be noted, yet within the current systematic uncertainties.

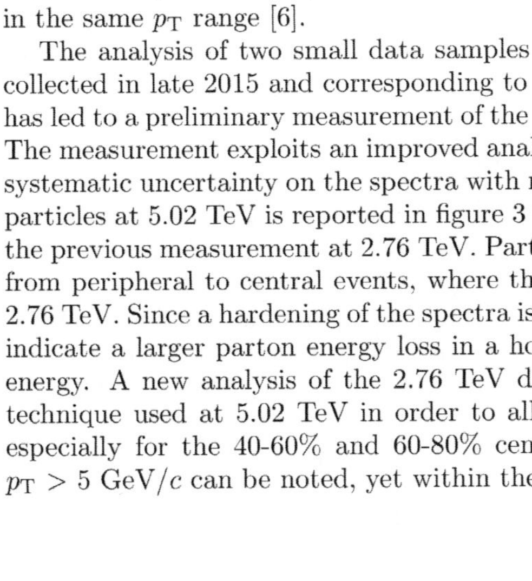

Figure 2. R_{AA} for various particle species in central Pb–Pb collisions at $\sqrt{s_{NN}} = 2.76$ TeV.

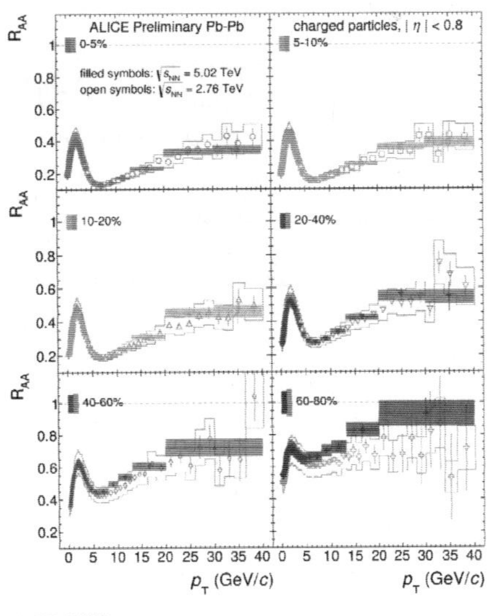

Figure 3. R_{AA} of inclusive charged particles in central Pb–Pb collisions at $\sqrt{s_{NN}} = 5.02$ TeV.

Similarly to Pb–Pb, no flavour dependence is observed at high-$p_{\rm T}$ for the nuclear modification factor $R_{\rm pPb}$ in p–Pb collisions at $\sqrt{s_{\rm NN}}$ = 5.02 TeV [8]. The $R_{\rm pPb}$ of π, K, p is consistent with unity above $p_{\rm T} \sim 8$ GeV/c, indicating that the suppression observed in central Pb–Pb collisions is due to parton energy loss in the medium created in the collision. At intermediate $p_{\rm T}$ (2 to 6 GeV/c), where also Ξ and Ω can be added to the picture [9], the Cronin peak exhibits mass ordering, reminiscent of the mass ordering observed for the $\langle p_{\rm T}\rangle$ [10].

3. Searches for collectivity in small systems

The role of radial flow and particle production mechanisms (fragmentation, recombination) in defining the spectral shapes in the intermediate-$p_{\rm T}$ range can be investigated by studying baryon-to-meson (B/M) ratios. Ratios of particles with a large mass difference, such as p/π and $\Lambda/{\rm K^0_S}$, or, on the contrary, with very similar masses such as p/ϕ, are good candidates in this respect. The comparison of the $\sqrt{s_{\rm NN}}$ = 2.76 TeV Pb–Pb results with theoretical models have shown that a good agreement with data is achieved over a broad $p_{\rm T}$ range when hydrodynamics is taken into account, as done in EPOS [11], thus suggesting that the so-called "baryon anomaly" observed in central Pb–Pb collisions may be a consequence of a strong radial flow. In addition, it was shown that models of parton recombination can also describe the data [12]. The p/π and $\Lambda/{\rm K^0_S}$ ratios have been measured as well in pp at 7 TeV and p–Pb at 5.02 TeV for various event classes [13]. Both ratios exhibit a depletion at low $p_{\rm T}$ and an enhancement at intermediate $p_{\rm T}$ going from low to high multiplicity within each system. Although the magnitudes of the depletion and enhancement are different across different systems, the similarity with the Pb–Pb case is striking. Is this an indication for the presence of collectivity in small systems?

In order to try to answer this question, the transverse momentum spectra of pions, kaons and protons in both pp and p–Pb have been fitted simultaneously with a Blast-Wave function [14], as done for the Pb–Pb case [15]. For each multiplicity class the fitting range is optimised in order to ensure a χ^2/ndf $<$ 1.Then the fit is used to quantify the freeze-out parameters and the evolution of the system with centrality in the T$_{kin}$ - $\langle\beta_{\rm T}\rangle$ plane, where T$_{kin}$ represents the kinetic freeze-out temperature and $\beta_{\rm T}$ the radial expansion velocity. In this picture, radial velocity increases while the freeze-out temperature decreases with increasing centrality in Pb–Pb. The resulting fit parameters are reported for the three systems in figure 4. The evolution with multiplicity is similar in pp and p–Pb, but at similar multiplicities the smaller system exhibits a larger $\beta_{\rm T}$. In p–Pb the fit to the highest-multiplicity class π, K and p spectra returns parameters that can also describe the Ξ and Ω spectra, contrary to what happens in Pb–Pb collisions [9]. In pp collisions, despite still large uncertainties on the parameters of the fit, Ξ and Ω spectra can also be described fairly well by the Blast-Wave model. The Blast-Wave fit to particle spectra is increasingly being used as a tool to search for radial flow-like collectivity. However some caution has to be applied in the interpretation of the results, especially when comparing the results of the fit across different systems or experiments, where the sensitivity to the fitting range may play a role. Moreover, it has to be noted that mechanisms such as Color Reconnection [16], as implemented in the PYTHIA8 event generator [17], can mimic the effects of a collective radial expansion. Finally, the spectra of resonances seem not to be successfully predicted by using the parameters from the simultaneous Blast-Wave fit to long-lived hadrons.

Short-lived resonances decay very early, with lifetimes that are of the same order of magnitude as the hadronic phase in Pb–Pb collisions. Only the ϕ $p_{\rm T}$ spectrum is satisfactorily described by the Blast-Wave model in central Pb–Pb collisions [6], whereas the model overshoots the yield of ${\rm K^{*0}}$ at low $p_{\rm T}$. This is presently interpreted as an effect of re-scattering in the hadronic phase. Re-scattering effects of resonances are expected to be stronger in central Pb–Pb collisions, where the medium is denser and longer lived, and to affect predominantly the low momentum part of the spectrum ($p_{\rm T}<$ 2 GeV/c [18]). This is reflected in the suppression observed for the $p_{\rm T}$-integrated yield ratios of ${\rm K^{*0}}$/K and ρ^0/π going from peripheral to central Pb–Pb [19]. ${\rm K^{*0}}$ and

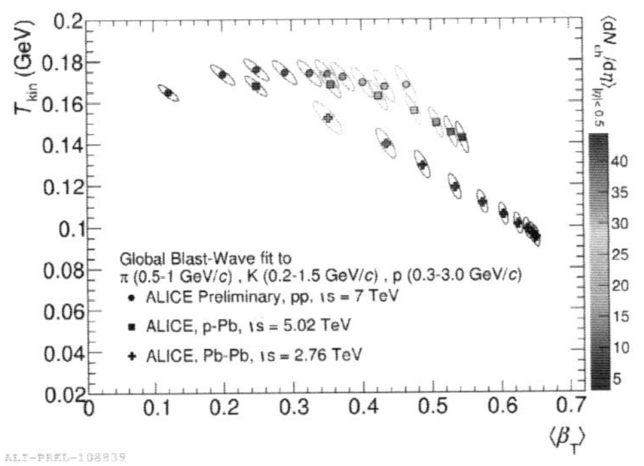

Figure 4. Blast-Wave parameters resulting from the simultaneous fit of identified particle spectra in pp, p–Pb and Pb–Pb collisions in various multiplicity or centrality bins. Colors indicate the scale in multiplicity.

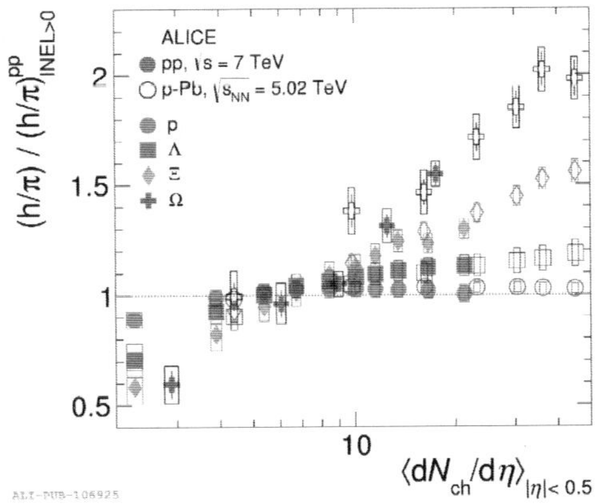

Figure 5. Particle yield relative to pion in pp at $\sqrt{s}$ = 7 TeV and p–Pb at $\sqrt{s_{\rm NN}}$ = 5.02 TeV as a function of the average charged particle multiplicity, normalised to the same ratio in multiplicity-integrated pp (INEL>0).

ρ^0 are indeed about ten times shorter-lived than the ϕ, whose ratio to K shows no significant variation over centrality. The decreasing trend of both resonances is qualitatively reproduced by the EPOS3 event generator when the core-corona development of the fireball is followed by a hadronic afterburner phase modelled via UrQMD [20]. Recent measurements of the $\mathrm{K}^{*0}/\mathrm{K}$ ratio in p–Pb and pp (see figure 1) also hint at the presence of a decreasing trend with multiplicity, although not significant given the present uncertainties.

4. Strangeness production in pp, p–Pb and Pb–Pb

The ratios of strange and multi-strange hadron production relative to pions in pp at $\sqrt{s}$ = 7 TeV, p–Pb at $\sqrt{s_{\rm NN}}$ = 5.02 TeV and Pb–Pb at $\sqrt{s_{\rm NN}}$ = 2.76 TeV are reported in figure 1 as a function of $\langle \mathrm{d}N_{\rm ch}/\mathrm{d}\eta \rangle$. In Pb–Pb the ratios follow a rather constant behaviour, saturating to values that are consistent with those predicted by Grand-Canonical thermal models [21, 22]. In pp and p–Pb, Ξ/π and Ω/π increase from low to high multiplicity events to values consistent with those measured in central and peripheral Pb–Pb collisions, respectively. A similar behaviour with multiplicity is observed in p–Pb for the resonant state of Ξ in the Ξ^{*0}/π ratio, suggesting that the strangeness content of the hyperon drives the increase and dominates over possible mass effects. A comprehensive set of measurements in pp collisions at $\sqrt{s}$ = 7 TeV [23] shows that while $\Lambda/\mathrm{K}^0_{\rm S}$ and p/π do not increase significantly, the Ξ/π and Ω/π increase with multiplicity following a trend compatible with that in p–Pb collisions across the same range of multiplicity. The comparison with other non-strange identified-particle ratios, such as the p/π ratio or $\Lambda/\mathrm{K}^0_{\rm S}$ can help in understanding if the increase is related to mass or to strangeness content. Both ratios are found flat with multiplicity. Figure 5 shows the pp and p–Pb hadron-to-pion ratios as a function of multiplicity normalised by the multiplicity-integrated ratio in pp, corresponding to the INEL>0 event class. While the p/π ratio stays practically constant, the increase of strange and multi-strange obeys a clear hierarchy with the strangeness content. PYTHIA8 [17] cannot reproduce the observed enhancement with multiplicity in pp. DIPSY [24] and EPOS LHC [11] predict a rising trend with multiplicity but miss the magnitude of the ratio and the flat trend of p/π.

Studies on multiplicity dependence of identified particle production continue on the data of pp collisions at $\sqrt{s} = 13$ TeV collected during the Run II of the LHC. An open question is whether the observed trend continues even further with multiplicity or if, instead, the strange to non-strange production saturates. In order to address this question, ALICE has collected a data sample with dedicated triggers to select high-multiplicity proton-proton events, that are being analysed at the moment of writing. The first results in minimum bias pp collisions at $\sqrt{s} = 13$ TeV show that the average charged particle multiplicity increases by about 15% from 7 TeV to the new collision energy [25]. No dependence on $\sqrt{s}$ is observed in the differential and integrated K/π ratio, nor for the integrated K^{*0}/K and ϕ/K ratios, continuing the trend from lower to LHC energies. Instead, Ξ/π and Ω/π exhibit a small increase from inelastic collisions at $\sqrt{s} = 7$ TeV to $\sqrt{s} = 13$ TeV. Whether this increase can be described as being only a consequence of the higher average multiplicity that is attained at the higher collision energy is a question that will be addressed with a multiplicity-differential study based on the 13 TeV data.

5. Outlook and conclusions

Several new ALICE results have been presented at this conference, including the preliminary measurements of the nuclear modification factor of charged particles in Pb–Pb collisions at $\sqrt{s_{NN}} = 5.02$ TeV and the measurements of the production of a comprehensive set of identified particles with different strangeness content. The results in pp collisions have revealed interesting features that are similar to those observed in p–Pb and Pb–Pb. While these features are usually interpreted as due to collectivity in Pb-Pb collisions, their origin in smaller systems is still to be fully understood (radial flow, color reconnection...). An enhancement of strangeness production has been observed towards high-multiplicity pp with respect to inelastic events, that is not satisfactorily explained by QCD-inspired models available. The analysis of the data collected in the LHC Run II at the highest collision energy ever reached in a collider may shed more light on the energy and multiplicity evolution of strangeness production.

References

[1] Rafelski J and Müller B 1982 *Phys. Rev. Lett.* **48**(16) 1066–1069
[2] Khachatryan V *et al.* (CMS) 2010 *JHEP* **09** 091
[3] Abelev B *et al.* (ALICE) 2013 *Phys.Lett.* **B719** 29–41
[4] Abelev B B *et al.* (ALICE Collaboration) 2014 *Int. J. Mod. Phys.* **A29** 1430044
[5] Adam J *et al.* (ALICE Collaboration) 2016 *Phys. Rev.* **C93** 034913
[6] Abelev B B *et al.* (ALICE Collaboration) 2015 *Phys. Rev.* **C91** 024609
[7] Gronefeld J M (ALICE Collaboration) 2016 *PoS (LHCP2016)* 126
[8] Adam J *et al.* (ALICE) 2016 *Phys. Lett.* **B760** 720–735
[9] Colella D (ALICE Collaboration) 2016 *these proceedings*
[10] Adam J *et al.* (ALICE) 2016 *Eur. Phys. J.* **C76** 245
[11] Pierog T, Karpenko I, Katzy J M, Yatsenko E and Werner K 2015 *Phys. Rev.* **C92** 034906
[12] Minissale V, Scardina F and Greco V 2015 *Phys. Rev.* **C92** 054904
[13] Souza R D D (ALICE Collaboration) 2016 *these proceedings*
[14] Schnedermann E, Sollfrank J and Heinz U W 1993 *Phys. Rev.* **C48** 2462–2475
[15] Abelev B *et al.* (ALICE Collaboration) 2013 *Phys.Rev.* **C88** 044910
[16] Bierlich C and Christiansen J R 2015 *Phys. Rev. D* **92**(9) 094010
[17] Sjostrand T, Mrenna S and Skands P Z 2008 *Comput. Phys. Commun.* **178** 852–867
[18] Bleicher M and Aichelin J 2002 *Phys. Lett.* **B530** 81–87
[19] Knospe A (ALICE Collaboration) 2016 *these proceedings*
[20] Knospe A G, Markert C, Werner K, Steinheimer J and Bleicher M 2016 *Phys. Rev.* **C93** 014911
[21] Stachel J, Andronic A, Braun-Munzinger P and Redlich K 2014 *J. Phys. Conf. Ser.* **509** 012019
[22] Cleymans J, Kraus I, Oeschler H, Redlich K and Wheaton S 2006 *Phys. Rev.* **C74** 034903
[23] Adam J *et al.* (ALICE) 2016 (*Preprint* 1606.07424)
[24] Bierlich C, Gustafson G, Lönnblad L and Tarasov A 2015 *JHEP* **03** 148
[25] Adam J *et al.* (ALICE Collaboration) 2016 *Phys. Lett.* **B753** 319–329

Strangeness in STAR experiment at RHIC

Shusu Shi (for the STAR collaboration)
Key Laboratory of Quarks and Lepton Physics (MOE) and Institute of Particle Physics, Central China Normal University, Wuhan, 430079, China
E-mail: shiss@mail.ccnu.edu.cn

Abstract. We present the recent results of strangeness production at the mid-rapidity in Au + Au collisions at RHIC, from $\sqrt{s_{\rm NN}}$ = 7.7 to 200 GeV. Elliptic asymmetry v_2 of multi-strange baryon Ω and ϕ mesons are similar to that of pions and protons in the intermediate p_T range (2 - 5 GeV/c) in $\sqrt{s_{\rm NN}}$ = 200 GeV Au + Au collisions, indicating that the major part of collective flow has been built up at partonic stage. The breaking of mass ordering between ϕ mesons and protons in the low p_T range (< 1 GeV/c) is consistent with a picture that ϕ mesons are less sensitive to later hadronic interaction. The nuclear modification factor $R_{\rm CP}$ and baryon to meson ratio change dramatically when the collision energy is lower than 19.6 GeV. It suggests a possible change of the created QCD medium properties at lower energies compared to those from high energies.

1. Introduction

The main motivations of the experiments of heavy-ion collision at RHIC are 1) generating the new form of matter Quark Gluon Plasma at high energy collisions and studying its properties, and 2) exploring the QCD phase structure by scanning the collision energy from 200 to 7.7 GeV. Strangeness production is regarded as a sensitive probe to the phase transition [1]. In addition, the hadronic interaction cross sections for strange hadrons, especially for multi-strange hadrons Ξ, Ω and ϕ mesons are expected to be small and their freeze-out temperatures are close to the quark-hadron transition temperature predicted by lattice QCD [2, 3, 4]. Hence, these hadrons are expected to provide information primarily from the partonic stage of the collision. They are good probes to study the QGP properties and explore the QCD phase structure. The STAR experiment has covered the beam energies of $\sqrt{s_{\rm NN}}$ = 7.7, 11.5, 14.5, 19.6, 27, 39, 62.4 and 200 GeV. The extracted baryonic chemical potental (μ_B) range based on a statistical model [5] from the 0-5% central collisions is $20 \leq \mu_B \leq 420$ MeV which covers a wide region of the QCD phase diagram. In these proceedings, we will focus on the following observables for strange and multi-strange hadrons: elliptic flow v_2, nuclear modification factor $R_{\rm CP}$ and baryon to meson ratio.

2. Results and Discussions

In the RHIC runs of year 2010 and 2011, about 730 million minimum bias events of Au + Au collisions at $\sqrt{s_{\rm NN}}$ = 200 GeV were recorded by STAR. Sufficient statistics of multi-strange hadrons and ϕ mesons allow for the precise measurements on v_2. Figure 1 shows the v_2 as a function of p_T for (a) pions, protons and (b) ϕ, Ω in Au + Au collisions at $\sqrt{s_{\rm NN}}$ = 200 GeV for 0-80% centrality [6]. A comparison between v_2 of charged pions and protons, consisting of up

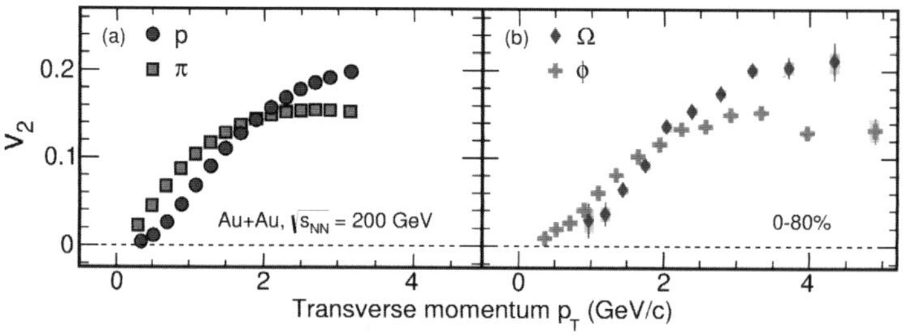

Figure 1. The v_2 as function of p_T for π (π^++π^-), p (p+$\bar{p}$) and ϕ, Ω (Ω^-+$\bar{\Omega}^+$) in Au+Au collisions at $\sqrt{s_{\rm NN}}$ = 200 GeV for 0-80% centrality [6].

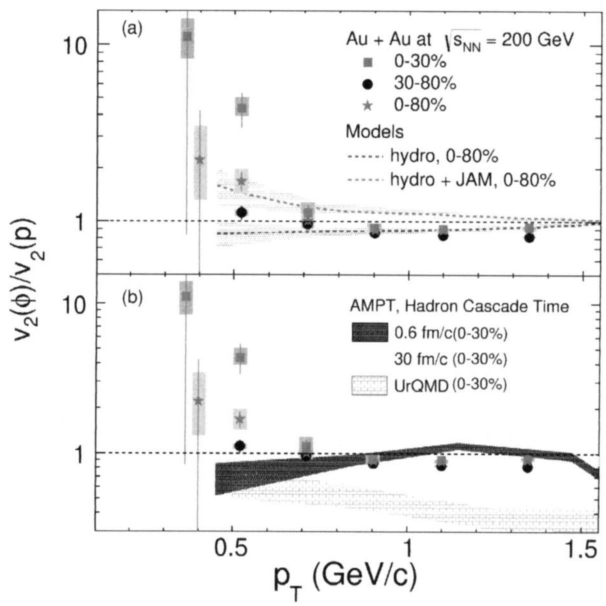

Figure 2. The ratio of $v_2(\phi)$ to $v_2(p)$ as function of p_T in Au + Au collisions at $\sqrt{s_{\rm NN}}$ = 200 GeV for 0-30% and 30-80% centrality. The bands in panel (a) and (b) represent the hydro and transport model calculations for $v_2(\phi)/v_2(p)$, respectively [6].

($u(\bar{u})$) and down ($d(\bar{d})$) light constituent quarks is shown in panel (a). Correspondingly, panel (b) shows a comparison of v_2 of ϕ and $\Omega(\Omega^-$+$\bar{\Omega}^+)$ containing $s(\bar{s})$ constituent quarks. This is the first time that high precision measurement of Ω baryon v_2 up to 4.5 GeV/c is available in experiments of heavy-ion collisions. In the low p_T region (p_T < 2.0 GeV/c), the v_2 of ϕ and Ω follows mass ordering. At intermediate p_T (2.0 < p_T < 5.0 GeV/c), a baryon-meson separation is observed. It is evident that the $v_2(p_T)$ of hadrons consisting only of strange constituent quarks (ϕ and Ω) is similar to that of light hadrons, pions and protons. However the ϕ and Ω do not participate vigorously in the hadronic interactions, because of the smaller hadronic cross sections compared to pions and protons. This suggests the major part of the collectivity is developed

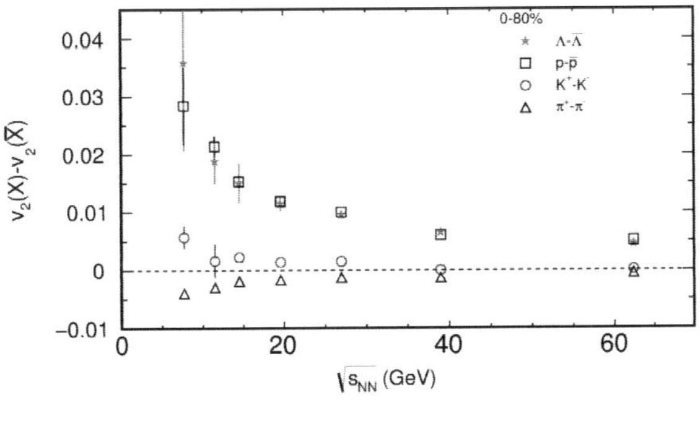

Figure 3. The difference in v_2 between particles (X) and their corresponding antiparticles ($\bar{X}$) as a function of beam energy for $0 - 80\%$ central Au + Au collisions [10, 11].

during the partonic phase in high energy heavy-ion collisions. Experimental results indicate that partonic collectivity has been built up in top energy heavy-ion collisions at RHIC. The ϕ mesons and protons show different sensitivity to the hadronic rescatterings. The ϕ mesons are less sensitive to the late hadron hadron interactions than light hadrons due to the smaller hadronic cross section. Hydrodynamical model calculations [7] predict that v_2 as a function of p_T for different particle species follows mass ordering, where the v_2 of heavier hadrons is lower than that of lighter hadrons. Hirano *et al.* predict the mass ordering of v_2 could be broken between ϕ mesons and protons at low p_T ($p_T <$ 1.5 GeV/c) based on a model with ideal hydrodynamics with hadron cascade process [7]. As the model calculations assign a smaller hadronic cross section for ϕ mesons compared to protons, the broken mass ordering is regarded as the different hadronic rescattering contributions on the ϕ meson and proton v_2. Figure 2 shows the ratios of ϕ v_2 to proton v_2 from model calculations and experimental data [6]. This ratio is larger than unity at $p_T \sim 0.5$ GeV/c for 0-30% centrality. It indicates breakdown of the expected mass ordering in that momentum range. This could be due to a large effect of hadronic rescatterings on the proton v_2. The data of 0-80% centrality around 0.5 GeV/c quantitatively agrees with hydro + hadron cascade calculations indicated by the shaded red band in panel (a) of Fig. 2, even though there is a deviation in higher p_T bins. A centrality dependence of $v_2(\phi)$ to $v_2(p)$ ratio is observed in the experimental data. The breakdown of mass ordering of v_2 is more pronounced in 0-30% central collisions than in 30-80% non-central collisions. This observation indirectly supports the idea that the ϕ meson has a smaller hadronic interaction cross section. The ratio of ϕ v_2 to proton v_2 was also studied by using the transport models AMPT [8] and UrQMD [9]. The panel (b) of Fig. 2 shows the $v_2(\phi)$ to $v_2(p)$ ratio for 0-30% centrality from AMPT and UrQMD models. The black shaded band is from AMPT with a hadronic cascade time of 0.6 fm/c while the yellow band is for a hadronic cascade time of 30 fm/c. Larger hadronic cascade time is equivalent to stronger hadronic interactions. It is clear that the $v_2(\phi)/v_2(p)$ ratio increases with increasing hadronic cascade time. This is attributed to a decrease in the proton v_2 due to an increase in hadronic rescattering while the ϕ meson v_2 is less affected. The ratios from the UrQMD model are much smaller than unity (shown as a brown shaded band in the panel (b) of Fig. 2). The UrQMD model lacks partonic collectivity, thus the ϕ meson v_2 is not fully developed. None of these models could describe the detailed shape of the p_T dependence.

A very interesting feature of the v_2 measurements from RHIC Beam Energy Scan program is the observation of an energy dependent difference in v_2 between particles and their corresponding antiparticles [10]. Figure 3 shows the difference in v_2 between particles and their corresponding antiparticles as a function of beam energy. The difference between baryon and anti-baryon is much more pronounced than difference between mesons. Proton versus anti-proton and Λ versus $\bar{\Lambda}$ show the same magnitude of difference. This difference naturally breaks the number of constituent quark scaling (NCQ) in v_2 which is regarded as an evidence of partonic collectivity in the top energy heavy-ion collisions at RHIC. It indicates the hadronic degrees of freedom play a more important role at lower collision energies.

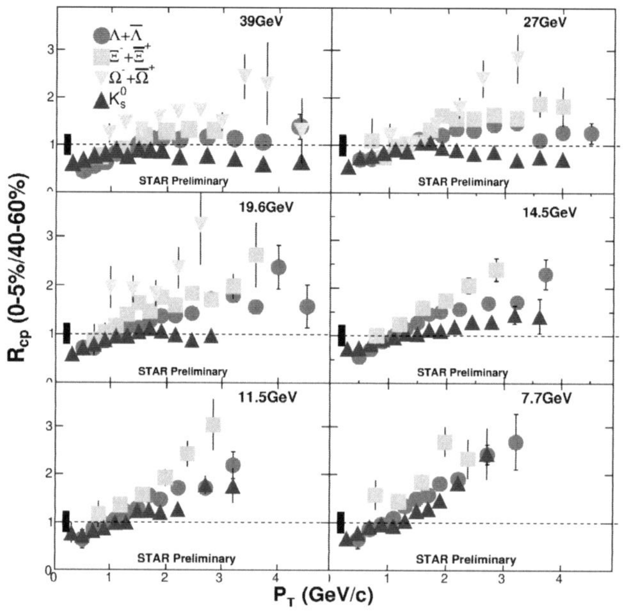

Figure 4. The nuclear modification factor $R_{\rm CP}$ as a function of p_T in Au + Au collisions at $\sqrt{s_{\rm NN}}$ = 7.7 - 39 GeV for K^0_S, Λ, Ξ and Ω at mid-rapidity $|y| < 0.5$ [12].

The $R_{\rm CP}$ ratio is defined as the ratio of particle yields in central collisions over those in peripheral ones scaled by number ($N_{\rm bin}$) of inelastic binary collisions. Here, $N_{\rm bin}$ is determined from Monte Carlo Glauber model calculations. The $R_{\rm CP}$ of K^0_S, Λ, Ξ and Ω in Au+Au 7.7 - 39 GeV collisions are presented in Fig. 4 [12]. The $R_{\rm CP}$ will be unity if nucleus-nucleus collisions are just simple superpositions of nucleon-nucleon collisions. Deviation of these ratios from unity would imply contributions from nuclear or medium effects. At collision energy $\geq$ 19.6 GeV, the $R_{\rm CP}$ of $K^0_S \leq 1$ at $p_T >$ 1.5 GeV/c and stays below $R_{\rm CP}$ of baryons. The high p_T suppression of K^0_S and baryon/meson separation is qualitatively consistent with the behavior observed at higher RHIC energies [13]. However, for collisions at 14.5 GeV and below, the data seem to be qualitatively different from those from higher energies. There is no suppression for K^0_S at p_T > 1.5 GeV/c, and at intermediate p_T the baryon/meson separation becomes less significant. It suggests that different properties of the system have been created in collisions at 14.5 GeV and below compared to those from high energies.

The enhancement of baryon to meson ratios at intermediate p_T compared to elementary collisions is interpreted as a consequence of hadron formation from parton recombination and parton collectivity [13]. Therefore, the baryon to meson ratios are expected to be sensitive to parton dynamics of the collision system. Figure 5 shows the ratio of $\bar{\Lambda}/K^0_S$ as a function of p_T in Au + Au collisions at 7.7 - 39 GeV at mid-rapidity $|y| < 0.5$ [12]. The ratios of $\bar{\Lambda}$ to K^0_S at intermediate p_T are close to each other at 27 and 39 GeV, and show a slight decrease at 19.6

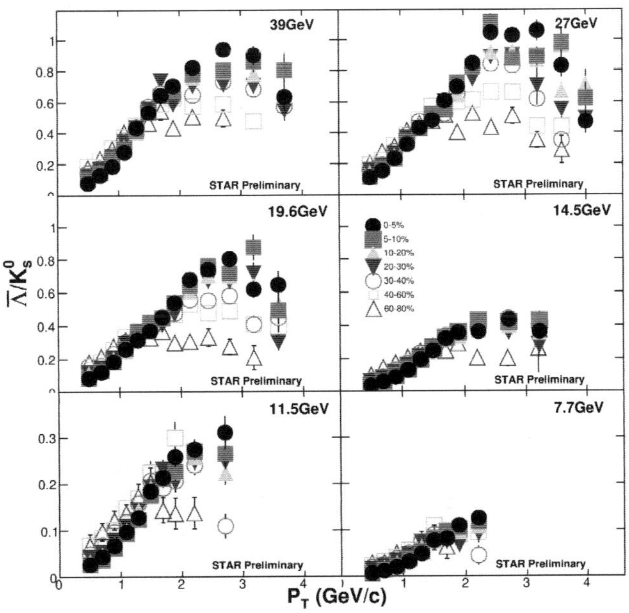

Figure 5. The ratio of $\bar{\Lambda}/K_S^0$ as a function of p_T in Au + Au collisions at $\sqrt{s_{\rm NN}}$ = 7.7 - 39 GeV at mid-rapidity $|y| < 0.5$ [12].

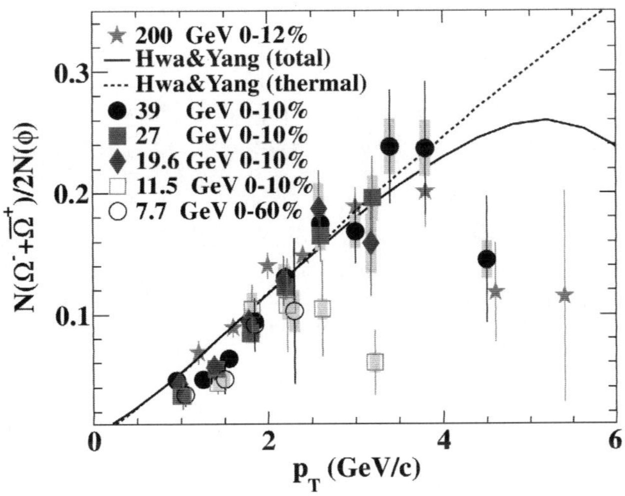

Figure 6. The ratio of $N(\Omega^- + \overline{\Omega}^+)$ to $(2N(\phi))$ as a function of p_T in Au+Au collisions at $\sqrt{s_{\rm NN}}$ = 7.7 to 200 GeV at mid-rapidity ($|y| < 0.5$) [14, 15]. The solid and dashed lines represent recombination model calculations for central collisions at $\sqrt{s_{\rm NN}}$ = 200 GeV with total and thermal strange quark contributions respectively [16].

GeV. There is a sudden decrease of intermediate p_T ratios between 19.6 and 14.5 GeV. Besides, the separation of central (0-5%) and peripheral (40-60%) collisions in the ratio becomes less obvious in collisions at 14.5 GeV and below. It suggests a possible change of underlying hadron formation mechanism and/or parton dynamics between these two energies. We use multi-strange hadrons, Ω to ϕ ratios, to further study the energy dependence of baryon to meson ratios.

In Fig. 6 [15] we present the results on $\Omega^-+\bar{\Omega}^+$ and ϕ yields ratio as a function of p_T for Au+Au collisions at 7.7 to 200 GeV. A model calculation by Hwa and Yang for Au+Au collisions at 200 GeV predicted that most of the Ω and ϕ yields up to the intermediate p_T region are from coalescence of thermal strange quarks [16]. The straight dotted line assumes that these thermal strange quarks have exponential p_T distributions. Deviations from the straight line at high p_T are attributed [16] to recombination with strange quarks from high p_T showers. The measured ratios from central Au+Au collisions at 19.6, 27 and 39 GeV follow closely the ratio from 200 GeV and are consistent with a picture of coalescence dynamics over a broad p_T range of 1-4 GeV/c. The ratios at 11.5 GeV seem to deviate from the trend observed at higher beam energies. In particular, the ratios appear to turn down around p_T of 2 GeV/c. The decrease in the Ω to ϕ ratios from central collisions at 11.5 GeV compared to those at 19.6 GeV or above may indicate a significant change in the hadron formation and/or in strange quark p_T distributions at the lower energy. Such a change may arise from a transition from hadronic to partonic dynamics with increasing beam energy.

3. Summary

In summary, the high precision v_2 data of multi-strange hadrons, namely for Ω baryon and ϕ meson strongly suggest the partonic collectivity has been built-up at top energy heavy-ion collisions at RHIC. The violation of mass ordering between ϕ mesons and protons at low p_T agrees with the expectation that ϕ mesons are less sensitive to late hadronic interactions. R_{CP} ratio for K^0_S increases with decreasing beam energies indicating that partonic energy loss effect is less important at lower energy. The separation of central and peripheral $\bar{\Lambda}/K^0_S$ ratio is diminishing for Au+Au collisions at $\sqrt{s_{\mathrm{NN}}} = 14.5$, 11.5 and 7.7 GeV and the decrease of $(\Omega^-+\bar{\Omega}^+)/2\phi$ ratios from central collisions at 11.5 GeV suggest the change of medium property and phase transition possibly happen between 19.6 GeV and lower energies.

4. Acknowledgments

This work was supported in part by the National Basic Research Program of China (973 program) under grant No. 2015CB8569 and National Natural Science Foundation of China under grant No. 11475070 and 11628508.

[1] J. Rafelski and B. Muller, Phys. Rev. Lett. **48**, 1066 (1982).
[2] J. Adams *et al.*, Nucl. Phys. **A 757** 102 (2005).
[3] A. Shor, Phys. Rev. Lett. **54**, 1122 (1985).
[4] H. van Hecke, H. Sorge and N. Xu, Phys. Rev. Lett. **81**, 5764 (1998).
[5] A. Andronic, *et al.*, Nucl. Phys. **A834**, 237(2010); A. Andronic, P. Braun-Munzinger and J. Stachel, Nucl. Phys. **A772**, 167(2006).
[6] L. Adamczyk *et al.* (STAR Collaboration), Phys. Rev. Lett. **116**, 062301 (2016).
[7] T. Hirano *et al.*, Phys. Rev. **C 77**, 044909 (2008); S. Takeuchi *et al.*, Phys. Rev. **C 92**, 044907 (2015).
[8] Z.-W. Lin *et al.*, Phys. Rev. **C 72**, 064901 (2005).
[9] S. A. Bass *et al.*, Prog. Part. Nucl. Phys. **41**, 255 (1998).
[10] L. Adamczyk *et al.* (STAR Collaboration), Phys. Rev. C **86**, 054908 (2012); Phys. Rev. C **88**, 014902 (2013); Phys. Rev. C **93**, 014907 (2016).
[11] S.S. Shi, Adv. High Energy Phys. **2016**, 1987432 (2016).
[12] M. U. Ashraf (for the STAR Collaboration), J. Phys. Conf. Ser. **668** 012095 (2016).
[13] M. M.Aggarwal *et al.* (STAR Collaboration), arXiv:1007.2613.
[14] X. Zhang (for the STAR Collaboration), J. Phys. Conf. Ser. **668** 012033 (2016).
[15] L. Adamczyk *et al.* (STAR Collaboration), Phys. Rev. C **93**, 021903 (2016).
[16] R. C. Hwa and C. B. Yang, Phys. Rev. C **66**, 025205 (2002); Phys. Rev. C **75**, 054904 (2007).

Collectivity and thermalization in p+p, p+A, d+Au and ^{3}He+Au collisions

Shengli Huang

Department of Physics and Astronomy, Station B 1807, Vanderbilt University, Nashville, TN 37235, USA

E-mail: shengli.huang@vanderbilt.edu

Abstract. In this proceeding, the recent measurements of the anisotropy(v_2, v_3) in small collision systems such as p+p, p+A, d+Au and ^{3}He+Au collisions at the LHC and RHIC energies are presented. In p+p collisions, the v_2 are measured by two and multi-particle $\Delta\phi$ correlations at $\sqrt{s_{NN}}$ = 2.76, 5, 7, and 13 TeV. In 200 GeV p+Au, d+Au and ^{3}He+Au collisions. the v_2 are measured with event plane methods. I also present the measurements of v_2 for K^0_S and $\Lambda/\overline{\Lambda}$ in high multiplicity p+p collisions, in which a particle mass ordering is observed. The v_2 of identified $\pi^\pm$, $K^\pm$ and(anti)protons in central ^{3}He+Au collisions are also measured and quark-number scaling is observed. These observations are similar to those seen in A+A collisions, and support the interpretation of a collective origin.

1. Introduction

Recently, in the high-multiplicity small collision systems such as p+p and $p/d/^3$He+A collisions, the near-side, long-range two particle angular correlation (so called 'ridge') and anisotropic flow (v_n) have been observed at the LHC [1, 2, 5, 3, 4] and RHIC [6, 7, 8]. Several physics models, which include initial-state gluon saturation [9], and hydrodynamic flow from a mini-QGP(Quark Gluon Plasma) [10] are potential explanations for these observations. While the former will not depend on the initial geometry, the hydrodynamic models have strong initial geometry dependence. Therefore, measuring the v_2 and v_3 in $p/d/^3$He+Au collisions will provide direct testing for these explanations since the initial geometry is quite different for these collisions [11]. If a mini-QGP exists in the small collision system, studying it will help us to further understand: 1)the condition for the QGP thermalization; 2)the role of the internal structure of proton on the initial geometry eccentricity [12]; and 3)the contributions from pre-equilibrium stage to the system evolution [13].

2. The v_2 measurements in p+Au, d+Au and ^{3}He+Au collisions at $\sqrt{s_{NN}}$ = 200 GeV at RHIC

The v_2 for inclusive charged hadrons produced at mid-rapidity $|\eta| < 0.35$ are measured in 0-5% central p+Au in the PHENIX experiment and compared with that from d+Au and ^{3}He+Au collisions, as shown in the Fig. 1. The v_2 in the p+Au is found to be smaller than that of d+Au and ^{3}He+Au. The eccentricity of p+Au is also smaller than that of d+Au and ^{3}He+Au. It indicates the anisotropy in the small collisions system depends on the initial geometry, which is supported by the hydrodynamic model calculation [13] as shown in the figure.

The v_3 for inclusive charged hadrons produced at mid-rapidity $|\eta| < 0.35$ in high-multiplicity ^{3}He+Au collisions at $\sqrt{s_{NN}} = 200$ GeV are also measured in the PHENIX experiment with respect to the Ψ_3 event planes. The v_3 results are shown in Fig. 2 with v_2 and different theory calculations. Four such predictions shown in Fig. 2 employ viscous hydrodynamics with η/s at or near the conjectured lower bound $1/4\pi$ [15] and one is from the AMPT (A-Multi-Phase-Transport-Model) framework [16]. The SONIC calculation [11] employs Glauber initial conditions, viscous hydrodynamics, and then at $T = 170$ MeV a transition to a hadronic cascade. The SUPERSONIC calculation [13] additionally includes pre-equilibrium dynamics that boosts the initial velocity fields at the earliest times. The impact of pre-equilibrium is modest on the v_2 values and the data agree with both calculations within uncertainties. The effect of pre-equilibrium on v_3 is significantly larger as the triangular flow takes longer to develop [11]. The SUPERSONIC prediction agrees well with the experimental data for $p_T < 1.2$ GeV/c, and then the data trends towards the SONIC prediction at higher p_T.

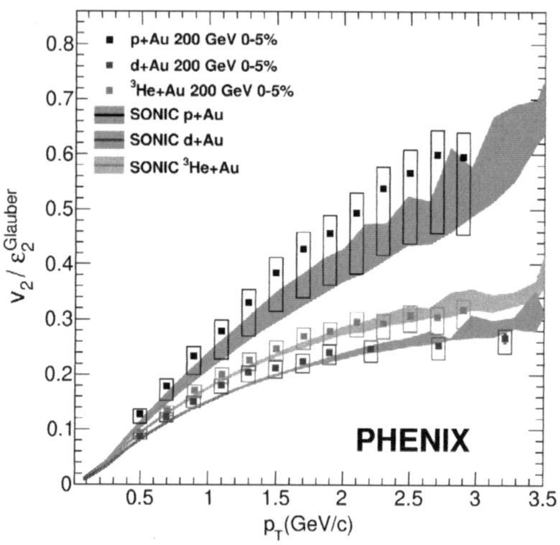

Figure 1: Measured v_2 for mid-rapidity charged tracks in 0-5% central p+Au,d+Au and ^{3}He+Au at $\sqrt{s_{NN}} = 200$ GeV using the event plane method. Also shown are SONIC model calculations [13]

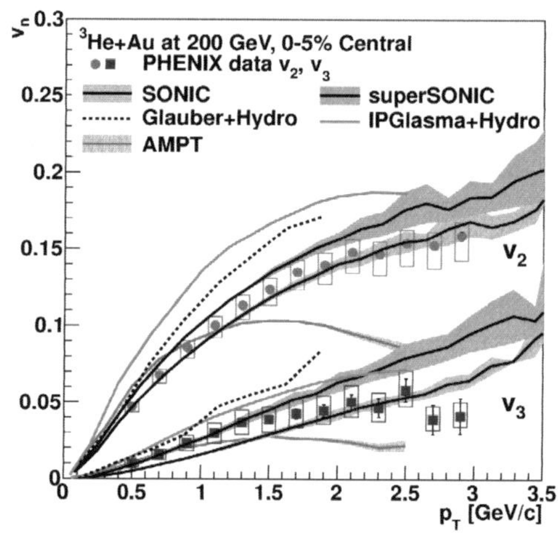

Figure 2: v_2 (circles) and v_3 (squares) as a function of p_T for inclusive charged hadrons at mid-rapidity in 0-5% central ^{3}He+Au collisions at $\sqrt{s_{NN}} = 200$ GeV; Also shown are various theoretical calculations.

3. The v_2 measurements in p+p collisions at $\sqrt{s_{NN}} = 2.76$, 5, 7, and 13 TeV at the LHC

The v_2 in p+p collisions at $\sqrt{s_{NN}} = 2.76$ and 13 TeV has been measured with long-range($|\Delta\eta| > 2$) two particle $\Delta\phi$ correlations by ATLAS [20]. A new method, called template fitting has been developed to separate the ridge and the back-to-back jet correlation in this long-rang two particles correlation, by assuming that the shape of the jet-induced correlations is invariant with event multiplicity and can be extracted from low-multiplicity events. The second order harmonic coefficients $v_{2,2}$ from template fitting and the v_2 extracted by assuming factorization are shown in the Fig 3. Both $v_{2,2}$ and v_2 show a very week multiplicity dependence and are nearly identical for collision energy from 2.76 to 7 TeV.

The anisotropy of charged particles are also measured in p+p collisions at 13 TeV by CMS [17]. A different subtraction method [4] is employed by CMS. The second-order Fourier coefficients

$V_{2\Delta}$ extracted from long-range two-particle $\Delta\phi$ correlations in the higher-multiplicity region are subtracted from the $V_{2\Delta}$ coefficients from $10 \leq N_{\rm trk}^{\rm offline} < 20$ with

$$V_{2\Delta}^{\rm sub} = V_{2\Delta} - V_{2\Delta}(10 \leq N_{\rm trk}^{\rm offline} < 20)\,\frac{N_{\rm assoc}(10 \leq N_{\rm trk}^{\rm offline} < 20)}{N_{\rm assoc}}\,\frac{Y_{\rm jet}}{Y_{\rm jet}(10 \leq N_{\rm trk}^{\rm offline} < 20)}. \quad (1)$$

Here, $Y_{\rm jet}$ represents the near-side jet yield obtained by integrating the difference of the short- and long-range event-normalized associated yields for each multiplicity class as shown for $105 \leq N_{\rm trk}^{\rm offline} < 150$ over $|\Delta\phi| < 1.2$. The ratio, $Y_{\rm jet}/Y_{\rm jet}(10 \leq N_{\rm trk}^{\rm offline} < 20)$, is introduced to account for the enhanced jet correlations resulting from the selection of higher-multiplicity events.

The $V_{2\Delta}$ is shown in the Fig 4 as a function of $N_{\rm trk}^{\rm offline}$ for charged particles. Before subtraction, the $V_{2\Delta}$ coefficients are found to be nearly constant as a function of multiplicity. After subtraction, $V_{2\Delta}$ exhibits an increase with multiplicity for $N_{\rm trk}^{\rm offline} \lesssim 100$. This jet subtraction procedure is also tested in PYTHIA simulations, where no jet modification from initial or final state effects is present. The $V_{2\Delta}$ after subtraction is found to be consistent with zero as shown in the Fig 4.

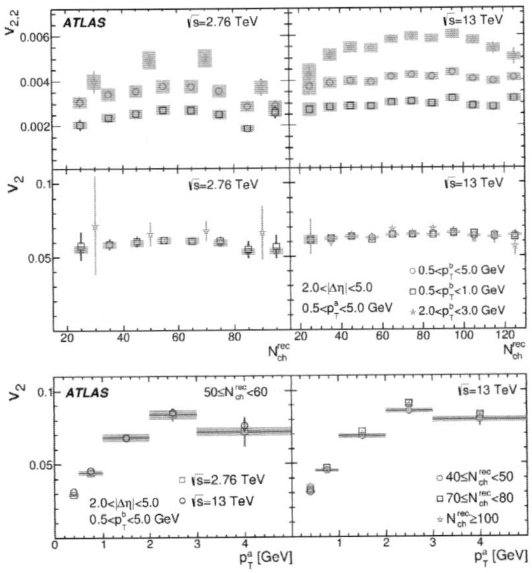

Figure 3: The integral $v_{2,2}$ and v_2 from template fitting as a function of multiplicity in in p+p collisions at $\sqrt{s_{NN}} = 2.76$ TeV(top left) and 7 TeV(top right). The v_2 as a function of p_T are also shown in these two collision energy(bottom left) and different multiplicity bins at 7 TeV.(bottom right)

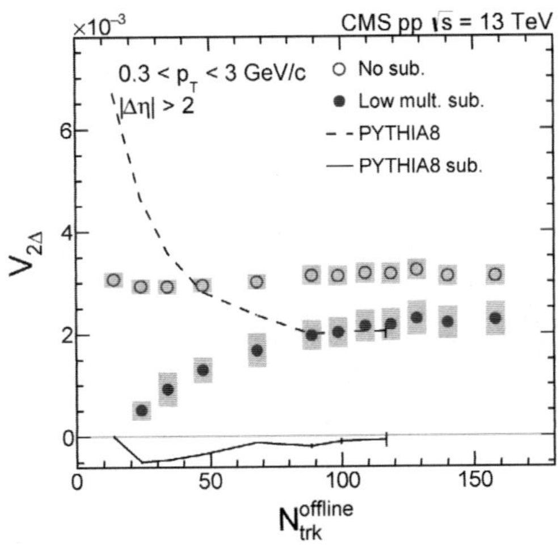

Figure 4: $V_{2\Delta}$ as a function of $N_{\rm trk}^{\rm offline}$ for charged particles, averaged over $0.3 < p_T < 3.0 GeV/c$, in p+p collisions at $\sqrt{s_{NN}} = 13$ TeV, before (open) and after (filled) subtraction of jet correlations, estimated from the $10 \leq N_{\rm trk}^{\rm offline} < 20$ range. Results from PYTHIA are shown as curves.

4. Fourier harmonics from multi-particle correlations

Different methods of jet subtraction will lead to quite different multiplicity dependence for v_2 measurements, as shown by results above. To avoid this uncertainty, CMS also measures the v_2 in p+p collisions by using a multi-particle cumulant method [21]. This technique has the advantage of suppressing short-range two-particle correlations such as jets and resonance decays. The corresponding cumulants, $c_n\{4\}$ and $c_n\{6\}$, are calculated as follows [21]:

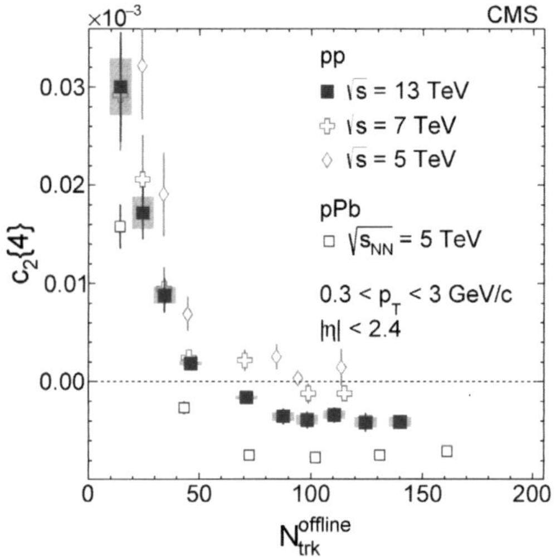

Figure 5: $c_n\{4\}$ as a function of $N_{\text{trk}}^{\text{offline}}$ for charged particles, averaged over $0.3 < p_T < 3.0 GeV/c$, in p+p collisions at $\sqrt{s_{NN}} = 5$, 7, and 13 TeV. The $c_n\{4\}$ from p+Pb collisions at 5 TeV are also shown for comparison

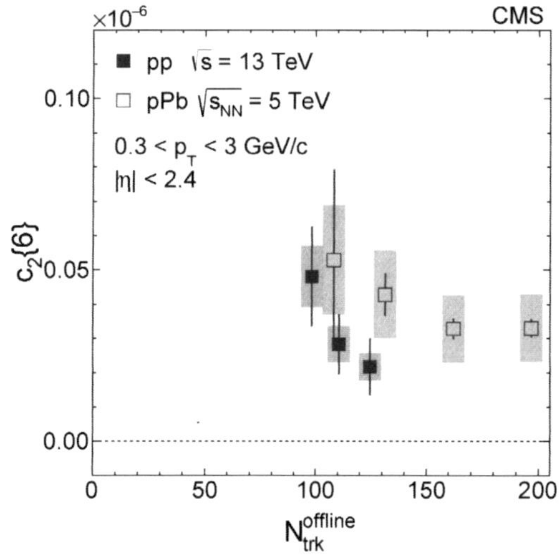

Figure 6: $c_n\{6\}$ as a function of $N_{\text{trk}}^{\text{offline}}$ for charged particles, averaged over $0.3 < p_T < 3.0 GeV/c$, in p+p collisions at $\sqrt{s_{NN}} = 13$ TeV and compare with that measured in p+Pb collisions at 5 TeV.

$$\begin{aligned} c_n\{4\} &= \langle\langle 4 \rangle\rangle - 2 \times \langle\langle 2 \rangle\rangle^2, \\ c_n\{6\} &= \langle\langle 6 \rangle\rangle - 9 \times \langle\langle 4 \rangle\rangle\langle\langle 2 \rangle\rangle + 12 \times \langle\langle 2 \rangle\rangle^3. \end{aligned} \tag{2}$$

Fig. 5 shows the four-particle cumulant $c_2\{4\}$ values for charged particles($0.3 < p_T < 3.0 GeV/c$) as a function of $N_{\text{trk}}^{\text{offline}}$ for p+p collisions at $\sqrt{s_{NN}} = 5$, 7, and 13 TeV. The p+Pb data at $\sqrt{s_{NN}} = 5$ TeV [4] are also plotted for comparison. The six-particle cumulant $c_2\{6\}$ values for p+p collisions at $\sqrt{s_{NN}} = 13$ TeV are shown in Fig. 6, comparing with p+Pb data at $\sqrt{s_{NN}} = 5$ TeV [4].

Similar to that found for p+Pb collisions, the $c_2\{4\}$ values decrease as a function of increasing multiplicity in p+p collisions for all three collision energies. In p+p collisions at 13 TeV, the $c_2\{4\}$ switches sign from positive to negative at $N_{\text{trk}}^{\text{offline}}$ above 60, and indicates a collective $v_2\{4\}$ signal [22]. An indication of energy dependence of $c_2\{4\}$ values is seen in Fig. 5, where $c_2\{4\}$ tends to be more positive for a given $N_{\text{trk}}^{\text{offline}}$ range at lower collision energies. It may be due to different average p_T values at different collisions energies or different multiplicity fluctuations. The positive $c_2\{6\}$ values are also observed in the p+p collisions at 13 TeV which is similar to what was observed in p+Pb collisions

5. The v_2 of identified particles in p+p and ^{3}He+Au collisions

The v_2 of K_S^0 and $\Lambda/\overline{\Lambda}$ particles are measured in the high multiplicity p+p collisions at 13 TeV by CMS [17]. After correction for jet correlations estimated from low-multiplicity data, the v_2^{sub} results as a function of p_T for $105 \leq N_{\text{trk}}^{\text{offline}} < 150$ are shown in Fig. 7 (top). The particle mass ordering of v_2 values is observed in the lower p_T region, while at higher p_T the ordering is reversed. The number of quark scaling as a function of KE_T/n_q is shown in Fig. 7 (bottom) with a dashed curve corresponding to a polynomial fitting to the K^0_S data. The ratio of v_2^{sub}/n_q results for K^0_S and $\Lambda/\overline{\Lambda}$ particles divided by this polynomial function fitting is also shown in Fig. 7 (bottom). An approximate scaling is seen for $KE_T/n_q \gtrsim 0.2GeV$ within about $\pm10\%$.

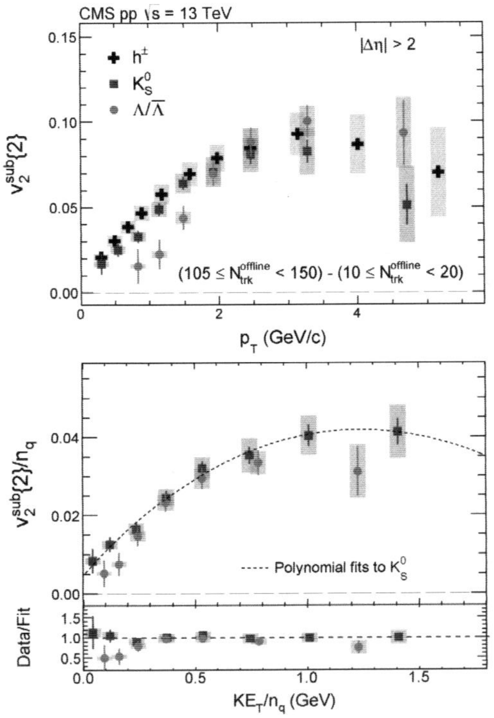

Figure 7: Top: the v_2^{sub} results of inclusive charged particles, K^0_S and $\Lambda/\overline{\Lambda}$ particles as a function of p_T in 13 TeV p+p for multiplicity as $105 \leq N_{trk}^{offline} < 150$, after subtracting jet correlations estimated from low-multiplicity data. Bottom: the v_2^{sub}/n_q for K^0_S and $\Lambda/\overline{\Lambda}$ as a function of KE_T/n_q. Ratios to a smooth fit function of data for K^0_S particles are also shown.

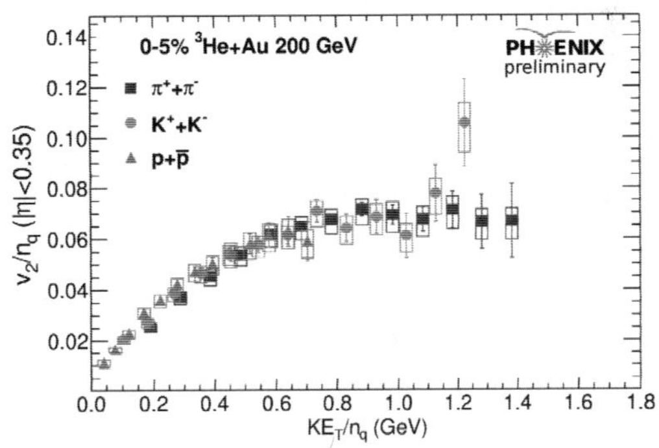

Figure 8: The scaling of number of quark of $v_2(p_T)$ for $\pi^{\pm}$, $K^{\pm}$ and (anti)protons in 0-5% central ^{3}He+Au collisions.

PHENIX measured the v_2 of $\pi^{\pm}$, $K^{\pm}$ and (anti)protons in the central ^{3}He+Au collisions at 200 GeV. Fig. 8 shows the number of quark scaling for v_2 of $\pi^{\pm}$, $K^{\pm}$ and (anti)protons as a function of KE_T/n_q (GeV). The number of quark scaling is found to hold in the small collisions system, which is consistent with viscous hydro calculations or quark coalescence models [18, 19].

6. Summary

I summarize the recent measurements of anisotropy(v_2, v_3) in small collision systems which include p+p, p+A, d+Au and ^{3}He+Au collisions. The v_2 and v_3 are measured with several different ways which include the long-range two particles $\Delta\phi$ correlations, event plane with large η gap and multi-particles cumulant methods. A sizable v_2 and v_3 are observed in the small collision systems and found to be related to the initial geometry. A particle mass ordering of v_2 values are also observed for the identified particles in small collisions. All these measurements are similar to what were observed in large A+A collisions system and indicate a mini-QGP has been generated in small collision systems.

References

[1] Chatrchyan, S *et al.* (CMS Collaboration) 2013 *Phys. Lett. B* **718** 795-814
[2] Aad, G *et al.* (ATLAS Collaboration) 2013 *Phys. Rev. Lett.* **110** 182302
[3] Aad, G *et al.* (ATLAS Collaboration) 2013 *Phys. Lett. B* **725** 60-78
[4] Chatrchyan, S *et al.* (CMS Collaboration) 2013 *Phys. Lett. B* **724** 213-240
[5] Abelev B *et al.* (ALICE Collaboration) 2013 *Phys. Lett. B* **719** 29-41
[6] Adare, A *et al.* (PHENIX Collaboration) 2013 *Phys. Rev. Lett.* **114** 212301
[7] Adare, A *et al.* (PHENIX Collaboration) 2015 *Phys. Rev. Lett.* **114** 192301
[8] Adare, A *et al.* (PHENIX Collaboration) 2015 *Phys. Rev. Lett.* **114** 142301
[9] Dusling, K *et al.* 2013 *Phys. Rev. D* **87** 094034
[10] Bozek, P *et al.* 2012 *Phys. Rev. C* **114** 014911
[11] Nagle, J. *et al.* 2014 *Phys. Rev. Lett.* **114** 112301
[12] Heikki Möntysaari, Björn Schenke, arxiv: 1603.04349
[13] Paul Romatschke, arxiv: 1502.04745
[14] Aidala, C *et al.* arXiv: 1311.3594
[15] Kovtun, P *et al.* 2005 *Phys. Rev. Lett.* **94** 111601
[16] Lin, Z.-W. Lin *et al.* 2005 *Phys. Rev. C* **72** 064901
[17] Chatrchyan, S *et al.* (CMS Collaboration) arXiv: 1606.06198
[18] Hwa, Rudolph C., Yang, C. B., 2003 *Phys. Rev. C* **67** 034902
[19] Fries, R. J. *et al.* 2003 *Phys. Rev. Lett.* **90** 202303
[20] Aad, G *et al.* (ATLAS Collaboration) 2016 *Phys. Rev. Lett.* **116** 172301
[21] Bilandzic, A *et al.* 2011 *Phys. Rev. C* **83** 044913
[22] Borghini, N *et al.* 2011 *Phys. Rev. C* **63** 054906

IOP Conf. Series: Journal of Physics: Conf. Series **779** (2017) 012010 doi:10.1088/1742-6596/779/1/012010

Λ production in p+p collisions at 40 and 158 GeV/c

Herbert Ströbele for the NA61 collaboration
Universität Frankfurt
E-mail: stroebel@ikf.uni-frankfurt.de

Abstract. The research programme of the NA61 collaboration covers a wide range of hadronic physics in the CERN SPS energy range. It encompasses measurements of hadron-hadron, hadron-nucleus as well as nucleus-nucleus collisions. The latter are analyzed to better understand the properties of hot and dense nuclear matter. In this contribution we present results on Λ production in proton-proton interactions at beam energies of 40 and 158 GeV/c.

1. Introduction
NA61 is a fixed target experiment at the CERN SPS pursuing an ambitious programme in hadron physics using proton and heavy ion beams, as well as in neutrino and cosmic ray physics by measuring hadron-nucleus cross sections as reference data for neutrino flux and atmospheric shower calculations. The subject of the hadron physics programme is the study of the energy region, where the onset of deconfinement is expected to occur. The STAR experiment at RHIC with its Beam Energy Scan programme is also studying this energy range. Next generation experiments at FAIR (CBM) in Darmstadt and NICA in Dubna are planned to follow up with dedicated accelerator facilities. Presently NA61 is the only experiment which does a two-dimensional scan (in energy and system size) of p+p and nuclear collisions. The NA61 detector has an acceptance for charged particles down to $p_T = 0$ for rapidities forward of mid-rapidity at all energies (20A GeV - 150A GeV). This and its particle identification capabilities allow to determine particle multiplicities in full phase space with small uncertainties introduced by extrapolations into unmeasured regions. In this contribution we present results on Λ production in p+p interactions at beam momenta of 158 and 40 GeV/c.

2. The experiment
The main components of the experimental setup are four large volume Time Projection Chambers (TPC) for charged particle tracking and identification by recording the energy loss in the TPC gas. A new small TPC (GTPC) on the beam line records fast forward going particles in p+p interactions [1]. Two time-of-flight walls each covering approximately two square meters improve the kaon identification capabilities. A small scintillator on the beam line downstream of the target is used to veto through-going beams and is used as minimum bias trigger. A hadron calorimeter serves as a projectile spectator detector and is used to select centrality in nuclear collisions.

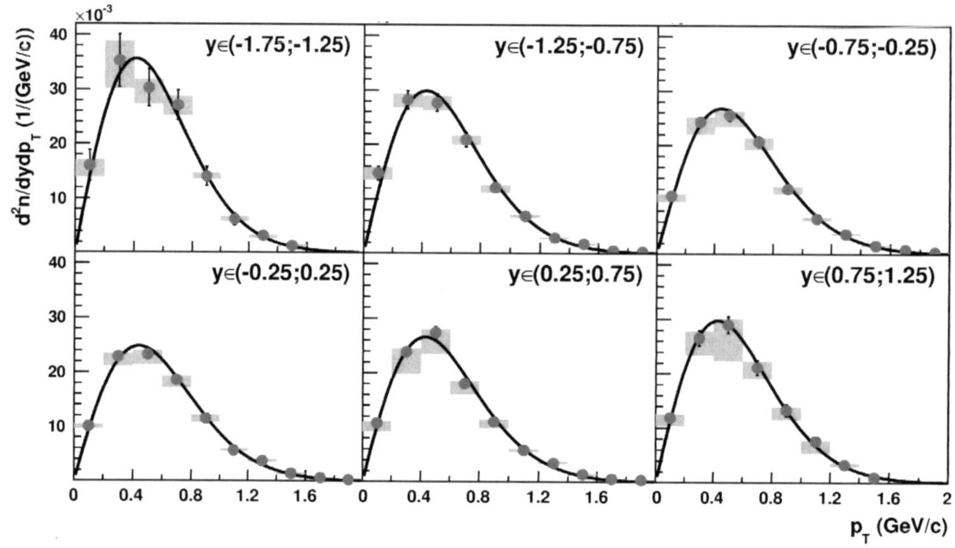

Figure 1. Transverse momentum distributions of Λ hyperons in p+p interactions at 158 GeV/c beam momentum. The rapidity intervals in which the dn/dp_T distributions are obtained are indicated in the individual figures. The lines are the result of fits of Eq.1 to the experimental points. The systematic uncertainties of the Λ yields are indicated by the vertical grey bands at the measured points.

3. Results

3.1. Λ production in p+p interactions at SPS energies

Candidates of charged decays of Λ hyperons are identified by the standard topological cuts applied to pairs of positively and negatively charged particles. The invariant masses of the resulting candidate decay particles give a distribution which consist of the Λ signal and a

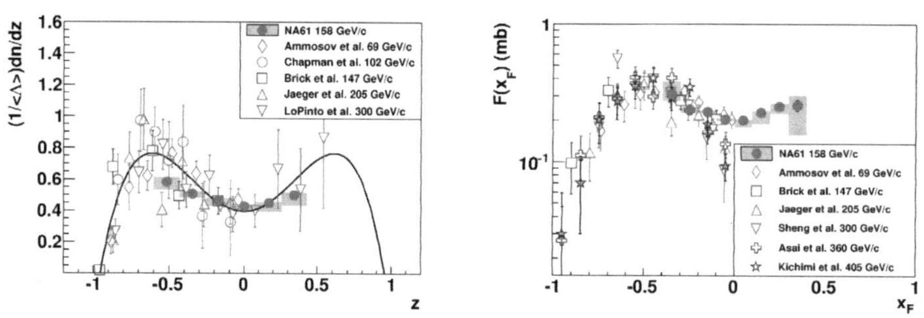

Figure 2. (left) Normalized rapidity ($z=y/y_{beam}$) distribution of Λ hyperons produced in p+p interactions at 158 GeV/c together with data from other experiments [3, 4, 5, 6, 7]. A symmetric 4th order polynomial (solid line) is used to estimate the systematic uncertainty of the mean Λ multiplicity (see [2]); (right) same for the invariant cross section as function of x_F.

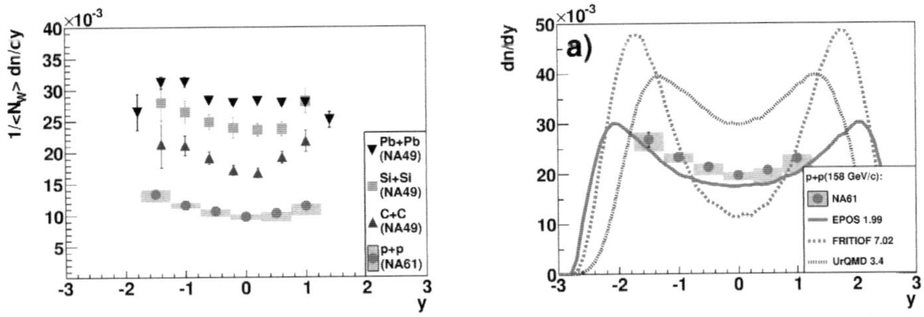

Figure 3. (left) Rapidity distribution of Λ hyperons divided by the number of wounded nucleons in inelastic p+p interactions at 158 GeV/c together with the data from Si+Si, C+C [8], and Pb+Pb [9] collisons; (right) comparison of the Λ rapidity distribution in p+p collisions to calculations by the EPOS [10], UrQMD [11, 12], and Fritiof [13] models.

structure-less background. The candidate decay particle pairs populate a large part of the two-dimensional longitudinal rapidity (y) or Feynman x (x_F) and transverse momentum (p_T) space. A grid in longitudinal and transverse momentum defines phase space cells in which the Λ yields are obtained from fits of signal and background functions to the corresponding invariant mass distributions. These yields are corrected for acceptance and for losses due to the topological and track selection cuts. The yields in phase space cells within a given longitudinal momentum interval form the transverse momentum distributions (dn/dp_T). Integration and extrapolation of the transverse momentum distributions provide the data points for dn/dy and dn/dx_F distributions. The transverse momentum distributions of Λ shown in Fig.1 are from inelastic p+p collisions at a beam momentum of 158 GeV/c [2]. They cover rapidities from

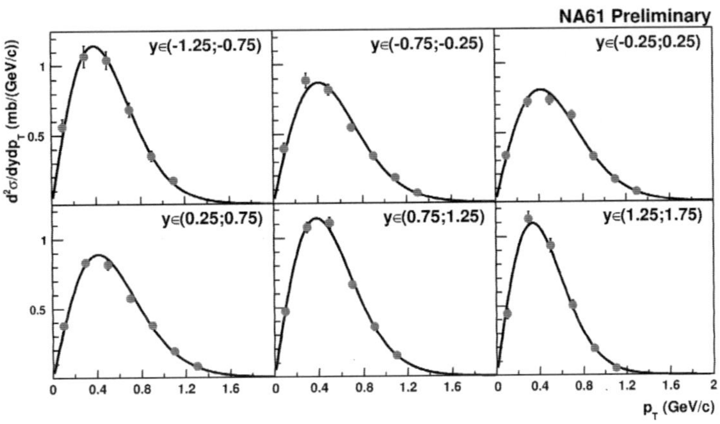

Figure 4. Transverse momentum distributions of Λ hyperons in p+p interactions at 40 GeV/c beam momentum. The rapidities at which the dn/dp_T distributions are obtained are indicated in the individual figures. The lines are the result of fits of Eq.2 to the experimental points.

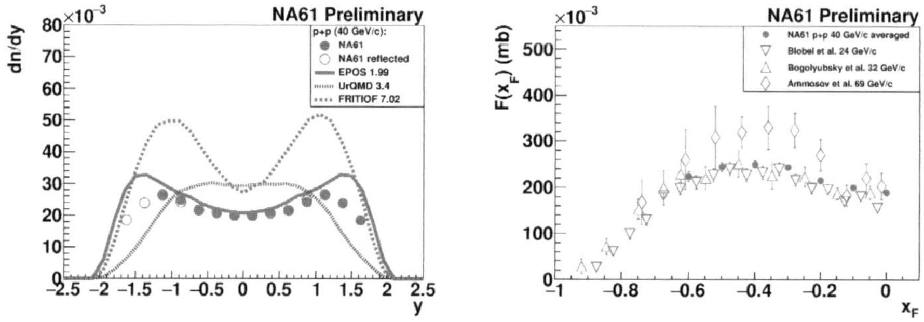

Figure 5. (left) Rapidity distribution of Λ hyperons produced in inelastic p+p interactions at 40 GeV/c together with calculations by the EPOS [10], UrQMD [11, 12], and Fritiof [13] models; (right) invariant cross section as function of x_F of Λ hyperons together with data from other experiments at similar energies [3, 14, 15].

-1.75 to 1.25. The solid lines represent fits of the function

$$\frac{1}{p_T}\frac{d^2n}{dydp_T} = \frac{1}{m_T}\frac{d^2n}{dydm_T} = e^{-m_T/\mathrm{T}}, m_T = \sqrt{p_T^2 + m_\Lambda^2} \tag{1}$$

to the data points which describe the data well. Integration of the dn/dydp_T distributions in the 6 rapidity intervals provide the 6 data points for the rapidity distribution in Fig.2(left). Also shown are results from other experiments. The mean Λ multiplicity was obtained by extrapolating the dn/dx_F distribution into the unmeasured region (see [2]). In order to compare with the other measurements at similar energies reduced rapidity (z=y/y_{beam}) is used for the abscissa. The NA61 data have statistical uncertainties smaller than the symbol size. The size of the systematic uncertainty is indicated by the grey areas. The world data bracket the high precision NA61 results quite well. The same conclusion can be drawn from Fig.2 (right) in which the invariant Λ cross section is plotted as function of the Feynman x variable (x_F) together again with the world data. The measurement by the NA61 collaboration has reduced significantly the uncertainty in the experimental knowledge about Λ production in p+p interactions at SPS energies. Another aspect of measurements of particle production in p+p interactions is their role as reference for studies of nuclear collisions. In Fig.3 (left) we therefore compare the NA61 data on Λ production with the reuslts from Pb+Pb, Si+Si and C+C collisons at the same c.m. energy per nucleon pair. The strangeness enhancement, well established in nuclear collisons, increases with system size in a large range of Λ rapidities. Finally, a comparison with the transport models EPOS and UrQMD as well as with the two-fireball excitation model FRITIOF (Fig.3 right) reveals that only EPOS reproduces the yield and the shape of dn/dy of the NA61 results on Λ production in p+p interactions.

New preliminary NA61 data on Λ production in p+p interactions at 40 GeV/c beam momentum are presented in the following paragraph. The analysis steps are essentially the same as for the 158 GeV/c data. The resulting dn/dp_T distributions are shown in Fig.4 together with fits of the function

$$u = A \cdot p_T \cdot e^{-m_T/\mathrm{T}} \tag{2}$$

with A and T as free parameters. The rapidity distribution is again obtained by integration of the p_T distributions (Fig.5). The statistical uncertainties are smaller than the symbol size. Systematic uncertainties are under study. Since the distribution must be symmetric, the data points can be mirrored at y=0 (open symbols). At 40 GeV/c the covered portion of the rapidity

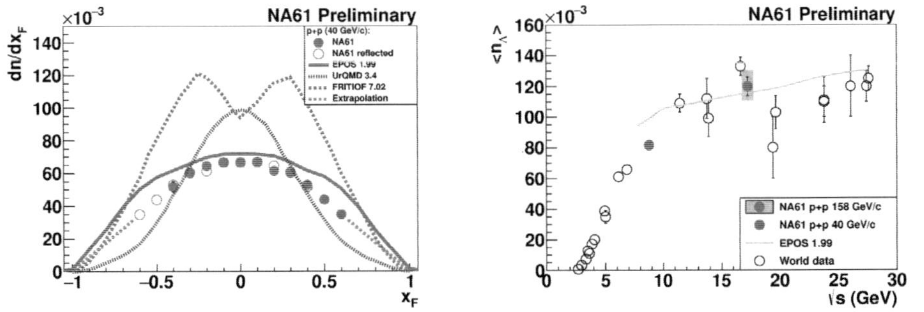

Figure 6. (left) Λ hyperon yields in p+p interactions at 40 GeV/c as function of x_F together with the same model calculations as shown Fig.5. In this form the longitudinal distribution can be extrapolated into the unmeasured region without significant uncertainty (see dotted line); (right) Λ multiplicity in p+p interactions as function of $\sqrt{s}$ together with world data and results from EPOS [10] calculations.

range is larger than at 158 GeV/c allowing to measure the falling of the leading baryon yield at large rapidities. The comparison to world data is again done using $z = y/y_{beam}$ as longitudinal variable (Fig.5 right). The same observations can be made as at 158 GeV/c. The preliminary NA61 data have small (statistical) uncertainties and are bracketed well by the world data.

Extrapolation of the dn/dx_F distribution into the unmeasured regions (see e.g. dotted lines in Fig.6) provides estimates of the Λ yields. The corresponding multiplicities (average yields per event) are compared to world data in Fig.6 (right) as function of c.m. energy ($\sqrt{s}$). The NA61 data follow the trend set by the world data and at the same time reduce uncertainties in the spectral yields significantly at the two energies. Above $\sqrt{s}$=19 GeV the experimental data have large uncertainties and lie consistently below the EPOS calculation. It is desirable to have new experimental data in this energy range. Below 10 GeV EPOS underestimates Λ yields. However this model is not intended to produce reliable results at this energy and below.

References

[1] Abgrall N et al. 2014 *J. Instrum.* **9** P06005
[2] Aduszkiewicz A et al. 2016 *Eur. Phys. J.* C **76** 198
[3] Ammosov V et al. 1976 *Nucl. Phys.* B **115** 269
[4] Chapman J et al. 1973 *Phys.Lett.* **47B** 465
[5] Brick D et al. 1980 *Nucl. Phys.* B **164** 1
[6] Jaeger K et al. 1975 *Phys. Rev.* D **11** 2405
[7] LoPinto F et al. 1980 *Phys. Rev.* D **22** 573
[8] Alt C et al. 2005 *Phys. Rev. Lett.* **94** 052301
[9] Alt C et al. 2008 *Phys. Rev.* C **78** 034918
[10] Werner K, Liu F-M and Pierog T 2006 *Phys. Rev.* C **74** 044902
[11] Bass S A et al. 1998 *Prog. Part. Nucl. Phys.* **41** 225
[12] Bleicher M et al. 1999 *J. Phys.* G **25** 1859
[13] Anderson B, Gustafson G and Hong Pi 1993 *Z. Phys.* C **57** 485
[14] Blobel V et al. 1974 *Nucl. Phys.* B **69** 454, Blobel V et al. 1978 *Nucl. Phys.* B **135** 379
[15] Bogolyubsky M et al. 1989 *Sov. J. Nucl. Phys.* **50** 424, *Yad. Fiz.* **50** 683

IOP Conf. Series: Journal of Physics: Conf. Series **779** (2017) 012011 doi:10.1088/1742-6596/779/1/012011

Collectivity in small systems: Initial correlations or final state flow?

Björn Schenke

Physics Department, Brookhaven National Laboratory, Upton, NY 11973, USA

E-mail: bschenke@bnl.gov

Abstract. I review recent progress in understanding correlation measurements in small collision systems, such as proton+lead and proton+proton collisions at the Large Hadron Collider (LHC) and proton+gold, deuteron+gold, and ^{3}He+gold collisions at the Relativistic Heavy Ion Collider (RHIC). I discuss two distinct theoretical approaches to describing the experimental data on multi-particle correlations. The first attributes the origin of the measured correlations to strong final state interactions, often described by hydrodynamics, the second employs the color glass condensate effective theory and is able to reproduce many features of the data from initial state effects only. I discuss how to distinguish which of the two sources of correlations dominates the experimental observables, and give an outlook on how to make progress on the theory side.

1. Introduction

Collective effects such as long range correlations in rapidity with distinct azimuthal structures have been observed in nuclear collision systems ranging from p+p to A+A collisions, at least for high enough multiplicity events [1]. Our experience with heavy ion collisions, where such correlations are understood as being generated via the response of the evolving hot and dense matter to the initial fluctuating geometry [2], led many to believe that small systems, such as p+A collisions, behave in a similar way, at least when large numbers of particles are produced. This is conceivable, at least for particles with low enough transverse momentum, which are numerous. Nevertheless, the applicability of viscous hydrodynamics may be questioned as the system becomes smaller, since quantities like the Knudsen number or the inverse Reynolds number become large [3]. This does not immediately exclude significant final state effects but questions the tool of hydrodynamics, which assumes small deviations from local thermal equilibrium and isotropy, for describing these effects. In the following I will briefly discuss the hydrodynamic description of small collision systems followed by the initial state picture. I will then address how to distinguish the two scenarios using experimental observables.

2. Hydrodynamics

Hydrodynamics, supplemented with an appropriate initial state, has been applied to compute particle production in small collision systems. Agreement with certain experimental data on spectra and correlations can be very good [4, 5, 6, 7, 8, 9, 10, 11]. In fact, one could say that hydrodynamics, a gradient expansion, does an "unreasonably good" job in describing the experimental data, because pressure anisotropies in the small systems can be very large. This was most recently summarized in [12], where it is argued that hydrodynamics working does not necessarily imply near-equilibrium behavior of the matter.

Assuming that at least for low momenta ($p_T \lesssim 2\,\mathrm{GeV}$) hydrodynamics provides a good description of the final state interactions, and thus can provide reliable calculations of the integrated azimuthal anisotropy coefficients v_n, the disagreement of hydrodynamic calculations using the very successful IP-Glasma initial state model with experimentally obtained v_n was surprising [13]. While the smaller system size in this framework [14] is more compatible with experimental Hanbury-Brown-Twiss (HBT) data [15] than calculations in the MC-Glauber model [16], the v_n from the latter agree much better with the data.

Detailed HBT calculations using IP-Glasma initial conditions are still outstanding, however, it is likely that a consistent description of all low p_T data is only possible with the IP-Glasma initial state when including one important modification: The inclusion of a more refined geometric substructure of the proton.

3. The fluctuating shape of the proton

The original IP-Glasma model, which has been successfully applied to heavy ion collisions, assumes individual nucleons to be round (up to small variations from color charge fluctuations). This feature led to extremely small anisotropy coefficients in p+A collisions. Because in the IP-Glasma model the interaction region is dominated by the shape of the smaller projectile, i.e., the proton, a round proton shape led to an approximately round distribution of deposited energy, which in turn led to almost isotropic expansion and small anisotropic flow coefficients. This was a strong first hint at the fact that p+A collisions provide data sensitive to the geometric shape of the proton.

In fact, it was recently found that HERA data on diffractive exclusive vector meson production in e+p collisions [17, 18, 19, 20] also requires the model description to include strong geometric fluctuations of the gluon distribution in a proton at high energies [21, 22]. In particular, the transverse momentum transfer $|t|$ dependent incoherent cross section is sensitive to fluctuations at various length scales. In the dipole picture employed in [21, 22] not including geometric fluctuations leads to an under-estimation of the incoherent cross section at the largest measured $|t|$ by almost an order of magnitude. Qualititively, the observed effect of fluctuations is in line with early discussions of incoherent diffraction [23].

Comparing a model that assumes gluons to be centered around three valence quarks, whose positions fluctuate randomly, to experimental data, allowed to constrain the shape and degree of fluctuations of the gluon distribution. Other models such as a stringy proton can also lead to good agreement with the data, yet all models have one thing in common - the requirement of strong geometric fluctuations.

The next step is now to use such constrained fluctuating protons as input for IP-Glasma + hydrodynamics calculations in p+A collisions, and to see if experimental observables can be systematically reproduced, including v_n coefficients and HBT measurements.

4. Initial state gluon correlations

Long range correlations are present in the initial gluon production from colliding color glass condensates (CGC) [1, 24]. They show a similar azimuthal structure as the experimental data and calculations in the CGC framework have been able to quantitatively reproduce a variety of data from small systems [25]. These calculations were done in the dilute-dense limit, where one projectile is assumed to be dilute, and one can constrain the calculation to two-gluon exchange. For very high multiplicity events that assumption may no longer hold and calculations have to take into account multiple gluon exchanges. This can be achieved in the classical Yang-Mills framework and calculations of azimuthal anisotropies of gluons in p+A collisions yield magnitudes of v_2 and v_3 comparable to the experimental data [26].

Of course one cannot directly compare gluons with final state hadrons that are measured in the experiments. First attempts to implement fragmentation using Lund strings [27] were presented in [28]. It was shown that 1) the azimuthal correlations survive the hadronization and 2) characteristic differences in the mean transverse momentum and v_2 as a function of hadron mass are comparable to those observed in the experimental data. This "mass-splitting" effect clearly comes from particle production from a common boosted source, which in this case is the moving string, in hydrodynamics a fluid cell on the freeze-out surface. Thus it is natural that both pictures yield similar features.

The initial state correlations are uncorrelated with the global geometry of the collision system and their strength decreases with the number of uncorrelated color field domains [26]. Thus, their importance decreases as the system becomes bigger, and azimuthal anisotropies in heavy ion collisions cannot be explained without final state interactions of the produced matter.

5. Distinguishing the two scenarios

At this point it is still unresolved what mechanism dominates the generation of the observed structures in the experimental data for different small systems, multiplicities and collision energies. Hydrodynamics has a good chance of providing a reasonable description for very high multiplicities and small transverse momentum particles, which also dominate in the integrated v_n. Above a p_T of approximately 1.5 GeV viscous effects are too large to get reliable results from hydrodynamics, which means that the tool starts to fail, but not that final state effects become irrelevant.

More generally though, the strong final state interaction interpretation faces various challenges, in particular with respect to jet quenching: There is a surprising lack of quenching of the pronounced back-to-back correlation of di-hadrons in small systems [29], which is present in larger systems [30]. Thus, a more detailed study of such effects could shed more light on whether strong final state effects are actually present.

A promising experiment driven venue for distinguishing the two scenarios is the systematic study of small systems with expected differences in the initial geometry: At RHIC different systems such as p+Au, d+Au and ^{3}He+Au have been analyzed [31]. Due to the nucleon content of the smaller projectile on average one expects a somewhat larger elliptic shape in d+Au and a larger triangular shape in ^{3}He+Au compared to p+Au collisions. The hydrodynamic framework thus predicts [32] a larger elliptic flow in d+Au collision, and triangular flow v_3 in ^{3}He+Au, consistent with the experimental data. Initial state frameworks have yet to perform such a systematic comparison across various collision systems.

On the theory side it is desirable to develop a combined framework that includes both initial correlations and final state interactions, potentially via a parton cascade simulation. This is work in progress and it will allow to vary the system size/multiplicity to systematically study which contribution dominates under which circumstances.

It should be mentioned that there is a third scenario which could be described as weak final state interactions that produce an anisotropic parton escape. This situation was observed within AMPT simulations [33, 34] and the effect could be dominant at multiplicities/system sizes between those where the other scenarios dominate.

6. Summary

The study of correlations in small collision systems provides a unique opportunity to learn about the many-body features of quantum chromo dynamics (QCD) at high energy. There is likely a window in multiplicity and transverse momentum that will allow us to study the geometric structure of gluons in the proton via p+A collisions. Furthermore, we can potentially access intrinsic correlations within the incoming nuclear wave functions which can lead to a deeper understanding of how QCD acts within high energy nuclei.

Acknowledgments
This work was supported under DOE Contract No. DE SC0012704. The author acknowledges a DOE Office of Science Early Career Award.

References

[1] Dusling K, Li W and Schenke B 2016 *Int. J. Mod. Phys.* **E25** 1630002 (*Preprint* 1509.07939)
[2] Gale C, Jeon S and Schenke B 2013 *Int. J. Mod. Phys.* **A28** 1340011 (*Preprint* 1301.5893)
[3] Niemi H and Denicol G S 2014 (*Preprint* 1404.7327)
[4] Werner K, Karpenko I, Pierog T, Bleicher M and Mikhailov K 2011 *Phys. Rev.* **C83** 044915 (*Preprint* 1010.0400)
[5] Werner K, Karpenko I and Pierog T 2011 *Phys. Rev. Lett.* **106** 122004 (*Preprint* 1011.0375)
[6] Bozek P 2012 *Phys.Rev.* **C85** 014911 (*Preprint* 1112.0915)
[7] Bozek P and Broniowski W 2013 *Phys.Lett.* **B718** 1557–1561 (*Preprint* 1211.0845)
[8] Bozek P and Broniowski W 2013 *Phys. Rev.* **C88** 014903 (*Preprint* 1304.3044)
[9] Kozlov I, Luzum M, Denicol G, Jeon S and Gale C 2014 (*Preprint* 1405.3976)
[10] Romatschke P 2015 *Eur. Phys. J.* **C75** 305 (*Preprint* 1502.04745)
[11] Shen C, Paquet J F, Denicol G S, Jeon S and Gale C 2016 (*Preprint* 1609.02590)
[12] Romatschke P 2016 (*Preprint* 1609.02820)
[13] Schenke B and Venugopalan R 2014 *Phys. Rev. Lett.* **113** 102301 (*Preprint* 1405.3605)
[14] Bzdak A, Schenke B, Tribedy P and Venugopalan R 2013 *Phys. Rev.* **C87** 064906 (*Preprint* 1304.3403)
[15] Abelev B B *et al.* (ALICE) 2014 *Phys. Lett.* **B739** 139–151 (*Preprint* 1404.1194)
[16] Bozek P and Broniowski W 2013 *Phys. Lett.* **B720** 250–253 (*Preprint* 1301.3314)
[17] Chekanov S *et al.* (ZEUS) 2002 *Eur. Phys. J.* **C24** 345–360 (*Preprint* hep-ex/0201043)
[18] Chekanov S *et al.* (ZEUS) 2003 *Eur. Phys. J.* **C26** 389–409 (*Preprint* hep-ex/0205081)
[19] Aktas A *et al.* (H1) 2003 *Phys. Lett.* **B568** 205–218 (*Preprint* hep-ex/0306013)
[20] Aktas A *et al.* (H1) 2006 *Eur. Phys. J.* **C46** 585–603 (*Preprint* hep-ex/0510016)
[21] Mantysaari H and Schenke B 2016 *Phys. Rev. Lett.* **117** 052301 (*Preprint* 1603.04349)
[22] Mantysaari H and Schenke B 2016 *Phys. Rev.* **D94** 034042 (*Preprint* 1607.01711)
[23] Miettinen H I and Pumplin J 1978 *Phys. Rev.* **D18** 1696
[24] Lappi T, Schenke B, Schlichting S and Venugopalan R 2016 *JHEP* **01** 061 (*Preprint* 1509.03499)
[25] Dusling K and Venugopalan R 2013 *Phys. Rev.* **D87** 054014 (*Preprint* 1211.3701)
[26] Schenke B, Schlichting S and Venugopalan R 2015 *Phys. Lett.* **B747** 76–82 (*Preprint* 1502.01331)
[27] Andersson B, Gustafson G, Ingelman G and Sjostrand T 1983 *Phys. Rept.* **97** 31–145
[28] Schenke B, Schlichting S, Tribedy P and Venugopalan R 2016 (*Preprint* 1607.02496)
[29] Khachatryan V *et al.* (CMS) 2016 *Submitted to: Physics Letters B* (*Preprint* 1606.06198)
[30] Adler C *et al.* (STAR) 2003 *Phys. Rev. Lett.* **90** 082302 (*Preprint* nucl-ex/0210033)
[31] Adare A *et al.* (PHENIX) 2015 *Phys. Rev. Lett.* **115** 142301 (*Preprint* 1507.06273)
[32] Nagle J L, Adare A, Beckman S, Koblesky T, Orjuela Koop J, McGlinchey D, Romatschke P, Carlson J, Lynn J E and McCumber M 2014 *Phys. Rev. Lett.* **113** 112301 (*Preprint* 1312.4565)
[33] Bzdak A and Ma G L 2014 *Phys. Rev. Lett.* **113** 252301 (*Preprint* 1406.2804)
[34] He L, Edmonds T, Lin Z W, Liu F, Molnar D and Wang F 2016 *Phys. Lett.* **B753** 506–510 (*Preprint* 1502.05572)

Hadron yields, the chemical freeze-out and the QCD phase diagram

A Andronic[1], P Braun-Munzinger[1,2], K Redlich[3], and J Stachel[2]

[1] Research Division and EMMI, GSI Helmholtzzentrum für Schwerionenforschung, Darmstadt, Germany

[2] Physikalisches Institut, Universität Heidelberg, Germany

[3] Institute of Theoretical Physics, University of Wrocław, Poland

E-mail: A.Andronic@gsi.de

Abstract. We present the status of the chemical freeze-out, determined from fits of hadron yields with the statistical hadronization (thermal) model, with focus on the data at the LHC. A description of the yields of hadrons containing light quarks as well as the application of the model for the production of the J/ψ meson is presented. The implications for the QCD phase diagram are discussed.

1. Chemical freeze-out of light-quark hadrons

Chemical freeze-out in nucleus-nucleus collisions is addressed in the statistical hadronization (thermal) model, based on the statistical operator for the hadron resonance gas. It describes a snapshot of the collision dynamics, namely assuming a rapid chemical freeze-out. The model is simple, but very powerful considering its small set of parameters (temperature T, baryochemical potential μ_B and volume V). Importantly, the model is a unique phenomenological approach linking the production of hadrons in nucleus-nucleus collisions to the QCD phase diagram [1] (see [2, 3] for early ideas).

All known hadrons, stable and decaying, are employed in the calculations. Even as the knowledge of the hadron spectrum is constantly improving, missing states could in principle still have an effect on T, as shown in lattice QCD studies [4]. As currently the prefered T values are around 160 MeV (see below) the relevance of higher-lying resonances is diminished. In the commonly-used grand canonical approach, chemical potentials μ ensure conservation on average of additive quantum numbers (baryon number, isospin, strangeness, charmness), fixed by "initial conditions".

The model, in its "standard" implementation (meaning the minimal set of parameters listed above), was sucessfully used to describe hadron production in heavy-ion collisions over a wide range of collision energies (see e.g. [5, 6]). Several versions with extended set of parameters have been proposed [7, 8]. In particular, a strangeness suppression factor γ_s is used as a fit parameter to test the possible departure from equilibrium of hadrons containing strange quark(s) (see, e.g. [6]). The question of flavor-dependent freeze-out is also addressed [9, 10, 11]. Beyond the sudden freeze-out concept, a hadronic phase with "chemical activity" with the UrQMD transport model [12] was proposed. Some times, possible repulsive interactions among hadrons are modeled in a hard-sphere excluded-volume approach. In the simplest implementation, identical (and moderate) values for the radii are used ($R_{meson} = R_{baryon} = 0.3$ fm), leaving T and μ_B unaffected in comparison with results for point-like hadrons [5]. Employing species-dependent hard-sphere interactions [13] leads to an extended parameter set, which, however, cannot be constrained by current knowledge of hadron-hadron interactions. As the results below demonstrate, the data do not require any of these extensions.

For small systems and/or low energies, a canonical treatment is needed [14], usually implemented only for strangeness. Recent such studies in p–Nb and Ar–KCl collisions [15] and in pp collisions [16, 17, 18] lead to values of T comparable to (or even larger than) those in (central) Au–Au or Pb–Pb collisions. The studies performed by ALICE in p–Pb [19] and pp [20] collisions revealed that in high-multiplicity events hadron production in these small systems resembles that in Pb–Pb collisions.

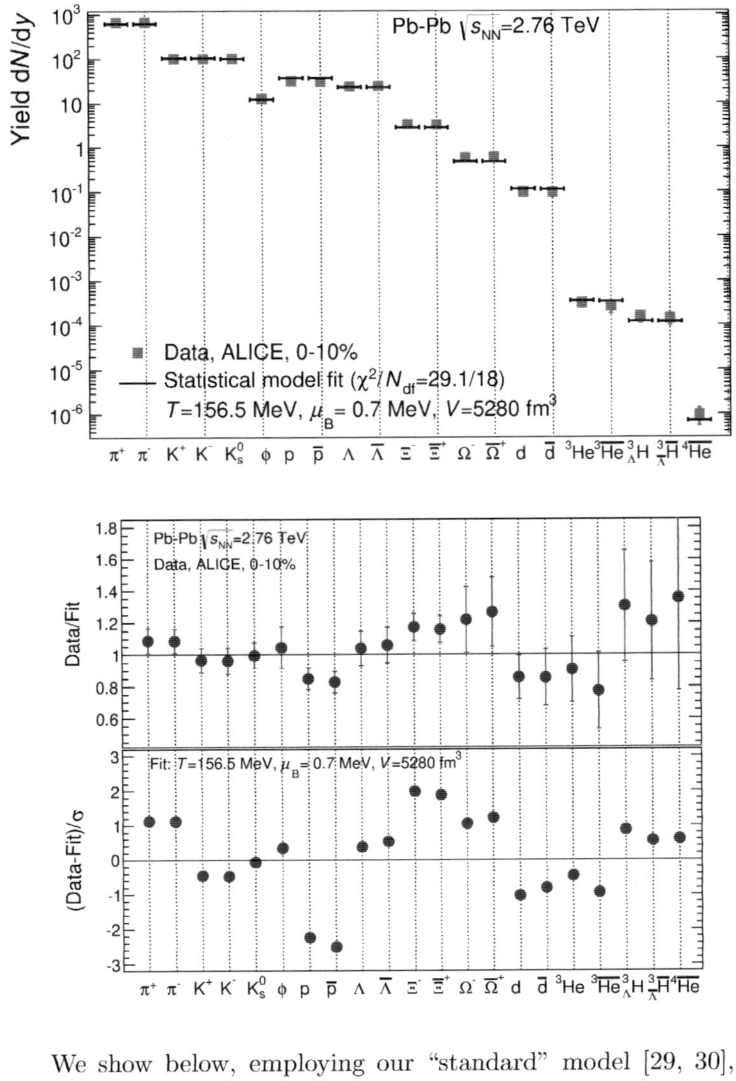

Figure 1. Hadron multiplicities in central (0-10%) Pb–Pb collisions at the LHC, ALICE data ([21, 22, 23, 24, 25, 26, 27]; the values for $\bar{\Lambda}$ [28] and anti-^{4}He are preliminary) and best fit (upper panel). The lower panels show the ratio of data value to fit (with error bars the total, statistical and systematic, uncertainties) and the difference between data and model fit in units of the experimental uncertainty.

We show below, employing our "standard" model [29, 30], the current description of the LHC data in central (0-10%) Pb–Pb collisions at $\sqrt{s_{NN}} = 2.76$ TeV. This model is based on the statistical operator for the hadron resonance gas which was shown earlier [31] to lead to an equation of state in good agreement with lattice QCD calculations. The best fit values and their uncertainties are: $T = 156.5 \pm 1.5$ MeV, $\mu_B = 0.7 \pm 3.8$ MeV, $V_{\Delta y=1} = 5280 \pm 410$ fm^3, achieved for $\chi^2_{min} = 29.1$ per 18 d.o.f., indicating a very good description of data, which extends over 9 orders of magnitude in hadron yields, see Fig. 1 and Fig. 2. Included in the calculations are the contribution in the yields of π, $K^{\pm}$, and K^0 from the decays of charmed hadrons, amounting to a relative contribution for the best fit of 0.7%, 2.9%, and 3.1%, respectively. While we used $R_{meson} = R_{baryon} = 0.3$ fm, point-like hadrons lead to a fit with same T and μ_B values, but a volume smaller by about 25%.

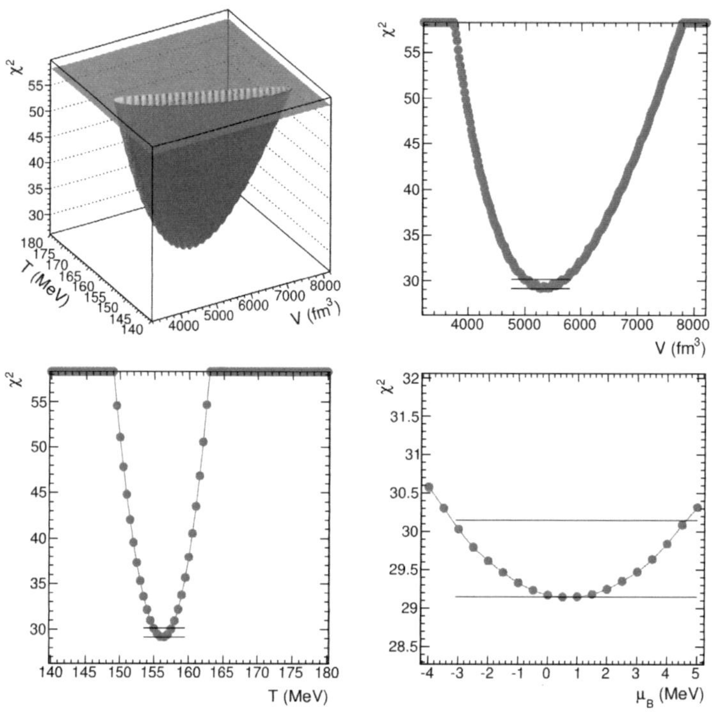

Figure 2. The variation of χ^2 as a function of the fit parameters. The upper-left panel shows χ^2 in the $T-V$ correlation, the other panels the variation of χ^2 for each parameter (the horizontal lines indicate the χ^2_{min} and $\chi^2_{min}+1$ values).

In Fig. 3 an illustration is shown of the success of the thermal model in reproducing over a broad energy range the production of (anti)protons and kaons relative to pions.

The value of the (pseudo-)critical temperature, T_c, at vanishing baryochemical potential (μ_B) is currently calculated in lattice QCD [35, 36] to be 155±9 MeV. The phenomenological QCD phase diagram is shown in Fig. 4. Remarkably, at low μ_B chemical freeze-out coincides with T_c, indicating hadron formation from deconfined matter. Each point corresponds to a fit of hadron yields in central Au–Au or Pb–Pb collisions at a given collision energy. The agreement between the results from several independent analyses [29, 32, 33, 34] is very good. Note that in some cases [32, 33, 34] an additional fit parameter, the strangeness suppression factor γ_s, is used to test possible departure from equilibrium of hadrons containing strange quark(s). Values of γ_s (slightly) below unity are found. The non-equilibrium model [7, 8] leads to rather different results (smaller T values for small μ_B), while the model with an extended hadronic phase [12] implies larger T values (neither of these 2 models are shown here).

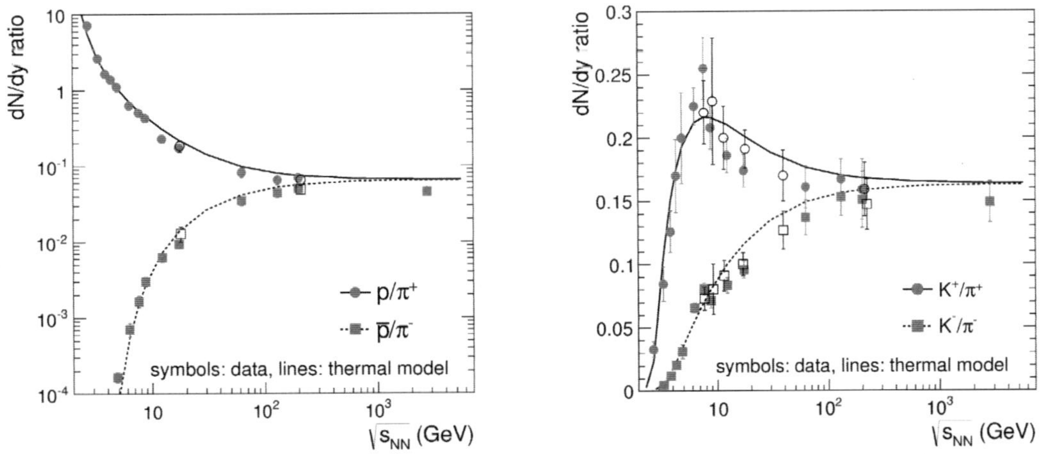

Figure 3. Collision energy dependence of ratios of yields of protons and antiprotons (left panel) and of kaons (right panel) to yields of pions. The symbols are data, the lines are thermal model calculations for energy-dependent parametrizations of T and μ_B (as in Ref. [29], but with $T_{lim} = 159$ MeV). Full symbols represent the data of NA49 and STAR (p, $\bar{p}$ from weak decays subtracted based on the thermal model); at 17 GeV open symbols represent the NA44 data, at 200 GeV BRAHMS data, and for lower energies STAR BES data (preliminary).

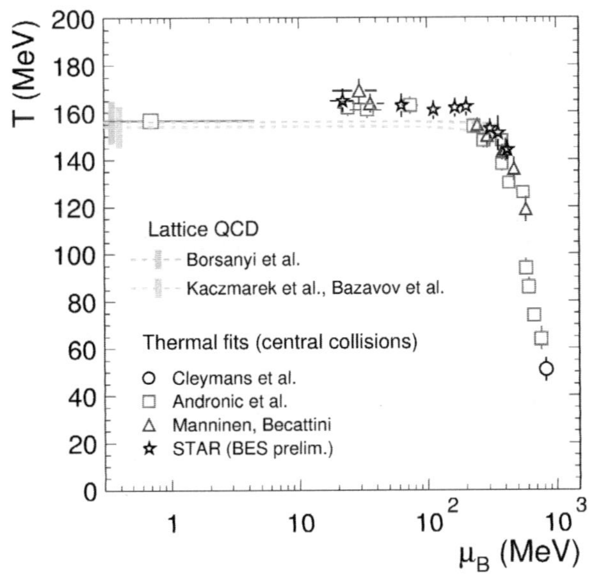

Figure 4. The phase diagram of strongly interacting matter with the points representing the thermal fits of hadron yields at various collision energies [14, 29, 32, 33, 34]. The crossover T values from lattice QCD [35, 36] are shown as bands at small μ_B values. The dashed lines represent lattice QCD calculations of the curvature parameter [37, 38].

2. Charmonium

The statistical hadronization model outlined above can be applied as well to production of hadrons with heavy quarks (HQ), charm and bottom [39, 40]. Given that the HQ themselves are produced predominantly in primary hard collisions (for charm, $t_{c\bar{c}} \sim 1/2m_c \simeq 0.1$ fm/c), the model describes, in a more explicit way than for lighter quarks, the hadronization stage. One additional input parameter, the HQ production cross section, is employed, as well as the assumption the HQ survive and thermalize in QGP (thermal, but not chemical equilibrium). We discuss here the current status of the model description of the J/ψ production at the LHC, quantified via the nuclear modification factor $R_{AA}^{J/\psi}$.

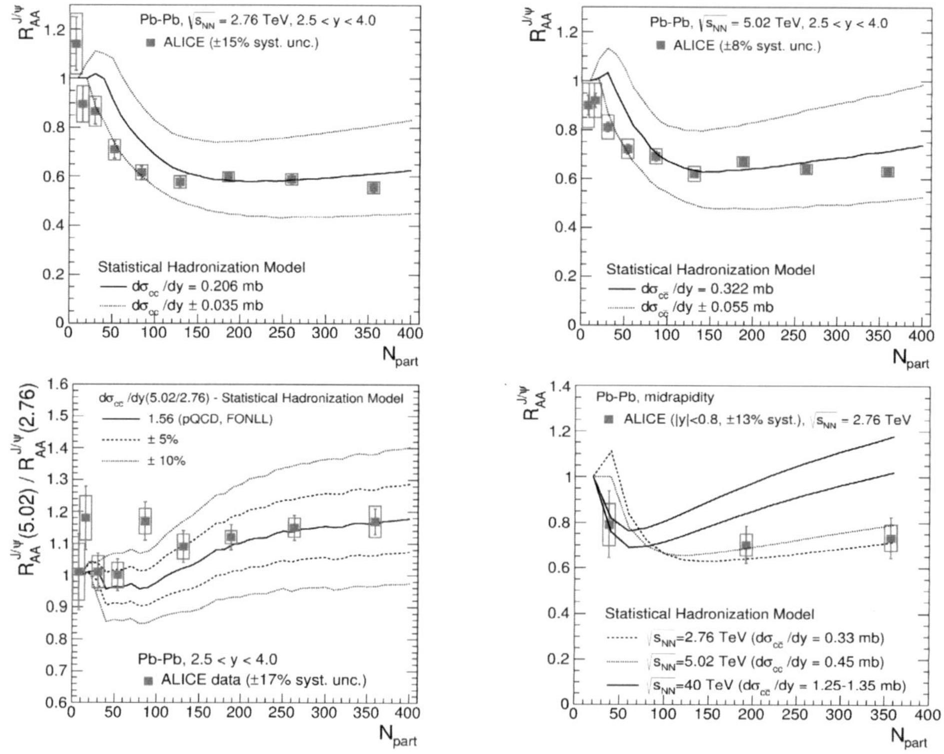

Figure 5. The centrality dependence of the nuclear modification factor of J/ψ production at the LHC. The model is compared to ALICE data [41, 42] at forward rapidity ($2.5 < y < 4$, upper panel for $\sqrt{s_{\rm NN}} = 2.76$ and 5.02 TeV and lower-left panel for the ratio of the 2 energies) and at midrapidity (lower-right panel, for $\sqrt{s_{\rm NN}} = 2.76$ TeV and with predictions for 40 TeV [43]).

Within the statistical hadronization approach a new charmonium production regime was predicted [39, 40] for LHC energies. Consequently, the measurement was expected to be decisive in clarifying the suppression via the Debye screening mechanism [44] and answering if (re)generation scenarios are viable production mechanisms. The data [41, 42] showed that statistical generation at the chemical freeze-out is likely the mechanism of J/ψ production, as demonstrated by the good agreement of model predictions and data, Fig. 5. The uncertainty in the model predictions is due to the $c\bar{c}$ production cross section, obtained by extrapolating the LHCb measurement [45] in pp collisions at $\sqrt{s}$=7 TeV using FONLL pQCD calculations [46] and a shadowing factor from EPS09 calculations [47] (0.71±0.10 for $2.5 < y < 4.0$). Transport models [48, 49] describe the data as well, assuming production during the whole lifetime of QGP. However, the tantalizing possibility of having the J/ψ meson as a probe of the QCD phase boundary is very appealing.

Acknowledgments This work is part of and supported by the DFG Collaborative Research Centre "SFB 1225 (ISOQUANT)". K.R. acknowledges support by the Polish Science Center (NCN), under Maestro grant DEC-2013/10/A/ST2/00106.

References

[1] Braun-Munzinger P and Wambach J 2009 *Rev. Mod. Phys.* **81** 1031–1050 (*Preprint* 0801.4256)

[2] Cabibbo N and Parisi G 1975 *Phys. Lett. B* **59** 67–69

[3] Hagedorn R 1985 *Lect. Notes Phys.* **221** 53–76

[4] Bazavov A, Ding H T, Hegde P, Kaczmarek O, Karsch F *et al.* 2014 *Phys. Rev. Lett.* **113** 072001 (*Preprint* 1404.6511)

[5] Andronic A, Braun-Munzinger P and Stachel J 2006 *Nucl. Phys. A* **772** 167–199 (*Preprint* nucl-th/0511071)

[6] Becattini F, Manninen J and Gazdzicki M 2006 *Phys. Rev. C* **73** 044905 (*Preprint* hep-ph/0511092)

[7] Letessier J and Rafelski J 2008 *Eur. Phys. J. A* **35** 221–242 (*Preprint* nucl-th/0504028)

[8] Petran M and Rafelski J 2013 *Phys. Rev. C* **88** 021901 (*Preprint* 1303.0913)

[9] Bazavov A, Ding H T, Hegde P, Kaczmarek O, Karsch F *et al.* 2013 *Phys. Rev. Lett.* **111** 082301 (*Preprint* 1304.7220)

[10] Bellwied R, Borsanyi S, Fodor Z, Katz S D and Ratti C 2013 *Phys. Rev. Lett.* **111** 202302 (*Preprint* 1305.6297)

[11] Chatterjee S, Godbole R and Gupta S 2013 *Phys. Lett. B* **727** 554–557 (*Preprint* 1306.2006)

[12] Becattini F, Steinheimer J, Stock R and Bleicher M 2016 (*Preprint* 1605.09694)

[13] Vovchenko V and Stöcker H 2015 (*Preprint* 1512.08046)

[14] Cleymans J, Oeschler H and Redlich K 1999 *Phys. Rev. C* **59** 1663 (*Preprint* nucl-th/9809027)

[15] Agakishiev G *et al.* (HADES Collaboration) 2016 *Eur. Phys. J. A* **52** 178 (*Preprint* 1512.07070)

[16] Vovchenko V, Begun V V and Gorenstein M I 2016 *Phys. Rev. C* **93** 064906 (*Preprint* 1512.08025)

[17] Das S, Mishra D, Chatterjee S and Mohanty B 2016 (*Preprint* 1605.07748)

[18] Cleymans J, Hippolyte B, Oeschler H, Redlich K and Sharma N 2016 (*Preprint* 1603.09553)

[19] Adam J *et al.* (ALICE Collaboration) 2016 *Phys. Lett. B* **758** 389–401 (*Preprint* 1512.07227)

[20] Adam J *et al.* (ALICE Collaboration) 2016 (*Preprint* 1606.07424)

[21] Abelev B *et al.* (ALICE Collaboration) 2012 *Phys. Rev. Lett.* **109** 252301 (*Preprint* 1208.1974)

[22] Abelev B *et al.* (ALICE Collaboration) 2013 *Phys. Rev. C* **88** 044910 (*Preprint* 1303.0737)

[23] Abelev B B *et al.* (ALICE Collaboration) 2013 *Phys. Rev. Lett.* **111** 222301 (*Preprint* 1307.5530)

[24] Abelev B B *et al.* (ALICE Collaboration) 2014 *Phys. Lett. B* **728** 216–227 (*Preprint* 1307.5543)

[25] Abelev B B *et al.* (ALICE Collaboration) 2015 *Phys. Rev. C* **91** 024609 (*Preprint* 1404.0495)

[26] Adam J *et al.* (ALICE Collaboration) 2016 *Phys. Lett. B* **754** 360–372 (*Preprint* 1506.08453)

[27] Adam J *et al.* (ALICE Collaboration) 2016 *Phys. Rev. C* **93** 024917 (*Preprint* 1506.08951)

[28] Schuchmann S and Appelshaeuser H 2015 *Modification of K^0_s and $\Lambda(\bar{\Lambda})$ transverse momentum spectra in Pb-Pb collisions at $\sqrt{s_{NN}}$ = 2.76 TeV with ALICE* Ph.D. thesis Frankfurt U.

[29] Andronic A, Braun-Munzinger P and Stachel J 2009 *Phys. Lett. B* **673** 142–145 [Erratum: Phys. Lett. B 678, 516 (2009)] (*Preprint* 0812.1186)

[30] Stachel J, Andronic A, Braun-Munzinger P and Redlich K 2014 *J. Phys. Conf. Ser.* **509** 012019 (*Preprint* 1311.4662)

[31] Andronic A, Braun-Munzinger P, Stachel J and Winn M 2012 *Phys. Lett. B* **718** 80–85 (*Preprint* 1201.0693)

[32] Manninen J and Becattini F 2008 *Phys. Rev. C* **78** 054901 (*Preprint* 0806.4100)

[33] Abelev B *et al.* (STAR Collaboration) 2010 *Phys. Rev. C* **81** 024911 (*Preprint* 0909.4131)

[34] Aggarwal M *et al.* (STAR Collaboration) 2011 *Phys. Rev. C* **83** 034910 (*Preprint* 1008.3133)

[35] Borsanyi S *et al.* (Wuppertal-Budapest Collaboration) 2010 *JHEP* **1009** 073 (*Preprint* 1005.3508)

[36] Bazavov A, Bhattacharya T, DeTar C, Ding H T, Gottlieb S *et al.* (HotQCD) 2014 *Phys. Rev. D* **90** 094503 (*Preprint* 1407.6387)

[37] Kaczmarek O, Karsch F, Laermann E, Miao C, Mukherjee S *et al.* 2011 *Phys. Rev. D* **83** 014504 (*Preprint* 1011.3130)

[38] Borsanyi S, Endrodi G, Fodor Z, Katz S, Krieg S *et al.* 2012 *JHEP* **1208** 053 (*Preprint* 1204.6710)

[39] Braun-Munzinger P and Stachel J 2000 *Phys. Lett. B* **490** 196–202 (*Preprint* nucl-th/0007059)

[40] Andronic A, Braun-Munzinger P, Redlich K and Stachel J 2007 *Nucl. Phys. A* **789** 334–356 (*Preprint* nucl-th/0611023)

[41] Abelev B B *et al.* (ALICE Collaboration) 2014 *Phys. Lett. B* **743** 314–327 (*Preprint* 1311.0214)

[42] Adam J *et al.* (ALICE Collaboration) 2016 (*Preprint* 1606.08197)

[43] Dainese A *et al.* 2016 (*Preprint* 1605.01389)

[44] Matsui T and Satz H 1986 *Phys. Lett. B* **178** 416

[45] Aaij R *et al.* (LHCb Collaboration) 2013 *Nucl. Phys. B* **871** 1–20 (*Preprint* 1302.2864)

[46] Cacciari M, Frixione S, Houdeau N, Mangano M L, Nason P *et al.* 2012 *JHEP* **1210** 137 (*Preprint* 1205.6344)

[47] Vogt R 2015 *Phys. Rev. C* **92** 034909 (*Preprint* 1507.04418)

[48] Zhao X and Rapp R 2011 *Nucl. Phys. A* **859** 114–125 (*Preprint* 1102.2194)

[49] Zhou K, Xu N, Xu Z and Zhuang P 2014 *Phys. Rev. C* **89** 054911 (*Preprint* 1401.5845)

IOP Conf. Series: Journal of Physics: Conf. Series **779** (2017) 012013 doi:10.1088/1742-6596/779/1/012013

Experimental Overview of the Search for Chiral Effects at RHIC

Gang Wang

Department of Physics and Astronomy, University of California, Los Angeles, California 90095, USA

E-mail: gwang@physics.ucla.edu

Abstract. In high-energy heavy-ion collisions, various novel transport phenomena in local chiral domains result from the interplay of quantum anomalies with magnetic field and vorticity, and could survive the expansion of the fireball and be detected in experiments. Among these phenomena are the chiral magnetic effect, the chiral vortical effect and the chiral magnetic wave, the experimental searches for which have aroused extensive interest. This review will describe the current status of experimental studies at Relativistic Heavy Ion Collider at BNL, and outline the future work in experiment needed to eliminate the existing uncertainties in the interpretation of the data.

1. Introduction

The quark-gluon plasma (QGP) created in high-energy heavy-ion collisions could specify its thermodynamic states by the axial chemical potential μ_5. μ_5 characterizes the imbalance of right-handed and left-handed fermions in a domain. A *chiral* domain that bears a nonzero μ_5 may be created locally in heavy-ion collisions on an event-by-event basis [1, 2, 3, 4, 5, 6]. In a noncentral collision, a strong magnetic field ($B \sim 10^{15}$ T) can be produced (mostly by energetic spectator protons) [2, 3], and will induce an electric current along $\vec{B}$ in chiral domains, $\vec{J_e} \propto \mu_5 \vec{B}$, which is called the chiral magnetic effect (CME) [1, 2]. On average, $\vec{B}$ is perpendicular to the reaction plane ($\Psi_{\rm RP}$) that contains the impact parameter and the beam momenta. Hence the CME will manifest a charge transport across the RP.

Considering the CME and other modes of collective motions, we Fourier decompose the azimuthal distribution of particles of given transverse momentum (p_T) and pseudorapidity (η):

$$\frac{dN_\alpha}{d\phi} \propto 1 + 2v_{1,\alpha}\cos(\Delta\phi) + 2v_{2,\alpha}\cos(2\Delta\phi) + ... + 2a_{1,\alpha}\sin(\Delta\phi) + ..., \qquad (1)$$

where ϕ is the azimuthal angle of a particle, and $\Delta\phi = \phi - \Psi_{\rm RP}$. The subscript α (+ or $-$) denotes the charge sign of the particle. Conventionally v_1 is called "directed flow" and v_2 "elliptic flow" [7]. a_1 ($a_{1,-} = -a_{1,+}$) quantifies the electric charge separation due to the CME.

Another complementary transport phenomenon to the CME has been found and named the chiral separation effect (CSE) [8, 9], in which chiral charges are separated along the axis of the magnetic field in the presence of finite density of vector charge: $\vec{J_5} \propto \mu_{\rm V}\vec{B}$. In a chirally symmetric phase, the CME and CSE form a collective excitation, the chiral magnetic wave

(CMW) [10, 11]. The CMW manifests a finite electric quadrupole moment of the collision system, where the "poles" ("equator") of the produced fireball acquire additional positive (negative) charge [10]. There are other chiral magnetic/vortical effects such as the chiral vortical effect (CVE) [12], the chiral electric separation effect (CESE) [13, 14] and the chiral vortical wave (CVW) [15]; see Ref [16] for a recent review on these effects.

2. Chiral Magnetic Effect

The experimental searches for the CME have been carried out extensively in the past decade at RHIC [17, 18, 19, 20, 21, 22] and the LHC [23]. From event to event, the signs of the μ_5 values are equally likely, and the signs of finite $a_{1,+}$ and $a_{1,-}$ will flip accordingly, leading to $\langle a_{1,+}\rangle = \langle a_{1,-}\rangle = 0$. One therefore has to search for the CME with charge-separation *fluctuations* perpendicular to the reaction plane, e.g., with a three-point correlator [24], $\gamma \equiv \langle\cos(\phi_\alpha + \phi_\beta - 2\Psi_{\rm RP})\rangle$, averaging over all particles in an event and over all events.

The STAR Collaboration first measured the γ correlator with the $1^{\rm st}$ and $2^{\rm nd}$ harmonic event planes for Au+Au and Cu+Cu collisions at 62.4 and 200 GeV with data from the 2004/2005/2007 RHIC runs [17, 18, 19]. All the results are in qualitative agreement with the CME: the opposite-charge ($\gamma_{\rm OS}$) and the same-charge ($\gamma_{\rm SS}$) correlations display the "right" ordering. The opposite-charge correlations in Cu+Cu are stronger than those in Au+Au, possibly reflecting the suppression of the correlations among oppositely moving particles in a larger system. Similar γ results for 200 GeV Au+Au and 2.76 TeV Pb+Pb were observed by the PHENIX Collaboration [22] and the ALICE Collaboration [23], respectively.

STAR has explored two different correlators, the modulated sign correlator [19] and the charge multiplicity asymmetry correlator [21], whose methodology is similar to the γ correlator, and yielded very similar results. PHENIX also employed a multiparticle charge-sensitive correlator [25], and their preliminary results also evidence the charge-separation effect. The background from conventional physics was studied with heavy-ion event generators MEVSIM [26], UrQMD [27] and HIJING [28] (with and without an elliptic flow afterburner implemented). None of these generators could achieve reasonable agreement with the data.

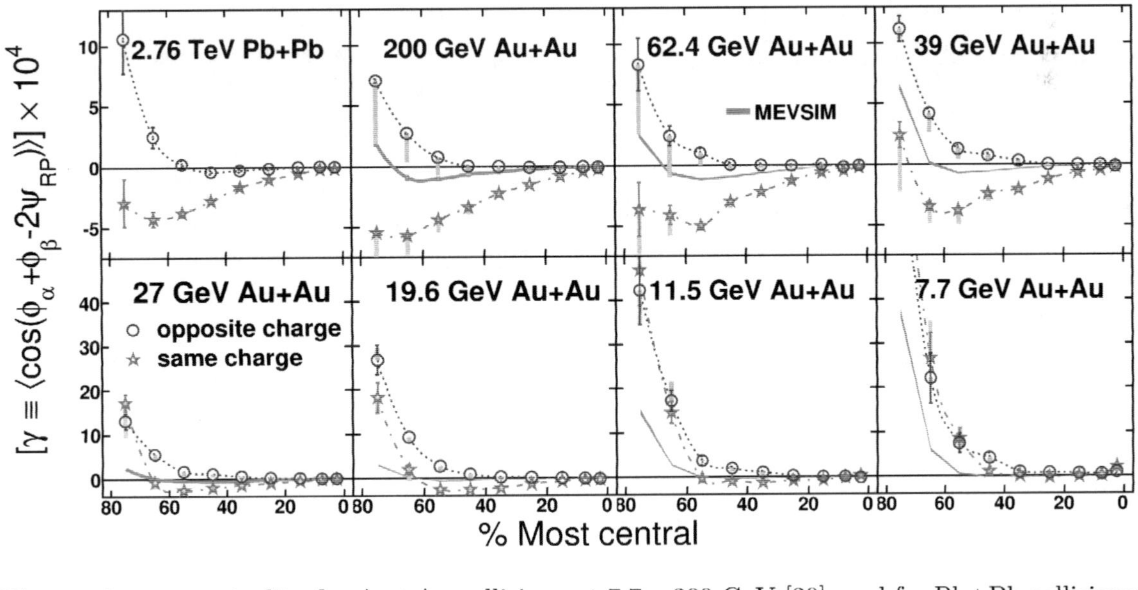

Figure 1. γ vs centrality for Au+Au collisions at $7.7-200$ GeV [20], and for Pb+Pb collisions at 2.76 TeV [23]. The gray bars reflect the conditions of $\Delta p_T > 0.15$ GeV/c and $\Delta\eta > 0.15$ applied to γ. Charge independent results from MEVSIM [26] are shown as gray curves.

To further understand the origin of the observed charge separation, Fig. 1 presents $\gamma_{\rm OS}$ and $\gamma_{\rm SS}$ correlators as functions of centrality for Au+Au collisions at $\sqrt{s_{NN}} = 7.7 - 200$ GeV measured by STAR [20], and for Pb+Pb collisions at 2.76 TeV by ALICE [23]. With decreased beam energy, both $\gamma_{\rm OS}$ and $\gamma_{\rm SS}$ tend to rise up starting from peripheral collisions, which seems to be charge independent and can be explained by momentum conservation and elliptic flow [19]. In most cases, the difference between $\gamma_{\rm OS}$ and $\gamma_{\rm SS}$ is still present with the expected ordering, manifesting extra charge-separation fluctuations perpendicular to the reaction plane. The difference between $\gamma_{\rm OS}$ and $\gamma_{\rm SS}$ seems to vanish at low collision energies, but the interpretation involves an ambiguity due to the flow-related background.

One useful tool to study the background is the two-particle correlator, $\delta \equiv \langle\cos(\phi_\alpha - \phi_\beta)\rangle$, which ideally should be protortional to $\langle a_{1,\alpha}a_{1,\beta}\rangle$, but in reality is dominated by backgrounds [20]. The background-subtracted correlator, H, can be obtained from the ensemble averages of several observables [20, 29]: $H^\kappa = (\kappa v_2\delta - \gamma)/(1+\kappa v_2)$. The major uncertainty in H, the coefficient κ, has been estimated with three approaches to be typically between 1 and 1.5, shown in Fig. 2 [30]. Figure 3 shows $(H_{\rm SS}^{\kappa=1} - H_{\rm OS}^{\kappa=1})$ as a function of beam energy for 30 – 60% Au+Au (Pb+Pb) collisions [20, 23]. $\Delta H^{\kappa=1.5}$ is depicted with the solid line. In both cases of κ, ΔH demonstrates a weak energy dependence above 19.6 GeV, and tends to diminish from 19.6 to 7.7 GeV, though the statistical errors are large for 7.7 GeV. This may be explained by the probable domination of hadronic interactions over partonic ones at low energies. A more definitive conclusion may be reached with a more accurate estimation of κ and with higher statistics at lower energies in the proposed phase II of the RHIC Beam Energy Scan program, as illustrated by the shaded band in Fig. 3.

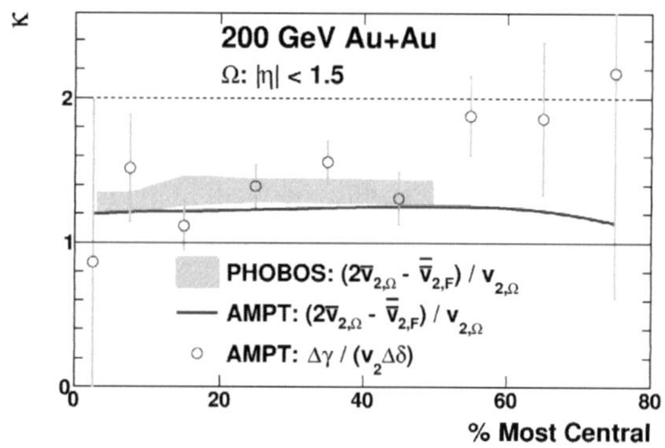

Figure 2. (Color online) Estimation of κ with three approaches for 200 GeV Au+Au [30].

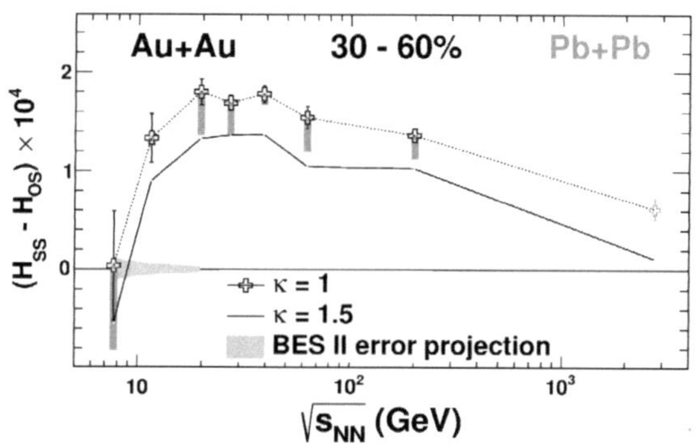

Figure 3. (Color online) ΔH as a function of beam energy for 30 – 60% Au+Au (Pb+Pb) collisions [20, 23]. The systematic errors (gray bars) bear the same meaning as in Fig. 1.

3. Chiral Magnetic Wave

The CMW will induce a finite electric quadrupole moment of the collision system, with additional positive (negative) charge at the "poles" ("equator") of the produced fireball [10]. This electric quadrupole, if boosted by radial flow, will lead to charge-dependent elliptic flow. Taking pions as an example, on top of the baseline $v_2^{\rm base}(\pi^\pm)$, the CMW will lead to [10]

$$v_2(\pi^\pm) = v_2^{\rm base}(\pi^\pm) \mp (\frac{q_e}{\bar{\rho}_e})A_{\rm ch}, \tag{2}$$

where q_e, $\bar{\rho}_e$ and $A_{\rm ch} = (N_+ - N_-)/(N_+ + N_-)$ are the quadrupole moment, the net charge density and the charge asymmetry of the collision event, respectively. Panel (a) of Fig. 4 shows the example of 30-40% 200 GeV Au+Au, where π^- v_2 increases with $A_{\rm ch}$ and π^+ v_2 decreases with a similar magnitude of the slope [31]. The v_2 difference between π^- and π^+ is fitted with a straight line in panel (b). The slope parameter r, or presumably $2q_e/\bar{\rho}_e$ from Eq. 2, is positive and qualitatively consistent with the expectation of the CMW picture.

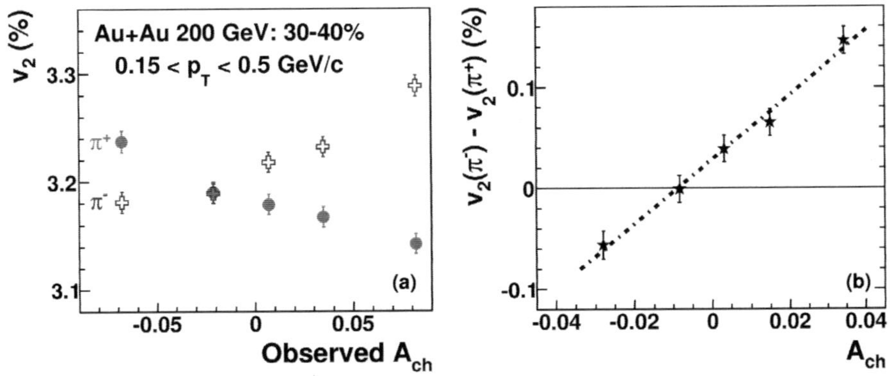

Figure 4. (a) pion v_2 as a function of observed $A_{\rm ch}$ and (b) v_2 difference between π^- and π^+ as a function of corrected $A_{\rm ch}$, for 30-40% Au+Au collisions at 200 GeV [31].

STAR retrieved the slope parameter r as a function of centrality for Au+Au collisions at $7.7 - 200$ GeV, as shown in Fig. 5 [31]. A similar rise-and-fall trend is observed in the centrality dependence of the slope parameter for all the beam energies except 11.5 and 7.7 GeV, where the slopes are consistent with zero with large statistical uncertainties. A comparison in the slope parameter r between STAR results for 200 GeV Au+Au [31] and ALICE results for 2.76 TeV Pb+Pb [32] shows a striking similarity. For Au+Au collisions at 200 GeV, the slopes extracted from UrQMD events are consistent with zero for the 10-70% centrality range, where the signal from the real data is prominent. Similarly, the AMPT event generator [34, 35] also yields null results. On the other hand, the simplified CMW calculations [33] demonstrate a centrality dependence of the slope parameter similar to the data.

It was pointed out in Ref. [36] that local charge conservation at freeze-out, when convoluted with the characteristic shape of $v_2(\eta)$ and $v_2(p_T)$, may provide a qualitative explanation for the finite v_2 slope. A realistic estimate of the contribution of this mechanism turns out to be smaller than the measurment by an order of magnitude [31]. A recent hydrodynamic study [37] suggested that simple viscous transport of charges, combined with certain specific initial conditions, might lead to a sizable contribution to the observed v_2 splitting of charged pions. However, certain predictions of this model (e.g. splitting for kaons) appear to be not in line with current experimental information [38].

4. Future measurements

The confirmation of the experimental observation of several chiral anomalous effects will bring forth an exciting program to directly study the non-perturbative sector of QCD. Future experimental measurements should aim at more detailed study of the observed signals as well as understanding the background effects. To disentangle the possible CME signal and the flow-related backgrounds, one can utilize experimental setups to either vary the backgrounds with the signal fixed, or vary the signal with the backgrounds fixed. The former approach was carried out by exploiting the prolate shape of the uranium nuclei [39]. However, it was found that the total multiplicity of detected hadrons is far less dependent on the number of binary collisions than expected [40], so it is very hard to isolate tip-tip collisions (that generate small v_2) from

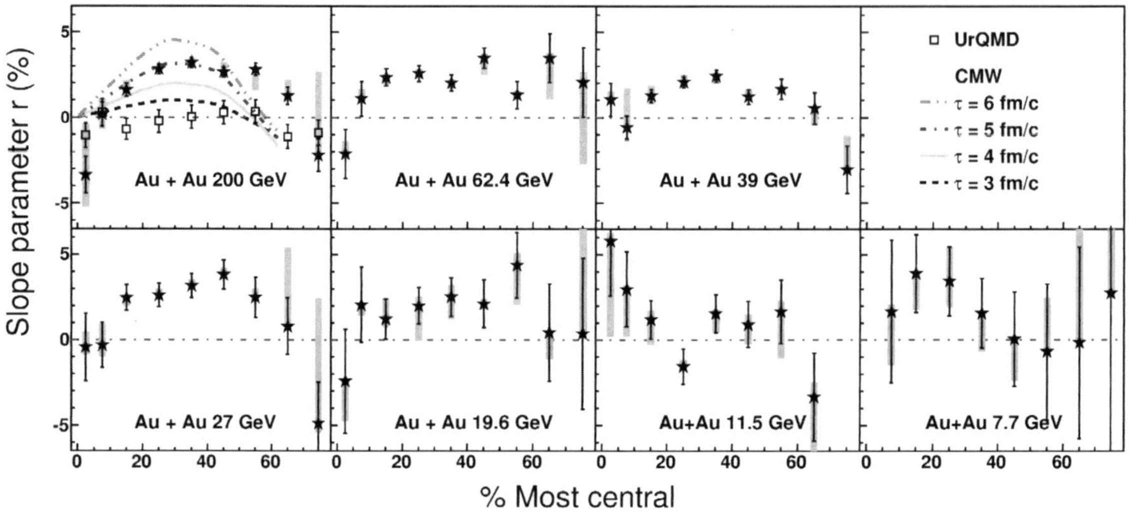

Figure 5. The slope parameter r as a function of centrality for Au+Au collisions at $7.7 - 200$ GeV [31]. The gray bands represent the systematic errors. For comparison, we also show the UrQMD calculations [27] and the calculations of the CMW [33] with different duration times.

body-body collisions (that generate large v_2). This significantly reduces the lever arm available to manipulate v_2 in order to separate flow backgrounds from the CME.

The latter approach (with the v_2-driven backgrounds fixed) can be realized, especially for mid-central/mid-peripheral events, with collisions of isobaric nuclei, such as $^{96}_{44}$Ru and $^{96}_{40}$Zr [39]. Ru+Ru and Zr+Zr collisions at the same beam energy are almost identical in terms of particle production [41], while the charge difference between Ru and Zr nuclei provides a handle on the initial magnetic field. Figure 6(a) shows the projection of $S \equiv N_{\rm part}\Delta\gamma$ for Ru + Ru and Zr + Zr collisions at 200 GeV, as functions of centrality, with the magnetic field and eccentricity obtained for case 1, and the background level $bg = 2/3$ [41]. Here "case 1" (case 2) refers to the information of the nucleus deformity based on e-A scattering experiments [42, 43] (comprehensive model deductions [44]). The statistical errors are estimated based on 400M events for each collision type. The systematic uncertainties in the projection are largely canceled out with the relative difference between Ru + Ru and Zr + Zr, R_S, shown in Fig. 6(b); in comparison, the relative difference in eccentricity, R_{ϵ_2}, is also shown. When combining the events of $20-60\%$ centralities, R_S is 5σ above R_{ϵ_2} for both parameter cases. Therefore, the isobaric collisions provide a unique test to pin down the underlying physics mechanism for the observed charge separation. As a by-product, v_2 measurements in central collisions will discern which information source (case 1 or 2) is more reliable regarding the deformity of the Ru and Zr nuclei.

References

[1] D. E. Kharzeev, L. D. McLerran and H. J. Warringa, Nucl. Phys. A **803**, 227 (2008).
[2] D. Kharzeev, Phys. Lett. B **633**, 260 (2006).
[3] D. Kharzeev and A. Zhitnitsky, Nucl. Phys. A **797**, 67 (2007).
[4] D. Kharzeev, A. Krasnitz and R. Venugopalan, Phys. Lett. B **545**, 298 (2002).
[5] I. Iatrakis, S. Lin and Y. Yin, Phys. Rev. Lett. **114**, 252301 (2015).
[6] K. Fukushima, D. E. Kharzeev and H. J. Warringa, Phys. Rev. Lett. 104, 212001 (2010).
[7] A. M. Poskanzer and S. Voloshin, Phys. Rev. C **58**, 1671 (1998).
[8] D. T. Son and A. R. Zhitnitsky, Phys. Rev. D **70**, 074018 (2004).
[9] M. A. Metlitski and A. R. Zhitnitsky, Phys. Rev. D **72**, 045011 (2005).
[10] Y. Burnier, D. E. Kharzeev, J. Liao and H.-U. Yee, Phys. Rev. Lett. **107**, 052303 (2011).

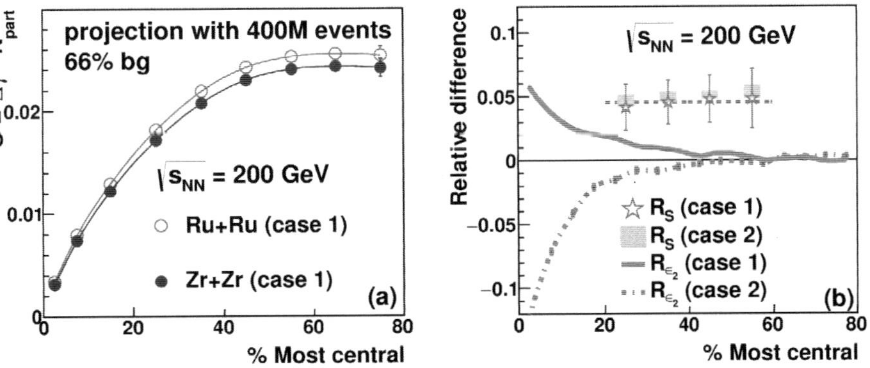

Figure 6. Projection of $S \equiv N_{\rm part}\Delta\gamma$ for Ru + Ru and Zr + Zr collisions at $\sqrt{s_{\rm NN}} = 200$ GeV for case 1 (a) and the relative difference (b) versus centrality, assuming the background level to be two thirds [41]. Also shown in panel (b) is the relative difference in the initial eccentricity from the Monte Carlo Glauber [45] simulation (pink solid and dashed lines).

[11] G. M. Newman, JHEP **0601**, 158 (2006).
[12] D. E. Kharzeev and D. T. Son, Phys. Rev. Lett. **106**, 062301 (2011).
[13] X. G. Huang and J. Liao, Phys. Rev. Lett. **110**, 232302 (2013).
[14] Y. Jiang, X. G. Huang and J. Liao, Phys. Rev. D **91**, 045001 (2015).
[15] Y. Jiang, X. G. Huang and J. Liao, Phys. Rev. D **92**, 071501 (2015).
[16] D.E. Kharzeev, J. Liao, S.A. Voloshin and G. Wang, Prog. Part. Nucl. Phys. **88**, 1 (2016).
[17] B. I. Abelev *et al.* [STAR Collaboration], Phys. Rev. Lett. **103**, 251601 (2009).
[18] 7. I. Abelev *et al.* [STAR Collaboration], Phys. Rev. C **81**, 54908 (2010).
[19] L. Adamczyk *et al.* [STAR Collaboration], Phys. Rev. C **88**, 064911 (2013).
[20] L. Adamczyk *et al.* [STAR Collaboration], Phys. Rev. Lett. **113**, 052302 (2014).
[21] L. Adamczyk *et al.* [STAR Collaboration], Phys. Rev. C **89**, 044908 (2014).
[22] N. N. Ajitanand, S. Esumi, R. A. Lacey (PHENIX Collaboration), in: Proc. of the RBRC Workshops, vol.**96**, 230 (2010): “P- and CP-odd ects in hot and dense matter”.
[23] B. I. Abelev *et al.* [ALICE Collaboration], Phys. Rev. Lett. **110**, 021301 (2013).
[24] S. A. Voloshin, Phys. Rev. C **70**, 057901 (2004).
[25] N. N. Ajitanand, R. A. Lacey, A. Taranenko and J. M. Alexander, Phys. Rev. C **83**, 011901 (2011).
[26] R. L. Ray and R. S. Longacre, arXiv:nucl-ex/0008009 (2000).
[27] S. A. Bass *et al.*, Prog. Part. Nucl. Phys. **41**, 255 (1998).
[28] M. Gyulassy and X.-N. Wang, Comput. Phys. Commun. **83**, 307 (1994); X.N. Wang and M.Gyulassy, Phys. Rev. D **44**, 3501 (1991).
[29] A. Bzdak, V. Koch and J. Liao, Lect. Notes Phys. **871**, 503 (2013) [arXiv:1207.7327 [nucl-th]].
[30] F. Wen, L. Wen and G. Wang, arXiv:1608.03205 (2016).
[31] L. Adamczyk *et al.* (STAR Collaboration), Phys. Rev. Lett. **114**, 252302 (2015).
[32] J. Adam *et al.* (ALICE Collaboration), Phys. Rev. C **93**, 044903 (2016).
[33] Y. Burnier, D.E. Kharzeev, J. Liao, H.-U. Yee, arXiv:1208.2537 (2012).
[34] B. Zhang, C.M. Ko, B.-A. Li and Z.-W. Lin, Phys. Rev. C **61**, 067901 (2000).
[35] Z.-W. Lin, C.M. Ko, B.-A. Li and B. Zhang, Phys. Rev. C **72**, 064901 (2005).
[36] A. Bzdak and P. Bozek, arXiv:1303.1138 (2013).
[37] Y. Hatta, A. Monnai and B. W. Xiao, Nucl. Phys. A **947**, 155 (2016).
[38] Q.-Y. Shou [STAR Collaboration], Nucl. Phys. A **931**, 758 (2014).
[39] S. A. Voloshin, Phys. Rev. Lett. **105**, 172301 (2010).
[40] L. Adamczyk *et al.* [STAR Collaboration], Phys. Rev. Lett. **115**, 222301 (2015).
[41] W.-T. Deng, X.-G. Huang, G.-L. Ma, G. Wang, arXiv:1607.04697 (2016).
[42] S. Raman, C. W. G. Nestor, Jr and P. Tikkanen, Atom. Data Nucl. Data Tabl. **78**, 1 (2001).
[43] B. Pritychenko, M. Birch, B. Singh and M. Horoi, Atom. Data Nucl. Data Tabl. **107**, 1 (2016).
[44] P. Moller, J. R. Nix, W. D. Myers and W. J. Swiatecki, Atom. Data Nucl. Data Tabl. **59**, 185 (1995).
[45] B. I. Abelev *et al.* [STAR Collaboration], Phys. Rev. C **79**, 034909 (2009).

IOP Conf. Series: Journal of Physics: Conf. Series **779** (2017) 012014 doi:10.1088/1742-6596/779/1/012014

From gluon topology to chiral anomaly: Emergent phenomena in quark-gluon plasma

Jinfeng Liao
Physics Department and Center for Exploration of Energy and Matter, Indiana University, 2401 N Milo B. Sampson Lane, Bloomington, IN 47408, USA.
RIKEN BNL Research Center, Bldg. 510A, Brookhaven National Laboratory, Upton, NY 11973, USA.

E-mail: liaoji@indiana.edu

Abstract. Heavy-ion collision experiments at RHIC and the LHC have found a new emergent phase of QCD, a strongly coupled quark-gluon plasma (sQGP) that is distinctively different from either the low temperature hadron phase or the very high temperature weakly coupled plasma phase. Highly nontrivial emergent phenomena occur in such sQGP and two examples will be discussed in this contribution: the magnetic component of sQGP that stems from topologically nontrivial configurations in the gluon sector; and the anomalous chiral transport that arises as macroscopic manifestation of microscopic chiral anomaly in the quark sector. For both examples, their important roles in explaining pertinent heavy-ion data will be emphasized.

1. Introduction

For the overwhelming part of the pursue of science in the past two thousand years or so since the early philosophers, people were trying to understand the structure of matter by *"reduction"*, i.e. by figuring out the most fundamental "building blocks" of all matter. That approach successfully led to the celebrated "Standard Model" and is leading further toward ideas beyond it. The opposite route, that is, to understand the integration of these "building blocks" into the various emergent forms and properties of matter, is no less fundamental and often proves equally nontrivial, as emphasized only relatively recently. In terms of the famous notion by P. W. Anderson, *"More Is Different"* [1]. A simple example to illustrate this notion, is perhaps the "lego matter": there are just a few different types of basic pieces with rather simple ways of coupling any two pieces together; given many thousands of lego pieces, one could imagine a simple/natural phase of lego matter being just a big random pile of them; there are however all kinds of ordered structures from a toy car to a splendid castle that could be constructed given enough of these pieces — these constructions are "emergent phases" of lego matter. The whole of condensed matter physics, is essentially studying the wealth of *emergent phenomena* based on the fundamental force of electromagnetism.

The world of nuclear matter is of course governed by a different type of fundament force, described by the theory of quantum chromodynamics (QCD). Nevertheless as recently emphasized by F. Wilcezk: "The study of the strong interactions is now a mature subject — we have a theory of the fundamentals (QCD)......The important questions involve 'emergent phenomena'." Indeed most of today's nuclear physics, in a way, may be regarded as the *"condensed matter physics of QCD"*, with several very active frontiers locating at different regimes on the QCD phase diagram in terms of temperature and baryon density. At very cold and dilute setting, the QCD matter is vacuum-like and one is at the *"force frontier"*, studying the emergent nonperturbative force in the vacuum-like environment that binds partons into observed hadrons by investigating spectroscopy, hadron scatterings/reactions, parton distributions, etc. Moving into the cold but extremely baryon-rich regime, the QCD matter takes the form from large nuclei to dense nuclear matter and eventually toward dense quark matter. This is the *"dense frontier"* where one investigates the many emergent phases that are crucial for understanding some of the most remarkable cosmic and astrophysical phenomena such as nucleosynthesis, supernovae and neutron stars. Finally upon increase of temperature to the extreme, the QCD matter enters an entirely new (and yet primordially old) phase as a quark-gluon plasma and one is at the *"hot frontier"*, creating the cosmic epoch QGP and measuring its properties via heavy-ion collisions.

For the rest of this contribution I will focus on the "hot frontier" and discuss a new emergent phase of QCD, the strongly coupled quark-gluon plasma (sQGP) that has been found in heavy-ion experiments at the Relativistic Heavy Ion Collider (RHIC) and the Large Hadron Collider (LHC) [2, 3]. This sQGP, spanning a temperature of range of about $1 \sim 3$ times the parton/hadron transition temperature $T_c \sim 165$MeV, is markedly different from either the low temperature hadron phase or the very high temperature weakly coupled plasma phase. In particular I will provide two nontrivial examples of emergent phenomena in sQGP that became appreciated only relatively lately. The first example involves the magnetic component of sQGP that stems from topologically nontrivial configurations known as chromo-magnetic monopoles in the gluon sector, while the second example is about the anomalous chiral transport that arises as macroscopic manifestation of microscopic chiral anomaly in the quark sector. In both cases, the importance in explaining pertinent heavy-ion data will be emphasized.

2. Magnetic component of sQGP

It was known long ago that the vacuum or vacuum-like state at $T \ll \Lambda_{QCD}$ is a complicated phase of QCD matter, with the remarkable phenomena of color confinement and spontaneous chiral symmetry breaking. It was also realized shortly after the birth of QCD that there should be *a natural/simple phase of QCD matter at asymptotically high temperature* $T \gg \Lambda_{QCD}$ where it should be much like a plasma of weakly coupled color charges (quarks and gluons). Indeed such early ideas motivated and eventually led to the successful heavy-ion experimental programs in the past decades. So where are we now? In short, the QGP is created, but turns out to be a rather different one than the originally expected asymptotically free QGP. Evidences from heavy-ion data as well as from lattice QCD simulations, particularly from its nearly perfect fluidity and extreme color-opaqueness, have been accumulated in support of this finding. In hindsight, a strongly coupled QGP may not be that surprising: after all, the reachable temperature in current heavy-ion collisions is only around $T \sim$ few times Λ_{QCD}. The measured properties of the sQGP, such as the shear viscosity (normalized by entropy density) $\frac{\eta}{s}$ and the (normalized) jet transport coefficient $\frac{\hat{q}}{T^3}$, clearly indicate nonperturbative dynamics in this temperature regime and calls for a deeper theoretical understanding. In addressing such problem, it was realized and proposed that this quark-gluon plasma is a special form of emergent plasma which is in a post-confinement regime *above but close to the parton/hadron phase boundary*, preserving essential ingredients that drive the system toward the confinement transition at T_c [4, 5, 6].

An obvious question is: what are the microscopic degrees of freedom in this near-T_c plasma? One hint comes from a comparison between how rapidly the thermal degrees of freedom (as measured by the normalized entropy density s/T^3) become liberated with increasing temperature and how rapidly the color degrees of freedom (as measured by Polyakov loop $< L >$, being zero in the confined phase while being unity in the fully deconfined phase) become liberated. Lattice data tell us that while the normalized entropy density increases by almost an order of magnitude

IOP Conf. Series: Journal of Physics: Conf. Series **779** (2017) 012014 doi:10.1088/1742-6596/779/1/012014

in a very narrow window just around T_c, the Polyakov loop stays well below unity even at $2T_c$. This implies that significant suppression on (chromo-)electric charges (i.e. the quarks and gluons) persists even into the $1 \sim 2\mathrm{T_c}$, as first emphasized with the name semi-QGP by Pisarski et al [7], and such non-perturbative suppression must originate from the same dynamics that enforces confinement below T_c. The "mismatch" between the liberation of thermal particles and the color charges in the plasma, must also indicate active alternative degrees of freedom that are not electrically charged and at play in the plasma. It was first conjectured by Shuryak and myself that a thermal *magnetic component* of relatively light and abundant (chromo-)magnetic monopoles must prevail in the near-T_c plasma [4, 5]. Indeed, if the confining regime below T_c would be a (dual) superconducting phase of magnetic monopoles, then it would be natural to expect a *normal phase* of the same thermal ensemble of monopoles above T_c. These monopoles are themselves magnetically charged and free from the Polyakov loop suppression. They should be the origin of the suppression of quarks/gluons in the semi-QGP regime and their Bose-Einstein condensation at T_c signals the confinement transition. This is the "magnetic scenario" for the sQGP. Evidences for such a picture were seen from lattice simulations for pure gauge theories where these monopoles are identified (albeit in certain gauge choice) and counted, with their densities and mutual-correlations being measured at temperatures around and above T_c [8, 9, 10].

It is natural to ask how the magnetic scenario may help understand the observed properties of the sQGP such as η/s and $\hat{q}/T^3$ and whether the magnetic scenario is indispensable to explaining certain heavy-ion data. As pointed out in [4, 5], by virtue of the Dirac condition the scatterings between electric and magnetic components are always strong and frequent, which is precisely what an abundant magnetic component has to offer for making a nearly perfect fluid with extreme color opaqueness. Based on such reasoning, a nontrivial prediction for the jet quenching phenomenon was first made in [12], namely the jet-medium interaction is strongly enhanced in the near-T_c regime due to the magnetic component. The analysis of the measured azimuthal anisotropy data for jet energy loss was shown to provide strong phenomenological evidence for such a near-T_c enhancement [12, 11] which implies a prominent peak around T_c for the normalized jet transport coefficients $\hat{q}/T^3$ in analogy to the "critical opalescence".

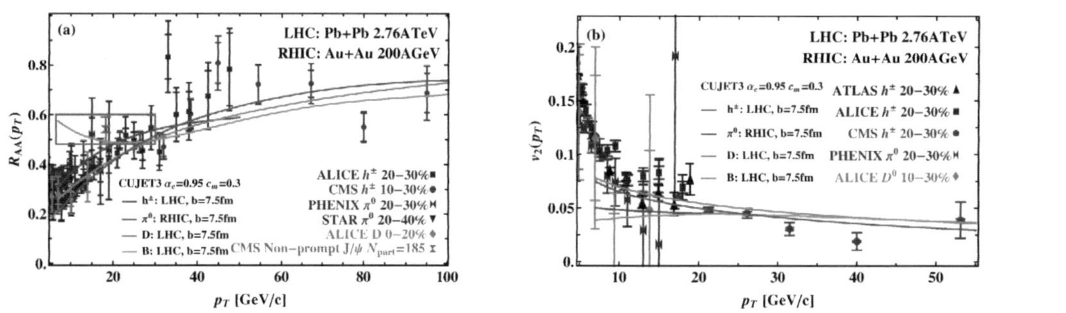

Figure 1. (a) $R_{AA}(p_T)$ and (b) $v_2(p_T)$ for inclusive neutral pions (π^0) and charged particles ($h^{\pm}$) in Au+Au $\sqrt{s_{NN}} = 200$ GeV and Pb+Pb $\sqrt{s_{NN}} = 2.76$ TeV collisions as well as that for open heavy flavors (D meson, red; B meson, green) at LHC semi-peripheral Pb+Pb $\sqrt{s_{NN}} = 2.76$ TeV, computed from CUJET3.0 with the impact parameter $b = 7.5$ fm, and compared with various available data from ALICE, ATLAS, CMS, PHENIX and STAR. See [13, 14] for details.

More recently, a major step forward was achieved for implementing the magnetic component of sQGP into a sophisticated and realistic modeling framework for heavy-ion collisions [13, 14]. This new comprehensive simulation framework, called CUJET3.0, (1) treats the radiative energy loss in the DGLV formalism; (2) convolutes energy loss with a bulk evolution of heavy-ion collisions from the state-of-the-art viscous hydrodynamic simulation (VISHNU hydro) that is data-validated; (3) implements nonpeturbative near-Tc enhancement of jet-medium coupling by introducing the monopole component in the plasma constituents as well as the semi-QGP suppression on quarks and gluons; (4) constrains the thermodynamic contents of the plasma constituents by current lattice QCD data. Using the CUJET3.0, jet quenching (leading-hadron) observables have been systematically investigated, including the nuclear modification factor R_{AA} and its azimuthal anisotropy v_2 at high transverse momenta for light and heavy flavors as well as for both RHIC and the LHC. Remarkably, this new framework with only one essential parameter successfully describes experimental data for seven sets of single hadron observables, including light hadrons' R_{AA} and v_2 at RHIC and LHC, D meson R_{AA} and v_2 at LHC, as well as B meson R_{AA} at LHC: see comparison in Fig. 1. This framework has also been used to predict the temperature dependence of sQGP transport properties $\hat{q}/T^3$ and η/s. The results show a strong non-monotonic structure, with a prominent peak for $\hat{q}/T^3$ and a strong minimum for η/s, both located in the near-Tc regime: see Fig. 2. The values of these coefficients from CUJET3.0 are in consistency with the phenomenological values inferred from experimental data.

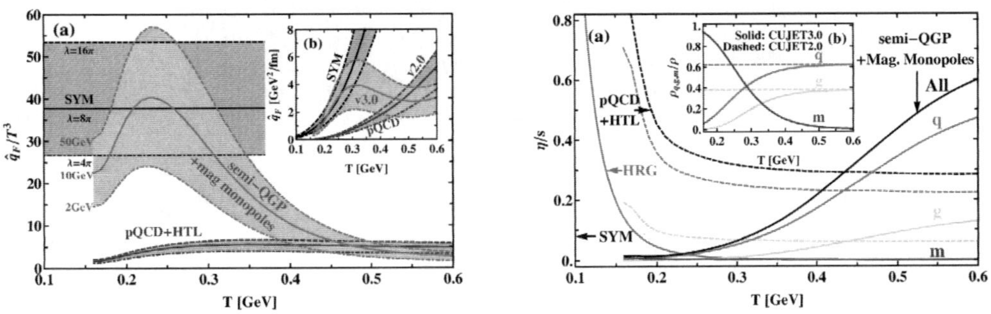

Figure 2. Temperature dependence of jet transport coefficient $\hat{q}/T^3$ (left) and shear viscosity per entropy density η/s (right), computed from CUJET3.0 (see [13, 14] for details).

3. Anomalous chiral transport

Let me now move to the discussion of an emergent phenomenon in the quark sector. In the hot QGP the chiral symmetry is resorted and the light flavor quarks (u and d), with $m_{u,d} \ll T$, can be considered chiral fermions to very good approximation. A most fundamental aspect in the quantum theory of chiral fermions is the famous chiral anomaly. An interesting question, is how such microscopic anomaly might manifest in macroscopic many-body setting. The answer by now is well known: from chiral anomaly emerges the anomalous chiral transport processes, with the most notable example of chiral magnetic effect (CME). (See reviews in e.g. [15, 16].)

As the QGP evolution in heavy-ion collisions is well described by hydrodynamics, let me rephrase this problem in the hydrodynamic context. The hydrodynamics by itself is precisely an emergent phenomenon from microscopic symmetries in the long-time and long-distance limit: spacetime translational invariance $\rightarrow$ conserved energy-momentum in microscopic scatterings $\rightarrow$ macroscopic hydro equations $\partial_\mu T^{\mu\nu} = 0$; phase invariance $\rightarrow$ conserved charges in microscopic scatterings $\rightarrow$ macroscopic hydro equations $\partial_\mu J^\mu = 0$. One may then ask: what happens in hydro for a "half" symmetry i.e. an anomaly that is conserved classically but broken quantum mechanically? The answer is a new type of anomalous hydrodynamics [17], where the chiral anomaly leads to new type of hydro equations for chiral currents: $\partial_\mu J^\mu_{R/L} = \pm C_A E^\mu B_\mu$ (where C_A the universal anomaly coefficient), and more importantly a constituent relation $J^\mu = n u^\mu + \nu^\mu_{NS} + \xi \omega^\mu + \xi_B B^\mu$ with the second term the normal Navier-Stocks viscous term

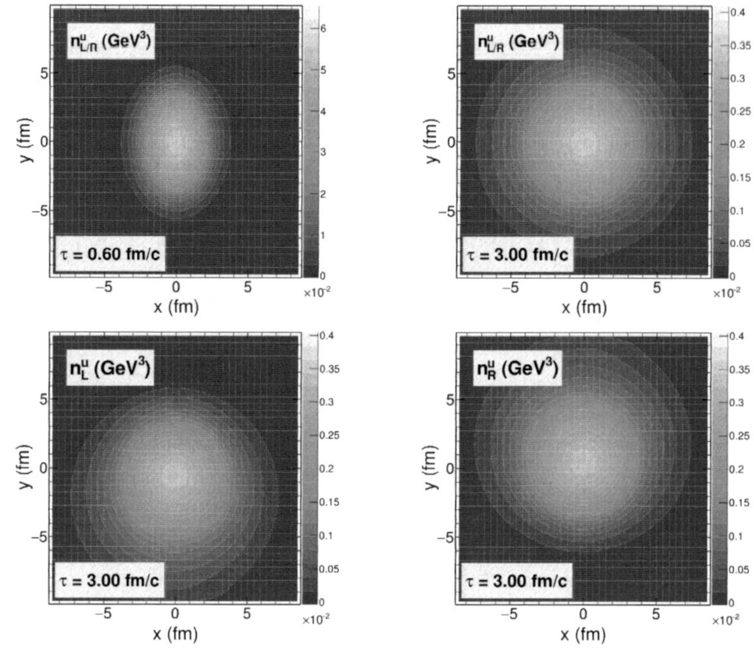

Figure 3. The charge densities at 3 fm/c starting from the same initial conditions (upper-left) in Anomalous-Viscous Fluid Dynamics for three cases: (upper-right) without $\vec{B}$ field; (lower-left) with $\vec{B} = B\hat{y}$ for left-handed density; (lower-right) with $\vec{B} = B\hat{y}$ for right-handed density.

while the last two terms the anomalous chiral transport currents corresponding to the chiral vortical effect and chiral magnetic effect respectively. This remarkable hydrodynamic framework that distinguishes "right" and "left", provides the necessary tool for quantitatively modeling anomalous chiral transport effects in heavy-ion collisions.

In heavy-ion collisions large electric and magnetic fields are present due to fast moving ions. Here let me refrain from discussing electric field (which also induces interesting effects e.g. [18]) but focus on magnetic field $\vec{B}$, which has been quantified in its magnitude as well as azimuthal orientation with event-by-event simulations [19]. The $\vec{B}$-driven CME could induce a charge separation effect along the out-of-plane direction, which can be measured with suitable charge-dependent azimuthal correlations. Such measurement faces the challenge of significant background contributions but a recent STAR analysis [20] has separated the potential signal via a two-component subtraction method [21]. Quantitative predictions for CME signal, based on realistic bulk evolution and integrating anomalous hydro framework, become crucial for comparison with data (see early attempts in e.g. [22, 23]). A tool for this purpose is now being built, the Anomalous-Viscous Fluid Dynamics (AVFD) simulation [24] which solves the anomalous hydro equations with 2nd-order viscous terms for chiral fermion currents in QGP with magnetic fields, on top of the data-validated VISHNew [25] bulk hydro evolution. The essential features of chiral density evolution in AVFD can be demonstrated in Fig. 3 where anomalous chiral transport is evident and differentiates the R/L fermions. Finally the quantitative predictions for CME-induced correlations from AVFD simulations, with realistic magnetic field magnitude and lifetime as well as estimated initial axial charge density, are presented in Fig. 4 and compared with pertinent STAR data. The main conclusion is that up to theoretical uncertainty, the CME predictions are quantitatively in the right ballpark of observed correlations. The residue difference might originate from a number of sources: theoretically

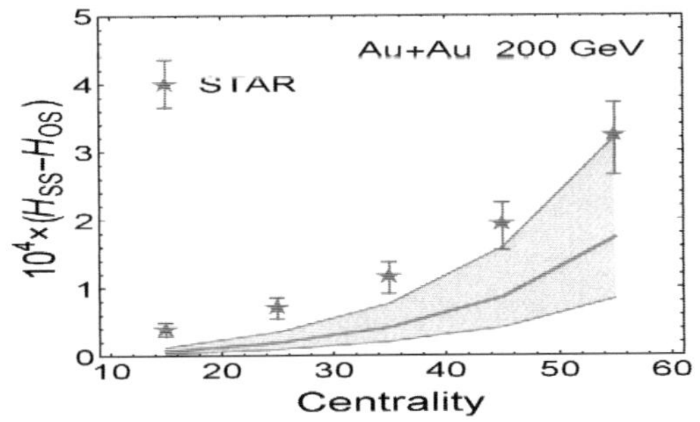

Figure 4. Quantitative predictions from Anomalous-Viscous Fluid Dynamics simulations for the CME-induced H-correlator, in comparison with STAR measurements [20]. The green bands reflect current theoretical uncertainty in the initial axial charge generated by gluonic field fluctuations.

various fluctuations and resonance decays need to be further included, while experimentally there are further backgrounds that shall be suitably suppressed in the H-correlators. These expected improvements together with anticipated isobaric collisions shall allow an ultimate conclusion.

Acknowledgments

The author thanks M. Gyulassy, Y. Jiang, S. Shi, J. Xu, Y. Yin for collaborations on the results reported here. The author is grateful to H. Huang, R. Pisarski, E. Shuryak, G. Wang and N. Xu for discussions. This material is based upon work supported by the U.S. Department of Energy, Office of Science, Office of Nuclear Physics, within the framework of the Beam Energy Scan Theory (BEST) Topical Collaboration. The work is also supported in part by the National Science Foundation (Grant No. PHY-1352368) and by the RIKEN BNL Research Center.

References

[1] P. W. Anderson, Science **177**, no. 4047, 393 (1972). doi:10.1126/science.177.4047.393
[2] M. Gyulassy and L. McLerran, Nucl. Phys. A **750**, 30 (2005) doi:10.1016/j.nuclphysa.2004.10.034
[3] E. V. Shuryak, Nucl. Phys. A **750**, 64 (2005) doi:10.1016/j.nuclphysa.2004.10.022
[4] J. Liao and E. Shuryak, Phys. Rev. C **75**, 054907 (2007) doi:10.1103/PhysRevC.75.054907
[5] J. Liao and E. Shuryak, Phys. Rev. Lett. **101**, 162302 (2008) doi:10.1103/PhysRevLett.101.162302
[6] J. Liao and E. Shuryak, Phys. Rev. Lett. **109**, 152001 (2012) doi:10.1103/PhysRevLett.109.152001
[7] R. D. Pisarski, Nucl. Phys. Proc. Suppl. **195**, 157 (2009). Y. Hidaka and R. D. Pisarski, Phys. Rev. D **78**, 071501 (2008); Phys. Rev. D **81**, 076002 (2010).
[8] M. N. Chernodub and V. I. Zakharov, Phys. Rev. Lett. **98**, 082002 (2007) doi:10.1103/PhysRevLett.98.082002
[9] A. D'Alessandro and M. D'Elia, Nucl. Phys. B **799**, 241 (2008) doi:10.1016/j.nuclphysb.2008.03.002
[10] C. Bonati and M. D'Elia, Nucl. Phys. B **877**, 233 (2013) doi:10.1016/j.nuclphysb.2013.10.004
[11] X. Zhang and J. Liao, Phys. Rev. C **89**, no. 1, 014907 (2014) doi:10.1103/PhysRevC.89.014907; Phys. Rev. C **87**, 044910 (2013) doi:10.1103/PhysRevC.87.044910
[12] J. Liao and E. Shuryak, Phys. Rev. Lett. **102**, 202302 (2009) doi:10.1103/PhysRevLett.102.202302
[13] J. Xu, J. Liao and M. Gyulassy, Chin. Phys. Lett. **32**, no. 9, 092501 (2015).
[14] J. Xu, J. Liao and M. Gyulassy, JHEP **1602**, 169 (2016) doi:10.1007/JHEP02(2016)169
[15] D. E. Kharzeev, J. Liao, S. A. Voloshin and G. Wang, Prog. Part. Nucl. Phys. **88**, 1 (2016) doi:10.1016/j.ppnp.2016.01.001 [arXiv:1511.04050 [hep-ph]].
[16] J. Liao, Pramana **84**, no. 5, 901 (2015); arXiv:1601.00381 [nucl-th].
[17] D. T. Son and P. Surowka, Phys. Rev. Lett. **103**, 191601 (2009) doi:10.1103/PhysRevLett.103.191601 [arXiv:0906.5044 [hep-th]].
[18] X. G. Huang and J. Liao, Phys. Rev. Lett. **110**, no. 23, 232302 (2013) doi:10.1103/PhysRevLett.110.232302
[19] J. Bloczynski, X. G. Huang, X. Zhang and J. Liao, Phys. Lett. B **718**, 1529 (2013)
[20] L. Adamczyk *et al.* [STAR Collaboration], Phys. Rev. Lett. **113**, 052302 (2014).
[21] A. Bzdak, V. Koch and J. Liao, Lect. Notes Phys. **871**, 503 (2013) [arXiv:1207.7327 [nucl-th]].
[22] Y. Hirono, T. Hirano and D. E. Kharzeev, arXiv:1412.0311 [hep-ph].
[23] Y. Yin and J. Liao, Phys. Lett. B **756**, 42 (2016) doi:10.1016/j.physletb.2016.02.065
[24] Y. Jiang, S. Shi, Y. Yin and J. Liao, to appear soon.
[25] C. Shen, Z. Qiu, H. Song, J. Bernhard, S. Bass and U. Heinz, Comput. Phys. Commun. **199**, 61 (2016).

IOP Conf. Series: Journal of Physics: Conf. Series **779** (2017) 012015 doi:10.1088/1742-6596/779/1/012015

Lattice QCD results on cumulant ratios at freeze-out

Frithjof Karsch

Fakultät für Physik, Universität Bielefeld, D-33615 Bielefeld, Germany;
Physics Department, Brookhaven National Laboratory, Upton, NY 11973, USA

E-mail: karsch@bnl.gov

Abstract. Ratios of cumulants of net proton-number fluctuations measured by the STAR Collaboration show strong deviations from a skellam distribution, which should describe thermal properties of cumulant ratios, if proton-number fluctuations are generated in equilibrium and a hadron resonance gas (HRG) model would provide a suitable description of thermodynamics at the freeze-out temperature. We present some results on 6th order cumulants entering the calculation of the QCD equation of state at non-zero values of the baryon chemical potential (μ_B) and discuss limitations on the applicability of HRG thermodynamics deduced from a comparison between QCD and HRG model calculations of cumulants of conserved charge fluctuations. We show that basic features of the μ_B-dependence of skewness and kurtosis ratios of net proton-number fluctuations measured by the STAR Collaboration resemble those expected from a $\mathcal{O}(\mu_B^2)$ QCD calculation of the corresponding net baryon-number cumulant ratios.

1. Introduction

A major goal in current experimental and theoretical studies of the thermodynamics of strong interaction matter is the exploration of its phase diagram. The hope is to find evidence for the existence of a second order phase transition point – the chiral critical point (CCP) – located at some value of the chemical potential, μ_B^{crit} [1]. This would be the starting point for a line of first order phase transitions at larger values of the baryon chemical potential μ_B.

At RHIC a dedicated research program – the beam energy scan (BES) – has been established that seeks evidence for the existence and location of the CCP. By varying the beam energy properties of matter in a regime of temperatures (T) up to about three times the transition temperature, $T_{pc} \sim 155$ MeV [2, 3], and baryon chemical potential up to $\mu_B \simeq 3T$ can be probed. It is generally expected that conserved charge fluctuations, which are generated close to, or at the freeze-out temperature, $T_f(\mu_B)$, can provide insight into the existence and location of the CCP. An important prerequisite for such studies, however, is to understand the thermodynamics of hot and dense matter in the crossover region and, in particular, close to freeze-out in QCD.

In the following we will point out the importance of characterizing this regime in terms of QCD rather than hadron resonance gas (HRG) model calculations, which are quite successful in approximating QCD thermodynamics at sufficiently low temperatures, but definitely fail to capture important aspects of QCD thermodynamics visible in conserved charge fluctuations at temperatures $T \gtrsim 160$ MeV.

2. Energy density in the crossover region

At small values of μ_B the QCD transition is not a true phase transition but a smooth transition from the low-T hadronic to the high-T partonic regime. This *crossover transition*

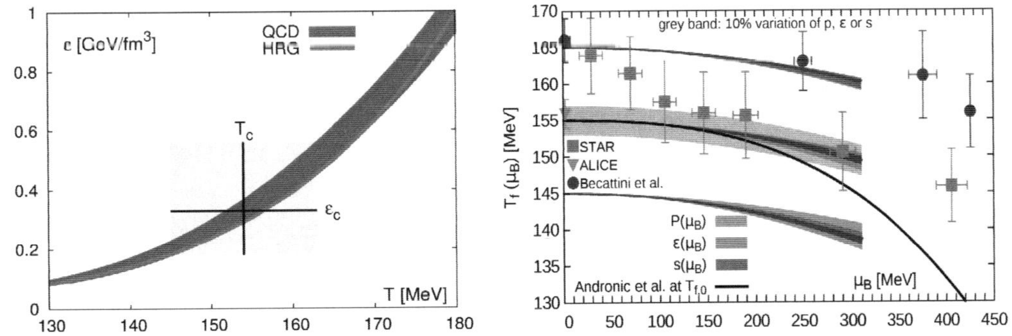

Figure 1. *Left*: Energy density of QCD versus temperature obtained from lattice QCD calculations (band) and HRG model calculations (line). *Right*: Freeze-out temperatures versus baryon chemical potential determined by STAR (boxes) [7] and ALICE (triangle) [8], and hadronization temperatures (circle) determined in Ref. [9]. The bands show lines of constant pressure, energy density and entropy density, respectively. They have been determined from a $\mathcal{O}(\mu_B^2)$ Taylor expansion of the QCD partition function. Also shown is a parametrization of the freeze-out line suggested by Andronic et al. (black line) [10] shifted to $T_f(0) = 155$ MeV.

does not happen at a well defined temperature. However, it can be characterized by pseudo-critical temperatures, i.e. temperatures that reflect characteristic features of e.g. fluctuation observables, which are guaranteed to converge to the true critical temperature in the chiral limit. One such observable is the chiral susceptibility, χ_q, the derivative of the chiral condensate with respect to quark mass. The location of the maximum of χ_q defines a pseudo-critical temperature. This has been obtained in lattice QCD calculations, $T_{pc} = 154(9)$ MeV [3].

In Fig. 1 (left) we show results for the energy density (ϵ) obtained from lattice QCD calculations with physical strange and light quark masses [6]. The temperature range covered by current uncertainties in T_{pc} corresponds to $\epsilon_c = (0.34 \pm 0.16)$ GeV/fm^3. Unfortunately, this energy density range is still quite large. In fact, it covers practically all energy scales of interest. While the central value corresponds to the close packing limit of nucleons with a radius $r_n \simeq 0.8$ fm, the lower limit is close to the energy density of nuclear matter ($\epsilon_{nm} \simeq 0.15$ GeV/fm^3) and the upper limit is larger than the energy density inside a nucleon ($\epsilon_n \simeq 0.45$ GeV/fm^3).

Using the Taylor expansion of the QCD equation of state one can follow lines of constant physics in the T-μ_B plane. In Fig. 1 (right) we show lines of constant pressure (p), energy density (ϵ) and entropy density (s) obtained in order μ_B^2. Corrections to this are small for $\mu_B/T \leq 2$. The three sets of curves, characterizing the current uncertainty band on T_{pc}, correspond to $\epsilon = 0.2$, 0.35 and 0.56 GeV/fm^3. Given the current uncertainties on T_{pc} it is not too surprising that all results on freeze-out parameters $(T_f(\mu_B), \mu_B)$ obtained from the analysis of particle yields at the LHC [8] and RHIC [7] and even the rather large hadronization temperatures extracted in Ref. [9] allow to state that "hadronization and freeze-out of hadrons occur close to, or in the QCD crossover region" (see Fig. 1 (right)) even though the temperatures in question are quite different, e.g. ranging from 155 MeV to 165 MeV at $\mu_B = 0$. They refer to environments, in which hadronization and freeze-out may take place, that are quite different, e.g. the energy and entropy density may differ by a factor 2. Lattice QCD calculations provide a wealth of other observables, e.g. higher order cumulants of conserved charge fluctuations, that are sensitive to the changes in physical properties of hot and dense matter that occur in this temperature interval. Understanding these changes also is of importance for our understanding of freeze-out conditions determined with the BES at RHIC as well as at the LHC.

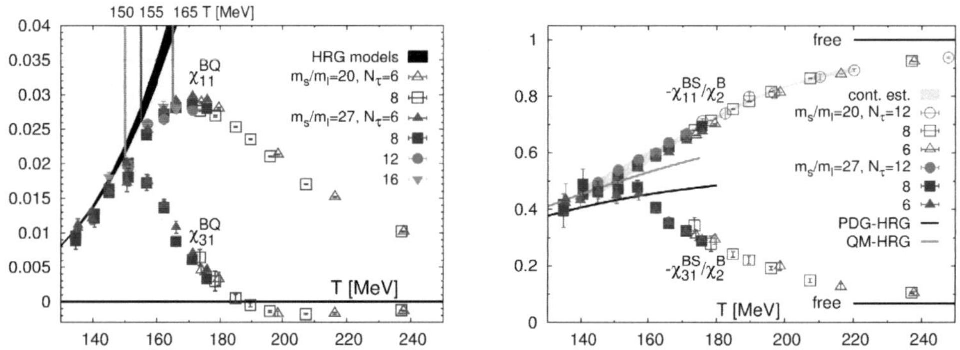

Figure 2. *Left*: Correlation between net baryon-number and net electric charge fluctuations (χ_{11}^{BQ}) versus temperature and the third moment of net baryon-number and net electric charge fluctuations (χ_{31}^{BQ}). The yellow band shows the crossover region defined by the uncertainty of the pseudo-critical temperature $T_{pc} = (154 \pm 9)$ MeV. *Right*: same as on the left but for correlations between net baryon-number and net strangeness fluctuations. These cumulants have been normalized with the quadratic fluctuations of net baryon-number.

3. Cumulants of conserved charge fluctuations

At $\mu_B = 0$ the energy density calculated in QCD as well as HRG models shown in Fig. 1 (left) varies smoothly as function of temperature when traversing the crossover region. Even quantitatively the latter is in quite good agreement with QCD calculations, although it has been pointed out that the QCD energy density is systematically larger, which may be taken as evidence for additional hadronic degrees of freedom contributing to bulk thermodynamics close to T_{pc} [4, 5]. Nonetheless, HRG model calculations seem to describe bulk thermodynamics quite well even at temperatures as large as $T \sim 180$ MeV. In fact, the situation is similar even for the specific heat [6]. Does this mean that the strongly interacting medium can be described in terms of hadronic degrees of freedom in the entire crossover region and even above?

Conserved charge fluctuations and correlations between them provide plenty of evidence that thermodynamics described in terms of hadronic degrees of freedom breaks down close to T_{pc}. In particular, higher order cumulants are quite different from conventional HRG model calculations. We show in Fig. 2 two examples of this. The left hand figure shows second and fourth order cumulants of net baryon-number and net electric charge fluctuations,

$$\chi_{11}^{BQ} = \left.\frac{\partial^2 p/T^4}{\partial\hat{\mu}_B\partial\hat{\mu}_Q}\right|_{\vec{\mu}=0} , \quad \chi_{31}^{BQ} = \left.\frac{\partial^4 p/T^4}{\partial\hat{\mu}_B^3\partial\hat{\mu}_Q}\right|_{\vec{\mu}=0} , \tag{1}$$

with $\hat{\mu}_X \equiv \mu_X/T$ and $\vec{\mu} = (\mu_B,\ \mu_Q,\ \mu_S)$. In the infinite temperature, ideal quark gas limit the net electric charge of 3-flavor QCD vanishes. Thus χ_{11}^{BQ} and χ_{31}^{BQ} will both approach zero while these correlations will keep increasing exponentially in HRG model calculations with point-like, non interacting hadrons.

At low temperature all hadronic degrees of freedom will either carry baryon number $B = 0$ or $B = \pm 1$. HRG model calculations thus give $\chi_{11}^{BQ} = \chi_{31}^{BQ}$. It is obvious from Fig. 2 (left) that this relation no longer holds for $T \gtrsim 150$ MeV. Also χ_{11}^{BQ} starts deviating from HRG model calculations at $T \gtrsim 155$ MeV. A similar pattern is found for second and fourth order cumulants of net baryon-number and net strangeness correlations shown in Fig. 2 (right). Various other combinations of 2nd and 4th order cumulants have been constructed that make it apparent that HRG models with point-like non-interacting hadrons are not suitable for describing QCD

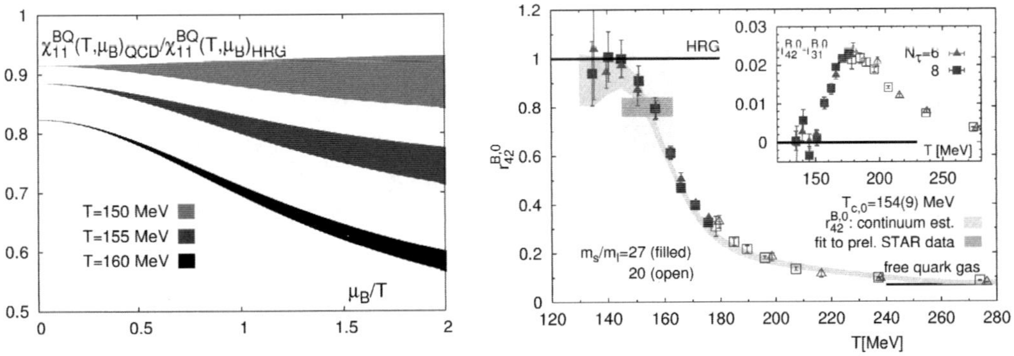

Figure 3. Correlations between net baryon-number and net electric charge as function of μ_B/T for three values of the temperature. Results from a QCD calculation at next-to-leading order Taylor expansion are normalized to a HRG model calculations truncated at the same order.

thermodynamics at temperatures $T \gtrsim T_{pc}$ [11]. In fact, this also has been verified for net baryon-number and net charm correlations [12] which strongly suggests that also the thermodynamics of open charm baryons cannot be described by HRG models above T_{pc}.

A consequence of the early deviation of 4th order cumulants from corresponding HRG calculations also is that differences between QCD and HRG calculations increase with increasing value of the chemical potential and will show up already in 2nd order cumulants. This is shown in Fig. 3 (left) for the correlation between net baryon-number and net electric charge evaluated up to $\mathcal{O}(\mu_B^2)$,

$$\chi_{11}^{BQ}(T,\mu_B) = \chi_{11}^{BQ} + \frac{1}{2}\chi_{31}^{BQ}\mu_B^2 \ . \tag{2}$$

Obviously QCD and HRG calculations differ significantly at $\mu_B/T = 2$ already for $T \gtrsim 155$ MeV.

4. Taylor expansions for skewness and kurtosis ratios

In HRG models with point-like non-interacting hadrons the distribution of net baryon number fluctuations is given by a skellam distribution. This leads to quite simple properties of higher order cumulants. In particular, the skewness ratio $S_B\sigma_B^3/M_B$ and the kurtosis ratio $\kappa_B\sigma_B^2$ should both be equal to unity at all values of μ_B. However, as discussed in the previous section this cannot be expected to hold in QCD at temperatures above T_{pc} where cumulants start to deviate significantly from HRG model calculations. The distribution of net baryon number fluctuations thus is e.g. not a simple skellam distribution. QCD calculations of the skewness and kurtosis ratios yield

$$\frac{S_B\sigma_B^3}{M_B} = \frac{\chi_3^B(T,\mu_B)}{\chi_1^B(T,\mu_B)} = \frac{\chi_4^B + s_1\chi_{31}^{BS} + q_1\chi_{31}^{BQ}}{\chi_2^B + s_1\chi_{11}^{BS} + q_1\chi_{11}^{BQ}} + \mathcal{O}(\mu_B^2) \equiv r_{31}^{B,0} + r_{31}^{B,2}\hat{\mu}_B^2 + \mathcal{O}(\mu_B^4) \tag{3}$$

$$\kappa_B\sigma_B^2 = \frac{\chi_4^B(T,\mu_B)}{\chi_2^B(T,\mu_B)} = \frac{\chi_4^B}{\chi_2^B} + \mathcal{O}(\mu_B^2) \equiv r_{42}^{B,0} + r_{42}^{B,2}\hat{\mu}_B^2 + \mathcal{O}(\mu_B^4) \ , \tag{4}$$

with s_1 and q_1 denoting the leading order expansion coefficients of μ_S and μ_Q in terms of μ_B that result from imposing the strangeness neutrality constraint $M_S = 0$ and a fixed electric charge to baryon number ratio, $M_Q/M_B = 0.4$. Results for the leading order expansion coefficients, defined in Eqs. 3 and 4, are shown in Fig. 3 (right). While at $T \simeq 150$ MeV the kurtosis ratio at $\mu_B = 0$ is still close to unity, $\kappa_B\sigma_B^2 \simeq 0.9$, it strongly deviates from unity for $T \simeq 160$ MeV,

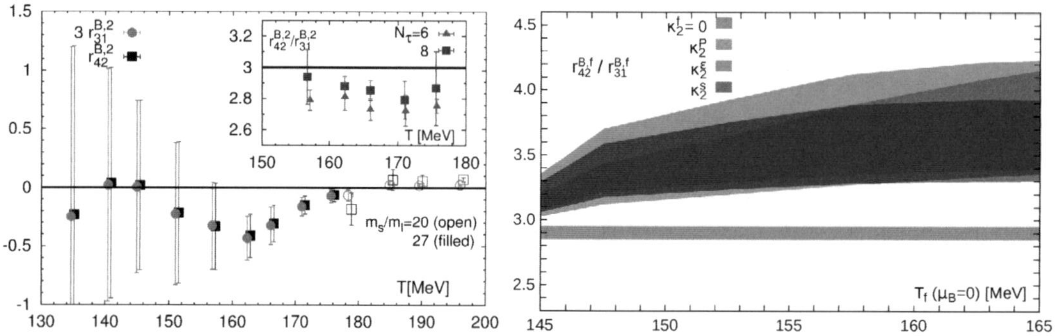

Figure 4. *Left*: The next-to-leading order (NLO) expansion coefficients of the skewness and kurtosis ratios at fixed temperature as introduced in Eqs. 3 and 4. *Right*: The ratio of the NLO expansion coefficients taking into account also a NLO expansion in terms of temperature along the line of constant pressure, energy density or entropy density, respectively.

giving $\kappa_B\sigma_B^2 \simeq 0.6$. In view of these differences it also is not obvious that the skewness and kurtosis ratios will still coincide in a QCD calculation. However, as can be seen from Eqs. 3 and 4 at $\mu_B = 0$ the leading order results will always be identical, if $\mu_Q = \mu_S = 0$, irrespective of the size of deviation from the skellam limit. For non-zero μ_Q and μ_S the skewness and kurtosis ratios will differ. Nonetheless, as can be seen from the insertion in Fig. 3 (right) this difference is small for all temperatures of interest. QCD thus predicts that the skewness and kurtosis ratios will approach each other in the limit $\mu_B \to 0$. This, however, changes for $|\mu_B| \neq 0$.

At next-to-leading order (NLO) the expansion coefficients $r_{31}^{B,2}$ and $r_{42}^{B,2}$ need to be calculated. This is computationally difficult, because 6th order cumulants need to be evaluated. Current results for $r_{31}^{B,2}$ and $r_{42}^{B,2}$ are shown in Fig. 4 (left). Although errors are still large, it is apparent that these expansion coefficients are negative for 150 MeV$\lesssim T \lesssim$175 MeV and that $r_{42}^{B,2}$ is about three times larger than $r_{31}^{B,2}$. This has been reported by us earlier [13]. It thus is expected that the skewness and kurtosis ratios, which are almost identical at $\mu_B = 0$, will start to differ for $\mu_B > 0$. As the expansion coefficients are negative, $\kappa_B\sigma_B^2$ will drop faster than $S_B\sigma_B^3/M_B$.

An additional subtlety in the analysis of the μ_B-dependence of the skewness and kurtosis ratios is that these need to be evaluated at a (freeze-out) temperature that changes with μ_B. This requires an additional Taylor expansion of the ratios introduced in Eqs. 3 and 4. Assuming that freeze-out happens along a line of constant physics described either by constant pressure, energy density or entropy density, i.e. the lines shown in Fig. 1, these contributions can be evaluated using the Taylor expansion of the pressure [16], changing $r_{nm}^{B,2}$ to $r_{nm}^{B,f}$. It turns out that the additional contributions increase the ratio $r_{42}^{B,f}/r_{31}^{B,f}$ somewhat as shown in Fig. 4 (right).

5. The STAR data on net proton-number skewness and kurtosis ratios

The STAR collaboration has measured the skewness ratio $S_P\sigma_P^3/M_P$ and kurtosis ratio $\kappa_P\sigma_P^2$ of net proton-number fluctuations [14, 15]. Obviously these cumulant ratios cannot directly be compared to QCD results on net baryon-number fluctuations. Already the strong sensitivity on the transverse momentum range used in the analysis, which is clearly visible in the data published so far by the STAR Collaboration (Fig. 5 (left) and (right), respectively), emphasizes that there is need for better understanding of various effects that enter the experimental analysis of higher order cumulants. Nonetheless, it is striking that the published as well as the new preliminary data on skewness and kurtosis ratios resemble all the features we expect to show up

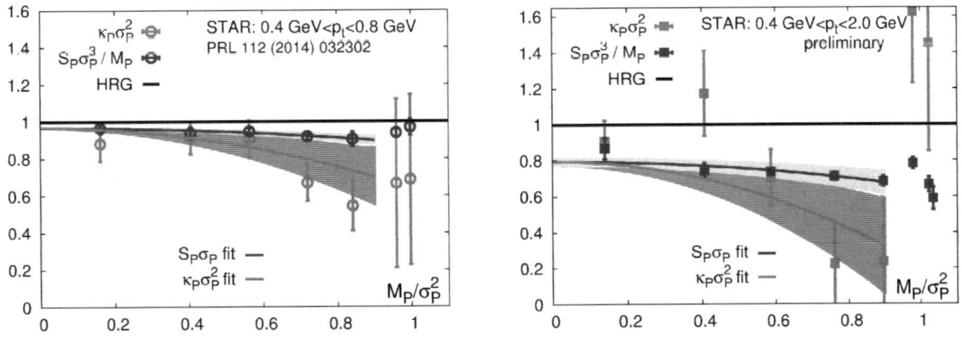

Figure 5. Skewness and kurtosis ratios of net proton-number fluctuations measured by the STAR collaboration in transverse momentum intervals 0.4 GeV $< p_t >$ 0.8 GeV (left) and 0.4 GeV $< p_t >$ 2.0 GeV (right), respectively. Data are plotted versus the ratio of mean (M_P) over variance (σ_P^2) of the net proton-number distribution, which also is measured at various beam energies. Curves show combined fits to the data for $M_P/\sigma_P^2 < 0.9$ or equivalently $\sqrt{s_{NN}} \le 19.6$ GeV/fm^3. They are constraint by demanding $S_P\sigma^3/M_P = \kappa_P\sigma_P^2$ at $M_P/\sigma_P^2 = 0$.

in equilibrium thermodynamics of QCD. The ratios (i) are smaller than unity, (ii) they seem to coincide in the limit $\mu_B \to 0$, (iii) they have a negative slope with increasing μ_B and (iv) the kurtosis ratio drops faster than the skewness ratio. In fact, a combined quadratic fit to these ratios, performed for all data obtained at beam energies $\sqrt{s_{NN}} \ge 19.6$ GeV/fm^3 and imposing the constraint $r_{42}^{B,0}/r_{31}^{B,0}$, yields for the ratio of slope parameters $r_{42}^{B,f}/r_{31}^{B,f} \sim 4 \pm 2$, which is in good agreement with the NLO QCD result shown in Fig. 4 (left). These fits are shown in Fig. 5.

Acknowledgments

This work has been partially supported through the U.S. Department of Energy under Contract No. DE-SC0012704 and within the framework of the Beam Energy Scan Theory (BEST) Topical Collaboration, and the German Bundesministerium für Bildung und Forschung (BMBF) under grant no. 05P15PBCAA.

References

[1] H. T. Ding, F. Karsch and S. Mukherjee, Int. J. Mod. Phys. E **24**, 1530007 (2015) [arXiv:1504.05274 [hep-lat]].
[2] Y. Aoki et al., JHEP **0906**, 088 (2009) [arXiv:0903.4155 [hep-lat]].
[3] A. Bazavov *et al.*, Phys. Rev. D **85**, 054503 (2012) [arXiv:1111.1710 [hep-lat]].
[4] A. Majumder and B. Muller, Phys. Rev. Lett. **105**, 252002 (2010) [arXiv:1008.1747 [hep-ph]].
[5] A. Bazavov *et al.*, Phys. Rev. Lett. **113**, 072001 (2014) [arXiv:1404.6511 [hep-lat]].
[6] A. Bazavov *et al.* [HotQCD Collaboration], Phys. Rev. D **90**, 094503 (2014) [arXiv:1407.6387 [hep-lat]].
[7] S. Das [STAR Collaboration], EPJ Web Conf. **90**, 08007 (2015) [arXiv:1412.0499 [nucl-ex]].
[8] M. Floris, Nucl. Phys. A **931**, 103 (2014) [arXiv:1408.6403 [nucl-ex]].
[9] F. Becattini, J. Steinheimer, R. Stock and M. Bleicher, arXiv:1605.09694 [nucl-th].
[10] A. Andronic, P. Braun-Munzinger and J. Stachel, Phys. Lett. B **673**, 142 (2009) Erratum: [Phys. Lett. B **678**, 516 (2009)] [arXiv:0812.1186 [nucl-th]].
[11] A. Bazavov *et al.*, Phys. Rev. Lett. **111**, 082301 (2013) [arXiv:1304.7220 [hep-lat]].
[12] A. Bazavov *et al.*, Phys. Lett. B **737**, 210 (2014) [arXiv:1404.4043 [hep-lat]].
[13] F. Karsch *et al.*, arXiv:1512.06987 [hep-lat].
[14] M. M. Aggarwal *et al.* [STAR Collaboration], Phys. Rev. Lett. **105**, 022302 (2010).
[15] X. Luo [STAR Collaboration], PoS CPOD **2014**, 019 (2015) [arXiv:1503.02558 [nucl-ex]].
[16] Bielefeld-BNL-CCNU Collaboration, in preparation.

Bulk properties of QCD matter from lattice simulations

Claudia Ratti
Department of Physics, University of Houston, Houston, TX 77204, USA
E-mail: cratti@uh.edu

Abstract. A review of the most recent results on QCD thermodynamics, obtained from lattice simulations, is presented. Particular focus is devoted to fluctuations of conserved charges and to their comparison with the experimental results from RHIC Beam Energy Scan.

1. Introduction

The deconfined phase of Quantum Chromodynamics (QCD), the Quark-Gluon Plasma (QGP) can be created in the laboratory in heavy ion collision experiments currently running at RHIC (Brookhaven National Laboratory) and at the LHC (CERN). These experiments allow to explore the QCD phase diagram and to extract the properties of this new phase of matter. The QGP turns out to be the most ideal fluid ever observed, which led to the idea that the system created in the collisions is strongly interacting and cannot be studied by means of perturbative techniques.

Lattice simulations are the best first-principle tool to address QCD in its non-perturbative regime. Given enough computer power, both statistical and systematic uncertainties can be kept under control. Due to a steady and continuous improvement in computer resources, numerical algorithms and our physical understanding which manifests itself in physical techniques (e.g. the Wilson-flow scale setting introduced in Ref. [1]), the lattice results which are being produced today reach an unprecedented level of accuracy, which allows a quantitative comparison to experimental observables for the first time.

The region of the phase diagram which can be explored experimentally corresponds to values of temperature T and chemical potential μ_B for which perturbative techniques are not applicable: this endorses lattice QCD as the main tool to investigate the properties of the matter created in heavy ion collisions. In the low temperature phase, interacting hadronic matter in the ground state can be well described in terms of a gas of non-interacting hadrons and resonances. Indeed the Hadron Resonance Gas (HRG) model, based on this idea, yields a very good description of the lattice results for QCD thermodynamics.

The availability of precise experimental data, as well as the accuracy of the lattice QCD results, allows a new synergy between fundamental theory and experiment, which leads to the determination of several properties of the QGP from first principles.

2. QCD equation of state

The QCD equation of state at zero chemical potential is now known with high accuracy: continuum extrapolated results are available for the pressure, energy density, entropy density, interaction measure and speed of sound of a system of 2+1 quark flavors with physical quark masses [2, 3, 4]. A selection of these results is shown in the left panel of Fig. 1, in which curves from the WB collaboration obtained with the 2stout action (gray) are compared to those from the HotQCD collaboration obtained with the HISQ action (colored). The agreement between the two sets of curves is a fundamental test of the validity of the lattice approach to solve QCD.

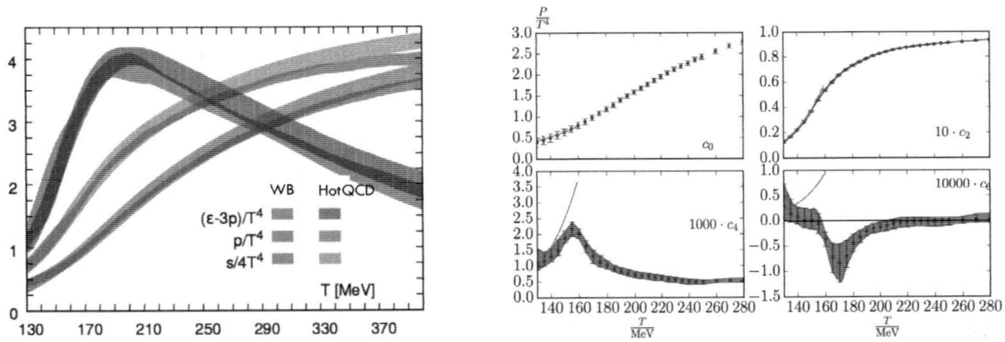

Figure 1. Left: Scaled pressure, entropy density and interaction measure as functions of the temperature. The gray curves have been obtained by the WB collaboration with the 2stout action [2, 3], the colored ones by the HotQCD collaboration with the HISQ action [4]. Right: Taylor coefficients c_0, c_2, c_4 and c_6 as functions of the temperature from the WB collaboration, obtained from imaginary μ_B simulations. The data are continuum extrapolated [14, 15, 16].

Lattice QCD simulations at finite chemical potential are unfortunately not possible, due to the sign problem. Different methods have been proposed in order to circumvent the problem: here I will focus on the Taylor expansion of thermodynamic observables around $\mu_B = 0$ [5, 6] (which can be considered as a truncated version of the multiparameter reweighting technique [7]) and on the analytical continuation from imaginary chemical potentials [8, 9, 10].

The Taylor expansion of the pressure in powers of μ_B/T can be written as:

$$\frac{p(\mu_B)}{T^4} = c_0(T) + c_2(T)\left(\frac{\mu_B}{T}\right)^2 + c_4(T)\left(\frac{\mu_B}{T}\right)^4 + c_6(T)\left(\frac{\mu_B}{T}\right)^6 + \mathcal{O}(\mu_B^8). \tag{1}$$

The coefficients $c_i(T)$ in the above expansion can be calculated on the lattice. Several results exist in the literature: $c_2 ... c_6$ have been calculated long ago on coarse lattices and with heavier than physical quark masses [11]; continuum extrapolated results for c_2 at the physical quark mass were published for the first time in Ref. [12]; results for c_4 at finite lattice spacing were shown in Ref. [13]. In Fig. 1 right I show the new results for c_2, c_4 and c_6 from the WB collaboration, in the continuum limit and for physical quark masses. These results have been obtained with the method of analytical continuation from imaginary chemical potential [14, 15, 16]. It is important to notice that the chemical potentials μ_B, μ_S and μ_Q are not independent of each other: μ_S and μ_Q are both functions of T and μ_B, such that the following experimental conditions are satisfied:

$$\rho_S = 0, \qquad \rho_Q = 0.4\rho_B, \tag{2}$$

where ρ_S, ρ_Q and ρ_B are the densities of strangeness, charge and baryonic number, respectively. The coefficients $c_2 ... c_6$ in Eq. (1) are full derivatives with respect to μ_B/T, namely they take into account the dependence of μ_S and μ_Q on μ_B and T.

IOP Conf. Series: Journal of Physics: Conf. Series **779** (2017) 012016 doi:10.1088/1742-6596/779/1/012016

3. QCD phase diagram

It is known from lattice QCD simulations that the deconfinement transition at $\mu_B = 0$ is an analytical crossover [17], taking place over a broad range of temperatures around $T_c \simeq 155$ MeV [18, 19, 20, 21]. The transition temperature is usually defined by locating the inflection point of relevant observables. As the chemical potential is increased, the position of the inflection point changes: it is possible to express the μ_B-dependence of T_c in the following way:

$$\frac{T_c(\mu_B)}{T_c(\mu_B = 0)} = 1 - \kappa \left(\frac{\mu_B}{T_c(\mu_B)} \right)^2 + \lambda \left(\frac{\mu_B}{T_c(\mu_B)} \right)^4 + \dots . \tag{3}$$

The coefficient κ is the curvature of the phase diagram. By looking at the chiral condensate, chiral susceptibility and strange quark susceptibility the WB collaboration finds a value of $\kappa = 0.0149 \pm 0.0021$ [22]; the strange quark chemical potential is fixed to impose strangeness neutrality. The phase diagram corresponding to this value of κ is showed in the left panel of Fig. 2, together with a compilation of freeze-out parameters obtained with different methods. The shaded band indicates the broadness of the QCD transition, while the dark blue band shows that it is possible to find a value for T_c with great accuracy when looking at a single observable (in this case the chiral condensate). Similar results have been obtained recently by two other groups: P. Cea *et al.* obtain a value of $\kappa = 0.020(4)$ by fixing $\mu_s = \mu_l$ [23], while Bonati *et al.* find $\kappa = 0.0135(20)$ both with $\mu_s = 0$ and $\mu_s = \mu_l$ [24]. In the right panel of Fig. 2, the phase diagram with the curvature from Ref. [23] is shown.

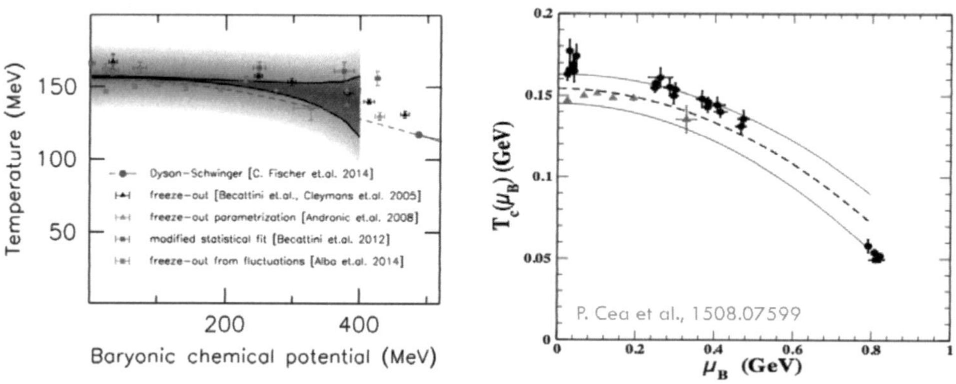

Figure 2. Left: The phase diagram based on the μ_B-dependent T_c from the chiral condensate, analytically continued from imaginary chemical potential [22]. The blue band indicates the width of the transition. The shaded black region shows the transition line obtained from the chiral condensate. The widening around 300 MeV is coming from the uncertainty of the curvature and from the contribution of higher order terms, thus the application range of the results is restricted to smaller values. We also show some selected non-lattice results: the Dyson-Schwinger result [25], and the freeze-out data of Refs. [26]-[32]. Right: analogous plot from Ref. [23].

4. Fluctuations of conserved charges

Fluctuations of conserved charges are the most prominent example of direct comparison between lattice QCD results and experimental data: they are fixed at the chemical freeze-out, therefore they can be used to gain information about this point in the evolution of a heavy ion collision [33, 34].

The definition of fluctuations is as follows:

$$\chi_{lmn}^{BQS}(T,\mu_B) = \frac{\partial^{l+m+n}p/T^4}{\partial(\mu_B/T)^l\partial(\mu_Q/T)^m\partial(\mu_S/T)^n} \tag{4}$$

and they can be related to the cumulants of the event by event distribution of the corresponding net conserved charge measured in experiments. In particular, it is possible to define volume-independent ratios which only depend on T and μ_B:

$$\begin{aligned} M/\sigma^2 &= \chi_1/\chi_2 \qquad & S\sigma^3/M = \chi_3/\chi_1 \\ S\sigma &= \chi_3/\chi_2 \qquad & \kappa\sigma^2 = \chi_4/\chi_2 \end{aligned} \tag{5}$$

where M is the mean of the experimental distribution, σ the variance, S the skewness and κ the kurtosis. The lattice QCD results for the above ratios will depend on T and μ_B: by comparing them to the experimental value it is therefore possible to extract the freeze-out temperature and chemical potential.

In the comparison between a first principle calculation in thermal equilibrium and the data from a heavy ion collision, we have to make sure that the experiment is measuring thermal fluctuations as well, and that other spurious sources of fluctuations are under control. The system created in a heavy ion collision is a canonical ensemble, in which charges like baryon number or electric charge are strictly conserved. However, if we look at a limited part of this system (which can be done due to the limited acceptance of the detector), we can treat it as a Grand Canonical Ensemble [35]. There are several effects which may generate non-thermal fluctuations: fluctuations in the initial volume, protons coming from the interaction with the beam pipe, effects due to cuts in rapidity and transverse momentum, and so on. A variety of effects has been identified and studied in the literature [36]-[42].

The WB collaboration showed that, by analyzing the fluctuations of electric charge and baryon numbers separately [44, 45], a consistency is found in the chemical freeze-out μ_B values [43], with the freeze-out temperature having an upper limit of $T_{ch} \leq 151\pm4$ MeV. More recently, the authors of Ref. [46] were able to obtain both the freeze-out temperature and the curvature of the freeze-out line. The value of the freeze-out temperature ($T_f = (147 \pm 2)$ MeV) is in agreement with the one obtained in Ref. [43]. The WB collaboration performed a combined fit of χ_1/χ_2 for electric charge and proton number and found the freeze-out temperature and chemical potential for the highest RHIC energies. These preliminary results are shown in the left panel of Fig. 3, together with the isentropic lines which match the freeze-out data, the contours for constant mean/variance of net-electric charge and net-baryon numbe4r from the lattice, and the results of a previous analysis based on the HRG model [32].

Strangeness is still missing from the picture mainly due to the lack of experimental data for the fluctuations of (multi-)strange baryons. Therefore, it is important to have a first-principle determination of the strangeness freeze-out temperature. The ALICE data for particle yields and ratios seem to indicate a tension between the freeze-out temperatures in the light and strange sectors [48, 49]. Several explanations have been proposed for this result [50]-[57], but so far none has been validated or excluded. A result from first principles would finally resolve this issue.

So far, the only feasible experimental measurement of fluctuations of strange particles is the net-kaon fluctuations, which will be finalized soon by the STAR collaboration. In Ref. [58] we showed that it is possible to isolate the kaon fluctuations in lattice QCD simulations: the experimental kaon distribution contains both primordial kaons and those produced by resonance decays. By using an HRG model with resonance decays, we showed that the χ_2/χ_1 of the full kaon distribution is effectively identical to the one obtained for primordial kaons in the Boltzmann approximation. The latter is given by a simple formula:

$$\frac{\chi_2^K}{\chi_1^K} = \frac{\cosh(\hat{\mu_S}+\hat{\mu_Q})}{\sinh(\hat{\mu_S}+\hat{\mu_Q})} \tag{6}$$

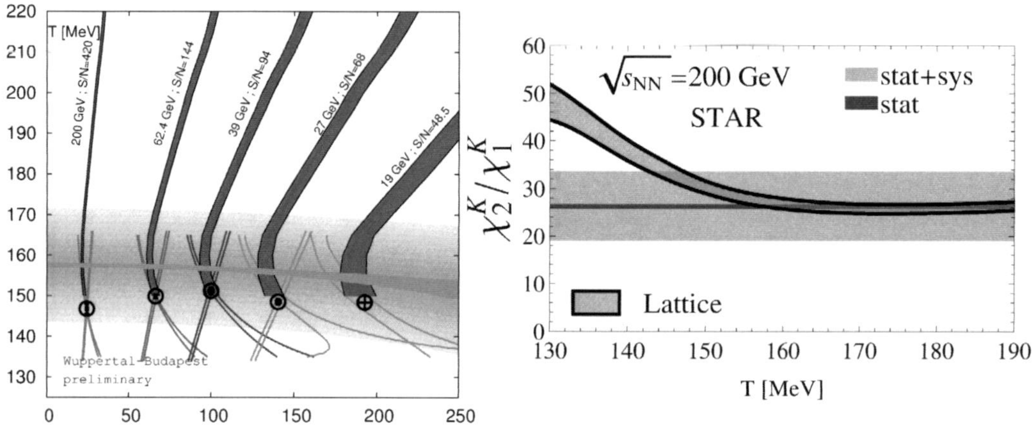

Figure 3. Left: Preliminary results for the freeze-out parameters from the WB collaboration. The colored full and dashed lines are the contours at constant mean/variance ratios of the net electric charge and net baryon number from lattice simulations. The contours that correspond to STAR data intersect in the freeze-out points of Ref. [32]. The red band is the QCD phase diagram shown in Fig. 2. Also shown are the isentropic contours that match the chemical freeze-out data [14, 15, 16]. Right: Example of comparison between χ_2^K/χ_1^K from lattice QCD and the preliminary STAR result at $\sqrt{s}$= 200 GeV shown at the Strangeness in Quark Matter 2016 conference [47]. The blue band corresponds to the statistical error, the orange one shows statistical and systematic errors summed in quadrature.

which can be easily calculated on the lattice. An example of lattice-to-experiment comparison for this observable is shown in the right panel of Fig. 3. When the STAR data for this observable are finalized, it will be possible to extract the freeze-out parameters for kaons from first principles.

5. Conclusions

In conclusion, lattice QCD simulation have reached a new level of accuracy in the last few years. Precise results are available for QCD thermodynamics at zero and small chemical potentials, which allow a quantitative comparison with experimental results for the first time. This will eventually enable us to achieve a comprehensive understanding of bulk properties of QCD matter from first principles.

Acknowledgements

This material is based upon work supported by the National Science Foundation under grant no. PHY-1513864 and by the U.S. Department of Energy, Office of Science, Office of Nuclear Physics, within the framework of the Beam Energy Scan Theory (BEST) Topical Collaboration. An award of computer time was provided by the INCITE program. This research used resources of the Argonne Leadership Computing Facility, which is a DOE Office of Science User Facility supported under Contract DE-AC02-06CH11357.

References

[1] S. Borsanyi *et al.*, JHEP **1209** (2012) 010
[2] S. Borsanyi *et al.*, JHEP **1011** (2010) 077
[3] S. Borsanyi, Z. Fodor, C. Hoelbling, S. D. Katz, S. Krieg and K. K. Szabo, Phys. Lett. B **730** (2014) 99
[4] A. Bazavov *et al.* [HotQCD Collaboration], Phys. Rev. D **90** (2014) 094503
[5] C. R. Allton *et al.*, Phys. Rev. D **66** (2002) 074507

[6] R. V. Gavai and S. Gupta, Phys. Rev. D **78** (2008) 114503
[7] Z. Fodor and S. D. Katz, Phys. Lett. B **534** (2002) 87
[8] P. de Forcrand and O. Philipsen, Nucl. Phys. B **642** (2002) 290
[9] M. D'Elia and M. P. Lombardo, Phys. Rev. D **67** (2003) 014505
[10] L. K. Wu, X. Q. Luo and H. S. Chen, Phys. Rev. D **76** (2007) 034505
[11] C. R. Allton *et al.*, Phys. Rev. D **71** (2005) 054508
[12] S. Borsanyi, G. Endrodi, Z. Fodor, S. D. Katz, S. Krieg, C. Ratti and K. K. Szabo, JHEP **1208** (2012) 053
[13] P. Hegde [BNL-Bielefeld-CCNU Collaboration], Nucl. Phys. A **931**, 851 (2014)
[14] R. Bellwied *et al.*, arXiv:1601.00466 [nucl-th].
[15] C. Ratti, arXiv:1601.02367 [hep-lat].
[16] J. Gunther, R. Bellwied, S. Borsanyi, Z. Fodor, S. Katz, A. Pasztor and C. Ratti, arXiv:1607.02493 [hep-lat].
[17] Y. Aoki, G. Endrodi, Z. Fodor, S. D. Katz and K. K. Szabo, Nature **443** (2006) 675
[18] Y. Aoki, Z. Fodor, S. D. Katz and K. K. Szabo, Phys. Lett. B **643** (2006) 46
[19] Y. Aoki, S. Borsanyi, S. Durr, Z. Fodor, S. D. Katz, S. Krieg and K. K. Szabo, JHEP **0906** (2009) 088
[20] S. Borsanyi *et al.* [Wuppertal-Budapest Collaboration], JHEP **1009** (2010) 073
[21] A. Bazavov *et al.*, Phys. Rev. D **85** (2012) 054503
[22] R. Bellwied, S. Borsanyi, Z. Fodor, J. Gnther, S. Katz, C. Ratti and K. Szabo, Phys. Lett. B **751** (2015) 559
[23] P. Cea, L. Cosmai and A. Papa, arXiv:1508.07599 [hep-lat].
[24] C. Bonati, M. D?Elia, M. Mariti, M. Mesiti, F. Negro and F. Sanfilippo, Phys. Rev. D **92** (2015) 5, 054503
[25] C. S. Fischer, J. Luecker and C. A. Welzbacher, Phys. Rev. D **90** (2014) 3, 034022
[26] J. Cleymans, B. Kampfer, M. Kaneta, S. Wheaton and N. Xu, Phys. Rev. C **71** (2005) 054901
[27] F. Becattini, J. Manninen and M. Gazdzicki, Phys. Rev. C **73** (2006) 044905
[28] A. Andronic, P. Braun-Munzinger, J. Stachel, Phys. Lett. B **673** (2009) 142 [Phys. Lett. B **678** (2009) 516]
[29] F. Becattini *et al.* Phys. Rev. Lett. **111** (2013) 082302
[30] J. Stachel, A. Andronic, P. Braun-Munzinger and K. Redlich, J. Phys. Conf. Ser. **509** (2014) 012019
[31] A. Andronic, Int. J. Mod. Phys. A **29** (2014) 1430047
[32] P. Alba *et al.* Phys. Lett. B **738** (2014) 305
[33] F. Karsch, Central Eur. J. Phys. **10** (2012) 1234
[34] S. Borsanyi, Z. Fodor, S. D. Katz, S. Krieg, C. Ratti and K. K. Szabo, Phys. Rev. Lett. **111** (2013) 062005
[35] V. Koch, Chapter of the book "Relativistic Heavy Ion Physics", R. Stock (Ed.), Springer, Heidelberg, 2010, p. 626-652. (Landolt-Boernstein New Series I, v. 23). (ISBN: 978-3-642-01538-0, 978-3-642-01539-7 (eBook)) [arXiv:0810.2520 [nucl-th]].
[36] V. Skokov, B. Friman and K. Redlich, Phys. Rev. C **88** (2013) 034911
[37] A. Bzdak and V. Koch, Phys. Rev. C **86** (2012) 044904
[38] S. Jeon and V. Koch, Phys. Rev. Lett. **85** (2000) 2076
[39] M. Kitazawa and M. Asakawa, Phys. Rev. C **85** (2012) 021901
[40] M. Nahrgang, M. Bluhm, P. Alba, R. Bellwied and C. Ratti, Eur. Phys. J. C **75** (2015) 12, 573
[41] P. Alba, R. Bellwied, M. Bluhm, V. M. Sarti, M. Nahrgang and C. Ratti, Phys. Rev. C **92** (2015) 6, 064910
[42] F. Becattini, M. Bleicher, T. Kollegger, T. Schuster, J. Steinheimer and R. Stock, Phys. Rev. Lett. **111** (2013) 082302
[43] S. Borsanyi, Z. Fodor, S. D. Katz, S. Krieg, C. Ratti and K. K. Szabo, Phys. Rev. Lett. **113** (2014) 052301
[44] L. Adamczyk *et al.* [STAR Collaboration], Phys. Rev. Lett. **112** (2014) 032302
[45] L. Adamczyk *et al.* [STAR Collaboration], Phys. Rev. Lett. **113** (2014) 092301
[46] A. Bazavov *et al.*, arXiv:1509.05786 [hep-lat].
[47] J. Xu, these proceedings.
[48] R. Preghenella (ALICE), Acta Phys. Polon. B43, 555 (2012).
[49] M. Floris, Nucl. Phys. A931, 103 (2014).
[50] F. Becattini, M. Bleicher, T. Kollegger, M. Mitrovski, T. Schuster, and R. Stock, Phys. Rev. C85, 044921 (2012).
[51] F. Becattini, M. Bleicher, T. Kollegger, T. Schuster, J. Steinheimer, and R. Stock, Phys. Rev. Lett. 111, 082302 (2013).
[52] J. Steinheimer, J. Aichelin, and M. Bleicher, Phys. Rev. Lett. 110, 042501 (2013)
[53] R. Bellwied, S. Borsanyi, Z. Fodor, S. D. Katz, and C. Ratti, Phys. Rev. Lett. 111, 202302 (2013).
[54] A. Bazavov et al., Phys. Rev. Lett. 113, 072001 (2014).
[55] J. Noronha-Hostler and C. Greiner, (2014), arXiv:1405.7298 [nucl-th]
[56] J. Noronha-Hostler and C. Greiner, Nucl. Phys. A931, 1108 (2014), arXiv:1408.0761 [nucl-th].
[57] S. Chatterjee, R. M. Godbole, and S. Gupta, Phys. Lett. B727, 554 (2013), arXiv:1306.2006 [nucl-th]
[58] J. Noronha-Hostler, R. Bellwied, J. Gunther, P. Parotto, A. Pasztor, I. Portillo Vazquez and C. Ratti, arXiv:1607.02527 [hep-ph].

Sub-threshold strangeness and charm production in UrQMD

J. Steinheimer[1] **and M. Bleicher**[1,2]

[1] Frankfurt Institute for Advanced Studies, Ruth-Moufang-Str. 1, 60438 Frankfurt am Main, Germany

[2] Institut für Theoretische Physik, Goethe Universität Frankfurt, Max-von-Laue-Strasse 1, D-60438 Frankfurt am Main, Germany

E-mail: steinheimer@fias.uni-frankfurt.de

Abstract. We present recent results on the sub-threshold production of strange and charmed hadrons in nuclear collisions. In particular we highlight how the excitation and decay of heavy baryonic resonances can be used to describe the production yield of ϕ mesons at the SIS18 accelerator and show how this production mechanism also consistently describes the ϕ nuclear transparency ratio. Including even more massive baryonic resonances in the model we are able to extend our approach and make, for the first time, realistic predictions on the sub-threshold production of J/Ψ and Λ_c in nuclear collisions at the SIS100 accelerator at FAIR. We find that even at a beam energy of $E_{\rm lab} = 6$ A GeV, charm physics is feasible at the CBM experiment, which opens up new opportunities to study charmed hadron properties at large baryon densities.

1. Introduction

As heavy flavors in nuclear collisions have to be newly produced (as $s + \overline{s}$ or $c + \overline{c}$ pair), it is possible to study near and sub-threshold production of strange and chamed hadrons in nuclear collisions at energies where systems of large net baryon density are created [1]. The properties of such hot and dense systems are in the focus of current and planned experimental programs at the GSI/FAIR [2], NICA and RHIC facilities. For the energies considered, microscopic transport models are usually employed and have a history of succesfully describing experimental observables. Several previous works found that the production rates and properties of Kaons, for example are a promising probe to extract their medium interactions in low energy nuclear collisions [3, 4, 5, 6, 7, 8, 9, 10, 11, 12, 13, 14, 15].

In particular strange particles, which are produced below their elementary production threshold have proven to show interesting properties. Here one tries to understand the large production probability for multi strange hadrons, the ϕ and the Ξ^-, as measured by experiments at the SIS18 accelerator [16, 17, 18, 19, 20]. Furthermore, charmed hadron production is considered to be also an excellent probe of the properties of hot and dense nuclear matter. Early works have argued that charmonium suppression in central nuclear collisions may serve as signal for the formation of a deconfined medium, the so called quark gluon plasma (QGP) [21]. In essence, it serves as a messenger of the properties of the deconfined stage. A focus of recent investigations was on charm production at ultrarelativistic energies, i.e. in experiments at the LHC and RHIC accelerators (see e.g. [22, 23, 24, 25, 26, 27, 28, 29, 30, 31, 32]).
In the physics program of the CBM experiment at FAIR the study of open charm and

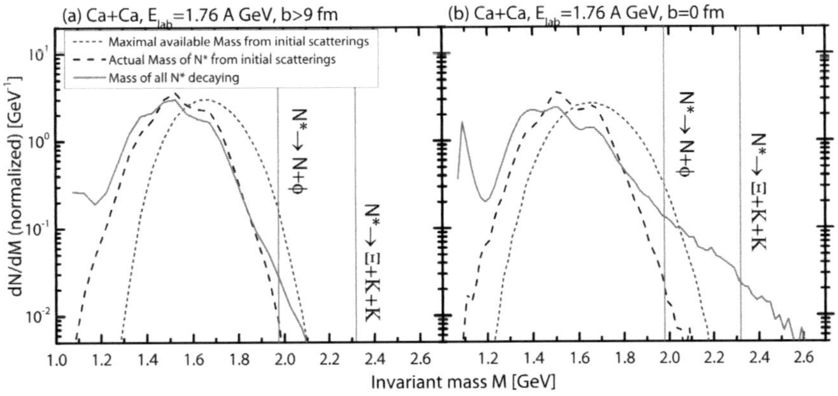

Figure 1. Shown (as blue short dashed line) is the distribution of the maximally available invariant mass for resonance creation from initial scatterings of target and projectile nucleons. Note that the peak of the distribution corresponds to the energy available in proton+proton collisions, at the same beam energy, while the Fermi momenta in the nucleus smear out the distribution. The black dashed line shows the actual invariant mass distribution of N* resonances from these initial nucleon+nucleon scatterings, which is shifted because the outgoing particles have finite momenta. The red solid line depicts the invariant mass distribution of all N* resonances which decay during the systems whole evolution. The left sub-figure presents results of very peripheral collisions of Ca+Ca at $E_{\rm lab} = 1.76$ A GeV, while he right sub-figure depicts results for head on collisions.

charmonium plays an essential role [2]. However, the FAIR project is planned to start with the SIS100 accelerator, which will be able to accelerate a beam of heavy ions only to an energy of $E_{\rm lab} \approx 11$ A GeV, an energy which is below the charm production threshold in elementary collisions. In order to verify if the planned CBM experiment at the FAIR facility is fit to do studies on open charm and charmonium production, it is of great importance to have reliable estimates on the production cross sections of these states.

To this aim we will show how the transport model UrQMD is extended to investigate near and sub-threshold strangeness and charm production.

2. The UrQMD model

In this work we employ the newest version of the UrQMD transport model [33, 34]. The UrQMD model is based on the propagation and scattering of hadrons and hadronic resonances. All the hadron properties, i.e. their masses, quantum numbers, widths as well as scattering cross sections are, where known, taken from the particle data book [35]. In recent publications we have presented new features that were implemented and which turned out to be essential for the description of strange particle production below their elementary energy threshold [36, 37]. These processes are the strangeness exchange reactions $Y + \pi \leftrightarrow N + \overline{K}$ as well as new decay channels for the most massive baryonic resonances included in UrQMD.

The direct resonance production cross section, in elementary p+p collisions, implemented in UrQMD follows from a phenomenological parametrization of measured experimental cross sections and phase space considerations. Here the cross section has the general form:

$$\sigma_{1,2\to 3,4}(\sqrt{s}) \propto (2S_3+1)(2S_4+1)\frac{\langle p_{3,4}\rangle}{\langle p_{1,2}\rangle}\frac{1}{s}\left|M(m_3,m_4)\right|^2 \ , \qquad (1)$$

where S_3 and S_4 are the spin of the outgoing particles, and $\langle p_{i,j}\rangle$ the average momentum of

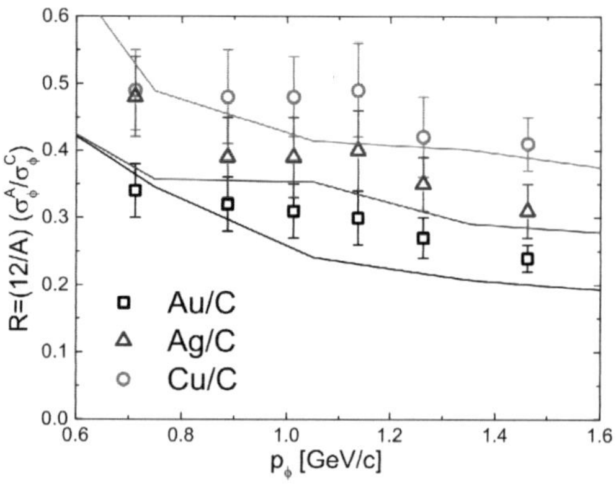

Figure 2. The transparency ratio R calculated within our approach (lines) compared to ANKE data from [38] (symbols).

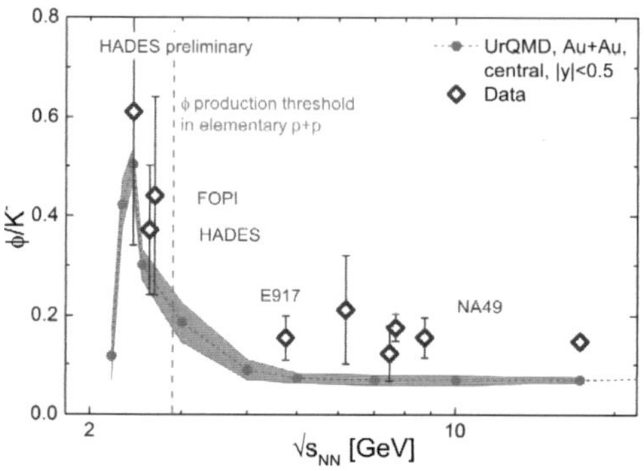

Figure 3. Excitation function of the ϕ/K ratio for central ($b < 5$ fm) Au+Au collisions, calculated with the UrQMD model. Experimental results are shown as symbols [17, 39, 20, 40, 41, 42].

the in- and out-going particles respectively. The matrix element $|M(m_3, m_4)|$ is assumed to not explicitly depend on the spins but only on the masses of the outgoing particles. It is given as:

$$|M(m_3, m_4)|^2 = A \frac{1}{(m_4 - m_3)^2 (m_4 + m_3)^2} . \tag{2}$$

The parameter A is determined by a fit to available data and is the same for any specific process, i.e. $N + N \to N + N^*$, for all N^*. The mass dependence of the production cross section is then essentially determined by phase space which has been shown to lead to a good description of measured resonance production cross sections.

It is important to note that in the current version the model does not include any additional medium effects which would change the hadron properties in nuclear collisions and we have not included any hadronic potentials. Thus, we only investigate basic effects like secondary scatterings as well as Fermi momenta on particle production.

3. Strange results

When discussing sub-threshold production of ϕ's in nuclear collisions one should note that there are two distinct mechanisms which allow for the production of hadrons with masses, higher than what is energetically forbidden in elementary reactions: One reason is that in a nucleus, the nucleons acquire Fermi momenta due to their bound state. Because of the Fermi momenta, the actual energy of two colliding nucleons will not be exactly the initial center of mass energy but a smeared out energy distribution. This allows for collisions of nucleons at energies higher than the initial center of mass energy. Furthermore energy can be accumulated due to secondary interactions of already excited states, produced earlier in the collision [43, 44].

To understand the deep sub-threshold production of multi-strange hadrons we will first investigate the production probability of resonance states with sufficiently high mass to decay into a ϕ in collisions of Ca+Ca. Figure 1 shows the calculated invariant mass distributions of N^* baryons produced in Ca+Ca collisions at a fixed target beam energy of $E_{lab} = 1.76$ A GeV [45]. It is clear from this figure that a substantial number of N^* resonances with masses larger than the apparent threshold energy is produced in the nuclear collision. We can utilize the decays of the most massive N^* resonances implemented in UrQMD, namely the $N^* \to N + \phi$

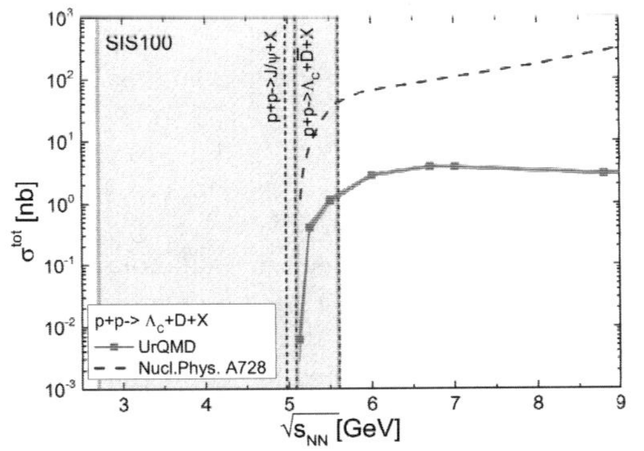

Figure 4. Comparison of the Λ_c cross section in p+p collsions from the UrQMD approach with previous theoretical estimates [46].

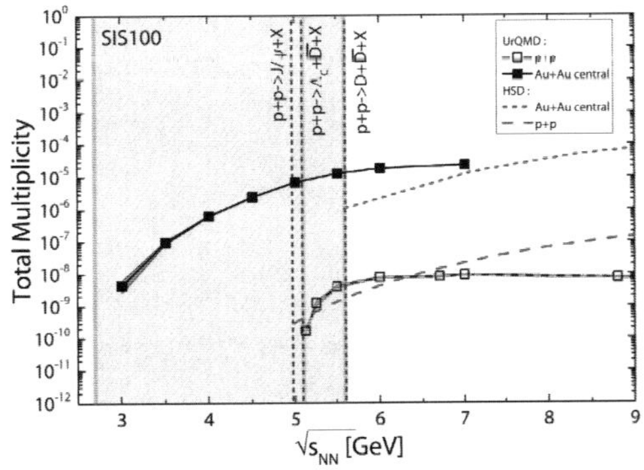

Figure 5. Comparison of the J/Ψ production yield in p+p and central A+A collisions from the UrQMD approach (symbols) with previous theoretical estimates (green short dashed) [47].

channel, in order to describe the production of ϕ mesons near and below their elementary threshold energies [37]. In particular we will use the N^* states with masses larger than 2 GeV, included in the UrQMD model, as their decay channels are experimentally not well constrained and they have a sufficiently large mass. One should note that, by introducing these new decay channels we also naturally, through detailed balance relations, introduce reactions of the kind $M + N \leftrightarrow N^* \leftrightarrow N + \phi$, where M could be any meson that couples to the N^* (e.g. η, ω, ρ or π). We are also able to describe ϕ absorption in cold nuclear matter, within our approach, as shown in figure 2, without the inclusion of an addition in-medium broadening of the ϕ.

Using ANKE data on near threshold ϕ production in $p + p$ collisions [48] we find, that a branching fraction of $\frac{\Gamma_{N+\phi}}{\Gamma_{\rm tot}} = 0.2\%$, for all the above mentioned N^* resonances, provides a very good description of the measured ϕ production cross section.

A ratio which has shown an interesting beam energy dependence, especially below the ϕ production threshold is the ϕ/K^- ratio, which is shown in figure 3 for nuclear collisions at different beam energies, measured by several experiments [17, 39, 20, 40, 41, 42]. Results from our simulations for most central ($b < 3.4$ fm) Au+Au collisions are shown as the red line.

From the comparison in figure 3, it is clearly visible that the qualitative and quantitative behavior of the data, a rapid increase of the ϕ/K^- ratio for sub-threshold energies, is nicely reproduced in our simulations. Also the value of the ratio is in nice agreement with the HADES Ar+KCl data with $E_{\rm lab} = 1.76$ A GeV as well as the FOPI results for Ni+Ni collisions at 1.91 A GeV. An interesting feature of the calculations is the peak in the ratio at a beam energy of approximately 1.2 A GeV. The experimental confirmation of this peak, by ϕ measurements at even lower than the current beam energies, would further support our approach for ϕ production in nuclear collisions.
However, we also observe that above the low SPS energy regime the present model underpredicts the ϕ/K^- ratio. This can be understood as a result of the above mentioned high threshold for ϕ production in the string break-up. Because string excitation dominate the particle production at beam energies above $\sqrt{s_{NN}} > 5$ GeV, the ϕ must always be produced together with a Kaon-Antikaon pair, which strongly suppresses the ϕ production.

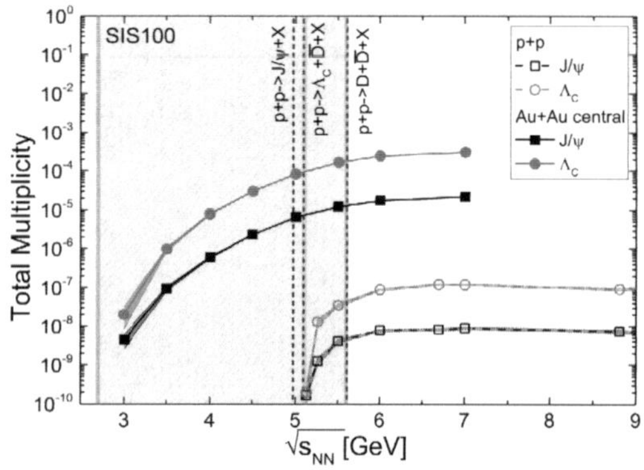

Figure 6. Production yields of J/Ψ and Λ_C in p+p and central Au+Au reactions as a function of the collision energy. The threshold energies of the corresponding channels in p+p reactions are again indicated as vertical lines. The grey area corresponds to the beam energy range expected for heavy ion collisions at the SIS100 accelerator.

4. Charming results

In the previous section we have shown that the near and sub-threshold production of strange hadrons can be well descibed by the use of heavy baryonic resonance decays. The resonances essentially serve as energy reservior and likely decay according to phase space dominance. In the following we will estimate also charm sub-threshold productio rates based on the same basic assumptions. Therefore we need to determine the branching fraction of the N* into the relevant charm channels, i.e. we need to determine the probability of $N^* \rightarrow N + J/\Psi$ and $N^* \rightarrow \Lambda_c + \overline{D}$.

To fix this crucial input we use the measured J/Ψ cross section in p+p collisions at $\sqrt{s_{pp}} = 6.7$ GeV. The resulting branching fraction, to describe the experimentally measured J/Ψ yield is $\Gamma_{N+J/\Psi}/\Gamma_{tot} = 2.5 \cdot 10^{-5}$. This branching ratio is two orders of magnitude smaller than the corresponding decay into a ϕ meson.

First we can compare the obtained Λ_c and J/Ψ production cross sections with previous near threshold estimates [46, 47], as shown in figures 4 and 5. Our estimates are in reasonable agreement (J/Ψ) or even very conservative estimates (Λ_c) as compared to previous work.

Figure 6 summarizes the results of the sub-threshold charm production in nuclear collisions, obtained with our model [49]. Even at a fixed target beam energy of $E_{\text{lab}} = 6$ A GeV we expect a yield of 3×10^{-7} J/Ψ and 4×10^{-6} Λ_c and $\overline{D}$ per central Au+Au event. For the highest available beam energy for heavy nuclei at the SIS100, $E_{\text{lab}} = 11$ A GeV, we expect that yield to be one order of magnitude larger. Hence we predict that a significant number of charmed hadrons can be measured at the CBM experiment, already with the SIS100 accelerator in place. This is of particular interest as the baryon number densities achieved at these low beam energies are very large, opening up the possibility to study charm production and propagation in a very dense hadronic environment. Of particular interest here is to understand and quantify the interaction of charmed hadrons with the nuclear environment and possible effects of chiral symmetry restoration of charmed hadron properties.

5. Summary

We have presented a method to describe the near and sub-threshold production of strange and charmed hadrons in a microscopic transport approach. By employing baryonic resonances as energy reservoir we are able to explain the measured beam energy dependence of the ϕ/K^- ratio. For the first time predictions for the sub-threshold production of charmed hadrons in nuclear collisions are presented showing that charm studies are feasible at the SIS100 accelerator planned for FAIR.

6. Acknowledgments

This work was supported by GSI and BMBF. The computational resources were provided by the LOEWE Frankfurt Center for Scientific Computing (LOEWE-CSC).

References

[1] J. Randrup and C. M. Ko, Nucl. Phys. A **343**, 519 (1980) [Erratum-ibid. A **411**, 537 (1983)].
[2] B. Friman *et al.*, Lect. Notes Phys. **814**, 1 (2011).
[3] J. Aichelin and C. M. Ko, Phys. Rev. Lett. **55**, 2661 (1985).
[4] A. Shor *et al.*, Phys. Rev. Lett. **63**, 2192 (1989).
[5] G. Hartnack, L. Sehn, J. Jaenicke, H. Stoecker and J. Aichelin, Nucl. Phys. A **580**, 643 (1994).
[6] X. S. Fang, C. M. Ko, G. Q. Li and Y. M. Zheng, Nucl. Phys. A **575**, 766 (1994)
[7] G. Q. Li, C. M. Ko and X. S. Fang, Phys. Lett. B **329**, 149 (1994).
[8] G. Q. Li and C. M. Ko, Phys. Lett. B **351**, 37 (1995)
[9] U. Mosel, Ann. Rev. Nucl. Part. Sci. **41**, 29 (1991).
[10] D. Miskowiec *et al.*, Phys. Rev. Lett. **72**, 3650 (1994).
[11] W. Cassing, E. L. Bratkovskaya, U. Mosel, S. Teis and A. Sibirtsev, Nucl. Phys. A **614**, 415 (1997)
[12] E. L. Bratkovskaya, W. Cassing and U. Mosel, Nucl. Phys. A **622**, 593 (1997)
[13] C. Hartnack, H. Oeschler and J. Aichelin, Phys. Rev. Lett. **90**, 102302 (2003)
[14] C. Hartnack, H. Oeschler and J. Aichelin, Phys. Rev. Lett. **96**, 012302 (2006)
[15] C. Hartnack, H. Oeschler, Y. Leifels, E. L. Bratkovskaya and J. Aichelin, Phys. Rept. **510**, 119 (2012)
[16] G. Agakishiev *et al.*, Phys. Rev. Lett. **114**, no. 21, 212301 (2015).
[17] G. Agakishiev *et al.* [HADES Collaboration], Phys. Rev. C **80**, 025209 (2009)
[18] G. Agakishiev *et al.* [HADES collaboration], Phys. Rev. Lett. **103**, 132301 (2009)
[19] E. E. Kolomeitsev, B. Tomasik and D. N. Voskresensky, Phys. Rev. C **86**, 054909 (2012)
[20] K. Piasecki *et al.* [FOPI Collaboration], Phys. Rev. C **91**, no. 5, 054904 (2015).
[21] T. Matsui and H. Satz, Phys. Lett. B **178**, 416 (1986).
[22] H. van Hees, M. Mannarelli, V. Greco and R. Rapp, Phys. Rev. Lett. **100**, 192301 (2008)
[23] R. Rapp and H. van Hees, arXiv:0903.1096 [hep-ph].
[24] M. He, R. J. Fries and R. Rapp, arXiv:1204.4442 [nucl-th].
[25] J. Aichelin, P. B. Gossiaux and T. Gousset, Acta Phys. Polon. B **43**, 655 (2012)
[26] J. Uphoff, O. Fochler, Z. Xu and C. Greiner, Phys. Rev. C **84**, 024908 (2011)
[27] J. Uphoff, O. Fochler, Z. Xu and C. Greiner, Phys. Lett. B **717**, 430 (2012)
[28] C. Young, B. Schenke, S. Jeon and C. Gale, Phys. Rev. C **86**, 034905 (2012)
[29] I. Vitev, A. Adil and H. van Hees, J. Phys. **G34**, S769 (2007)
[30] P. B. Gossiaux, J. Aichelin, T. Gousset and V. Guiho, J. Phys. **G37**, 094019 (2010)
[31] P. Gossiaux, J. Aichelin and T. Gousset, Prog. Theor. Phys. Suppl. **193**, 110 (2012)
[32] T. Lang, H. van Hees, J. Steinheimer and M. Bleicher, arXiv:1211.6912 [hep-ph].
[33] S. A. Bass *et al.*, Prog. Part. Nucl. Phys. **41**, 255 (1998) [Prog. Part. Nucl. Phys. **41**, 225 (1998)]
[34] M. Bleicher *et al.*, J. Phys. **G25**, 1859 (1999)
[35] K. A. Olive *et al.* [Particle Data Group Collaboration], Chin. Phys. C **38**, 090001 (2014).
[36] G. Graef, J. Steinheimer, F. Li and M. Bleicher, Phys. Rev. C **90**, 064909 (2014).
[37] J. Steinheimer and M. Bleicher, J. Phys. G **43**, no. 1, 015104 (2016).
[38] M. Hartmann *et al.*, Phys. Rev. C **85**, 035206 (2012)
[39] M. Lorenz [HADES Collaboration], Nucl. Phys. A **931**, 785 (2014).
[40] B. Holzman *et al.* [E917 Collaboration], Nucl. Phys. A **698**, 643 (2002)
[41] S. V. Afanasev *et al.* [NA49 Collaboration], Phys. Lett. B **491**, 59 (2000).
[42] J. Adams *et al.* [STAR Collaboration], Phys. Lett. B **612**, 181 (2005)
[43] C. Spieles, A. Jahns, H. Sorge, H. Stoecker and W. Greiner, Mod. Phys. Lett. A **8**, 2547 (1993).
[44] G. Zeeb, M. Reiter and M. Bleicher, Phys. Lett. B **586**, 297 (2004)
[45] J. Steinheimer, M. Lorenz, F. Becattini, R. Stock and M. Bleicher, Phys. Rev. C **93**, no. 6, 064908 (2016).
[46] W. Liu, C. M. Ko and S. H. Lee, Nucl. Phys. A **728**, 457 (2003)
[47] O. Linnyk, E. L. Bratkovskaya and W. Cassing, Int. J. Mod. Phys. E **17**, 1367 (2008)
[48] Y. Maeda *et al.* [ANKE Collaboration], Phys. Rev. C **77**, 015204 (2008)
[49] J. Steinheimer, A. Botvina and M. Bleicher, arXiv:1605.03439 [nucl-th].

Experimental overview on hadronic resonance production in high-energy nuclear collisions

Yosuke Watanabe

Center for Nuclear Study, the University of Tokyo

E-mail: wyosuke@cns.s.u-tokyo.ac.jp

Abstract. I summarize recent results of low-mass vector meson measurements in high-energy nuclear collisions.

1. Introduction

Hadronic resonances are important tools to study the properties of the hot and dense medium created in high-energy nuclear collisions. Table 1 summarizes the properties of low-mass vector mesons discussed in this article, ρ, ω, ϕ and K^* [1]. Due to the variations in the lifetimes and the decay modes, these resonances are sensitive to different aspects of the space-time evolution of the medium.

For example, the ρ meson has a short lifetime of 1.3 fm and is expected to decay inside the hot medium. Dilepton decays of the ρ meson directly carry the information about its in-medium properties, because leptons do not suffer from final state interactions with the medium. On the other hand, hadronic decays, such as $\rho \to \pi^+\pi^-$, are also sensitive to the later stage of the system evolution through the final state interactions of the daughter particles.

In contrast to the ρ meson, the ϕ meson has a long lifetime and decays after freezout of the hot medium. Since the hadronic-interaction cross section of the ϕ meson is small, the ϕ meson is expected to be frozen at an early stage and carry the information from the quark-gluon plasma (QGP).

2. Dilepton decays

In the hot and dense medium created in heavy-ion collisions, chiral symmetry is expected to be restored, and consequently, the spectral shapes of low-mass vector mesons could undergo modifications from the shapes in vacuum. Dilepton decays are suitable for the detection of such spectral modifications.

Dilepton measurements in the low-mass region below 1 GeV/c^2 are performed by SPS experiments, such as CERES and NA60 [2–7]. These experiments observed an enhancement from cocktail calculations of known hadronic sources at dilepton invariant mass (m_{ll}) around 500 MeV/c^2. This excess is identified as the contributions from the pion annihilation into dileptons, mediated by the ρ meson ($\pi\pi \to \rho \to \gamma^* \to ll$), inside the hot medium. Two main approaches are taken to describe the in-medium ρ spectral shapes: "broadening" and "mass shift". There seems now to be consensus that the "broadening" models are successful in describing the excess in the invariant mass spectra at SPS energies [8–13].

Table 1. Properties of low mass vector mesons [1].

	ρ	ω	ϕ	K^*
Mass (MeV/c^2)	775	783	1019	896
Lifetime (fm/c)	1.3	23	46	4.2
Decay modes	l^+l^-, $\pi^+\pi^-$	l^+l^-	l^+l^-, K^+K^-	$K^+\pi^-$

The connection between the ρ broadening models and the chiral symmetry restoration is currently being studied theoretically. For example, Ref. [14] demonstrates that the broadened ρ spectra get degenerated with its chiral partner, the a_1 meson, at high temperature, which is an indication of the chiral symmetry restoration.

Many experiments are ongoing to perform further studies of the in-medium ρ spectral shape at different collision energies. Here, most recent results from PHENIX, STAR, ALICE and HADES are summarized.

2.1. RHIC and LHC

At RHIC energies, PHENIX and STAR are performing dielectron measurements at midrapidity in heavy-ion collisions. There used to be discrepancies between PHENIX [15] and STAR [16, 17] in Au-Au collisions at $\sqrt{s_{NN}} = 200$ GeV.

Recently, PHENIX published new dielectron results, which solves the discrepancies. Those new results are obtained with the data sets taken in the 2010 RHIC run [18], in which PHENIX installed a new detector, called hadron-blind detector (HBD), to improve the signal-to-background ratio of the dielectron measurements. The enhancement factor in the mass region 0.3–0.76 GeV/c^2 reported by PHENIX is $2.3 \pm 0.4(stat) \pm 0.4(syst) \pm 0.2(model)$ [18], which is consistent with $1.76 \pm 0.06(stat) \pm 0.26(syst) \pm 0.29(model)$ reported by STAR [16, 17]. These two results are consistent with each other including the centrality and p_T dependence of the excess.

The PHENIX and the STAR results are well described by the ρ broadening models, which were successful at SPS energies. STAR is further studying the excess as a function of collision energy down to $\sqrt{s_{NN}} = 20$ GeV [19] and also in a heavier system, U-U collisions [20]. Currently, all the available experimental data at RHIC are well described by the same ρ broadening models.

At this conference, ALICE presented dielectron spectra below $m_{ll} < 1$ GeV/c^2 in Pb-Pb collisions at $\sqrt{s_{NN}} = 2.76$ TeV for the first time [21]. Figure 1 shows the dielectron spectrum (blue points) compared to the cocktail calculations of hadronic sources (black solid line) for the 0-10% centrality bin. They are applying single p_T cut of 0.4 GeV/c, which is stronger than the cut of 0.2 GeV/c used by the RHIC experiments. No significant enhancement from the cocktail is observed in the ALICE data, which could be attributed to the large experimental uncertainties and large contributions from heavy-flavor decays. The upper limits on the excess do not contradict with the RHIC experiments and also with the real photon measurements performed by ALICE.

One of the difficulties of dilepton measurements at high energies are in the huge contributions from open-charm semileptonic decays, $c\bar{c} \to l^+l^-X$, as is shown as a magenta line in Fig. 1. The large model dependence of $c\bar{c}$ calculations is recently discussed by PHENIX [18]. The enhancement factor of the new PHENIX data is $2.3 \pm 0.4(stat) \pm 0.4(syst) \pm 0.2(model)$ when PYTHIA [22] scaled by N_{coll} is used for the $c\bar{c}$ calculation, while the enhancement factor becomes $1.7 \pm 0.3(stat) \pm 0.3(syst) \pm 0.2(model)$ when MC@NLO [23, 24] is used instead of PYTHIA In addition, the simple N_{coll} scaling of PYTHIA or MC@NLO does not take into account the energy loss of charm quarks inside the QGP. Thus, further studies of the $c\bar{c}$ contribution both in elementary and heavy-ion collisions are essential to define the excess precisely.

2.2. *HADES*

Dielectron spectra at lower energies are systematically studied in various collision systems, such as *pp*, *p*-Nb, Ar-KCl and Au-Au, by HADES [25]. Figure 2 shows the excess spectrum in Au-Au collisions at 1.23A GeV presented at this conference. In the figure, the data points are compared to theoretical model calculations both including the in-medium broadening of the ρ meson [26, 27]. These models also describe the dielectron excess in a smaller collision system, Ar-KCl collisions [28].

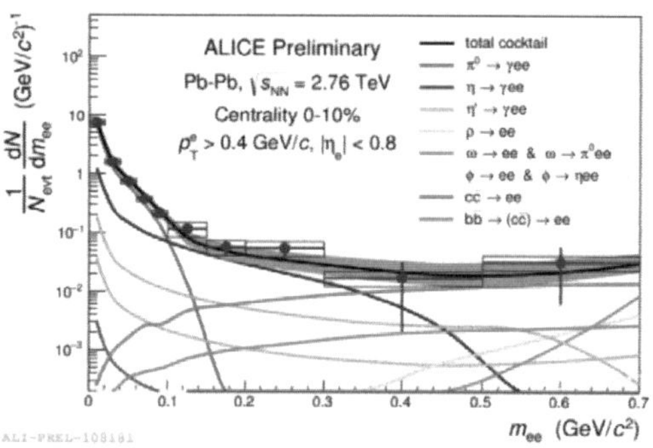

Figure 1. Dielectron spectra in the most central Pb-Pb collisions measured by ALICE [21]. The solid lines show the cocktail calculations of hadronic sources.

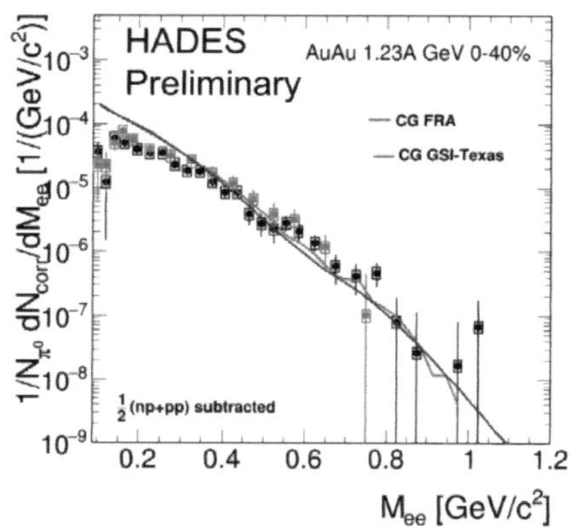

Figure 2. Dielectron excess spectra in Au-Au collisions measured by HADES [25]. The data points are compared to the model calculations (solid lines) including the in-medium broadening of the ρ meson.

3. Hadronic decays

The reconstruction of low-mass vector mesons with their hadronic decays is affected not only by the chiral symmetry restoration but also by the final state interactions of daughter particles. Therefore, to understand their spectral shapes and yields, a full description of the medium evolution is required.

The detection of spectral shape modifications using hadronic decays is experimentally challenging. The background contributions in the invariant mass spectra are usually subtracted by fitting the spectra with a combined function describing the background and the signal peak. In this way, the strongly modified signal is indistinguishable from the background. Small modifications can be detected as a change in the signal peak, however, such modifications are not detected, for example, in a K^* measurement by ALICE [29].

In this section, recent results on the yields of ρ and K^* are summarized.

3.1. *ρ meson*

Figure 3 shows the $\rho \to \pi^+\pi^-$ invariant mass spectra in Pb-Pb collisions reported by ALICE at this conference [30]. Even in the most central and also in the very low p_T bin, the ρ meson peak is clearly visible.

Figure 4 shows the ρ/π ratio as a function of $(dN_{ch}/d\eta)^{\frac{1}{3}}$ in pp and Pb Pb collisions. The ρ/π ratio is suppressed in central collisions compared to peripheral and pp collisions. This result indicates that rescattering processes dominate over regeneration processes during the hadronic phase. Figure 4 also shows the EPOS3 calculations with and without UrQMD afterburner, which describes the interactions in the hadronic phase [31]. While EPOS3 without UrQMD is not able to describe the decreasing trend of the data points, EPOS3 with UrQMD qualitatively describes the trend. Thus, the description of the hadronic phase is crucial to understand the measured yields.

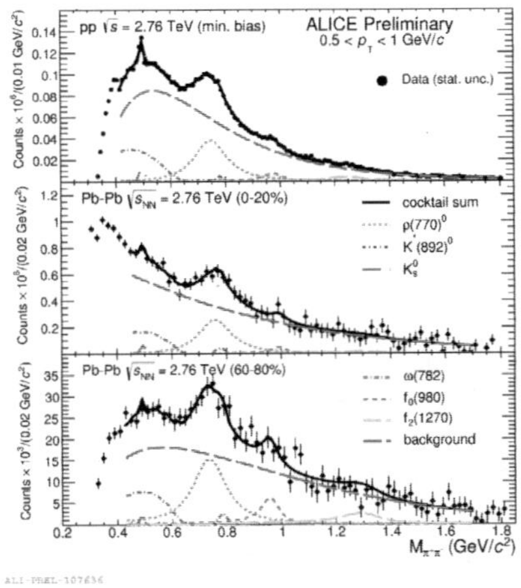

Figure 3. The $\pi^+\pi^-$ invariant mass spectra in pp (top), in the central Pb-Pb collisions (middle) and in the peripheral Pb-Pb collisions (bottom) measured by ALICE [30].

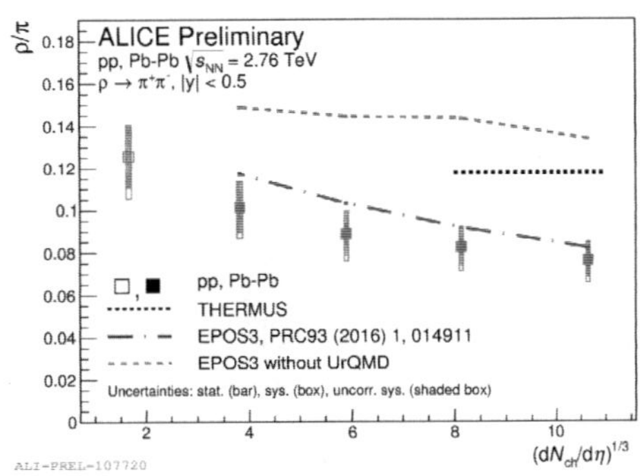

Figure 4. The ρ/π ratio as a function of $(dN_{ch}/d\eta)^{\frac{1}{3}}$ in pp and Pb-Pb collisions measured by ALICE [30].

3.2. K meson*

Figure 5 shows the K^*/K ratio (and the ϕ/K ratio) as a function of $(dN_{ch}/d\eta)^{\frac{1}{3}}$ in various collisions systems [29, 30, 32, 33]. A similar suppression pattern as the ρ/π ratio is seen for Pb-Pb collisions. The suppression pattern is again qualitatively described by the EPOS3 calculation with the UrQMD afterburner [31].

In the same figure, the multiplicity dependence of the K^*/K ratio in pp and p-Pb collisions is also shown. Similarly to the Pb-Pb collision results, the ratio decreases towards higher multiplicity bins. These results could be suggestive of the presence of rescattering effects even in such small collision systems.

4. ϕ meson

The ϕ meson is expected to be a penetrating probe of the hadronic phase due to its long lifetime and also due to its small hadronic interaction cross sections. In this section, as an example of the measurements using its penetrating nature, the elliptic flow (v_2) measurements by STAR are presented.

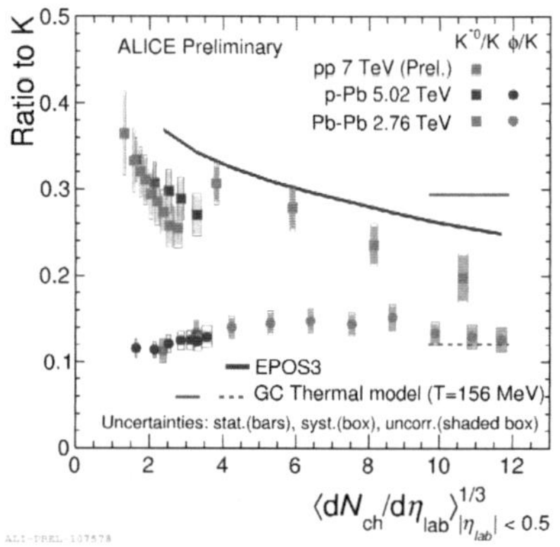

Figure 5. The K^*/K ratio as a function of a cube root of $dN_{ch}/d\eta$ in *pp*, *p*-Pb and Pb-Pb collisions measured by ALICE [29, 30, 32, 33].

4.1. Elliptic flow (v_2)

Hydrodynamical models predict the mass ordering of v_2 at low p_T, e.g. $v_2^{\pi} > v_2^{K} > v_2^{p}$. However, this mass ordering can be broken between proton and the ϕ meson due to large rescattering effects in the hadronic phase only on the proton v_2 [34].

Figure 6 shows the ϕ v_2 divided by the proton v_2 as a function of p_T in Au-Au collisions measured by STAR [35, 36]. Although the ϕ meson is heavier than proton, the ratio is above unity, an indication of violation of the mass ordering, in the p_T region below 0.5 GeV/c. In the top panel of the figure, the data points are compared to the hydrodynamical model with and without a hadron cascade model (JAM). By taking into account the hadron interactions, the breakdown of the mass ordering is well described by the hydro model.

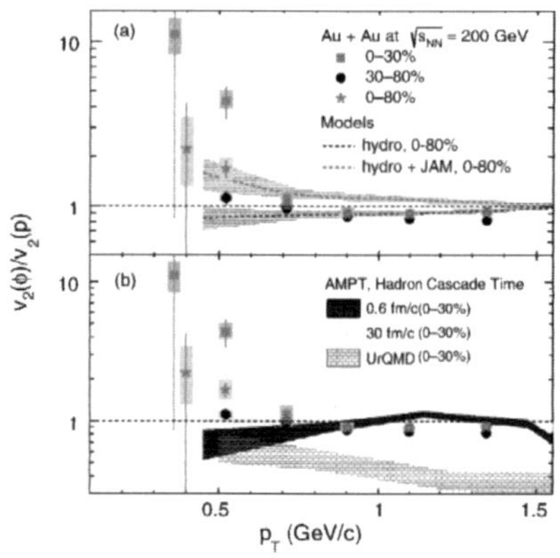

Figure 6. The ratio of ϕ v_2 and proton v_2 as a function of p_T for 0-10%, 30-80% and 0-80% centrality bin in Au-Au collisions measured by STAR. The bands in the top and bottom panels show the hydro models and transport model calculations, respectively [35, 36].

5. Summary

A broad range of new experimental results on low-mass vector meson measurements is presented at this conference.

Dilepton decays of low-mass vector mesons are studied by various experiments over wide ranges of energies. So far, all the available data sets are well described by theoretical models including the broadening of the ρ meson. At high energies, the understanding of the open-charm contributions are essential to study the in-medium ρ spectra in a more precise way.

In hadronic decays, the suppression of the K^* meson and the ρ meson are observed in central heavy-ion collisions. These results are indicative of the significant role of the hadronic phase in the understanding of the resonance production. Similar suppression pattern is also observed in small collision systems, such as pp and p-Pb collisions.

Lastly, the breakdown of the mass ordering between proton v_2 and ϕ v_2 is observed at top RHIC energies. This result is consistent with a view that the ϕ meson has a small hadronic-interaction cross section and is not strongly affected by the final state interactions with the hadronic medium.

References

[1] Olive K A *et al.* (Particle Data Group) 2014 *Chin. Phys.* **C38** 090001
[2] Agakichiev G *et al.* (CERES) 1995 *Phys. Rev. Lett.* **75** 1272–1275
[3] Agakishiev G *et al.* (CERES/NA45) 1998 *Phys. Lett.* **B422** 405–412 (*Preprint* nucl-ex/9712008)
[4] Adamova D *et al.* 2008 *Phys. Lett.* **B666** 425–429 (*Preprint* nucl-ex/0611022)
[5] Adamova D *et al.* (CERES/NA45) 2003 *Phys. Rev. Lett.* **91** 042301 (*Preprint* nucl-ex/0209024)
[6] Arnaldi R *et al.* (NA60) 2006 *Phys. Rev. Lett.* **96** 162302 (*Preprint* nucl-ex/0605007)
[7] Arnaldi R *et al.* (NA60) 2009 *Eur. Phys. J.* **C61** 711–720 (*Preprint* 0812.3053)
[8] van Hees H and Rapp R 2006 *Phys. Rev. Lett.* **97** 102301 (*Preprint* hep-ph/0603084)
[9] van Hees H and Rapp R 2008 *Nucl. Phys.* **A806** 339–387 (*Preprint* 0711.3444)
[10] Ruppert J, Gale C, Renk T, Lichard P and Kapusta J I 2008 *Phys. Rev. Lett.* **100** 162301 (*Preprint* 0706.1934)
[11] Dusling K, Teaney D and Zahed I 2007 *Phys. Rev.* **C75** 024908 (*Preprint* nucl-th/0604071)
[12] Bratkovskaya E L, Cassing W and Linnyk O 2009 *Phys. Lett.* **B670** 428–433 (*Preprint* 0805.3177)
[13] Linnyk O, Bratkovskaya E L, Ozvenchuk V, Cassing W and Ko C M 2011 *Phys. Rev.* **C84** 054917 (*Preprint* 1107.3402)
[14] Hohler P M and Rapp R 2014 *Phys. Lett.* **B731** 103–109 (*Preprint* 1311.2921)
[15] Adare A *et al.* (PHENIX) 2010 *Phys. Rev.* **C81** 034911 (*Preprint* 0912.0244)
[16] Adamczyk L *et al.* (STAR) 2014 *Phys.Rev.Lett.* **113** 022301 (*Preprint* 1312.7397)
[17] Adamczyk L *et al.* (STAR) 2015 (*Preprint* 1504.01317)
[18] Adare A *et al.* (PHENIX) 2016 *Phys. Rev.* **C93** 014904 (*Preprint* 1509.04667)
[19] Ruan L 2014 *Nucl. Phys.* **A931** 185–193 (*Preprint* 1407.8153)
[20] Yang S (STAR) 2016 *Nucl. Phys.* **A956** 429–432
[21] See A. Caliva, "Measurement of dielectrons in pp, p–Pb and Pb–Pb collisions with ALICE at the LHC", in these proceedings
[22] Sjostrand T, Eden P, Friberg C, Lonnblad L, Miu G, Mrenna S and Norrbin E 2001 *Comput. Phys. Commun.* **135** 238–259 (*Preprint* hep-ph/0010017)
[23] Frixione S and Webber B R 2002 *JHEP* **06** 029 (*Preprint* hep-ph/0204244)
[24] Frixione S, Nason P and Webber B R 2003 *JHEP* **08** 007 (*Preprint* hep-ph/0305252)
[25] See R. Holzmann, "Particle production investigated with HADES at SIS18", in these proceedings
[26] Galatyuk T, Hohler P M, Rapp R, Seck F and Stroth J 2016 *Eur. Phys. J.* **A52** 131 (*Preprint* 1512.08688)
[27] Endres S, van Hees H, Weil J and Bleicher M 2015 *Phys. Rev.* **C92** 014911 (*Preprint* 1505.06131)
[28] Agakishiev G *et al.* (HADES) 2011 *Phys. Rev.* **C84** 014902 (*Preprint* 1103.0876)
[29] Abelev B B *et al.* (ALICE) 2015 *Phys. Rev.* **C91** 024609 (*Preprint* 1404.0495)
[30] See A. Knospe, "Recent Hadronic Resonance Measurements at ALICE", in these proceedings
[31] Knospe A G, Markert C, Werner K, Steinheimer J and Bleicher M 2016 *Phys. Rev.* **C93** 014911 (*Preprint* 1509.07895)
[32] Adam J *et al.* (ALICE) 2016 *Eur. Phys. J.* **C76** 245 (*Preprint* 1601.07868)
[33] See F. Bellini, "Strangeness in ALICE at LHC", in these proceedings
[34] Hirano T, Heinz U W, Kharzeev D, Lacey R and Nara Y 2008 *Phys. Rev.* **C77** 044909 (*Preprint* 0710.5795)
[35] Adamczyk L *et al.* (STAR) 2016 *Phys. Rev. Lett.* **116** 062301 (*Preprint* 1507.05247)
[36] See S. Shi, "Strangeness in STAR at RHIC", in these proceedings

IOP Conf. Series: Journal of Physics: Conf. Series **779** (2017) 012019 doi:10.1088/1742-6596/779/1/012019

A new detector at RHIC, sPHENIX goals and status

Rosi Reed for the sPHENIX Collaboration
Lehigh University, Bethlehem, PA USA
E-mail: `RosiJReed@Lehigh.edu`

Abstract.
The study of heavy-ion collisions, which can create a new form matter, a nearly ideal strongly interacting fluid where quarks and gluons are no longer confined into nucleons, called Quark Gluon Plasma (QGP), is on the frontier of QCD studies. The Relativistic Heavy Ion Collider (RHIC) at Brookhaven National Lab (BNL) has had a long and successful program of QGP study since 2000, with many upgrades that have increased the delivered luminosity considerably in the last decade. The sPHENIX proposal is for a second generation experiment at RHIC, which will take advantage of the increased luminosity, and allow measurements of jets, jet correlations and Upsilons (Υs), with a kinematic reach that will overlap with measurements made at the Large Hadron Collider (LHC). Complementary measurements at RHIC and at the LHC probe the QGP at different temperatures and densities, which are necessary to determine the temperature dependence of transport coefficients of the QGP. The sPHENIX detector will have large acceptance electromagnetic and hadronic calorimetry, as well as precision tracking, and high rate capability which are necessary for precision jet and Υ observables. The experiment will enable a program of systematic measurements at RHIC, with a detector capable of acquiring a large sample of events in p+p, p+A, and A+A collisions. This proceedings outlines the key measurements enabled by the new detector, and status of the project itself.

1. Introduction

In relativistic heavy-ion collisions, the extreme temperature and baryon density cause a new state of matter, called the Quark Gluon Plasma (QGP), to be formed. In order to study these collisions and determine the dynamical changes in the QGP in terms of quasiparticles and excitations as a function of the temperature, one needs to probe the medium at a variety of length scales. Particle jets, formed when two quarks or gluons (partons) undergo a hard scatter and fragment and then hadronize into a collimated spray of particles, are one such probe. It is especially important to understand how transport coefficients, such as $\hat{q}$ and $\hat{e}$, which are the transverse momentum squared and longitudinal momentum gained per unit length by by a single parton without radiation, evolve with temperature. Techniques for measuring jet observables in heavy ion collisions have been developed at the Large Hadron Collider (LHC), where the cross-section for hard processes increased much faster with $\sqrt{s_{NN}}$ than the cross-section for soft processes, giving us a wealth of observables.

In order to fully constrain models, these measurements need to be repeated at Relativistic Heavy Ion Collider (RHIC) energies. Heavy quarkonia, specifically bottomonium, formed by a bottom and anti-bottom quark, is useful as a scale sensitive gauge of the Quantum Chromodyamic (QCD) coupling strength. The three lightest states of the Υ meson have proven to be a useful metric, and has been measured at the LHC and at RHIC [1, 2, 3] . However, the

Υ(2S) and Υ(3S) states have not been separated at RHIC energies.

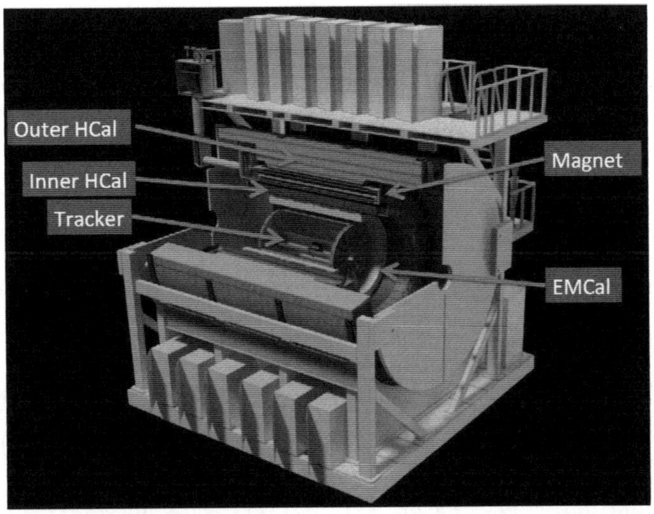

Figure 1. A schematic of the sPHENIX detector showing the different tracking and calorimeter systems.

With these observables in mind, the 2015 Nuclear Science Long Range Plan explicitly states the need for a state-of-the-art" jet and Υ detector at RHIC, called sPHENIX in its recommendations [4]. This detector will allow us to probe the inner workings of QGP by resolving its properties over a range of length scales. The complementarity of this facility with the Large Hadron Collider (LHC) is essential to fully quantify the key signatures of Quantum Chromodynamics (QCD) at extreme temperature and density. As RHIC completes its scientific mission, sPHENIX will play a crucial role in understanding the microscopic properties of the QGP [5]. Fully reconstructed jets at RHIC probe the medium near the critical temperature (T_c), where the coupling between the partons that ultimately fragment into jets and the medium is the strongest. A detector capable of detailed measurements of the modifications to the jet structure due to parton-QGP interactions at these collision energies will greatly enhance our understanding of energy loss in the QGP. Capabilities for heavy flavor jet measurements will provide the ability to study the flavor dependence of energy loss, which will clarify the roles of radiative and collisional energy losses in the medium. In addition, the high resolution tracking of the sPHENIX detector should be capable of separating the three lowest mass states of the upsilon (Υ) meson, which can be used to study the density dependence of the color screening.

Fifteen years of accelerator development has drastically increased the luminosity that RHIC can deliver, combined with a high data acquisition rate of 15 kHz will allow sPHENIX to collect large statistics data sets, which will increase the kinematic reach of previous measurements and the accessibility of rare probes. To fully utilize information gathered from experiments at both the LHC and RHIC, one should measure the same observables in the same kinematic range. The increased kinematic range and statistics from sPHENIX will provide sufficient overlap in observables between RHIC and LHC to constrain models and improve our understanding of the properties of the QGP at different temperatures.

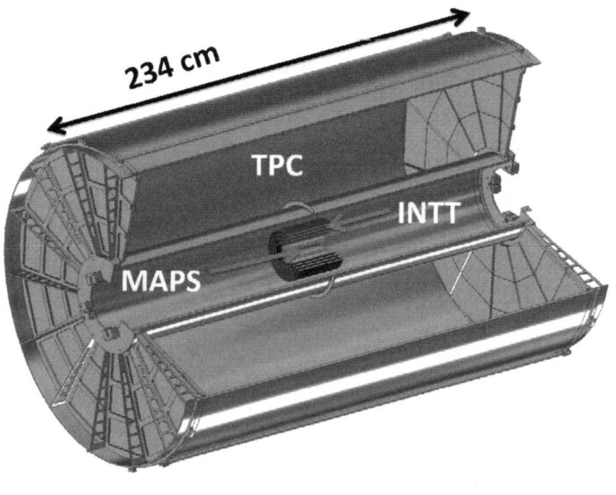

Figure 2. Schematic of the sPHENIX tracking system, indicating the three different subsystems and their relative positions. The 3-layer Monolithic Active Pixel Sensors (MAPS) detector extends from 2.3 cm $< r <$ 3.9 cm from the center of the beam-pipe. The four layer Intermediate Silicon Strip Tracker (INTT) extends from 6 cm$< r <$12 cm. The outer tracker is a compact TPC extending from 20 cm $< r <$78 cm.

2. The sPHENIX Design

The sPHENIX detector shown in Figure 1 will be housed where PHENIX was located on the RHIC ring and will utilize the existing PHENIX infrastructure. The PHENIX detector is currently being removed after its last run during 2016. The sPHENIX detector will be fully installed and ready for beam in 2022 and will have at least two years of data taking. The detector would then be available for use at an electron ion collider (EIC) in the years following the heavy-ion program. The sPHENIX detector is comprised of a tracking system surrounded by calorimetry based around the 1.5 Tesla BaBar solenoid magnet. The calorimetry design includes an electromagnetic calorimeter (EMCal) and two hadronic calorimeters (HCal) ??. The inner and outer HCals are located inside and outside the solenoid magnet respectively. The hadronic calorimeter serves as the flux return for the magnet. The solenoid magnet has already been obtained by Brookhaven National Lab (BNL) from the BaBar experiment for use in sPHENIX and has undergone a successful round of low power cold tests. The magnet has a diameter of 2.8 m and is 3.8 m long. The sPHENIX detector will have full azimuthal coverage and span $-1.1 < \eta < 1.1$ in pseudorapidity.

2.1. Tracking System

The sPHENIX physics program requires a high resolution tracking system in order to separate the Υ states and for high precision jet structure measurements. The tracker system has three subsystems which will provide primary and secondary vertex reconstruction, pattern recognition, and good momentum resolution.

The three subsystems are:

(i) Monolithic Active Pixel Sensors (MAPS) three layers identical to the ALICE Inner Tracking System ITS (r = 2.3 cm, 3.1 cm, and 3.9 cm) [6]

(ii) Intermediate Silicon Strip Tracker (INTT) four layer Si strip detector (r = 6 cm, 8 cm, 10 cm, and 12 cm)

(iii) Compact Time Projection Chamber (TPC) (20 cm $< r <$78 cm)

All subsystems cover $|\eta| < 1.1$ and 2π in azimuth. The INTT will be provided to sPHENIX by RIKEN, the MAPS detector received a recent boost by the approval of a LANL LDRD (Laboratory Directed Research and Development program) grant in support of its development for sPHENIX. A schematic of the tracking system can be seen in Figure 2.

Figure 3. The left figure shows a cross section of the calorimeters and the solenoidal magnet. The middle figure shows two 1D EMCal towers and their light guides. The right figure shows unwrapped HCal tiles and their embedded fibers.

2.2. Electromagnetic Calorimeter

A cross-section of the sPHENIX calorimeters is shown on the left side of Figure 3, with the solenoidal magnet between the inner and outer Hadronic Calorimeter. The EMCal is designed to measure electrons and photons through electromagnetic showers. The electron identification is important for measuring the different Υ states, as the cleanest measures of the Υ meson are through the dilepton channel. Direct photon measurements at high $p_{T,\gamma}$ will be used to identify photon-jet events. These are especially crucial for understanding partonic energy loss as the correlation between the photon and the hard scattered partons that fragment into jets is better than between the final state jets and the partons. Additionally, photon-jet pairs come preferentially from photon-quark pairs, which removes some of the flavor dependence. The segmentation of the current EMCal design is 0.025 x 0.025 for $\Delta\eta \times \Delta\phi$, which results in a total of 96×256 readout channels. Each tower is assembled by threading scintillator fibers through screens that are then positioned into a mold. Tungsten powder and epoxy are poured into the mold. After the tower has hardened the mold is removed and a light guide is attached as shown in the middle of Figure 3. The signal is read out by silicon photo-multipliers (SiPMs). The Moliere radius for the EMCal design is approximately 2.3 cm. 1D projective modules as shown in Figure 3 have been assembled and studied in prototype testing. However, 2D projective towers, which would improve the electron-pion separation, are also under investigation.

2.3. Hadronic Calorimeter

The purpose of the Hadronic Calorimeters is to measure the energy from hadrons produced in the collisions. The combination of the energy measured in the HCals and the EMCal is used

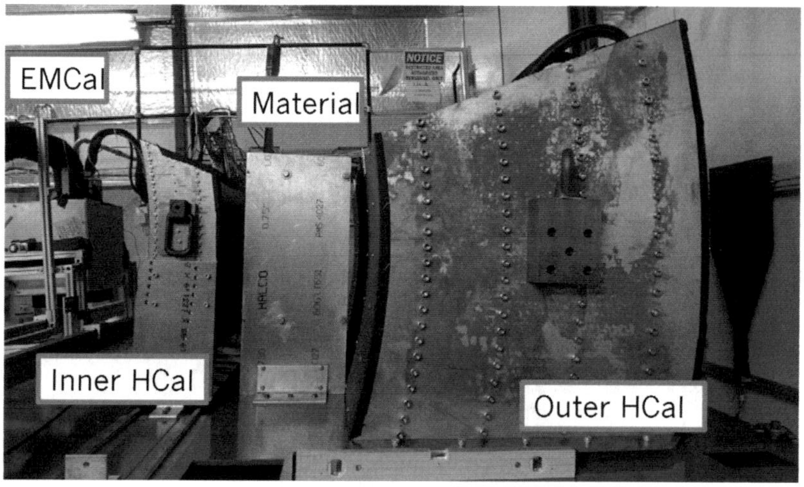

Figure 4. The calorimeter test beam set up used at FNAL for prototype tests held in April 2016. The material between the inner and outer HCal was placed in order to mock up the material budget of the solenoidal magnet.

to reconstruct the energy of jets, which are reconstructed from the calorimeters directly. To provide a jet energy resolution of approximately 20%, a key ingredient for all jet observables, the required single particle energy resolution for the HCal is $\sigma_E/E < 100\%/\sqrt{E}$. The HCals are comprised of alternating layers of steel plates and scintillator tiles. The scintillator tiles are 7mm thick polystyrene with a 1mm wave length shifting fiber embedded into them. An unwrapped outer HCal tile is shown on the right in Figure 3. This shows the routing pattern of the fiber used in the HCAL prototype. The ends of the fiber come together at the edge of tile and each tile illuminates a single SiPM. The SiPMs of five tiles are read out by a single pre-amplifier board as a single tower. This results in 3072 (2 x 24 x 64) total readout channels. The plates are tilted such that a straight line from the center of the detector will hit four tiles. Due to the difference in size, this results in a stronger tilt angle for the inner HCal than for the outer HCal. The plates in the inner HCal are tilted in the opposite direction from the outer HCal.

The EMCal, inner and outer HCal prototypes were constructed at BNL and were tested at the test beam facility at Fermilab in April 2016. A photo of the prototype is included in Figure 4. The energy resolution of these detectors is being studied at beam energies ranging from 2 to 64 GeV and particle species of pions, electrons and muons. Analysis of the data collected during the beam test is ongoing.

3. Conclusions

The scientific case has been established for the need to build the proposed sPHENIX detector as demonstrated in the Nuclear Science Long Range Plan recommendations [4]. At present a total of 58 institutions have joined the sPHENIX collaboration. Open meetings for the collaboration, topical groups and detector subsystems are held regularly and participation by new members and those interested in joining the sPHENIX collaboration is welcomed. Early results from the April 2016 test beam study for the calorimeter prototypes are promising, and should be completed soon. Another prototype test for the calorimeters is planned in early 2017. The proposed tracking system will meet the sPHENIX science goals. The project is on track to have sPHENIX fully installed and ready to take data by 2022. The new detectors and high luminosity from RHIC will allow sPHENIX to make high statistics measurements over a larger kinematic range than previous RHIC experiments. This is essential for measuring important observables such as jet structure, γ-jet and heavy flavor jets as well as separating the three Υ states. sPHENIX will play a crucial role as RHIC completes its scientific mission to understand the properties of the sQGP that it first created over a decade ago.

References

[1] S. Chatrchyan *et al.* [CMS Collaboration], "Observation of sequential Upsilon suppression in PbPb collisions," Phys. Rev. Lett. **109**, 222301 (2012) doi:10.1103/PhysRevLett.109.222301

[2] A. Adare *et al.* [PHENIX Collaboration], "Measurement of $\Upsilon(1S+2S+3S)$ production in $p+p$ and Au+Au collisions at $\sqrt{s_{NN}} = 200$ GeV," Phys. Rev. C **91**, no. 2, 024913 (2015) doi:10.1103/PhysRevC.91.024913

[3] L. Adamczyk *et al.* [STAR Collaboration], "Suppression of Υ production in d+Au and Au+Au collisions at $\sqrt{s_{NN}}$=200 GeV," Phys. Lett. B **735**, 127 (2014) Erratum: [Phys. Lett. B **743**, 537 (2015)] doi:10.1016/j.physletb.2014.06.028, 10.1016/j.physletb.2015.01.046

[4] Reaching For the Horizon, The 2015 Long Range Plan for Nuclear Science, `http://science.energy.gov/~/media/np/nsac/pdf/2015LRP/2015_LRPNS_091815.pdf` (DOE,NSF).

[5] A. Adare *et al.*, "An Upgrade Proposal from the PHENIX Collaboration," arXiv:1501.06197 [nucl-ex].

[6] B Abelev *et al.* [The ALICE Collaboration] "Technical Design Report for the Upgrade of the ALICE Inner Tracking System" J. Phys. G **41**, 087002 (2014). doi:10.1088/0954-3899/41/8/087002

Experimental summary for heavy flavor production

Rongrong Ma
Building 510A, Physics department, Brookhaven National Laboratory, Upton, NY 11973, USA
E-mail: marr@bnl.gov

Abstract. Measurements of heavy flavor production in heavy-ion collisions have played an important role in understanding the properties of the quark-gluon plasma created in such collision. Due to their large masses, heavy flavor quarks present unique sensitivity to the kinematics as well as the dynamics of the hot and dense medium. In this article, a selection of recent measurements on heavy flavor production in p+p, p+A and A+A collisions at both RHIC and LHC energies will be presented. The measurements in p+p collisions serve as benchmarks to fundamental theories, and as references to similar studies in A+A collisions where the hot medium effects are present. On the other hand, the measurements in p+A collisions can help to quantify the cold nuclear matter effects which are also in effect in A+A collisions and thus need to be taken into account when interpreting the measurements in heavy-ion collisions. The experimental results from A+A collisions are discussed and compared to theoretical calculations, which can shed lights on the understanding of the quark-gluon plasma.

1. Introduction

In relativistic heavy-ion collisions, a new state of matter, usually referred to as the "Quark-Gluon Plasma (QGP)", has been created in laboratory at RHIC since more than fifteen years ago [1]. Since then, understanding the properties of such nuclear matter under extreme conditions has been the central focus of the heavy-ion physics. Among various probes, heavy flavor quarks are playing an essential role in studying the kinematics and dynamics of the QGP [2, 3].

Open heavy flavor quarks (charm and bottom) are predominately produced during the initial hard scatterings at the early stage of the heavy-ion collisions. They consequently interact with the medium throughout the entire evolution of the medium, and thus encode information from various stages of the nuclear matter. The interaction with the medium leads to substantial energy loss, both radiative and collisional, for heavy quarks [4, 5]. By studying the dependence of energy loss on collision setup, e.g. energy, geometry, system, etc, as well as comparing the measurements to that of light hadrons and among different heavy flavor hadrons, one can better understand the properties of the QGP and extract key parameters for the plasma [6].

Quarkonium is another probe sensitive to the partonic nature of the QGP. Dissociation of the J/ψ meson in the medium due to the color-screening of the potential between charm and anti-charm quarks by the surrounding partons was proposed as an evidence for the formation of the deconfined nuclear matter [7]. However, regeneration of charm and anti-charm quarks, abundantly produced in the medium, significantly alters the observed J/ψ suppression pattern [8, 9]. Other effects, such as cold nuclear matter effects, co-mover absorption, formation time, feed-down, etc, need to be taken into account as well when interpreting the measurements of J/ψ suppression in heavy-ion collisions [9]. On the other hand, the Υ mesons suffer much less from the regeneration contribution due to the much smaller cross-section for bottom quark production in the medium compared to charm quarks. The three different Υ states are supposed to dissociate at different temperatures due to different binding energies, and measurements of this sequential suppression can help deduce the temperature of the QGP [10].

2. Heavy flavor production in p+p and p+A collisions

Due to the large masses, the heavy flavor production can be calculated analytically using perturbative-QCD (pQCD), which makes it a well-controlled probe to the QGP. The differential cross-section of J/ψ in p+p collisions at $\sqrt{s} = 500$ GeV measured by the STAR experiment for $0 < p_{\mathrm{T}} < 20$ GeV/c is shown in Fig. 1. A similar measurement for D^0 meson in p+p collisions

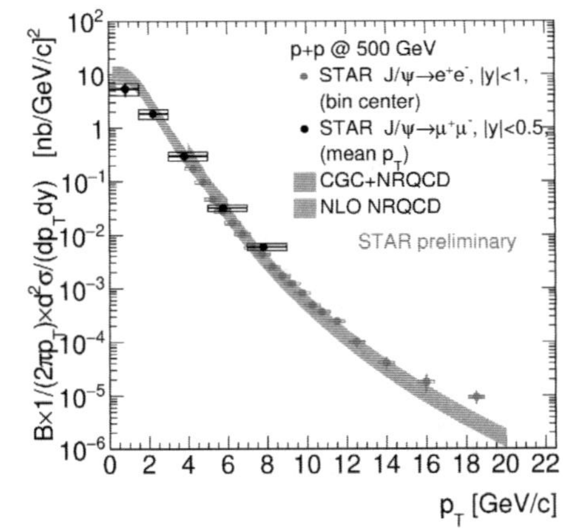

Figure 1: J/ψ cross-section in p+p collisions at $\sqrt{s} = 500$ GeV measured by the STAR experiment for $0 < p_{\mathrm{T}} < 20$ GeV/c.

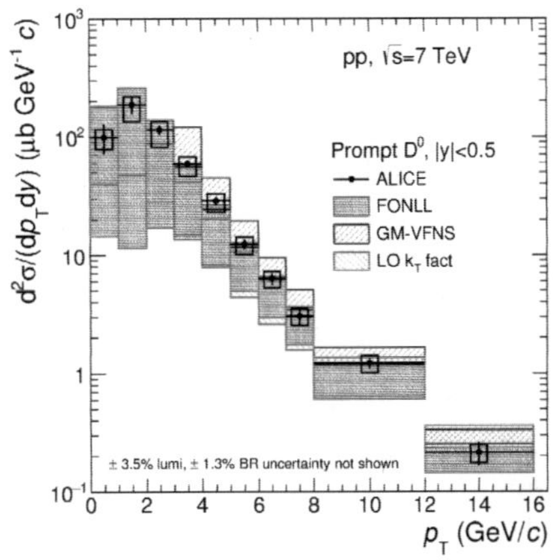

Figure 2: D^0 cross-section in p+p collisions at $\sqrt{s} = 7$ TeV measured by the ALICE experiment for $0 < p_{\mathrm{T}} < 16$ GeV/c [11].

at $\sqrt{s} = 7$ TeV by the ALICE experiment is shown in Fig. 2 covering the range of $0 < p_{\mathrm{T}} < 16$ GeV/c [11]. For both cases, QCD-inspired models can describe the data fairly well within the uncertainties from both model calculations and experimental measurements [11, 12, 13].

Figure 3a shows the measurements of nuclear modification factors (R_{AA}) for D^0 mesons as a function of p_{T} in minimum-bias p+Pb collisions at $\sqrt{s_{\mathrm{NN}}} = 5.02$ TeV by the ALICE collaboration [11]. R_{pPb} is consistent with unity at high p_{T} and shows a hint of nuclear showing at low p_{T}. Model calculations incorporating only Cold Nuclear Matter (CNM) effects agree with data reasonably well within the large uncertainties [11]. The ALICE collaboration also measured D^0-hadron correlations. The yields of associated hadrons (upper) and the widths of the correlation peaks (lower) as a function of D^0 p_{T} on the near-side are shown for both p+p collisions at $\sqrt{s} = 7$ TeV (circles) and p+Pb collisions at $\sqrt{s_{\mathrm{NN}}} = 5.02$ TeV (squares) in Fig. 3b [14]. They are seen to be consistent indicating that the fragmentation processes to D^0 mesons are largely unaltered in p+Pb collisions.

The modifications to J/ψ and Υ productions in minimum-bias p+Pb collisions at $\sqrt{s_{\mathrm{NN}}} = 5.02$ TeV are measured as a function of rapidity, as shown in Fig. 4a and Fig. 4b, respectively. They are significantly suppressed at forward rapidity while remain nearly unchanged at backward rapidity. This is consistent with nuclear shadowing, but there seems also room left for additional CNM effects, such as energy loss.

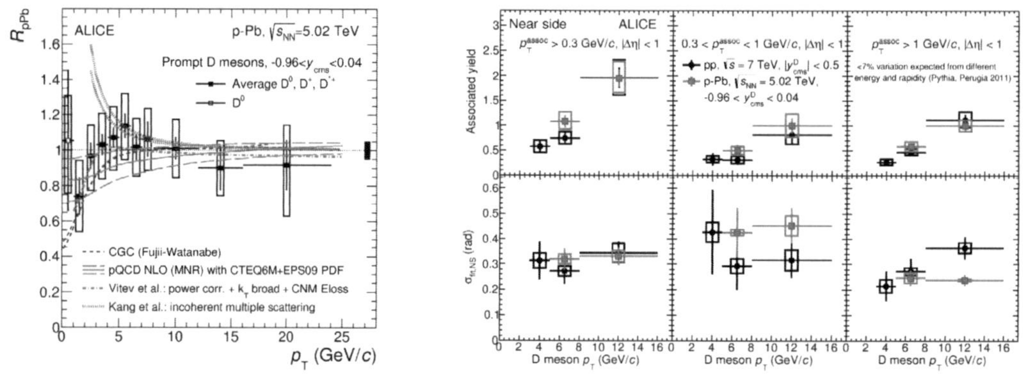

(a) (b)

Figure 3: Measurements of D^0 meson production in minimum-bias p+Pb collisions at $\sqrt{s_{NN}}$ = 5.02 TeV by the ALICE experiment. (a) D^0 R_{pPb} as a function of p_T along with model calculations [11]. (b) Yields of associated hadrons (upper) and the widths of the correlation peaks (lower) extracted from D^0-hadron correlations on the near-side (squares). Similar measurements in p+p collisions at $\sqrt{s}$ = 7 TeV (circles) are shown for comparison [14].

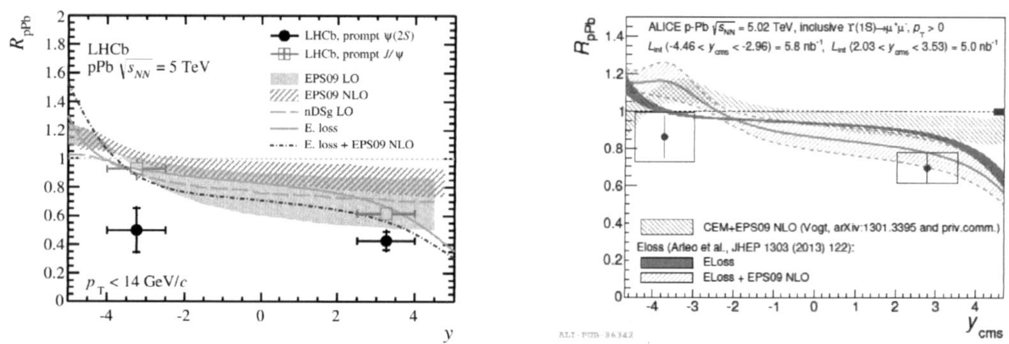

(a) J/ψ R_{pPb} as a function of rapidity by the LHCb experiment [15].

(b) Υ(1S) R_{pPb} as a function of rapidity by the ALICE experiment [16].

Figure 4: Measurements of quarkonium production in minimum-bias p+Pb collisions at $\sqrt{s_{NN}}$ = 5.02 TeV.

3. Heavy flavor production in A+A collisions

3.1. Open heavy flavor

Figure 5 shows measurements of D^0 R_{AA} as a function of p_T in 0-10% central heavy-ion collisions at RHIC and LHC energies [17, 18]. Compared to scaled p+p reference, D^0 production is significantly suppressed, up to a factor of 5-6, at high p_T, indicating strong interactions between charm quarks and the hot medium. A bump structure at lower p_T is observed at all three collision energies, which could be explained by the coalescence of charm quarks with the light quarks in the medium. Unlike the case at RHIC, the bump structure at the LHC energies does not exceed unity, probably because of the stronger shadowing effects. Several models incorporating energy loss mechanisms differently can qualitatively describe the data.

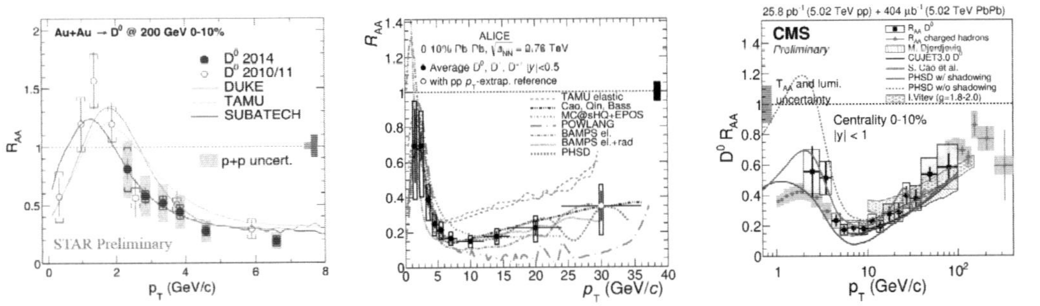

(a) Au+Au collisions at $\sqrt{s_{\mathrm{NN}}}$ = 200 GeV by STAR [17].

(b) Pb+Pb collisions at $\sqrt{s_{\mathrm{NN}}}$ = 2.76 TeV by ALICE [18].

(c) Pb+Pb collisions at $\sqrt{s_{\mathrm{NN}}}$ = 5.02 TeV by CMS.

Figure 5: Measurements of D^0 R_{AA} in 0-10% central heavy-ion collisions at RHIC and LHC.

The production of non-prompt J/ψ from B-meson decays at forward rapidities is measured in minimum-bias Cu+Au collisions at $\sqrt{s_{\mathrm{NN}}} = 200$ GeV by the PHENIX experiment. With the help of Monte Carlo models for decay kinematics, the B-meson R_{AA} is extracted as a function of B-meson p_{T} as shown in Fig. 6a, which is seen to be systematically higher than D^0 R_{AA} at the same p_{T}. Such a comparison is also done in Fig. 6b for Pb+Pb collisions at $\sqrt{s_{\mathrm{NN}}} = 2.76$

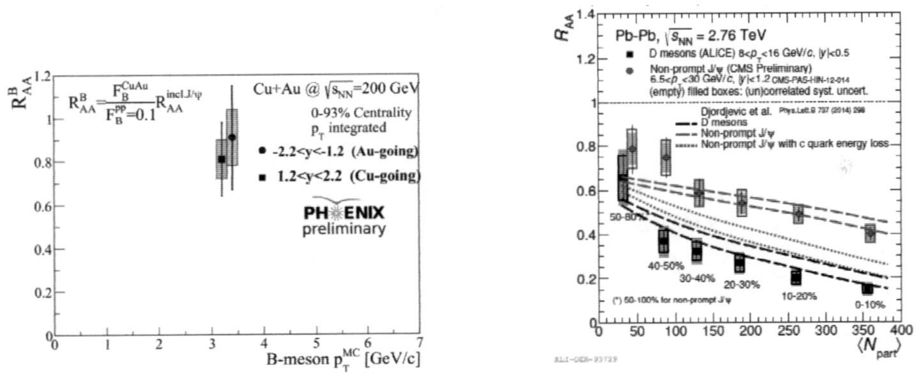

(a) (b)

Figure 6: (a) B-meson R_{AA} extracted from measurements of non-prompt J/ψ in Cu+Au collisions at $\sqrt{s_{\mathrm{NN}}} = 200$ GeV from the PHENIX experiment. (b) R_{AA} of D^0 and non-prompt J/ψ as a function of N_{part} [19] in Pb+Pb collisions at $\sqrt{s_{\mathrm{NN}}} = 2.76$ TeV [20].

TeV at the LHC [20], and similarly the R_{AA} of non-prompt J/ψ is significantly higher than that of D^0 mesons. Model calculations [21] taking into account both the expected mass hierarchy of the radiative energy loss, i.e. $\Delta E_c > \Delta E_b$, as well as spectrum shape can describe the data fairly well, as shown in Fig. 6b. Other calculations, such as the MC@sHQ+EPOS2 model [22] and the TAMU elastic model [23], are also compatible with such a hierarchy.

3.2. Quarkonium

A first measurement of J/ψ suppression within $0 < p_{\mathrm{T}} < 10$ GeV/c at mid-rapidity via the di-muon channel is done by the STAR experiment for 0-40% central Au+Au collisions at $\sqrt{s_{\mathrm{NN}}}$

IOP Conf. Series: Journal of Physics: Conf. Series **779** (2017) 012020 doi:10.1088/1742-6596/779/1/012020

= 200 GeV, as shown in Fig. 7a, along with similar measurements in Pb+Pb collisions at $\sqrt{s_{NN}}$ = 2.76 TeV [24, 25]. At low p_T, the J/ψ R_{AA} is higher at the LHC which can be explained by a larger regeneration contribution due to the larger cross-section for charm quark production, while at high p_T the J/ψ R_{AA} is lower at the LHC due probably to the larger dissociation rate because of the higher medium temperature. Similar measurements at forward rapidity are

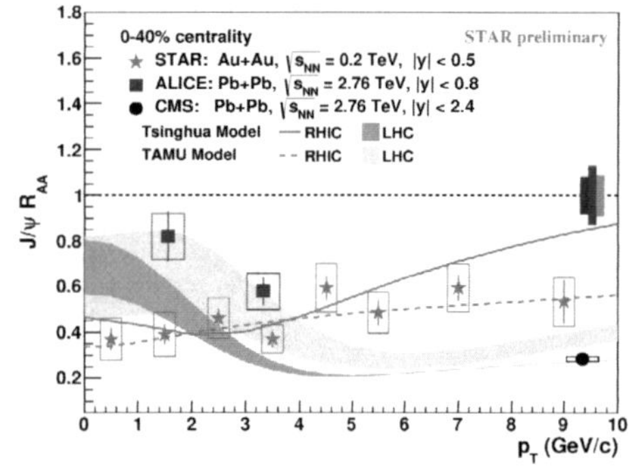

(a) J/ψ at mid-rapidity in Au+Au collisions at $\sqrt{s_{NN}}$ = 200 GeV and Pb+Pb collisions at $\sqrt{s_{NN}}$ = 2.76 TeV.

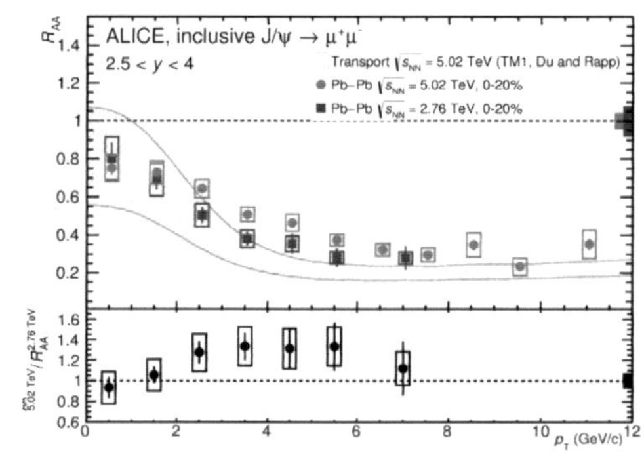

(b) J/ψ at forward-rapidity in Pb+Pb collisions at $\sqrt{s_{NN}}$ = 2.76 and 5.02 TeV [26].

Figure 7: Measurements of J/ψ R_{AA} as a function of p_T at RHIC and LHC energies.

shown in Fig. 7b for 0-20% central Pb+Pb collisions at $\sqrt{s_{NN}}$ = 5.02 TeV [26], which are seen to be systematically higher than that at $\sqrt{s_{NN}}$ = 2.76 TeV for $2 < p_T < 6$ GeV/c. One possible explanation is that the regeneration contribution dominates the dissociation in this kinematic range. Transport models from Tsinghua [8] and TAMU [9] groups include both regeneration and dissociation contributions, and can qualitatively describe the data.

Suffering less from regeneration contribution compared to J/ψ, Υ mesons are considered a cleaner probe. Figure 8a shows the ratio of $\Upsilon(2S+3S)/\Upsilon(1S)$ measured in Au+Au collisions

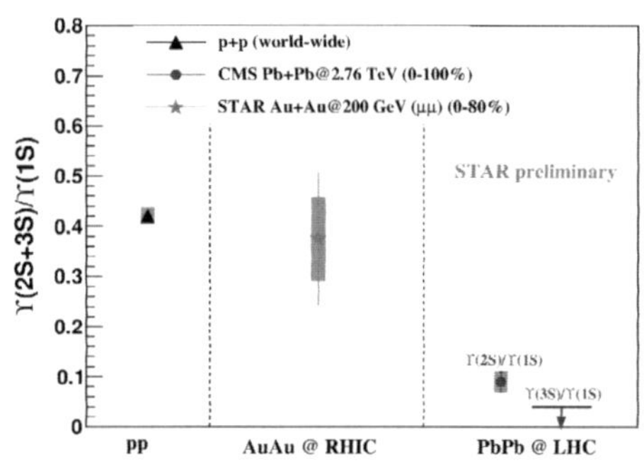

(a)

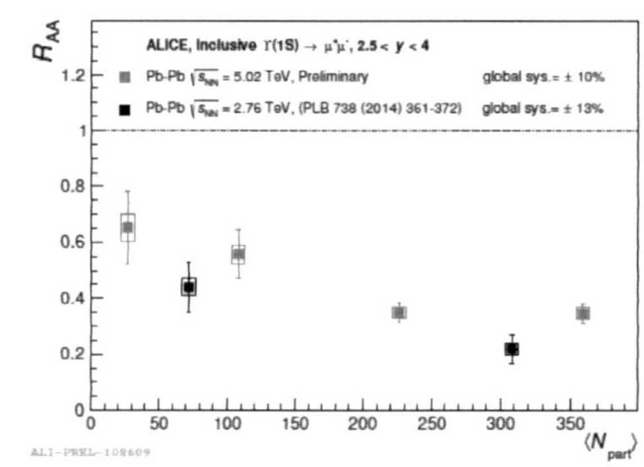

(b)

Figure 8: (a) Measurements of ratio of excited Υ states to ground state in Au+Au collisions at $\sqrt{s_{NN}}$ = 200 GeV, compared with that from world-wide p+p data [27] and in Pb+Pb collisions at $\sqrt{s_{NN}}$ = 2.76 TeV. (b) Measurements of Υ R_{AA} as a function of N_{part} in Pb+Pb collisions at $\sqrt{s_{NN}}$ = 2.76 and 5.02 TeV.

at $\sqrt{s_{NN}}$ = 200 GeV. Compared to similar measurements at higher collision energies at the LHC, there is a hint of less relative suppression of excited Υ states to the ground state at RHIC. Υ R_{AA} as a function of N_{part} are shown in Fig. 8b for Pb+Pb collisions at $\sqrt{s_{NN}}$ = 2.76 and 5.02 TeV, where the larger R_{AA} values for 5.02 TeV collisions could be due to a larger regeneration contribution as well since the bottom quark production becomes sizable at such a high collision energy. Future measurements of feed-down contributions to the Υ states in the heavy-ion collisions are critical for a thorough understanding of different mechanisms in play.

4. Summary

Measurements of heavy flavor production in heavy-ion collisions have revealed many important aspects of the QGP properties in the last two decades. The strong interaction of heavy quarks with the medium leads to considerable energy loss for both charm and bottom quarks. The suppression seems to be larger for D^0 meson compared to non-prompt J/ψ coming from B-meson decays, which is consistent with the expected mass hierarchy [5]. These measurements can help constrain the transport coefficient ($\hat{q}$) of the medium. On the other hand, the color-screening mechanism responsible for quarkonium dissociation in the medium seems to hold at both RHIC and LHC even though an additional contribution from regeneration also plays a critical role in interpreting the suppression patterns at different collision energies. This provides potentially a strong evidence for the QGP formation. With increased luminosity at RHIC and new beam energies at LHC, heavy flavor measurements will continue to contribute significantly towards a comprehensive understanding of the QGP properties.

References

[1] Adams J *et al.* (STAR) 2005 *Nucl. Phys.* **A757** 102–183 (*Preprint* nucl-ex/0501009)
[2] Frawley A D, Ullrich T and Vogt R 2008 *Phys. Rept.* **462** 125–175 (*Preprint* 0806.1013)
[3] Andronic A *et al.* 2016 *Eur. Phys. J.* **C76** 107 (*Preprint* 1506.03981)
[4] Braaten E and Thoma M H 1991 *Phys. Rev.* **D44** R2625
[5] Dokshitzer Y L and Kharzeev D E 2001 *Phys. Lett.* **B519** 199–206 (*Preprint* hep-ph/0106202)
[6] Averbeck R 2013 *Prog. Part. Nucl. Phys.* **70** 159–209 (*Preprint* 1505.03828)
[7] Matsui T and Satz H 1986 *Phys. Lett.* **B178** 416–422
[8] Zhou K, Xu N, Xu Z and Zhuang P 2014 *Phys. Rev.* **C89** 054911 (*Preprint* 1401.5845)
[9] Zhao X and Rapp R 2010 *Phys. Rev.* **C82** 064905 (*Preprint* 1008.5328)
[10] Mocsy A and Petreczky P 2008 *Phys. Rev.* **D77** 014501 (*Preprint* 0705.2559)
[11] Adam J *et al.* (ALICE) 2016 (*Preprint* 1605.07569)
[12] Ma Y Q, Wang K and Chao K T 2011 *Phys. Rev. Lett.* **106** 042002 (*Preprint* 1009.3655)
[13] Ma Y Q and Venugopalan R 2014 *Phys. Rev. Lett.* **113** 192301 (*Preprint* 1408.4075)
[14] Adam J *et al.* (ALICE) 2016 (*Preprint* 1605.06963)
[15] Aaij R *et al.* (LHCb) 2016 *JHEP* **03** 133 (*Preprint* 1601.07878)
[16] Abelev B B *et al.* (ALICE) 2015 *Phys. Lett.* **B740** 105–117 (*Preprint* 1410.2234)
[17] Adamczyk L *et al.* (STAR) 2014 *Phys. Rev. Lett.* **113** 142301 (*Preprint* 1404.6185)
[18] Adam J *et al.* (ALICE) 2016 *JHEP* **03** 081 (*Preprint* 1509.06888)
[19] Miller M L, Reygers K, Sanders S J and Steinberg P 2007 *Ann. Rev. Nucl. Part. Sci.* **57** 205–243 (*Preprint* nucl-ex/0701025)
[20] Adam J *et al.* (ALICE) 2015 *JHEP* **11** 205 (*Preprint* 1506.06604)
[21] Djordjevic M, Djordjevic M and Blagojevic B 2014 *Phys. Lett.* **B737** 298–302 (*Preprint* 1405.4250)
[22] Nahrgang M, Aichelin J, Gossiaux P B and Werner K 2014 *Phys. Rev.* **C89** 014905 (*Preprint* 1305.6544)
[23] He M, Fries R J and Rapp R 2013 *Nucl. Phys.* **A910-911** 409–412 (*Preprint* 1208.0256)
[24] Adam J *et al.* (ALICE) 2015 *JHEP* **07** 051 (*Preprint* 1504.07151)
[25] Chatrchyan S *et al.* (CMS) 2012 *JHEP* **05** 063 (*Preprint* 1201.5069)
[26] Adam J *et al.* (ALICE) 2016 (*Preprint* 1606.08197)
[27] Zha W, Yang C, Huang B, Ruan L, Yang S, Tang Z and Xu Z 2013 *Phys. Rev.* **C88** 067901 (*Preprint* 1308.4720)

IOP Conf. Series: Journal of Physics: Conf. Series **779** (2017) 012021 doi:10.1088/1742-6596/779/1/012021

Summary of recent experimental results on strangeness production

Alexander Kalweit

European Organization for Nuclear Research (CERN), Geneva, Switzerland

E-mail: `Alexander.Philipp.Kalweit@cern.ch`

Abstract. This article summarises the highlights of the recent experimental findings on strangeness production presented at the 16th edition of the *International Conference on Strangeness in Quark Matter* in Berkeley. Results obtained by eight large experimental collaborations (ALICE, ATLAS, CMS, HADES, LHCb, NA-61, PHENIX, STAR) spanning a large range in centre-of-mass energy and a variety of collision systems were presented at the conference. The article does not aim at being a complete review, but rather at connecting the experimental highlights of the different collaborations and at pointing towards questions which should be addressed by these experiments in future.

1. Introduction

Historically, the interest in strangeness production in heavy-ion collisions dates back to the original proposal of strangeness enhancement as a signature of QGP formation [1]. In a more modern view, the formation of strange particles in relativistic nucleus-nucleus (AA) collisions is described together with all other light flavour hadrons (hadrons consisting of u, d, and s valence quarks) in the framework of statistical thermal models and hydrodynamics (see e.g. [2, 3]).

2. Nucleus-nucleus collisions at LHC and RHIC

Light flavour hadrons of low transverse momentum define the collective behaviour of the fireball created in the collision of ultra-relativistic heavy ions. Even at the highest LHC energies about 98% of all particles exhibit transverse momenta $p_{\mathrm{T}} < 2$ GeV/c [4]. They are thus produced in a regime which is not easily accessible to perturbation theory approaches in QCD. However, the assumption that the medium created in the collision is in local thermodynamic equilibrium allows to describe the main features of bulk particle production with two basic concepts. Firstly, the p_{T}-integrated particle yields follow the expectations from statistical-thermal models (chemical equilibrium). Secondly, the spectral shapes and azimuthal anisotropies can be explained by a common hydrodynamic expansion (kinetic equilibrium). One of the main goals in the current investigation of light flavour hadron production is to determine up to which precision this baseline model holds and to carefully investigate the deviations from it for the search of new phenomena.

The validity of the hydrodynamic picture has recently been confirmed at the highest available centre-of-mass energy by the ALICE collaboration with a measurement of the v_2, v_3, and v_4 flow harmonics for charged particles in Pb-Pb collisions at $\sqrt{s_{\mathrm{NN}}} = 5.02$ TeV [5]. One of the most distinct predictions of hydrodynamics is a mass ordering in the p_{T}-dependent flow coefficients resulting from the interplay of radial and anisotropic flow: the common expansion velocity β

IOP Conf. Series: Journal of Physics: Conf. Series **779** (2017) 012021 doi:10.1088/1742-6596/779/1/012021

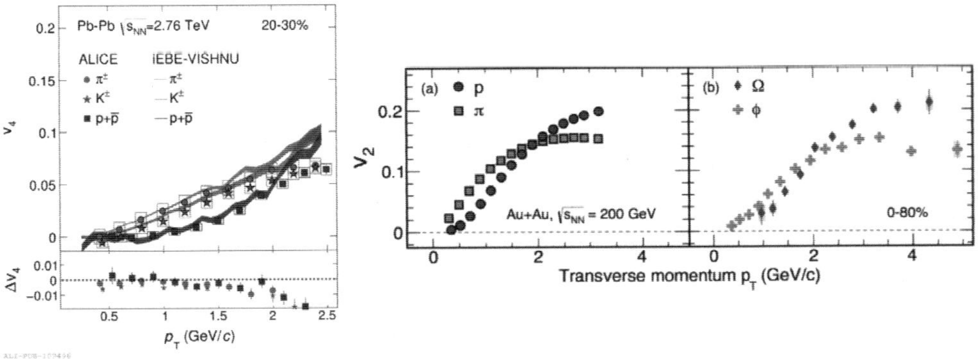

Figure 1. The p_T-differential v_4 for π, K and p measured in 20-30% Pb-Pb collisions at 2.76 TeV compared to expectations from hydrodynamics (left). The v_2 as function of p_T for π, p (panel a) and ϕ, Ω (panel b) from minimum bias Au+Au collisions at 200 GeV for 0-80% centrality (right). Figures taken from [6] (left) and from [7] (right).

leads to stronger blue-shift for heavier particles following $p = m \cdot \beta\gamma$ where m corresponds to the particle mass and γ to the Lorentz factor. Figure 1 (left) shows a striking confirmation of this prediction for the fourth order flow harmonic which was recently measured by ALICE [6]. The STAR Collaboration confirmed that the mass ordering for elliptic flow even holds for the triple-strange Ω-Baryon as shown in Fig. 1 (right). As the hadronic cross-sections for Ω and ϕ are significantly smaller than for the non-strange particles p and π, the results also indicate that a major part of the collectivity is already built up at the partonic stage.

The success of thermal-statistical models in describing particle yields over a wide range of energies in central heavy-ion collisions is also confirmed at LHC energies. The latest comparison of the complete set of particles available at LHC energies with the model prediction is shown in Fig. 2. The yields of light flavour hadrons are described over nine orders of magnitude within 20% with a common chemical freeze-out temperature of $T_{ch} \approx 156$ MeV. Most notably, also the yields of light (anti-)(hyper-)nuclei show agreement with equilibrium thermal model predictions. The largest deviations are observed for protons ($\approx 2.8\sigma$) which would prefer a lower T_{ch} and for Ξ ($\approx 2.0\sigma$) which would prefer a higher T_{ch}. The question must be answered if these deviations are due to physical processes which are not taken into account in the current versions of thermal-statistical models or if the deviations are not significant and can be neglected by invoking the Occam's razor principle, giving preference to a model as simple as possible. One open issue was related to the p/π-ratio at top RHIC energies measured by STAR, that was found significantly higher than the ALICE measurement and whose value was thus indicating chemical freeze-out temperatures of approximately 163 MeV. This topic is now re-addressed with new data on proton production in which feed-down corrections are improved with the help of the STAR heavy-flavour tracker. The first results presented at the conference indeed show a lower p/π-ratio for prompt protons which is more consistent with the ALICE and PHENIX results [8]. A careful future investigation of this topic is needed, because hadron resonance gas calculations on fluctuation observables start to deviate significantly from Lattice QCD for temperatures above 160 MeV [9].

On the theoretical side, two main ideas are on the market as an explanation for the different p and Ξ yield with respect to the thermal model at LHC energies: firstly, the concept that the chemical freeze-out temperatures for s quarks could be higher than for u and d quarks and secondly the presence of inelastic collisions in the hadronic phase [11]. A potential difference in the chemical freeze-out temperature was originally motivated by Lattice QCD

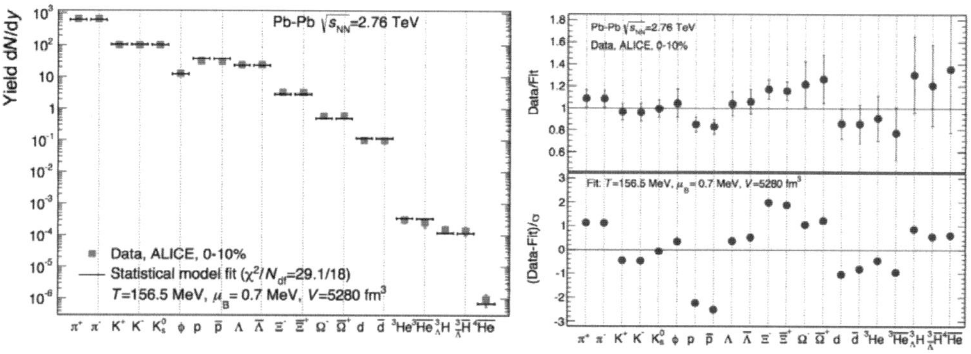

Figure 2. Comparison of grand-canonical thermal model predictions with ALICE data for central Pb–Pb collisions. Figure taken from [10].

results from the Wuppertal-Budapest collaboration [12]. In order to substantiate this claim, freeze-out parameters have previously been extracted from measurements of net-charge and net-baryon number fluctuations [13, 14] and now first attempts have been made for net-strangeness fluctuations, but as shown in Fig. 3, the current experimental precision on net-kaon fluctuations is not sufficient to clarify if the freeze-out temperature for strange particles is higher than for non-strange particles. On the other hand, the dynamical picture of a freeze-out of a fraction of the partons in the system is not yet clear and also a clear ordering in the deviation of the particle yields from the thermal model is not observed: in principle, the triple strange Ω-Baryon should show a larger deviation from a *common* chemical freeze-out temperature, but the data does not show the expected ordering ($\Omega > \Xi$, $\phi > \Lambda$, K) of the deviations. New data with increased precision for multi-strange particles from the LHC run 2 and the upgraded detectors at RHIC is needed in order to investigate the exact magnitude of the deviations and their ordering.

The effects of inelastic collisions in the hadronic phase are studied with UrQMD afterburners added to the thermal-statistical fits [11]. A significant improvement of the χ^2/n_{dof} is observed in this case together with an increase in T_{ch}. The increase in T_{ch} leads at LHC energies to regimes of chemical freeze-out temperatures in which the hadron resonance gas already deviates significantly from Lattice QCD [9]. Future experimental work is needed to investigate precisely the centrality dependence of particle ratios such as p/π and Ξ/π. While the plain statistical-thermal model predicts that particle ratios are saturated above approximately 60% centrality, the lifetime of the hadronic phase is expected to increase with increasing centrality and thus also its influence on the particle ratios [15]. While the dependence of particle ratios on centrality in Pb-Pb collisions is not significant within the current systematic uncertainties, a careful experimental study on the correlated fraction of the systematic uncertainty across different centralities might allow further insights.

Also the deviation of the K^{0*} yield from the thermal model expectation is commonly attributed to the influence of the hadronic phase, but this time to elastic scattering processes of the decay daughters instead of inelastic collisions. The magnitude of the effect is expected to increase with the cross-section of the decay daughters and to decrease for larger lifetimes of the resonance. This picture is now further substantiated by a new ALICE measurement of the ρ/π ratio which shows a suppression pattern that follows the ordering expected from the lifetimes [4]. On the other hand, the results on light (anti-)nuclei production could indicate a very short lifetime of the hadronic phase: despite their low binding energy, which is much smaller than the chemical freeze-out temperature of about 156 MeV, their production yields

IOP Conf. Series: Journal of Physics: Conf. Series **779** (2017) 012021 doi:10.1088/1742-6596/779/1/012021

are in agreement with thermal model expectations. In addition, the shapes of the p_T-spectra and the elliptic flow pattern are in agreement with a hydrodynamic expansion described by the blast-wave model [16, 17] and thus show no difference with respect to non-composite particles. Either, this indicates a production at the phase boundary without subsequent interaction, or a formation by final state coalescence (which takes into account the complex space-time dynamics), or an isentropic expansion which forces the system to keep the entropy-per-baryon constant after chemical freeze-out [18, 19]. Simple coalescence models assuming a point-like source without radial expansion fail to describe the data.

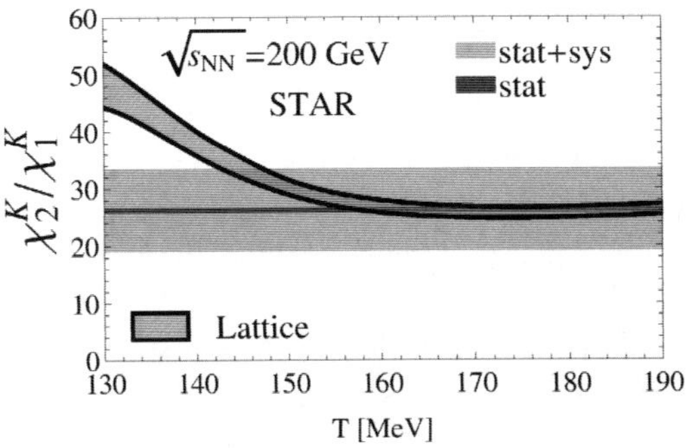

Figure 3. Comparison of the net-kaon fluctuation measurement of the STAR collaboration with Lattice QCD predictions. Figure taken from [20, 21].

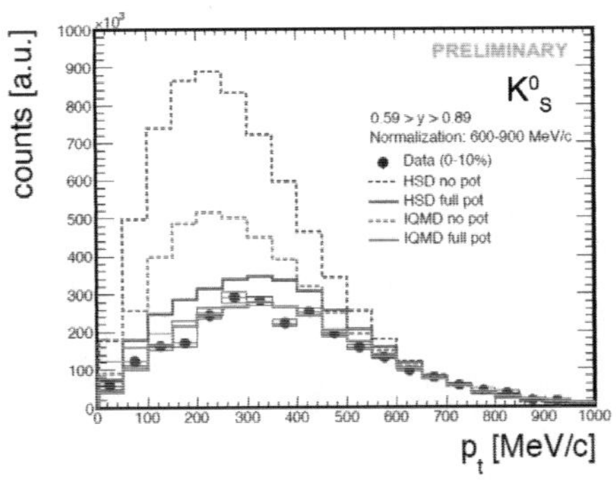

Figure 4. Comparison of HADES kaon spectra with transport code calculations Figure taken from [22].

3. Strangeness production at low energies

In the SPS energy domain, the NA-61/SHINE experiment is on the verge of reaching its originally anticipated goals. The measurement of excitation functions in pp collisions is nearly completed and while data on Ar+Sc collisions is analysed, a beam time for Xe+La is scheduled to complete the system size scan. The data shows that the horn-like structure in the K^+/π^+-ratio around $\sqrt{s_{NN}} \approx 6 - 10$ GeV (often discussed in the context of thermal models [23] or as a potential signature for the onset of deconfinement [24]) is a genuine heavy-ion effect as the pp data follows a flat behaviour. The experiment delivers important input for two external communities. Firstly, with data from proton beams on a carbon target the differential cross-sections for pion production are measured for the precise calculation of the flux of decay neutrinos. Secondly, data from a pion beam on a carbon target improves the modelling of cosmic ray air showers.

At even lower centre-of-mass energies, the HADES experiment has completed a first comprehensive set of measurements on strange particle production in Au-Au collisions at $\sqrt{s_{NN}} = 2.41$ GeV. Feed-down contributions from ϕ decays explain the different slopes of the K^+ and the K^- spectra which were previously attributed to a sequential freeze-out. At such low energies, all strange particles are produced below the nucleon-nucleon threshold leading to a strong sensitivity to medium effects and multi-particle collisions. The data thus helps to constrain better the poorly known kaon-nucleon potential by comparisons with transport codes as shown in Fig. 4. The preliminary statistical-thermal fit shows surprisingly that the larger Au+Au system does not follow the trend of the chemical freeze-out curve in contrast to the smaller Ar+KCl system. A future measurement of the Ξ/π-ratio in the Au+Au system is eagerly awaited in order to see if it also drastically deviates from the thermal model baseline as in the Ar+KCl system [25].

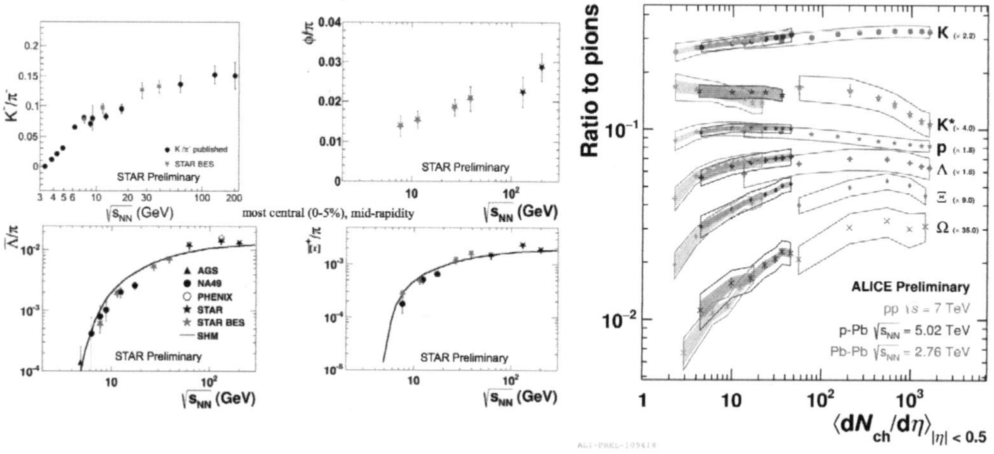

Figure 5. Yields of strange particles as a function of beam energy (left) and as a function of system size at LHC energies (right). Figure taken from [26] (left) and from [4] (right).

4. Strangeness production in small systems

Results from the RHIC beam energy scan together with older experimental data nicely demonstrate a smooth evolution of strangeness production (or the particle chemistry in general) across different centre-of-mass energies. One of the most interesting results presented at the conference is that such a smooth evolution of strangeness production is also observed across different collision systems at LHC energies if the multiplicity dependence within the small collision systems is taken into account as shown in Fig. 5. In contrast to p-Pb or Pb–Pb data, the availability of the pp measurements now also allows to compare to a complete set of QCD-inspired event generators. While the DIPSY and EPOS models describe qualitatively the enhanced production of strangeness with increasing event multiplicity, they will need at least further tuning for a correct quantitative description [4]. The evolution of the spectral shapes with event multiplicity follows a pattern which is consistent with a common radial expansion. A simultaneous blast-wave fit to the spectra of π/K/p (ALICE) or Λ and K^0_S (CMS) shows a good description of the spectra in the relevant p_T-range. The trend of the freeze-out parameters T_{kin} (kinetic freeze-out temperature) and $\langle\beta_T\rangle$ (average radial expansion velocity) is very similar for pp, p–Pb, and Pb–Pb collisions. However, $\langle\beta_T\rangle$ is larger for pp and p–Pb than Pb-Pb collisions at comparable multiplicities [4, 27]. This could indicate a stronger radial flow due to larger pressure gradients, if a hydrodynamical description applies to such small systems [28].

5. Summary and roadmap for the future

A large variety of high quality data on strangeness production from a pp, pA, and AA collisions at various centre-of-mass energies has been presented at the SQM 2016 conference in Berkeley. In the near future, experimental research should address the following five open problems:

(i) At the highest RHIC and LHC (run 2) energies, the centrality dependence of yield ratios such as p/π and Ξ/π needs to be measured with good precision including the correlated part of the systematic uncertainty across different centrality bins. This will allow to address the question of potentially different chemical freeze-out temperatures and the influence of inelastic scatterings in the hadronic phase.

(ii) Measurements of the event-by-event fluctuations of the conserved quantities in QCD over the entire beam energy range will allow to investigate further a potential flavour dependence of chemical freeze-out and effects related to critical chiral dynamics.

(iii) With the silicon vertex detectors operational at RHIC, feed-down corrections for the light flavour hadron yields can be done more precisely. The results should be updated and the thermal-statistical fit should be repeated which might resolve the current discrepancy between the chemical freeze-out temperature at top RHIC energies being slightly higher than the critical temperature obtained in Lattice QCD.

(iv) The set of strange hadron measurements in Au+Au collisions at SIS-18 energies needs to be completed by the Ξ/π-ratio which showed a surprising deviation in Ar+KCl with respect to the thermal model [25].

(v) The multiplicity dependence of particle yield ratios in pp collisions at LHC energies should be repeated at 13 TeV centre-of-mass energy in order to disentangle the evolution of particle chemistry with $\sqrt{s}$ from the evolution with event multiplicity.

Even after 25 years of research, the investigation of light flavour hadron production in general and of strange hadrons in particular, remains a highly interesting topic of modern research.

Acknowledgements
The author would like to thank A. Andronic, F. Bellini, R. Bellwied, P. Braun-Munzinger, D. Chinellato, B. Doenigus, R. Holzmann, F. Karsch, H. Oeschler, C. Ratti, K. Safarik, T. Scheib, M. Schmelling, J. Schukraft, S. Shi, J. Steinheimer, H. Stroebele, Y. Zhou.

References

[1] Rafelski J and Muller B 1982 *Phys. Rev. Lett.* **48** 1066 [Erratum: Phys. Rev. Lett.56,2334(1986)]
[2] Braun-Munzinger P, Redlich K and Stachel J 2003 In Quark Gluon Plasma 3, eds. R.C. Hwa and Xin-Nian Wang, World Scientific Publishing
[3] Voloshin S A, Poskanzer A M and Snellings R 2008
[4] Bellini F *these proceedings and slides presented at SQM2016*
[5] Adam J *et al.* (ALICE) 2016 *Phys. Rev. Lett.* **116** 132302
[6] Adam J *et al.* (ALICE) 2016 (*Preprint* 1606.06057)
[7] Adamczyk L *et al.* (STAR) 2016 *Phys. Rev. Lett.* **116** 062301
[8] Mizuno S *these proceedings and slides presented at SQM2016*
[9] Karsch F *these proceedings and slides presented at SQM2016*
[10] Andronic A *these proceedings and slides presented at SQM2016*
[11] Becattini F, Steinheimer J, Stock R and Bleicher M 2016 (*Preprint* 1605.09694)
[12] Bellwied R, Borsanyi S, Fodor Z, Katz S D and Ratti C 2013 *Phys. Rev. Lett.* **111** 202302
[13] Bazavov A *et al.* 2012 *Phys. Rev. Lett.* **109** 192302
[14] Borsanyi S, Fodor Z, Katz S D, Krieg S, Ratti C and Szabo K K 2014 *Phys. Rev. Lett.* **113** 052301
[15] Becattini F, Grossi E, Bleicher M, Steinheimer J and Stock R 2014 *Phys. Rev.* **C90** 054907
[16] Adam J *et al.* (ALICE) 2016 *Phys. Rev.* **C93** 024917
[17] Lea R (*Preprint* 1606.09379)
[18] Kapusta J I 1980 *Phys. Rev.* **C21** 1301–1310
[19] Andronic A, Braun-Munzinger P, Stachel J and Stoecker H 2011 *Phys. Lett.* **B697** 203–207
[20] Ratti C *these proceedings and slides presented at SQM2016*
[21] Noronha-Hostler J *et al.* 2016 (*Preprint* 1607.02527)
[22] Scheib T *these proceedings and slides presented at SQM2016*
[23] Andronic A, Braun-Munzinger P and Stachel J 2009 *Phys. Lett.* **B673** 142–145 (*Preprint* 0812.1186)
[24] Alt C *et al.* (NA49) 2008 *Phys. Rev.* **C77** 024903 (*Preprint* 0710.0118)
[25] Agakishiev G *et al.* (HADES) 2016 *Eur. Phys. J.* **A52** 178
[26] Shi S *these proceedings and slides presented at SQM2016*
[27] Khachatryan V *et al.* (CMS) 2016 *Submitted to: Phys. Lett. B* (*Preprint* 1605.06699)
[28] Shuryak E and Zahed I 2013 *Phys. Rev.* **C88** 044915

IOP Conf. Series: Journal of Physics: Conf. Series **779** (2017) 012022 doi:10.1088/1742-6596/779/1/012022

SQM2016: Theory Summary

Vincenzo Greco

Department of Physics and Astronomy, University of Catania, Via S. Sofia 64, I-95125 Catania, IT

E-mail: greco@lns.infn.it

Abstract. This is an overview of the main theoretical developments presented at SQM2016 held in Berkeley, California (USA) in June 2015.

1. Introduction

During the *"Strangeness in Quark Matter 2016"* (SQM2016) conference, more than 50 talks were presented about the theoretical effort to understand the many facets of the heavy-ion collisions and the properties of the hot QCD matter. Here in a few pages we overview the main aspects that have been subject of intense discussion focusing on the theoretical plenary talks.

2. Collectivity in Small Systems

The study of several observables and their critical assessment impinges on the understanding of the initial state space-momentum distribution and correlations. At this conference it was presented a main advancement showing that strong eccentric initial fluctuations of the proton shape, once coupled to viscous hydrodynamics, can account for the v_2 observed at $p + Pb$ collision at LHC [1]. A main novelty is that proton shape fluctuations can be constrained by exclusive vector meson production in lepton-proton scatterings at HERA lepton-proton scattering, creating a nice link with high-energy hadronic physics. It remains the issue of the validity of hydrodynamics for a system with large initial Kundsen number. Although the system may undergo "hydrodynamization" in a short time scale as indicated by AdS/CFT approaches [2], it would be important to see if kinetic theory also leads to a similar conclusions. However still such an approach has to be contrasted with the relevance of local initial momentum correlations (on a $1/Q_s$ scale) that lead also to final anisotropies, but not directly related to the space eccentricities [3].

3. Sub-threshold Strangeness and Charm production

The large ϕ/K^- and Ξ^-/Λ ratios reported by HADES and FOPI [4, 5] at subthreshold energy have not been correctly predicted by microscopic transport model and the thermal fits are also not convincing. It has been shown, employing uRQMD augmented by $N^* \to \phi, \Xi^-$ decays, that the Fermi motion plus secondary interaction are a fast way to generate collisions pairs with invariant mass that goes above the threshold. Such a mechanism appears to correctly reproduce the low $\sqrt{s_{NN}}$ enhancement measured experimentally [7], predicting a peak at $\sqrt{s_{NN}} = 1.25\,\text{GeV}$. The quantitative prediction is reasonably regulated by the ANKE pp data for the ϕ production in the $2.5 - 3$ GeV energy range [6]. A similar study for subthreshold charm production has also been presented, showing that $J/\Psi, \Lambda_c, \overline{D}$ may have the chance to be abundant already at SIS100 energy range, $E_{lab} = 5 - 11$ AGeV. Finally it is becoming clear that fast resonance excitation and decays allow for an effective thermalization of hadron yields in a time scale comparable with the nuclei passing time, thanks to the large density of states [8].

4. Thermal model, freeze-out, susceptibilities and phase diagram

The statistical thermal hadronization model (SHM) is able to correctly predict the hadronic yields, and surprisingly recently also light nuclei like $^3He,^3_\Lambda He$ and their anti-nuclei. Considering that this involves a span of about 9 orders of magnitude and is valid at least over the entire range of collisions from SPS to LHC energy [10], this result constitutes one of the bulk achievements of the field that easily at a glance shows that a matter hadronizing with a temperature $T \simeq 155 - 165\, MeV$ has been created [9]. At first glance it always appears quite surprising that the hadronic system can achieve a nearly full equilibrium in a short time. At this conference it has been shown that a microscopic mechanism like the Hagedorn state decays, at the energy densities relevant for QCD transition, can lead to a system nearly born in chemical equilibrium [11].

It is however known that there is a tension in the data on $p, \overline{p}, \Lambda$ production at LHC. This is certainly very interesting and according to some of the talks may signal hadronic chemical re-interaction [12, 13] that shifts the apparent temperature. While this could be a quite natural explanation, an important warning comes from the most recent lattice QCD results on susceptibilities, which has been an hot topic of the conference. From lQCD study of the susceptibilities, involving the strange quantum number and in particular the Koch' ratio [14], it appears clear that the Hadron Resonance Gas (HRG) model, underlying the SHM, from the PDG is not able to reproduce the Koch's ratio,χ_{11}^{BS}/χ_2^S, as evaluated from lQCD for $T \geq 160\,$MeV. A new result presented also shows that even a HRG based on Quark Model that appears to be in a quite good agreement with several lQCD calculations, misses for example the χ_4^S/χ_2^S already at $T > 140\,$MeV, probably because of lack of $s = 1$ meson states [15, 16].

A fundamental part of the phase diagram is certainly the Equation of State (EoS) at finite μ_B/T. The Wuppertal-Budapest Collaboration has shown to be able to evaluate it up to $\mu_B/T < 2.5$ and up to term of the $(\mu_B/T)^6$ order [15].

A key part of the relativistic HIC program is certainly the search for the QCD critical point. It has been shown that kurtosis $\kappa\sigma^2$ of net protons measured by STAR and χ_4/χ_2 in lQCD appear to be in good agreement within the present uncertainties in the region $\sqrt{s_{NN}} > 19\,$GeV corresponding to $\mu_B/T < 2$, where lQCD results are reliable [14]. It has to be mentioned that in this region lQCD does not see a critical point. Dynamical studies of critical fluctuations and of resonance decay impact have been presented, assuming a critical point. They show a robust remnant of critical fluctuations [17], but further advancements are needed before a reliable comparison can be done. Soon the entire range accessible with BES-I and II at RHIC will be accessible through Taylor expansion in lQCD. A proposal to face the search for the critical point by looking the high energy collisions versus the rapidity has been presented in a parallel session[18].

5. Heavy Flavor

The heavy flavor attracted quite some interest already more than a decade ago because of the large suppression observed with a call for large non-perturbative effects at least for charm quarks with $p_T \leq 10\,$GeV. A main difference from the physics of the bulk QGP matter comes from the fact that the initial p_T distribution can be evaluated in a pQCD scheme and in particular the FONLL has been show to be in agreement with the experimental data within the inherent uncertainties. Furthermore one expects marginal thermal production, which is confirmed, within current error bars, also by the observed N_{coll} scaling. A main challenge, currently, is the simultaneous description of both the $R_{AA}(p_T)$ and the $v_2(p_T)$. It has been pointed out that several ingredients contribute to this, but the main impact can come from the T dependence of the transport coefficient. In particular a constant drag coefficient γ leads to a more efficient building-up of v_2 especially in the low-intermediate $p_T \leq 4\,\text{GeV}$. There a weak dependence of the drag as predicted by Quasi-Particle models [20, 19] or by T-matrix approach [21] leads to v_2 about a factor 2-3 larger w.r.t. the T^2 dependence as in pQCD ($\alpha_s = cost.$) or AdS/CFT. A similar trend is observed also for v_3 for heavy quarks at intermediate-high p_T, as discussed in [22].

Another source of uncertainty comes from the microscopic heavy quark dynamics. A real heavy quark would undergo a Brownian motion well described by Langevin transport equations. A charm quark can be envisaged to be marginally heavy, if one considers that while for the thermal production m_c has to be compared with T, for the scatterings other scales enters the problem, gT and/or πT, that are comparable to the charm mass at temperatures $T \sim 300-400\,\text{MeV}$, reached in particular at LHC energy. A direct comparison between Langevin and Boltzmann dynamics shows non sizeable differences for bottom quarks, while for charm it can lead to differences that range between $10-30\%$ in the intermediate p_T region for spectra and elliptic flow, depending on the specific model for the in-medium scatterings considered [23]. The effect is predicted to be larger for $c-\bar{c}$ angular correlations. In the low p_T limit the differences between the two approaches become marginal for the predictions on the charm spectra (or R_{AA}) that especially at LHC energy appear to reach thermalization when a thermal model is able to correctly describe $R_{AA}(p_T)$ for $p_T < 2\,\text{GeV}$.

The comparison to data favours a space-diffusion coefficient $2\pi T D_s$ that rises with temperature T in agreement with lQCD calculation, although such a statement suffers from the systematic errors in both lQCD and the phenomenological models. In the latter the main sources of difference may come from the background bulk expansion and from details in the hadronization mechanism. However, in general, it appears that models including quark coalescence are able to provide predictions closer to the experimental data at both RHIC and LHC energy.

An important advancement presented has been the Linearized Boltzmann Transport (LBT), from the LBL-CCNU Collaboration, that is able to treat in the same framework both the light and heavy hadrons suppression [24], confirming in the heavy quark sector the necessity to have a modification of the the diffusion coefficient w.r.t. the T^2 dependence, as pointed out in [19]. It is interesting to notice that a similar dependence for the diffusions coefficient appears to be necessary also for the $R_{AA}-v_2$ for high p_T minijets as pointed out by the CUJET approach [25]. In the latter the microscopic mechanism of a minimum in $2\pi T D_s$ would be driven by the presence of monopoles in the bulk medium, while for heavy quarks in the low p_T region it would be consistent with the dynamics implied by the remnants of confinement according to TAMU T-matrix approach [26, 27].

A novel aspect of the heavy quarks physics that has been started to be investigated is the impact of the strong initial magnetic field on their dynamical evolution. Some first work shows a modification of the parallel and transverse HQ diffusion coefficient that becomes significant for value of the magnetic field $eB > T^2$ [28, 29], which however may marginally occur in HIC at current energy.

On the other hand it has been shown initial B entails a sizeable directed flow (v_1) of charm quarks (CQs) much larger than light quarks due to a combination of several favorable conditions for CQs, mainly: (i) unlike light quarks formation time scale of CQs, $\tau_f \simeq 0.1\text{fm/c}$ is comparable to the time scale when B is around its maximum value and (ii) the kinetic relaxation time of CQs is similar to the QGP lifetime [30]. The effect is also odd under charge exchange allowing to distinguish it from the vorticity of the bulk matter due to the initial angular momentum conservation.

6. Quarkonia

The medium modification of the quarkonia production is certainly a key probe of the deconfined matter. During the years and going to collisions at increasing energies it has become clear that for charmonia more than the suppression it is the regeneration that plays a key role for the understanding of its production especially at top LHC energy. The large amount of data collected as a function of rapidity and centralities clearly show this. The theoretical approaches based on a transport evolution including both regeneration and dissociation are able to predict the J/Ψ production fairly well [31, 32, 33]. However the impact of the uncertainties in the charm cross sectio still leaves open a key question: is charmonia yield consistent with a regeneration at the chemical freeze-out temperature as for light hadron production? or it is necessary to have a dynamical description of suppression and regeneration to understand the production in the QGP matter? From the point of view of the SHM a clear prediction would be a strong reduction of the ratio $\Psi(2S)/J/\Psi$ with respect to pp and pA [34]. From the point of view of the dynamical regeneration approach a signature of the dynamical nature of the production is the anomalous regeneration of $\Psi(2S)$ observed in central $Pb+Pb$ collisions at high $p_T > 3\,\mathrm{GeV}$, a behavior that is predicted in such a picture due to the delayed recombination into $\Psi(2S)$ w.r.t. J/Ψ and a larger inherited radial flow [35]. This implies a quite different behavior of $R_{AA}(p_T)$ for the J/Ψ w.r.t. $\Psi(2S)$ with the latter showing a peak at $p_T \simeq 4\,\mathrm{GeV}$. Upcoming experimental data at low p_T will allow to clarify this issue. Certainly it remains important to measure the v_2 of the J/Ψ with significantly reduced statical and systematic errors. This would allow, independently of the measurement of the charm cross section, the dynamics of the J/Ψ production to be clarified, as anticipated in [36].

An important theoretical development toward a quantum treatment of the quarkonia suppression has been presented. In fact most of the approaches treat the several quarkonia states as independent states by means of decays width, while they are different excited quantum states of the same system. In the framework of a Schroedinger-Langevin equation that encodes decoherence dynamics [37]. A first comparison show a reasonable agreement with data on bottomonia suppression except for most central collisions. However before full realistic calculation, the impact of realistic 3D potential extracted from lQCD and the impact of the assumptions for the initial quantum states of quarkonia as well as the relation to lQCD spectral functions, still have to be investigated [38].

7. Chiral Anomaly

The Chiral Magnetic Effect (CME) is a remarkable phenomenon emerging from a highly nontrivial interplay of QCD chiral symmetry, axial anomaly, and gluonic topology. The heavy ion collisions supply a hot chiral-symmetric QGP under a strong magnetic field, hence gluonic topological fluctuations could generate a chirality imbalance observable experimentally as a charge separation along the direction of the magnetic field. Experimentally a signal of such a separation has been measured especially in the energy range of BES-RHIC at $\sqrt{s_{NN}} = 7.7 - 62.4\,\mathrm{GeV}$ [39], although currently it seems that models with magnetic field-independent flow backgrounds can also be tuned to reproduce the current measurements.

Indeed on the theoretical side a transport theory that includes the dynamics of chiral currents and the associated anomaly, is needed. This is leading to the development of a chiral viscous hydrodynamics and some attempts to develop a chiral kinetic transport theory; the latter would allow studying the pre-thermal glasma stage that is likely to be the most pertinent one given the short life-time of the magnetic field. Such efforts will supply the necessary theoretical tools to predict the size and the evolution of the CME with energy and centrality. Also it can allow us to study in a self-consistent way several observables related to CME and to see if from the experimental data a coherent picture for a fundamental effect emerges, which however till now is associated to a weak signal over a large background. Nicely, a first attempt to describe the centrality dependence of the charge separation within chiral hydro appears to be consistent with the dynamics implied by the CME [40].

8. Global Polarization

An aspect of relativistic HIC that has been mostly overlooked is that in non-central collisions due to the large orbital momentum one is likely to create a system with a huge vorticity. One can envisage that particles emerging from such a highly vorticous fluid are globally polarized with their spins on average pointing along the system angular momentum. To study such a dynamics the development of a self-consistent theory that treat vorticity in a relativistic framework is needed. This has been recently investigated in the context of the QGP physics assuming a thermal fluid deriving a general formula that relates the vorticity of a fluid to its polarization for particles of spin 1/2 [41]. This has allowed embedding the polarization dynamics into the ECHO-QGP viscous hydro code for RHIC's studying in particular the Λ polarization that is accessible experimentally [42, 43]. Distinctive feature of thermodynamic polarization would be a C-even effect, that particle and antiparticle have the same polarization, unlike e.m. induced polarization. The predictions appear to follow a similar increase of the polarization with decreasing beam energy. Interestingly in experiments the polarization of the order of a few percent appears to be even larger than the prediction [43]. However the effect of resonance decays, hadronic rescattering, impact of the Pauli Blocking [44] have to be investigated. Certainly the developments of such studies has opened a new direction and can allow us to have an insight on the vorticity of the matter created in the early stage of the collision, adding a new dimension in the HIC phase space.

9. Summary

In Summary, there is a clear progress in the understanding of the QCD matter at high temperature created in relativistic heavy-ion collisions. Certainly in the last decade an understanding of the main features of the collision dynamics and its collective expansion has been achieved, and the surprising behavior of several observables in pA from strangeness production to anisotropic flows will at the end allow us to have deeper understanding of the matter created in such collisions and the properties of hot QCD. At the some time one should appreciate that many questions were not even conceivable a decade ago, but are now under active investigation thanks to the progress that the field has achieved in the understanding of the bulk properties of the matter created.

10. Acknowledgments

V.G. is supported under QGPDyn ERC Grant no. 259684.

References

[1] H. Mntysaari and B. Schenke, Phys. Rev. Lett. **117** (2016) no.5, 052301
[2] B. Schencke, 2016 *contribution to this conference*
[3] T. Lappi, B. Schenke, S. Schlichting and R. Venugopalan, JHEP **1601** (2016) 061
[4] G. Agakishiev *et al.* [HADES Collaboration], Phys. Rev. C **80** (2009) 025209
[5] G. Agakishiev *et al.* [HADES Collaboration], Phys. Rev. Lett. **103** (2009) 132301
[6] M. Hartmann *et al.*, Phys. Rev. C **85** (2012) 035206
[7] J. Steinheimer, 2016 *contribution to this conference*
[8] J. Steinheimer, M. Lorenz, F. Becattini, R. Stock and M. Bleicher, Phys. Rev. C **93** (2016) no.6, 064908
[9] A. Andronic, P. Braun-Munzinger and J. Stachel, Phys. Lett. B **673** (2009) 142 Erratum: [Phys. Lett. B **678** (2009) 516]
[10] A. Andronic, 2016 *contribution to this conference*
[11] M. Beitel, C. Greiner and H. Stoecker, Phys. Rev. C **94** (2016) no.2, 021902: C. Greiner, 2016 *contribution to this conference*
[12] F. Becattini, E. Grossi, M. Bleicher, J. Steinheimer and R. Stock, Phys. Rev. C **90** (2014) no.5, 054907

[13] R. Stock, 2016 *contribution to this conference*
[14] F. Karsch, 2016 *contribution to this conference*
[15] C. Ratti, 2016 *contribution to this conference*
[16] J.-H. Noronha, 2016 *contribution to this conference*
[17] M. Bluhm, 2016 *contribution to this conference*
[18] J. Kapusta, 2016 *contribution to this conference*
[19] S. K. Das, F. Scardina, S. Plumari and V. Greco, Phys. Lett. B **747** (2015) 260
[20] H. Berrehrah, E. Bratkovskaya, W. Cassing, P. B. Gossiaux, J. Aichelin and M. Bleicher, Phys. Rev. C **89** (2014) no.5, 054901
[21] H. van Hees, M. Mannarelli, V. Greco and R. Rapp, Phys. Rev. Lett. **100** (2008) 192301
[22] Caio A.G. Prado, 2016 *contribution to this conference*
[23] S. K. Das, F. Scardina, S. Plumari and V. Greco, Phys. Rev. C **90** (2014) 044901
[24] S. Cao, T. Luo, G. Y. Qin and X. N. Wang, Phys. Rev. C **94** (2016) no.1, 014909
[25] J. Xu, J. Liao and M. Gyulassy, JHEP **1602** (2016) 169
[26] S. Y. F. Liu and R. Rapp, arXiv:1609.04877 [hep-ph].
[27] S. Y. F. Liu, 2016 *contribution to this conference*
[28] Ho-Yunh Yee, 2016 *contribution to this conference*
[29] K. Fukushima, K. Hattori, H. U. Yee and Y. Yin, Phys. Rev. D **93** (2016) no.7, 074028
[30] S. K. Das, S. Plumari, S. Chatterjee, J. Alam, F. Scardina and V. Greco, arXiv:1608.02231 [nucl-th].
[31] K. Zhou, 2016 *contribution to this conference*
[32] Y. p. Liu, Z. Qu, N. Xu and P. f. Zhuang, Phys. Lett. B **678** (2009) 72
[33] X. Zhao and R. Rapp, Nucl. Phys. A **859** (2011) 114
[34] A. Andronic, F. Beutler, P. Braun-Munzinger, K. Redlich and J. Stachel, Phys. Lett. B **678** (2009) 350
[35] X. Du, 2016 *contribution to this conference*
[36] V. Greco, C. M. Ko and R. Rapp, Phys. Lett. B **595** (2004) 202
[37] P. B. Gossiaux and R. Katz, Nucl. Phys. A **956** (2016) 737
[38] J. P. Gossiaux, 2016 *contribution to this conference*
[39] D. E. Kharzeev, J. Liao, S. A. Voloshin and G. Wang, Prog. Part. Nucl. Phys. **88** (2016)
[40] J. Liao, 2016 *contribution to this conference*
[41] F. Becattini, V. Chandra, L. Del Zanna and E. Grossi, Annals Phys. **338** (2013) 32
[42] F. Becattini *et al.*, Eur. Phys. J. C **75** (2015) no.9, 406
[43] Y. Karpenko, 2016 *contribution to this conference*
[44] X. G. Huang, 2016 *contribution to this conference*

Open heavy-flavour measurements in p–Pb and Pb–Pb collisions with ALICE at the LHC

Cristina Terrevoli, for the ALICE Collaboration
Università degli Studi di Padova & INFN Padova, via Marzolo 8, 35131, Padova, Italy
E-mail: cristina.terrevoli@cern.ch

Abstract. Heavy flavours are sensitive probes of the hot and dense QCD medium formed in high-energy heavy-ion collisions. Measurements of their production in p–Pb collisions are crucial for the interpretation of heavy-ion results, by investigating the cold nuclear matter effects. The open heavy-flavour production studied with ALICE at the LHC in p–Pb collisions at $\sqrt{s_{\rm NN}}$ = 5.02 TeV and in Pb–Pb collisions at $\sqrt{s_{\rm NN}}$ = 2.76 TeV are presented. Emphasis is given to the recent measurements of $\rm D^0$ production cross section down to $p_{\rm T}$=0, the nuclear modification factor of heavy-flavour hadron decay electrons in p–Pb collisions, the nuclear modification factor of D-meson, and heavy-flavour hadron decay electron elliptic flow in Pb–Pb collisions, as a function of centrality.

1. Introduction

The main goal of the ALICE experiment [1] is the characterization of the Quark-Gluon Plasma (QGP), the hot and dense matter created in high-energy nuclear collisions. Heavy quarks (charm and beauty) are unique probes of the QGP because they are dominantly produced in hard partonic scattering processes occurring in the initial stage of the collisions, and thus probe the entire evolution of the system. The nuclear modification factor $R_{\rm AA}$ and the elliptic flow v_2 are among the key observables for the QGP characterization. The nuclear modification factor is sensitive to the quark in-medium energy loss, providing a test of its colour-charge and parton-mass dependence. Theoretical calculations predict a hierarchy in partonic energy loss: $\Delta E_{\rm g}>$ $\Delta E_{\rm light_q}>$ $\Delta E_{\rm c}>$ $\Delta E_{\rm b}$. This can be studied with the $R_{\rm AA}$ of light and heavy-flavored hadrons. The v_2 is the second coefficient of the Fourier expansion of the $p_{\rm T}$-dependent particle azimuthal distribution with respect to the reaction plane. It is related to the azimuthal anisotropy of particle production in non-central collisions and represents an effective tool to investigate to what extent heavy quarks participate in the collective expansion in the medium. The measurement of the heavy-flavour production in p–Pb collisions provides insight into the role of cold nuclear matter effects (CNM). In this contribution, the ALICE measurements of D-meson and heavy-favour hadron decay lepton production in Pb-Pb and p-Pb collisions, focusing on the results from recent publications, are presented.

2. Open heavy-flavour reconstruction in ALICE

The ALICE detector provides precise tracking, vertexing and charged particle identification over a broad momentum range. D mesons are reconstructed via the $\rm D^0$, $\rm D^{*+}$, $\rm D^+$ and $\rm D_s^+$ hadronic decay channels at mid-rapidity ($|y_{\rm lab}|$<0.5) through the topological selection of the reconstructed decay vertices displaced by a few hundred μm from the interaction vertex, in $1< p_T< 24$ GeV/c [2, 3]. In addition, the prompt D^0 production was measured in p–Pb collisions down to $p_T = 0$ using an analysis technique based on the estimation and subtraction of the combinatorial background, without reconstructing its decay vertex [4]. The open heavy-flavour production is also accessible via semi-leptonic decays of charm and beauty hadrons, both at mid-rapidity (electrons, $|y_{lab}|<0.7$) and at forward rapidity (muons, $2.5 < y_{lab} < 4$). Muons are selected and identified with the muon tracking and trigger chambers, with acceptance and geometrical cuts and MC cocktail methods to subtract the background from K and π decay muons [5]. Electrons are identified through a combination of electron identification strategies with different detectors. The background is subtracted via an invariant mass reconstruction of the couples of e^-e^+ pairs, or cocktail method based on data [6, 7].

3. Highlights in p–Pb collisions

The initial-state effects are expected to have a small impact on D-meson production at high p_T, but they can induce a modification of the D-meson cross section with p_T below a few GeV/c. For this reason, a measurement of the D-meson production down to $p_T = 0$ provides important information. Below 1-2 GeV/c, the D-meson decay topology can not be efficiently resolved because of the small Lorentz boost. Furthermore, the selection criteria based on secondary-vertex displacement tend to select non-prompt D mesons from beauty-hadron decays with higher efficiency, increasing the systematic uncertainty on the subtraction of the beauty feed-down contribution. Using an analysis technique based on particle identification and on the estimation and subtraction of combinatorial background, via event mixing, like-sign pairs, track rotation and side-band fit, it was possible to measure the D^0-meson yield down to $p_T= 0$ in p–Pb collisions, with reduced feed-down systematic uncertainties and large efficiency [4]. In Fig. 1, on the left, our most precise measurement of prompt D^0 p_T-differential cross section is presented. The results are obtained with the background subtraction method for $0 <p_T< 2$ GeV/c [4] and with the vertexing method for $p_T> 2$ GeV/c [2]. The total cross section for prompt D^0-meson production per unit of rapidity in $-0.96<y_{cms}<0.04$ is:

$$d\sigma^{\text{prompt } D^0}_{\text{p–Pb}, 5.02\,\text{TeV}}/dy = 79.0 \pm 7.3\,(\text{stat.})^{+\ 7.1}_{-13.4}\,(\text{syst.}) \pm 2.9\,(\text{lumi.}) \pm 1.0\,(\text{BR})\ \text{mb}.$$

The R_{pPb} for D mesons in p–Pb collisions at $\sqrt{s_{NN}} = 5.02$ TeV is shown in Fig. 1 (middle): this is the average R_{pPb} of D^0, D^{*+}, D^+ mesons in the interval $1 <p_T<24$ GeV/c [2], together with the D^0 R_{pPb} in $0 <p_T< 1$ GeV/c. The data are compared with results of theoretical calculations including only CNM effects [8, 9, 10, 11] and it is consistent with unity within the uncertainties. The p_T-integrated nuclear modification factor of prompt D^0 mesons in $-0.96<y_{cms}<0.04$ is: $R^{D^0}_{pPb} = 0.89 \pm 0.11$ (stat.)$^{+0.13}_{-0.18}$ (syst.). On the right the heavy-flavour decay electron R_{pPb} is shown. It is consistent with unity within the uncertainties, but also consistent with an enhancement in the region $1<p_T<6$ GeV/c, as observed in d–Au collisions at $\sqrt{s_{NN}}$=200 GeV [6]. Data are compared with theoretical models that include initial-state effects [11, 10] or radial flow [12].

4. Highlights in Pb–Pb collisions

The left panel of Fig. 2 shows the R_{AA} of prompt non-strange D mesons, D^0, D^{*+}, D^+, in comparison with that of D_s^+ [13]. The maximum suppression of the D-meson R_{AA} (factor 5-6) is observed at p_T=10 GeV/c in central Pb–Pb collisions. The R_{pPb} shown in Fig. 1 in the middle, which is consistent with unity, indicates that the suppression in central Pb–Pb collisions is induced by final-state effects due to quark energy loss in the medium. If hadronization via recombination occurs at low p_T, the relative abundance of D_s^+ with respect to non-strange D mesons should be larger in Pb–Pb than in pp collisions, due to the expected strangeness abundance. An indication of less suppression for D_s^+ is observed at low p_T, and is in agreement

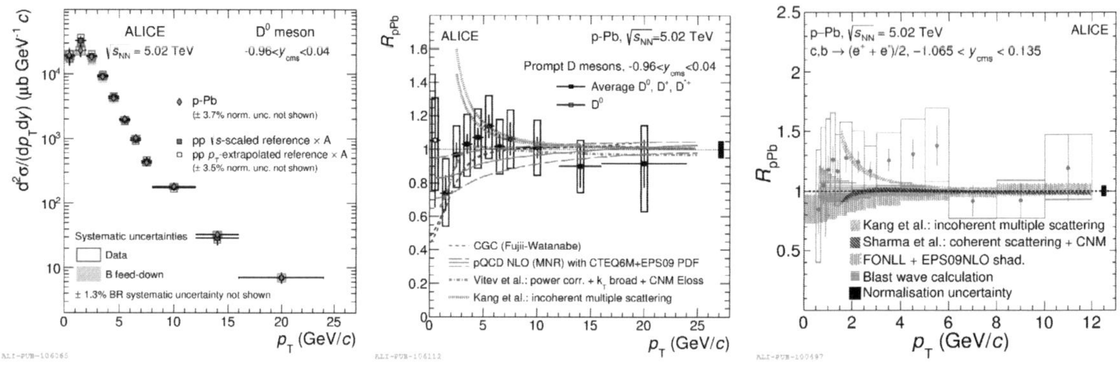

Figure 1. Measurements in p–Pb collisions at $\sqrt{s_{\rm NN}} = 5.02$ TeV. Left:$p_{\rm T}$-differential production cross section of D^0 mesons in 0<$p_{\rm T}$<16 GeV/c [4]. Middle:$R_{\rm pPb}$ of D mesons (with D^0 in 0<$p_{\rm T}$<1 GeV/c) [4]. Right: heavy-flavour electrons $R_{\rm pPb}$ [6]. Data are compared with models.

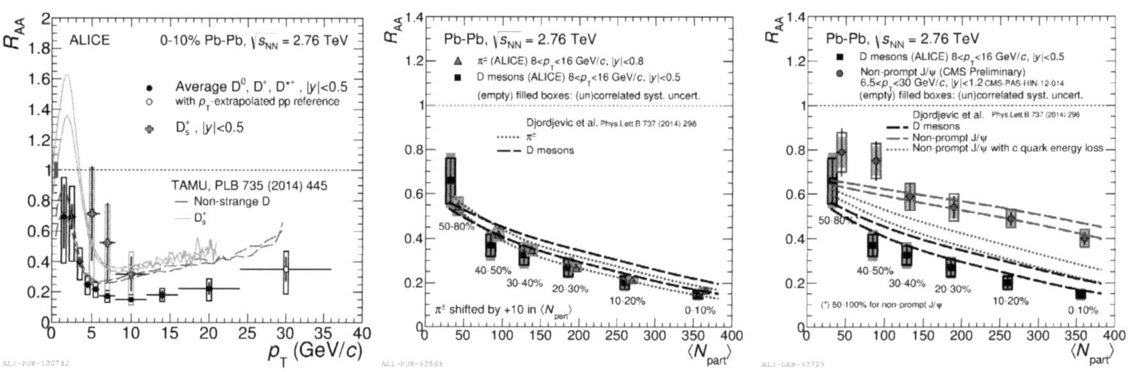

Figure 2. Left: $R_{\rm AA}$ in the 0-10% centrality class, of prompt non-strange D mesons and $R_{\rm AA}$ of D$_{\rm s}^{+}$ compared with TAMU model [13]. Middle and Right: $R_{\rm AA}$ of D mesons in 8< $p_{\rm T}$<16 GeV/c as a function of centrality in comparison with model [14]. Middle: compared with charged pions. Right: compared with non-prompt J/ψ in 6.5<$p_{\rm T}$<30 GeV/c, measured by CMS [15]

with TAMU models [13] within uncertainties. In the middle panel of Fig. 2 the $R_{\rm AA}$ of prompt D mesons in the transverse momentum region $8 < p_{\rm T} < 16$ GeV/c is shown as a function of centrality in comparison with the $R_{\rm AA}$ of pions [3]. The two $R_{\rm AA}$ are compatible within uncertainties. This observation is described by a model that takes into account the mass dependences of energy loss and the softer $p_{\rm T}$ spectrum and fragmentation of gluons and light quarks with respect to charm quarks [14]. The comparison of the $R_{\rm AA}$ of D mesons and of J/ψ from B-hadron decays (measured by the CMS Collaboration [15]) is displayed in the right panel of Fig. 2. It shows a stronger suppression for charm than for beauty hadrons at high $p_{\rm T}$ in central Pb–Pb collisions, consistent with the expectation $R_{\rm AA}$(D) <$R_{\rm AA}$(B). The two measurements are described by the predictions based on a pQCD model including mass-dependent radiative and collisional energy loss [14]. In this model the difference in the $R_{\rm AA}$ of charm and beauty mesons is mainly due to the mass dependence of quark energy loss, as demonstrated by comparing with the curve in which the non-prompt J/ψ $R_{\rm AA}$ is calculated assuming that b quarks have the same energy loss as c quarks. The elliptic flow of heavy-flavour hadron decay electrons at

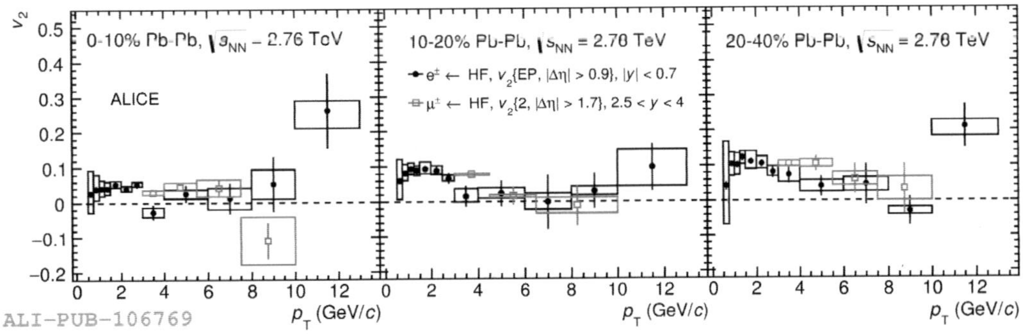

Figure 3. Elliptic flow of heavy-flavour hadron decay electrons at mid-rapidity and muons at forward rapidity in Pb–Pb collisions at $\sqrt{s_{NN}}$=2.76 TeV for three centrality intervals [7, 5].

mid-rapidity [7] and muons at forward rapidity [5], measured with the event-plane method and with two-particle Q-cumulant, respectively, in Pb–Pb collisions at $\sqrt{s_{NN}} = 2.76$ TeV is shown in Fig. 3 for three centrality classes. The measurements in the two rapidity regions are comparable in magnitude and the v_2 increases from central to semi-central collisions. In semi-central (20-40%) Pb–Pb collisions a positive v_2 is observed with a significance of 3 in $2 < p_T < 3$ GeV/c for electrons and in $3 < p_T < 5$ GeV/c for muons. These results suggest the significant interaction of heavy quarks, mainly charm in these p_T intervals, with the medium. Furthermore, the measured v_2 of prompt D mesons (not shown) is larger than zero in semi-central (30-50%) Pb–Pb collisions with a significance of 5.7 in $2 < p_T < 6$ GeV/c [16].

5. Conclusions

The ALICE results on open heavy-flavour production show a R_{pPb} consistent with unity and with models including CNM effects. The first measurement of D^0 meson production down to p_T=0 in p–Pb collisions was presented. The strong suppression observed in Pb–Pb collisions for p_T>3 GeV/c is due to final-state effects, and can be described by models with collisional and radiative energy loss mechanisms. The positive elliptic flow indicates that heavy quarks participate to the collective expansion at low p_T.

References

[1] ALICE Collaboration 2008 *JINST* **3** S08002
[2] ALICE Collaboration 2014 *Phys. Rev. Lett.* **113** no. 23 232301
[3] ALICE Collaboration 2016 *JHEP* **1603** 081
[4] ALICE Collaboration 2016 *arXiv:1605.07569* [nucl-ex], submitted to *JHEP*
[5] ALICE Collaboration 2016 *Phys. Lett. B* **753** 41
[6] ALICE Collaboration 2016 *Phys. Lett. B* /bf 754 81-93
[7] ALICE Collaboration 2016 arXiv:1606.00321 [nucl-ex] submitted to *JHEP*
[8] Fujii H and Watanabe K 2013 *Nucl. Phys. A* **920** 78-93
[9] Eskola K, Paukkunen K and Salgado C 2009 *JHEP* **04** 065
[10] Sharma R, Vitev I and Zhang C 2009 *Phys. Rev. C* **80** 054902
[11] Kang Z, Vitev I, Wang E, Xing H and Zhang C 2015 *Phys. Lett. B* **740** 23-29
[12] Sickles M, 2014 *Phys. Lett. B* **731** 51- 56
[13] ALICE Collaboration 2015 *JHEP* **1603** 082
[14] M. Djordjevic, 2014 *Phys. Rev. Lett. B* **737** 286-298
[15] CMS Collaboration 2012 *JHEP* **1205** 063
[16] ALICE Collaboration 2014 *Phys. Rev. C* **90** 034904

Heavy-flavour production in pp collisions and correlations in pp and p-Pb collisions measured with ALICE at the LHC

Fabio Colamaria, on behalf of the ALICE Collaboration
INFN, Sezione di Bari, Via E. Orabona 4, 70125 Bari, Italy
E-mail: fabio.colamaria@ba.infn.it

Abstract. Thanks to its excellent tracking and particle identification performance, the ALICE detector is capable of measuring D mesons at mid-rapidity via their hadronic decay channels down to very low transverse momentum. We show an extension of the prompt D^0 p_T-differential cross section measurement down to zero transverse momentum, which allows us to determine the p_T-integrated charm production cross section at mid-rapidity in pp collisions at $\sqrt{s} = 7$ TeV. We also present measurements of azimuthal correlations of prompt D^0, D^+ and D^{*+} mesons with charged hadrons in pp collisions at $\sqrt{s} = 7$ TeV and p–Pb collisions at $\sqrt{s_{NN}} = 5.02$ TeV and compare the results with expectations from models.

1. Introduction

The study of heavy-flavour production in pp collisions at LHC energies allows us to test perturbative QCD (pQCD) calculations and it provides a reference for studies in heavy-ion collisions. At low p_T, heavy-flavour production is dominated by low-x gluons. Measurements in this region can thus provide constraints on the pQCD calculations and gluon parton distribution functions at low-x values. An analysis technique without selections on the displaced vertex topology allowed measurements of the p_T-differential prompt D^0 production cross section down to $p_T = 0$ GeV/c [1] in pp collisions at $\sqrt{s} = 7$ TeV with ALICE, extending the p_T reach of previous measurements [2]. This also allowed an evaluation of the total charm production cross section with a reduced uncertainty with respect to the previous ALICE measurement [3]. More differential measurements of charm production provided by ALICE are now also available [4, 5]. In particular, the analysis of angular correlations between heavy-flavour particles and charged particles is a tool to characterize the heavy-quark fragmentation process and is sensitive to their production mechanism. In p–Pb collisions, this measurement allows us to investigate possible effects due to the presence of a nucleus in the collision (cold nuclear matter effects) on the heavy-quark production and hadronisation.

2. Prompt D^0 production cross section down to $p_T = 0$ GeV/c

In ALICE [6], D mesons are reconstructed from their hadronic decay channels ($D^0 \to K^-\pi^+$, $D^+ \to K^-\pi^+\pi^+$, $D^{*+} \to D^0\pi^+ \to K^-\pi^+\pi^+$, $D_s^+ \to \phi\pi^+ \to K^-K^+\pi^+$ and their charge conjugates). The standard analysis [2] is based on (i) the selection of D-meson decay products by exploiting particle identification and track-quality cuts, and ii) a selection exploiting the

D-meson decay length of a few hundred μm. At very low D-meson p_T, this technique is not efficient due to the small Lorentz boost of the D mesons and the degraded track impact parameter resolution. A different technique, relying only on the particle identification without any topological selection, is used to evaluate the D^0 meson p_T-differential production cross section down to $p_T = 0$ GeV/c. For each p_T interval, the D^0-meson raw yield is extracted from an analysis of the D^0 candidate invariant-mass distribution. The central value of the yield is obtained as the arithmetic average of the raw yields obtained with four different approaches used for the evaluation of the background distribution (like-sign pairs, event mixing, track rotation and side-band fit). With this analysis technique prompt and feed-down D^0 mesons are reconstructed with the same efficiency, differently from the standard approach, which favours D^0 candidates from beauty-hadron decays which on average are more displaced. As a consequence, the feed-down subtraction uncertainty on the p_T-differential cross section is reduced. The left panel of Fig. 1 compares the p_T-differential production cross section for D^0 mesons at mid-rapidity in pp collisions at $\sqrt{s} = 7$ TeV obtained with the two analysis techniques [1]. The results are compatible within the uncertainties in the common p_T range. The analysis without secondary vertex reconstruction, however, enables the study of the $0 < p_T^D < 1$ GeV/c interval, and provides smaller uncertainty for the $1 < p_T^D < 2$ GeV/c interval. In the right panel of Fig. 1 the most precise measurement of the cross section is shown, exploiting for each p_T interval the measurement with the smallest total uncertainty. ALICE measurements are compared to predictions from FONLL [7], GM-VFNS [8] and k_t-factorization [9] pQCD calculations, which are observed to be in agreement with data results within the theoretical and experimental uncertainties. The resulting total $c\bar{c}$ production cross section in pp collisions at $\sqrt{s} = 7$ TeV is $\sigma_{c\bar{c}}^{tot} = 7.96 \pm 0.65(\text{stat.})^{+0.87}_{-1.57}(\text{syst.})^{+2.34}_{-0.35}(\text{extr.}) \pm 0.28(\text{lumi.}) \pm 0.10(\text{BR}) \pm 0.03(\text{FF})$ mb, with reduced systematic and extrapolation uncertainties with respect to the previous measurement [3].

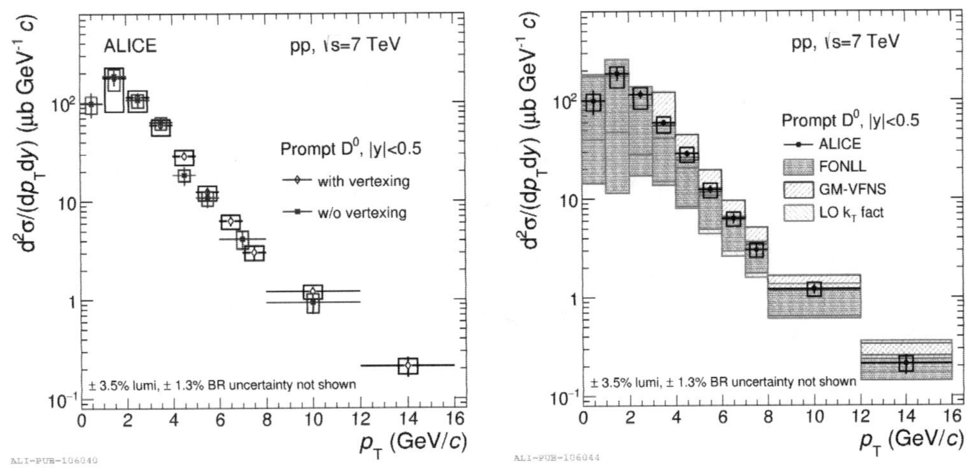

Figure 1. Left panel: comparison of the p_T-differential production cross section for D^0 mesons with $|y| < 0.5$ in pp collisions at $\sqrt{s} = 7$ TeV obtained with the two analysis techniques [1]. Right panel: comparison of the most precise measurements of the cross section in each p_T interval with predictions by pQCD calculations [1].

3. D meson-charged particle angular correlations

Azimuthal correlation distributions of D^0, D^+ and D^{*+} mesons (trigger particles) with charged particles (associated particles) in $|\Delta\eta| < 1$ are evaluated in pp collisions at $\sqrt{s} = 7$ TeV and p–Pb collisions at $\sqrt{s_{\mathrm{NN}}} = 5.02$ TeV, for different ranges of the D-meson ($3 < p_{\mathrm{T}}^{\mathrm{D}} < 16$ GeV/c in pp, $5 < p_{\mathrm{T}}^{\mathrm{D}} < 16$ GeV/c in p–Pb) and associated particle p_{T} (starting from $p_{\mathrm{T}}^{\mathrm{assoc}} > 0.3$ GeV/c). The contribution of D-meson combinatorial background is removed by subtracting the correlation distribution evaluated from the sidebands of the D-meson invariant mass distribution. An event-mixing correction is applied to account for detector inhomogeneities and limited acceptance. The distributions are corrected for inefficiencies in the reconstruction and selection of trigger and associated particles. The contribution of D mesons from beauty-hadron decays is subtracted, exploiting templates of the angular correlations of feed-down D mesons and charged particles obtained from PYTHIA simulations [10], normalized to the expected feed-down contribution evaluated from FONLL calculations [7]. The azimuthal correlation distributions of D^0, D^+ and D^{*+} mesons are compatible within uncertainties. A weighted average of the three D-meson measurements is thus performed to reduce the statistical uncertainty. The per-trigger azimuthal correlation distributions are fitted with two Gaussian functions and a constant (baseline), allowing us to extract quantitative observables such as the near-side associated yield, near-side peak width and baseline. Figure 2 compares the baseline-subtracted D meson-charged particle azimuthal correlation distributions extracted in pp collisions with predictions by PYTHIA6 [10], PYTHIA8 [11] and POWHEG+PYTHIA6 [12, 13] simulations, as a function of the D-meson p_{T}, for different associated particle p_{T} ranges [5]. The measured correlation pattern is reproduced by the simulations within uncertainties, though a hint for a more pronounced near-side peak in data than in models is visible in the $8 < p_{\mathrm{T}}^{\mathrm{D}} < 16$ GeV/c range. A more quantitative comparison of the near-side peak properties can be performed considering the observables extracted from the fit to the correlation distributions. The trends of the near-side associated yields and the near-side peak widths extracted in pp and p–Pb collisions versus the D-meson p_{T} are compared in Fig. 3 for different associated particle p_{T} ranges. Compatible values of the near-side observables are obtained in pp and p–Pb collisions. No modifications of the near-side peaks due to cold nuclear matter effects are observed in p–Pb collisions with the current uncertainties. Predictions by POWHEG+PYTHIA6 simulations, including nuclear shadowing effects for the nucleon parton distribution functions, are also in agreement with the measurements.

4. Summary

An extension of the p_{T}-differential prompt D^0 production cross section measurement down to $p_{\mathrm{T}} = 0$ GeV/c was presented and was found to be compatible with previous ALICE measurements in the common p_{T} range. It allowed to reduce the uncertainties on the total charm production cross section. The D meson-charged particle azimuthal correlation distributions, measured in pp and p–Pb collisions, as well as observables describing near-side peak properties extracted from their fits, are in agreement between each other within uncertainties, and are well described by PYTHIA and POWHEG+PYTHIA Monte Carlo simulations.

References

[1] ALICE Collaboration 2016 *arXiv:1605.07569 [nucl-ex]*, submitted to *JHEP*
[2] ALICE Collaboration 2012 *JHEP* **1201** 128
[3] ALICE Collaboration 2012 *JHEP* **1207** 191
[4] ALICE Collaboration 2015 *JHEP* **1509** 148
[5] ALICE Collaboration 2016 *arXiv:1605.06963 [nucl-ex]*
[6] ALICE Collaboration 2014 *Int. J. Mod. Phys.* A **29** 1430044
[7] Cacciari M, Frixione S, Houdeau N, Mangano M, Nason P, Ridolfi G 2012 *JHEP* **1210** 137
[8] Kniehl B A, Kramer G, Schienbein I, Spiesberger H 2012 *Eur. Phys. J.* C **72** 2082
[9] Maciula R, Szczurek A 2013 *Phys. Rev.* D **87** no.9 094022
[10] Sjöstrand T, Mrenna S, Skands P Z 2006 *JHEP* **0605** 026
[11] Sjöstrand T, Mrenna S, Skands P Z 2008 *Comput. Phys. Commun.* **178** 852
[12] Nason P 2004 *JHEP* **0411** 040
[13] Frixione S, Nason P, Oleari C 2007 *JHEP* **0711** 070

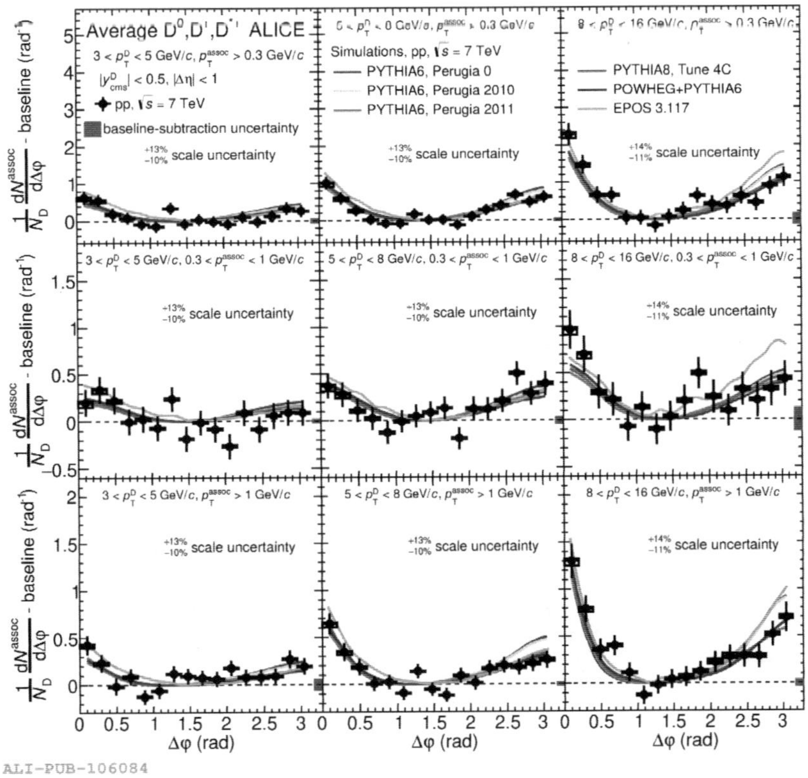

Figure 2. Comparison of D meson-charged particle azimuthal correlation distributions in pp collisions with predictions from Monte Carlo simulations, for different D-meson and associated particle p_{T} ranges [5].

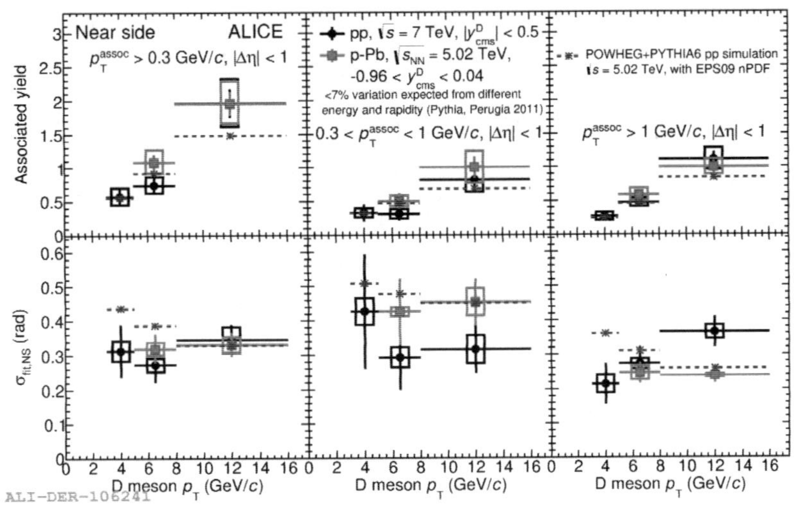

Figure 3. Comparison of near-side associated yields and widths extracted in pp and p–Pb collisions and predicted by PYTHIA+POWHEG simulations as a function of the D-meson p_{T} for different associated particle p_{T} ranges [5].

IOP Conf. Series: Journal of Physics: Conf. Series **779** (2017) 012025 doi:10.1088/1742-6596/779/1/012025

Heavy-flavour multiplicity dependence in p-Pb collisions with ALICE at the LHC

Jan Wagner for the ALICE Collaboration
Research Division and ExtreMe Matter Institute, GSI Helmholtzzentrum für Schwerionenforschung GmbH, Darmstadt, Germany

E-mail: j.wagner@cern.ch

Abstract. The nuclear modification factor Q_{pPb} of D^0, D^+ and D^{*+} mesons is shown in the transverse-momentum (p_{T}) range $1 < p_{\mathrm{T}} < 24\,\mathrm{GeV}/c$ as a function of centrality in p-Pb collisions at $\sqrt{s_{\mathrm{NN}}} = 5.02\,\mathrm{TeV}$ as measured with ALICE at the LHC. The Q_{pPb} measured in different centrality intervals agree with each other for all D-meson species. Also presented are the self-normalized yields of D mesons in the same p_{T} range and of electrons from semi-leptonic heavy-flavour hadron decays in the p_{T} range $0.5 < p_{\mathrm{T}} < 8\,\mathrm{GeV}/c$ as a function of the charged-particle multiplicity. Both measurements agree with each other within their uncertainties. The self-normalized yields increase with the charged-particle multiplicity independent of the p_{T} interval.

1. Introduction

Heavy-flavour production is studied in p-Pb collisions to gain insight into potential Cold Nuclear Matter (CNM) effects present in Pb-Pb and p-Pb collisions. The p_{T}-differential heavy-flavour production cross section is compared to a baseline measurement obtained in pp collisions, where the heavy-flavour cross sections are in good agreement with perturbative QCD calculations. ALICE has measured the nuclear modification factor (R_{pPb}) for D mesons [1] and electrons from heavy-flavour hadron decays [2], indicating insignificant CNM effects for $p_{\mathrm{T}} > 2\,\mathrm{GeV}/c$ in agreement with various theoretical models. The dependence of these effects on the centrality is an important aspect to understand and constrain these model calculations further. The centrality-dependent nuclear modification factor (Q_{pPb}) is defined as:

$$Q_{\mathrm{pPb}} = \frac{(\mathrm{d}N^{\mathrm{D}}/\mathrm{d}p_{\mathrm{T}})^{\mathrm{cent}}_{\mathrm{pPb}}}{\langle T_{\mathrm{pPb}}\rangle \times (\mathrm{d}\sigma^{\mathrm{D}}/\mathrm{d}p_{\mathrm{T}})_{\mathrm{pp}}} \tag{1}$$

where $(\mathrm{d}N^{\mathrm{D}}/\mathrm{d}p_{\mathrm{T}})^{\mathrm{cent}}_{\mathrm{pPb}}$ is the yield of prompt D mesons in p-Pb collisions in a given centrality class, $(\mathrm{d}\sigma^{\mathrm{D}}/\mathrm{d}p_{\mathrm{T}})_{\mathrm{pp}}$ is the cross section of prompt D mesons in pp collisions at the same collision energy and $\langle T_{\mathrm{pPb}}\rangle$ is the average nuclear overlap function in a given centrality class calculated in a Glauber Monte Carlo approach [3].

Another interesting aspect is the interplay between hard processes such as heavy-quark production and the underlying event charged-particle distribution driven by soft processes. This can be studied using self-normalized yields which are yields obtained in a multiplicity interval and normalized to the multiplicity integrated yield. Due to the cancellation of correlated systematic uncertainties self-normalized yields are a good tool to study high multiplicity events. Unlike for $Q_{\rm pPb}$, in which particle production in samples of 20% of the analyzed events is examined, the self-normalized yields explore events to extremely high multiplicities corresponding to only 5% of the analyzed events in p-Pb collisions [4]. In p-Pb collisions multiple binary collisions take place, for which a linear scaling of the self-normalized heavy-flavour yields as a function of the charged-particle multiplicity is expected. However, at very high multiplicities other effects might play a role such as Multiple Parton Interactions (MPIs). MPIs can be an explanation for a faster-than-linear scaling which is observed in measurements of self-normalized D-meson yields in pp collisions [5].

2. Analysis

The centrality and charged-particle multiplicity are estimated in ALICE using various approaches. For the centrality estimation, the zero degree calorimeter (ZNA) and the VZERO scintillator hodoscope (V0A) are used at backward rapidity i.e. in direction of the Pb beam [3]. For the estimation of the charged-particle multiplicity two approaches in different rapidity regions are used. At backward rapidity the V0A is used. At mid-rapidity tracklets, which are connected hits in both Silicon Pixel Detector (SPD) layers pointing to the vertex, are used.

Charm quark production is measured via the reconstruction of D mesons from their decay kaons and pions in an invariant-mass analysis [1]. The combinatorial background is reduced with $p_{\rm T}$-dependent topological cuts. The feed-down contribution from B-meson decays is subtracted to obtain the prompt D-meson yield.

Another approach to measure heavy-flavour production is to determine inclusively the yield of electrons from semi-leptonic heavy-flavour hadron decays [2]. The main contribution to the background is from electrons from Dalitz decays and photon conversion. Since these processes produce low-mass electron-positron pairs the corresponding electron-background yield can be measured via an invariant-mass analysis. The combinatorial background is estimated using the invariant-mass distribution of uncorrelated same-signed electron or positron pairs. The remaining electron background from weak kaon decays ($\rm K_{e3}$) and J/ψ decays is estimated by Monte Carlo simulations using measured cross sections as input. The inclusive measurement of electrons gives access to beauty physics since the contribution of electrons from semi-leptonic beauty-hadron decays is more than 50% for $p_{\rm T} > 4\,{\rm GeV}/c$ [6].

The D-meson cross sections are measured as a function of centrality and multiplicity and are published in [4]. The measurement of self-normalized electron yields is performed as a function of the charged-particle multiplicity.

3. Results

The measurement of the $Q_{\rm pPb}$ for different centrality classes is biased due to the method of centrality estimation. The least bias is reached with a hybrid approach using the ZNA as described in [3]. No significant centrality-dependent modification of D-meson production is observed in p-Pb collisions with respect to pp collisions as shown in Figure 1. These results are consistent with those obtained for charged hadrons at high $p_{\rm T}$ [3].

ALICE has measured a linear dependence of the self-normalized yields of D mesons in p-Pb collisions at $\sqrt{s_{\rm NN}} = 5.02\,{\rm TeV}$ as a function of charged-particle multiplicity estimated at backward rapidity and a faster-than-linear increase as a function of the charged-particle multiplicity estimated at mid-rapidity [4].

A new measurement of the self-normalized yields of electrons from heavy-flavour hadron decays as a function of the charged particle multiplicity is presented here. The charged-particle multiplicity is estimated in two different rapidity regions (at mid-rapidity and backward rapidity), similar to the analysis of the multiplicity dependence of the self-normalized D-meson

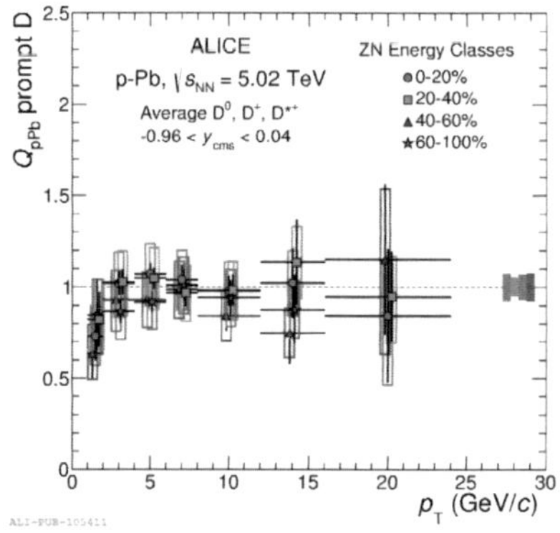

Figure 1. Average Q_{pPb} of D^0, D^+ and D^{*+} mesons as a function of p_{T} and event centrality [3].

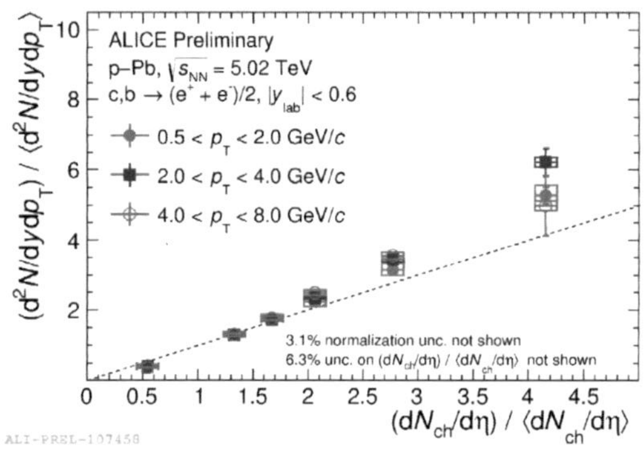

Figure 2. Self-normalized yields of electrons from heavy-flavour hadron decays for different electron p_{T} as a function of the charged-particle multiplicity estimated at mid-rapidity.

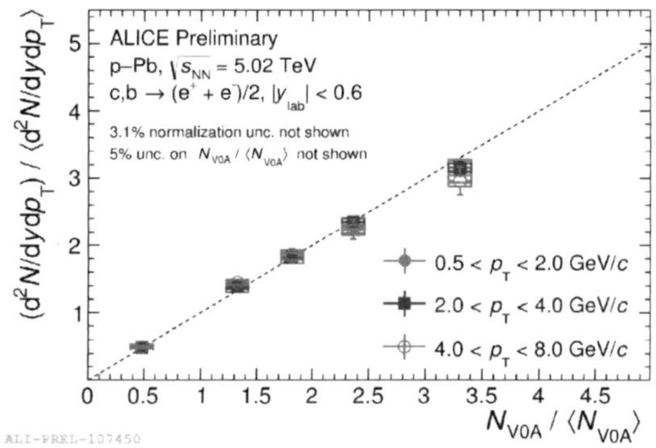

Figure 3. Self-normalized yields of electrons from heavy-flavour hadron decays for different electron p_{T} as a function of the charged-particle multiplicity estimated at backward rapidity.

yields. In Figures 2 and 3 the self-normalized yields of electrons from semi-leptonic heavy-flavour hadron decays are shown as a function of the charged-particle multiplicities estimated at mid-rapidity and backward rapidity, respectively. The self-normalized yields of electrons and D mesons as a function of the charged-particle multiplicity estimated at mid-rapidity are plotted in Figure 4 in different kinematic ranges in each panel to account for the decay kinematics. The self-normalized yields of electrons and D mesons are compatible with each other within their uncertainties and they do not depend on the transverse momentum. Also shown are results from model calculations using EPOS 3 [7] for D mesons considering two approaches, where the simulated initial conditions are evolved using hydrodynamics in one of the cases. The calculations using a hydrodynamic evolution are in good agreement with the faster-than-linear increase of the D-meson measurement, suggesting that collective effects could play a role in particle production in high-multiplicity p-Pb collisions.

The self-normalized yields of electrons include a significant contribution from electrons from beauty-hadron decays for $p_{\mathrm{T}} > 4\,\mathrm{GeV}/c$ [6]. No p_{T} dependence is observed for the self-normalized yields of electrons as shown in Figures 2 and 3. The measurements give a hint that the production mechanisms of charm and beauty as a function of the multiplicity are similar.

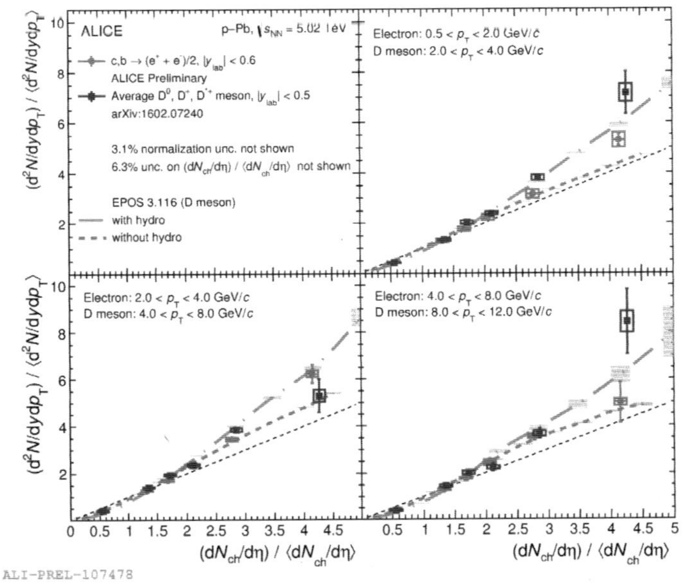

Figure 4. Comparison of self-normalized yields of D mesons averaged over all species and electrons from heavy-flavour hadron decays as a function of the charged-particle multiplicity estimated at mid-rapidity. Also shown are the results of EPOS 3 model calculations [7] without hydrodynamic evolution as a dashed line and including a hydrodynamic evolution shown as a dot-dashed line.

4. Summary

No centrality dependence is observed for the Q_{pPb} of D mesons in p-Pb collisions at $\sqrt{s_{\mathrm{NN}}} =$ 5.02 TeV within the uncertainties of the measurement. ALICE has measured the self-normalized yields of D mesons and electrons from semi-leptonic heavy-flavour hadron decays as a function of the charged-particle multiplicity estimated at mid-rapidity and backward rapidity. A linear increase of the self-normalized yields is observed for D mesons and electrons from semi-leptonic heavy-flavour hadron decays as a function of the charged-particle multiplicity at backward rapidity, compatible with each other in all p_{T} intervals within uncertainties. A trend for a faster-than-linear increase of the self-normalized yields of electrons as a function of the charged-particle multiplicity estimated at mid-rapidity is observed in agreement with the measurement of D mesons [4]. The faster-than-linear increase gives a hint that other effects besides binary scaling, such as MPIs, might play a role in heavy-flavour production in p-Pb collisions. Model calculations with MPIs can explain a faster-than-linear increase of self-normalized D mesons in pp collisions [5]. No significant difference between various p_{T} intervals of the self-normalized yields of electrons has been observed, suggesting similar production mechanisms for charm and beauty quarks as a function of the multiplicity in p-Pb collisions.

Acknowledgments

This work was supported in part by the National Research Foundation of Korea (NRF), Basic Science Research Program through the National Research Foundation of Korea funded by the Ministry of Education, Science and Technology (Grant number: 2013S1A2A2035612).

References

[1] Abelev B B *et al.* (ALICE Collaboration) 2014 *Phys. Rev. Lett.* **113** 232301
[2] Adam J *et al.* (ALICE Collaboration) 2016 *Phys. Lett.* **B754** 81–93
[3] Adam J *et al.* (ALICE Collaboration) 2015 *Phys. Rev.* **C91** 064905
[4] Adam J *et al.* (ALICE Collaboration) 2016 *JHEP* **08** 078
[5] Adam J *et al.* (ALICE Collaboration) 2015 *JHEP* **09** 148
[6] Abelev B *et al.* (ALICE Collaboration) 2013 *Phys. Lett.* **B721** 13–23
[7] Werner K *et al.* 2014 *Phys. Rev.* **C89** 064903

D^0 Meson Production in Heavy Ion Collisions in CMS experiment

Jing Wang for the CMS Collaboration
Massachussetts Institute of Technology, 77 Massachussetts Ave, Cambridge, MA, US
E-mail: jing.wang@cern.ch

Abstract. The measurement of heavy flavour production is a powerful tool to study the properties of the high-density QCD medium created in heavy-ion collisions as heavy quarks are sensitive to the transport properties of the medium and may interact with the QCD matter differently from light quarks. In particular, the comparison between the nuclear modification factors of light and heavy flavoured particles provides insights into the expected flavour dependence of in-medium parton energy loss. With the CMS detector, the D^0 meson production is studied in pp and PbPb collisions at 2.76 and 5.02 TeV. In this talk, the nuclear modification factor of D^0 meson are presented and compared to the charged hadron nuclear modification factor and theoretical calculations.

1. Introduction

Heavy quarks are considered to be produced mainly via initial hard scatterings in heavy ion collisions. Therefore, they propagate through the medium and carry information about the system at early stage. That makes heavy flavor hadrons good probes of the properties of QCD matter and important tools to study the mechanism of interaction with the matter. The interactions with the medium, the energy loss of a parton, happens via elastic and inelastic collisions, and via gluon brhemstrahlung. At small angles, the radiative energy loss is thought to be suppressed, the manifestation of the 'dead cone effect' [1]. If only this effect would be dominant, the nuclear modification factor of heavy flavor hadrons should be larger (less suppression) than that of light hadrons. This effect is also expected to have a strong p_T dependence, with the highest contribution at low transverse momentum of the heavy meson.

The measurements of the D^0 meson production were conducted by the Compact Muon Solenoid (CMS) [2] detector in PbPb collisions at 2.76 TeV [3] and 5.02 TeV [4]. The talk presented the analysis technique for extracting the D^0 signal, and the D^0 nuclear modification factor

$$R_{AA} = \frac{1}{T_{AA}} \frac{dN^{AA}/dp_T}{d\sigma^{pp}/dp_T} \tag{1}$$

results.

2. Analysis technique

Prompt D^0 production in PbPb collisions at 2.76 TeV was studied with CMS detector during LHC Run I in 2011, and that in both pp and PbPb collisions at 5.02 TeV was studied during Run II in 2015. The decay channel $D^0 \rightarrow K\pi$ is analyzed in the rapidity range $|y| < 1$. A detailed description of the CMS experiment can be found in Ref. [2]. Specifically, the silicon tracker, located in the 3.8 T magnetic field of the superconducting solenoid, is used to measure charged particles within the pseudorapidity range $|\eta| < 2.5$.

For 2.76 TeV, events selected online by a minimum bias trigger are used for the analysis. For 5.02 TeV, minimum bias events (about 2 billion events for pp and 150 million events for PbPb) are used for the results at low p_T (< 20 GeV/c), while events selected by a dedicated D^0 High-Level-Trigger (HLT) are used for the result at $p_T > 20$ GeV/c. The D^0 HLT path starts from events tagged by energy deposited in the calorimeters. A full track reconstruction is performed for these events, and pairs of tracks with $p_T > 2$ GeV/c (pp) or 8 GeV/c (PbPb) are analyzed. An event is selected if the pair passed a loose selection on their common vertex displacement. This D^0 trigger worked well and allows reaching very high p_T D^0.

Offline, D^0 candidates are reconstructed by combining pairs of oppositely charged particles and requiring an invariant mass within 0.2 GeV/c^2 of the nominal D^0 mass, as given by the particle data group (PDG) [5]. Topological selections based on pointing angle, 3D decay length normalized by its error and D^0 vertex probability are applied to reduce combinatorial background. The pointing angle is defined as the angle between D^0 momentum and the vector from primary vertex and secondary vertex in transverse plain.

Figure 1 shows the invariant mass distribution for selected D candidates, for $5 < p_T < 6$ GeV/c, for pp and PbPb collisions at 5.02 TeV. The signal shape is modeled by two Gaussians with same mean but different widths. The combinatorial background, generated by randomly combining pairs of tracks not originating from a D^0 meson decay, is modeled by a third order polynomial. With the choice of not using particle identification, one track is assigned the mass of a K, and the other the mass of a π. An additional Gaussian function is used to describe the invariant mass shape of D^0 candidates with incorrect mass assignment from the exchange of K and π designation.

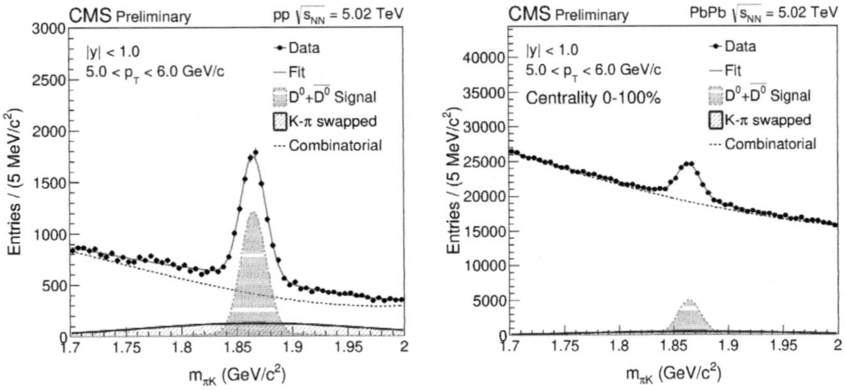

Figure 1. Invariant mass distributions of D^0 candidates and their charge conjugates for selected p_T intervals of pp (left) and PbPb (right) at 5.02 TeV in centrality 0-100%

The raw yields are then corrected for detector acceptance and reconstruction efficiency effects, calculated using PYTHIA, and PYTHIA events embedded in HYDJET PbPb events. In addition, the fraction of D^0 comming from c-quark fragmentation, are determined by fitting data distribution of the distance of closest approach (DCA) of D^0, with a prompt and nonprompt shape from Monte Carlo.

Systematic uncertainties from various sources are taken into account. Systematic uncertainty from signal extraction is estimated by varying signal and background fitting function. Systematic uncertainty from D^0 meson selection is obtained by comparing data and MC driven efficiencies varying cut values. Tracking efficiency systematic uncertainty is estimated by 2- and 4- prongs decays of D^0. Systematic uncertainty from prompt fractions is obtained by comparing the results of default method and an alternative method based on FONLL predictions.

3. Results

Figure 2 shows the prompt D^0 R_{AA} in PbPb collisions at 2.76 TeV as a function of p_T for centrality 0-20% and the comparison with the result of ALICE [6]. The pp reference used for this result was built using both FONLL [7] and data extrapolation: in the range $p_T < 16$ GeV/c range, the ALICE collaboration measurement of prompt D^0 cross-section at 7 TeV [8] was rescaled to 2.76 TeV using FONLL input, following the same procedure used in Ref. [9].

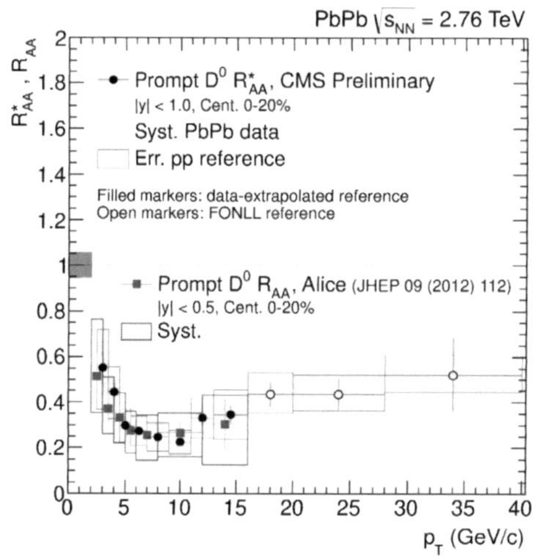

Figure 2. Prompt D^0 R_{AA} as a function of p_T for centrality 0-20% in PbPb collisions at 2.76 TeV by CMS collaboration. The results from ALICE collaboration are also shown for comparison [6]

Prompt D^0 R_{AA} as a function of p_T in PbPb collisions at 5.02 TeV for centrality classes 0-100% (left) and 0-10% (right) are shown in Figure 3. The measurements are compared with predictions from some theoretical calculations [10, 11, 12, 13, 14, 15, 16, 17, 18], and most of them describe data well. Comparison with charge particles R_{AA} in PbPb collisions at 5.02 TeV in the same centrality classes [19] is also shown in Figure 3.

4. Summary

The measurements of prompt D^0 nuclear modification factors in 5.02 TeV PbPb collisions from CMS collaboration are presented. D^0 production presents strong suppression. No significant difference of R_{AA} is observed between D^0 and charged hadrons, in the kinematic region studied. These results put strong constraints on theoretical calculations in a wide kinematic range. With LHC Run II data, there are various exciting studies concerning heavy flavors. For c quark related studies, there is an effort to reach prompt D^0 R_{AA} at very low p_T, and D^0 v_n measurements are

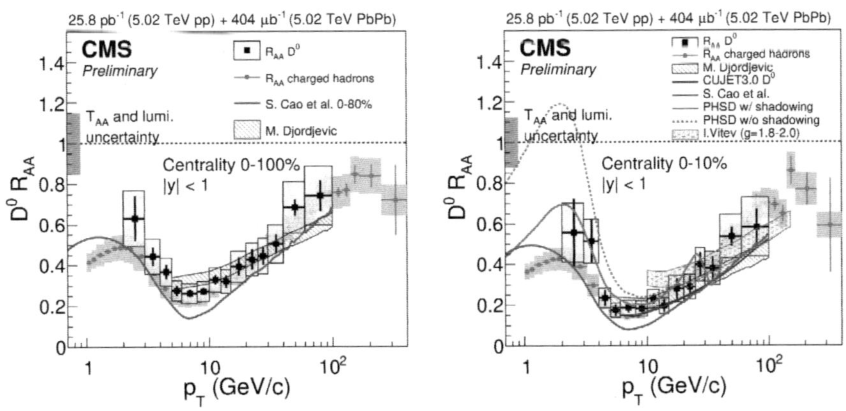

Figure 3. Prompt D^0 R_{AA} as function of p_T in PbPb collisions at 5.02 TeV for centrality classes 0-100% (left) and 0-10% (right)

also conducted. Also, the results of measurements of B mesons nuclear modification factor are expected to come soon.

References

[1] Armesto N, Salgado C A and Wiedemann U A 2004 *Phys. Rev. D* **69**(11) 114003 URL `http://link.aps.org/doi/10.1103/PhysRevD.69.114003`

[2] Chatrchyan S *et al.* (CMS Collaboration) 2008 *JINST* **3** S08004

[3] 2015 Nuclear Modification Factor of prompt D^0 in PbPb Collisions at $\sqrt{s_{\rm NN}} = 2.76$ TeV Tech. Rep. CMS-PAS-HIN-15-005 CERN Geneva URL `http://cds.cern.ch/record/2055466`

[4] 2016 D^0 meson nuclear modification factor in PbPb collisions at $\sqrt{s_{\rm NN}} = 5.02$ TeV Tech. Rep. CMS-PAS-HIN-16-001 CERN Geneva URL `http://cds.cern.ch/record/2157844`

[5] Beringer J *et al.* (Particle Data Group) 2012 *Phys.Rev.* **D86** 010001

[6] Abalev B e a (ALICE Collaboration) 2012 *JHEP* **09** 112 (*Preprint* `arXiv:nucl-ex/1203.2160`)

[7] Cacciari M, Greco M and Nason P 1998 *Journal of High Energy Physics* **1998** 007 URL `http://stacks.iop.org/1126-6708/1998/i=05/a=007`

[8] Abelev B *et al.* (ALICE Collaboration) 2012 *JHEP* **1201** 128 (*Preprint* `arXiv:hep-ex/1111.1553`)

[9] Averbeck R, Bastid N, del Valle Z C, Crochet P, Dainese A and Zhang X 2011 (*Preprint* `arXiv:hep-ph/1107.3243`)

[10] Xu J, Liao J and Gyulassy M 2016 *JHEP* **02** 169 (*Preprint* `arXiv:hep-ph/1508.00552`)

[11] Xu J, Buzzatti A and Gyulassy M 2014 *JHEP* **08** 063 (*Preprint* `arXiv:hep-ph/1402.2956`)

[12] Jiechen X, Jinfeng L and Gyulassy M 2015 *Chinese Physics Letters* **32** 092501 URL `http://stacks.iop.org/0256-307X/32/i=9/a=092501`

[13] Djordjevic M and Djordjevic M 2015 *Phys. Rev. C* **92**(2) 024918 URL `http://link.aps.org/doi/10.1103/PhysRevC.92.024918`

[14] Cao S, Luo T, Qin G Y and Wang X N (*Preprint* `arXiv:nucl-th/1605.06447v1`)

[15] Song T, Berrehrah H, Cabrera D, Torres-Rincon J M, Tolos L, Cassing W and Bratkovskaya E 2015 *Phys. Rev. C* **92**(1) 014910 URL `http://link.aps.org/doi/10.1103/PhysRevC.92.014910`

[16] Song T, Berrehrah H, Cabrera D, Cassing W and Bratkovskaya E 2016 *Phys. Rev. C* **93**(3) 034906 URL `http://link.aps.org/doi/10.1103/PhysRevC.93.034906`

[17] Kang Z B, Lashof-Regas R, Ovanesyan G, Saad P and Vitev I 2015 *Phys. Rev. Lett.* **114**(9) 092002 URL `http://link.aps.org/doi/10.1103/PhysRevLett.114.092002`

[18] Chien Y T, Emerman A, Kang Z B, Ovanesyan G and Vitev I 2016 *Phys. Rev. D* **93**(7) 074030 URL `http://link.aps.org/doi/10.1103/PhysRevD.93.074030`

[19] 2016 Measurement of the charged particle nuclear modification factor in PbPb collisions at $\sqrt{s_{\rm NN}} = 5.02$ TeV Tech. Rep. CMS-PAS-HIN-15-015 CERN Geneva URL `https://cds.cern.ch/record/2155301`

IOP Conf. Series: Journal of Physics: Conf. Series **779** (2017) 012027 doi:10.1088/1742-6596/779/1/012027

Recent Heavy-Flavor Results Utilizing the FVTX in PHENIX at RHIC

Xuan Li[1] for the PHENIX Collaboration
1) P-25, MS H846, Physics Divison, LANL, Los Alamos, NM, 87545, USA
E-mail: xuanlipx@rcf.rhic.bnl.gov

Abstract. Heavy-flavor and quarkona production are important hard probes to test Quantum Chromodynamics (QCD) and measure the properties of the Quark Gluon Plasma (QGP) created in high energy heavy ion collisions. The new PHENIX mid- and forward-rapidity silicon vertex detectors provide precise Distance of Closest Approach (DCA) determination, allowing charm and bottom yields identification, and improve di-muon mass resolution for charmonia studies. The relative $\psi(2S)$ over $J/\psi(1\text{S})$ yields measured in p+p, p+Al and p+Au data at $\sqrt{s} = 200$ GeV help us to understand final state effects on the charmonia production. Direct access to B-meson production has been enabled via measurements of non-prompt $J/\psi \rightarrow \mu^+ + \mu^-$ in p+p collisions at $\sqrt{s} = 510$ GeV and Cu+Au collisions at $\sqrt{s} = 200$ GeV. The fraction of B-meson decays in J/ψ yields in the p+p measurement agrees with higher energy results from Tevatron and the LHC. This fraction is enhanced in Cu+Au collisions suggesting the B-meson yield is not modified by the medium whereas J/ψ is strongly suppressed.

1. Introduction

Because of their high mass ($m_{\rm c,b} \gg \Lambda_{\rm QCD}$), heavy quarks are produced in the early stage of high energy collisions. Since the heavy-flavor quarks do not vanish or change into other flavors during the hard scattering and fragmentation processes, they can be treated as hard probes to study the full evolution of the medium created by relativistic heavy ion collisions. Heavy-flavor particles that traverse the medium in heavy ion collisions experience both cold and hot nuclear matter effects. To understand the hot nuclear matter effects in heavy ion collisions, it is necessary to study the same measurements in p+p and p+A collisions as well.

The Forward Silicon Vertex Tracker (FVTX) [1] installed in PHENIX since 2012 at RHIC dramatically improves the tracking quality in the rapidity range of $1.2 < |y| < 2.2$, with full azimuthal angle coverage. Better mass resolution of di-muon pairs reconstructed through the FVTX now allows separation of $\psi(2S)$ from $J/\psi(1\text{S})$ in the FVTX rapidity range. The mid-rapidity $\psi(2S)$ measured in $\sqrt{s} = 200$ GeV d+Au collisions is more suppressed compared with the J/ψ measurement in the same data set [2]. This result suggests that the final state effect is not negligible, as both J/ψ and $\psi(2S)$ are expected to undergo the same initial state effects. Extension to forward/backward rapidity measurements will further establish how final state effects impact the charmonia with different bound state energies.

The PHENIX mid-rapidity charm and bottom separated single electron $R_{\rm AA}$ measured in 200 GeV Au+Au collisions suggests that bottom production is less suppressed than charm production in the electron $p_{\rm T} < 4$ GeV/c region [3]. This measurement is consistent with models, which expect mass dependency of quark energy loss [4] and flavor dependent dissociation and formation time [5]. D and B mesons measured in the forward/backward rapidity may probe a different QGP evolution stage in heavy ion collisions from the mid-rapidity region. Decay particles with a non-zero displaced vertex can be identified with the help of the FVTX. Direct access to B-mesons in the forward/backward rapidity is first studied at PHENIX in both 510 GeV p+p collisions and 200 GeV Cu+Au collisions through the B $\rightarrow J/\psi \rightarrow \mu^+\mu^-$ decay channel with the branching ratio around 5×10^{-5}.

2. The double ratio of $\psi(2S)$ over J/ψ measured in p+p, p+Al and p+Au collisions at $\sqrt{s}$ = 200 GeV

Yields of J/ψ and $\psi(2S)$ have been measured in p+p, p+Al and p+Au collisions at $\sqrt{s} = 200$ GeV from the RHIC 2015 run. The double ratio of $\psi(2S)$ to J/ψ, which is defined as the ratio of $\psi(2S)$ yields over J/ψ yields measured in p+Al and p+Au collisions over the same ratio measured in p+p collisions, is used to characterize relative nuclear modifications between $\psi(2S)$ and J/ψ.

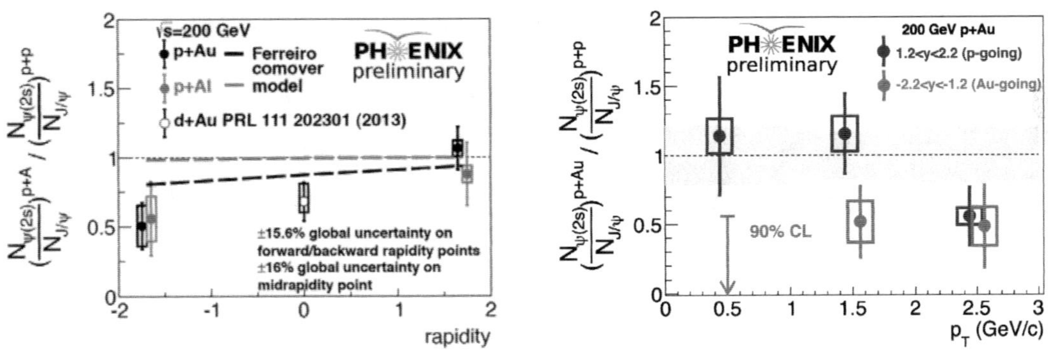

Figure 1. Left: The rapidity dependent double ratio of $\psi(2S)$ to J/ψ measured between p+Au, p+Al and p+p collisions at 200 GeV in comparison with the co-mover model calculation [6]. Right: The p_T dependent double ratio of $\psi(2S)$ to J/ψ measured in p+Au collisions at 200 GeV.

The forward rapidity (p-going) $\psi(2S)$ to J/ψ double ratios measured in both p+Au and p+Al collisions are near unity. This means $\psi(2S)$ to J/ψ have similar nuclear modifications in the proton going direction, where particle densities are smaller. In the backward rapidity, significant decreasing of the $\psi(2S)$ to J/ψ double ratios is measured in p+Au and p+Al collisions indicating the loosely bounded $\psi(2S)$ is suppressed more than J/ψ in the Au/Al going direction, where particle densities are larger. Comparison with co-mover model calculations [6] are shown in the left panel of Figure 1. Reasonable agreement between data and the co-mover model on the rapidity dependence means the charmonia break up with co-mover mechanism might be the dominant contributor to the $\psi(2S)$ and J/ψ differential suppression. The p_T dependent $\psi(2S)$ to J/ψ double ratio is studied in p+Au collisions as shown in the right panel of Figure 1. Absence of low p_T $\psi(2S)$ in the backward rapidity may be caused by the longer time interaction with soft co-movers.

3. The fraction of J/ψ from B-meson decay in $\sqrt{s}$ = 510 GeV p+p collisions and $\sqrt{s}$ = 200 GeV Cu+Au collisions

The fraction of J/ψ from B-meson decay is first studied in the RHIC 2012 510 GeV p+p collisions and 200 GeV Cu+Au collisions. Identification of J/ψ from B-meson decay at forward/backward rapidity $(1,2 < |y| < 2.2)$ relies on the Distance of Closet Approach (DCA) measurement with the FVTX and VTX detectors at PHENIX. Prompt particles (such as prompt J/ψ) show up as symmetric DCA distributions, while the DCA distribution of long lived decay particles (such as B-meson decays) produces a long tail for negative DCA and an asymmetric distribution.

The analysis use muons from unlike-sign di-muons in the J/ψ mass region. The DCA shapes of muons which decay from prompt J/ψ and B-meson decayed J/ψ are determined from a PYTHIA+GEANT+reconstruction simulation and samples of identified prompt hadrons. Combinatorial, detector mis-matching and heavy-flavor continuum backgrounds are determined separately using real data. A fit of the DCA distribution can simultaneously determine the prompt J/ψ and non-prompt J/ψ from B-meson decays.

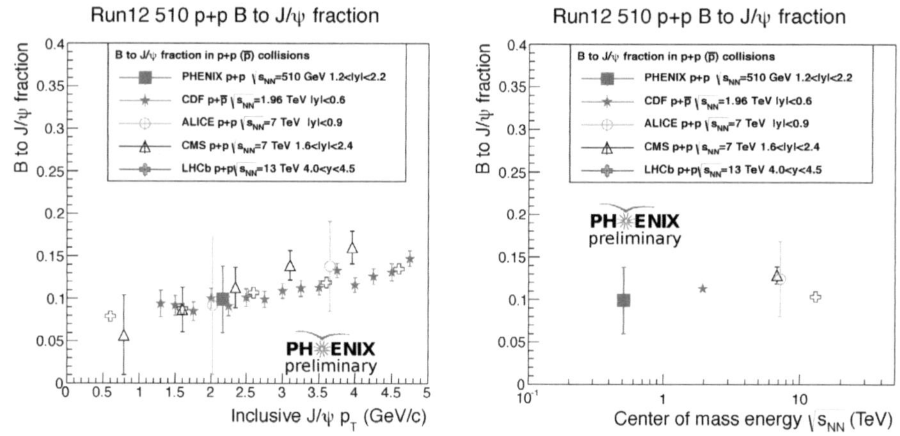

Figure 2. The fraction of J/ψ from B-meson decays measured in 510 GeV $p{+}p$ collisions at PHENIX and comparison with the global data from CDF [7], ALICE [8], CMS [9] and LHCb [10] experiments versus inclusive J/ψ p_{T} (left) and versus center of mass energy (right).

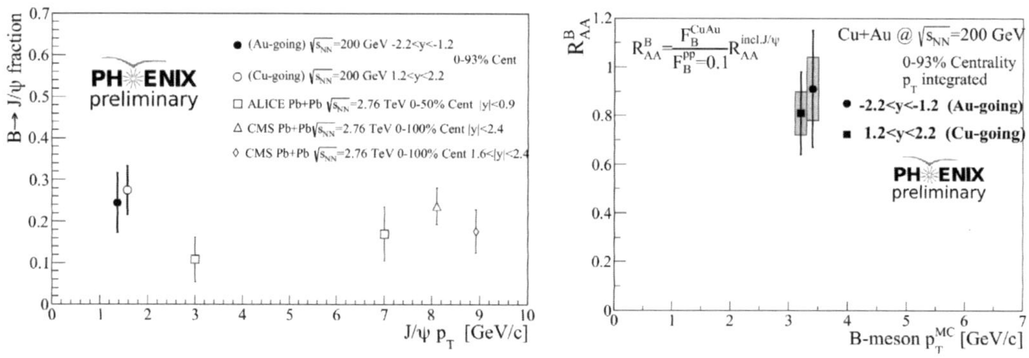

Figure 3. Left: The fraction of J/ψ from B-meson decay versus inclusive J/ψ p_{T} measured in 200 GeV Cu+Au collisions in comparison with the results measured in 2.76 TeV Pb+Pb collisions from the ALICE [11] and CMS [12] experiments.

The detector acceptance*efficiency corrected fraction of J/ψ from B-meson decay measured in 510 GeV $p{+}p$ collisions is shown in Figure 2. This result is compared with other measurements within the same inclusive J/ψ p_T region ($p_T < 5$ GeV/c) at different rapidities and center of mass energies. The PHENIX preliminary result follows the p_T dependence of the global data. The B-meson decayed J/ψ fraction does not have a strong energy dependence in p+p or p+$\bar{p}$ collisions for the J/ψ $p_T < 5$ GeV/c region.

Figure 3 shows the PHENIX preliminary result of J/ψ from B-meson decay fraction in Cu+Au collisions. The fractions in the Cu-going direction and Au-going direction are consistent with each other within large uncertainties. The J/ψ from B-meson decay fraction measured by PHENIX in Cu+Au collisions at $\sqrt{s_{NN}} = 200$ GeV is larger than results measured at the LHC in 2.76 TeV Pb+Pb collisions, indicating that B-mesons have weaker nuclear modification compared to the one from J/ψ at RHIC energy. To better understand this, the nuclear modification factor R_{CuAu} of J/ψ from B-mesons decay is determined. As the 200 GeV p+p analysis is ongoing, we assume the fraction is 0.1 in 200 GeV p+p collisions according to the scaling behavior shown in the right panel of Figure 2. We calculate the R_{CuAu} of J/ψ from B-mesons decay (shown in the right panel of Figure 3) by applying the measured B-meson decay J/ψ fraction in Cu+Au collisions and the assumed value in p+p collisions to the measured inclusive J/ψ R_{CuAu}. The result is consistent with no nuclear modification of momentum and centrality integrated B-meson yields. The FVTX detector provides almost uniform acceptance for a broad momentum range of B-mesons starting from zero p_T. This measurement is dominated by low p_T B-mesons.

4. Summary and Outlook

The differential suppression of the $\psi(2S)$ to J/ψ double ratio in 200 GeV p+Al and p+Au data shows strong evidence of final state effects in the backward rapidities. The co-mover model provides a reasonable description. First measurements of J/ψ from B-meson decay in forward/backward rapidity at low p_T are achieved for both 510 GeV p+p and 200 GeV Cu+Au collisions at PHENIX. No significant center of mass energy dependence for low p_T J/ψ from B-meson decay in p+p collisions is observed for energies of 510 GeV and above. No significant nuclear modification of integrated p_T and centrality B-mesons is observed in 200 GeV Cu+Au data, which is consistent with a hard scattering process and no additional production in the medium.

Large data sets in various types of heavy ion collisions collected by PHENIX provide opportunities to further study the heavy-flavor productions in forward/backward rapidities. The $\psi(2S)$ to J/ψ double ratio in 200 GeV Au+Au collisions will access the hot nuclear matter effects such as color screening in the QGP. The method to study the DCA of muons in the B-meson decays to J/ψ fraction analysis will be applied to other data sets and to the study of B- and D-meson semi-leptonic decays to muons in the forward/backward rapidities. Studies in p+Au and Au+Au collisions will improve the understanding of the cold and hot nuclear matter effects.

5. References

[1] Aidala C *et al.* 2014 *Nuclear Instruments and Methods in Physics Research Section A: Accelerators, Spectrometers, Detectors and Associated Equipment* **755** 44 – 61
[2] Adare A *et al.* 2013 *Phys. Rev. Lett.* **111** 202301
[3] Adare A *et al.* 2016 *Phys. Rev. C* **93** 034904
[4] Dokshitzer Y and Kharzeev D 2001 *Phys. Lett. B* **519** 199 – 206
[5] Adil A and Vitev I 2007 *Phys. Lett. B* **649** 139–146
[6] Ferreiro E G 2015 *Phys. Lett. B* **749** 98–103
[7] Acosta D *et al.* 2005 *Phys. Rev.* **D71** 032001
[8] Abelev B *et al.* 2012 *JHEP* **11** 065
[9] Khachatryan V *et al.* 2011 *Eur. Phys. J.* **C71** 1575
[10] Aaij R *et al.* 2015 *JHEP* **10** 172
[11] Adam J *et al.* 2015 *JHEP* **07** 051
[12] Chatrchyan S *et al.* 2012 *JHEP* **05** 063

IOP Conf. Series: Journal of Physics: Conf. Series **779** (2017) 012028 doi:10.1088/1742-6596/779/1/012028

$D_s^{\pm}$ meson production in Au+Au collisions at $\sqrt{s_{\rm NN}}$ = 200 GeV in STAR

Long Zhou (for the STAR collaboration)
#96 Jinzai road, University of Science and Technology of China, Hefei, Anhui, China
E-mail: zhoulong@mail.ustc.edu.cn

Abstract. In this article, we report the measurements of the nuclear modification factor (R_{AA}) and elliptic flow (v_2) for $D_s^{\pm}$ in Au+Au collisions at $\sqrt{s_{\rm NN}}$ = 200 GeV from the STAR experiment. These results are compared with the results of other open charm mesons to study the hadronization mechanism of the charm quarks and disentangle the transport properties of quark-gluon plasma and hadronic phase [1]. We found that the nuclear modification factor for D_s are systematically higher than unity and D^0 R_{AA}. The ratio of D_s/D^0 for 10-40% central Au+Au collisions is also higher than that in p+p collisions as predicted by PYTHIA. The D_s/D^0 ratio is also compared to that in Pb+Pb collisions at $\sqrt{s_{\rm NN}}$ = 2.76 TeV measured by the ALICE experiment. Our results indicate an enhancement of D_s meson production in Au+Au collisions.

1. Introduction
Heavy quarks are a unique probe to study the medium produced in ultra-relativistic heavy-ion collisions. Because of the large masses, their production is dominated by initial hard scatterings and is well described by perturbative QCD (pQCD) [2]. In particular the $D_s(c\bar{s}/\bar{c}s)$ production in heavy-ion collisions is expected to be affected by both the strangeness enhancement and the primordial charm quark production. The modifications of D_s meson spectrum and its elliptic flow(v_2) in ultra-relativistic heavy-ion collisions are identified as a quantitative probe of key properties of the hot nuclear medium [1]. Theoretical calculations predict a remarkable enhancement of the D_s meson production as the result of a strong charm quark coupling to the Quark-Gluon Plasma (QGP) and subsequent recombination with equilibrated strange quarks when compared to non-strange D mesons [1]. Like multi-strange hadrons, D_s mesons are expected to freeze out at pseudo-critical temperature (T_{pc}), while the non-strange D meson picks up significant additional v_2 during the hadronic phase. Therefore, the elliptic flow of D_s meson is considered to be a better measure of the partonic contribution to the charm hadron v_2 than that of D^0 or $D^{\pm}$ [1].

The Heavy Flavor Tracker (HFT), installed at the STAR experiment since 2014, is designed to extend STAR's capability of measuring open charm mesons via the topological reconstruction of displaced decay vertices. The HFT detector consists of 4 layers of silicon detectors. The outer layer is the Silicon Strip Detector (SSD). The Intermediate Silicon Tracker (IST), consisting of one layer of single-sided strips, is located inside the SSD. Two layers of the Silicon Pixel Detector (PXL) are inside the IST. The PXL detectors have the resolution necessary for a precise measurement of the displaced vertices. The track pointing resolution of the HFT detector is about 46 μm for kaons at 750 MeV/c [3].

2. Data Analysis

About 750 million minimum bias events taken during the 2014 Au+Au run at $\sqrt{s_{\rm NN}} = 200$ GeV by the STAR experiment were used in this analysis. The identification of daughter particles was done at mid-rapidity $|y| < 1$ using the Time Projection Chamber (TPC) and the Time-of-Flight (TOF) detector. The HFT detector is used to reconstruct the decay vertices. We reconstruct D_s meson through the hadronic decay channel: $D_s \to \phi(1020) + \pi \to K^+ + K^- + \pi$, whose branching ratio is $(2.27 \pm 0.08)\%$. Topological and kinematic cuts are applied to reduce the combinatorial background. Figure 1 shows the reconstructed D_s signal for 0-80% centrality.

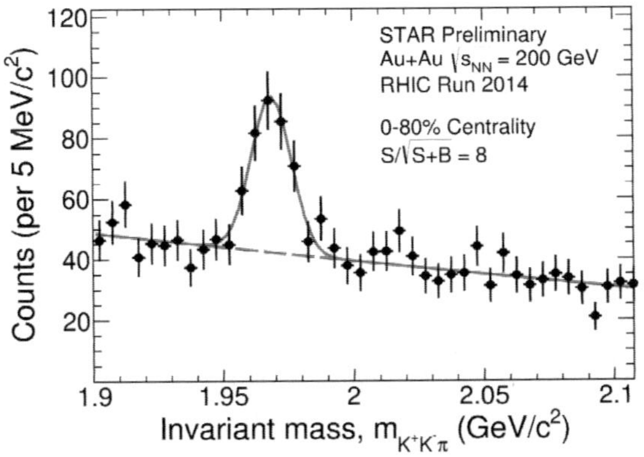

Figure 1. (Color online) The invariant mass distribution for D_s meson in 0-80% central Au+Au collisions at $\sqrt{s_{\rm NN}} = 200$ GeV

3. Physics Results

The invariant yield of D_s (average yield of D_s^+ and D_s^-) are calculated in two p_T bins in 10-40% central Au+Au collisions. The results are depicted in Fig. 2, where the invariant yield of D^0 from the same centrality bin [4] is also shown. The D_s/D^0 ratio as a function of p_T is shown in Fig. 3 for 10-40% central Au+Au collisions at $\sqrt{s_{\rm NN}} = 200$ GeV, along with a similar ALICE measurement for 0-10% central Pb+Pb collisions at $\sqrt{s_{\rm NN}} = 2.76$ TeV [6]. The two measurements follow the same trend within the large uncertainties. On the other hand, the Au+Au measurement is systematically higher than the ratio in p+p collisions predicted by PYTHIA [5].

Medium effects on D_s meson production are quantified as the nuclear modification factor, i.e. $R_{AA}(p_T) = (dN_{AA}/dp_T)/(N_{coll}dN_{pp}/dp_T)$, where dN_{AA}/dp_T is the differential yield in Au+Au collisions and dN_{pp}/dp_T is the corresponding yield in p+p collisions. N_{coll} is the number of binary collisions, derived from the Glauber model [15]. Since there is no measurement of D_s spectrum in p+p collisions at $\sqrt{s} = 200$ GeV, the charm quark cross-section measured by the STAR experiment in p+p collisions at the same collision energy [7] was used to derive the p_T spectrum for D_s meson. The fragmentation factor ($c \to D_s$) used to convert the charm cross-section to D_s cross-section is 0.09 ± 0.01 [8, 9, 10]. The R_{AA} of D_s at mid-rapidity ($|y| < 1.0$) in Au+Au collisions at $\sqrt{s_{\rm NN}} = 200$ GeV for 10-40% centrality bin is presented in Fig. 5. Model calculations of R_{AA} for D_s(red band) and D^0(purple dashed-dotted lines) [1] are also shown in Fig. 5. STAR measurements are consistent with model calculations within uncertainties. The R_{AA} of D_s is systematically higher than that of D^0 [4] but the enhancement is not statistically significant.

The elliptic flow can be used to probe the dynamics of early stages of heavy-ion collisions [11]. In Fig. 6 , the measurement of D_s v_2 for 0-80% centrality bin in Au+Au collisions at $\sqrt{s_{\rm NN}} = 200$ GeV is shown. Within large uncertainties, the D_s meson v_2 is consistent with v_2 of D^0 [13] and ϕ [14].

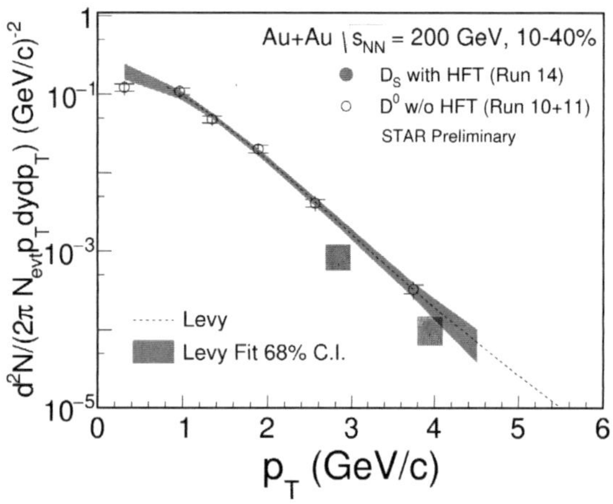

Figure 2. (Color online) The invariant yields of D_s in 10-40% central Au+Au collisions. The solid boxes represent the systematic uncertainties, while the vertical bars depict the statistical errors.

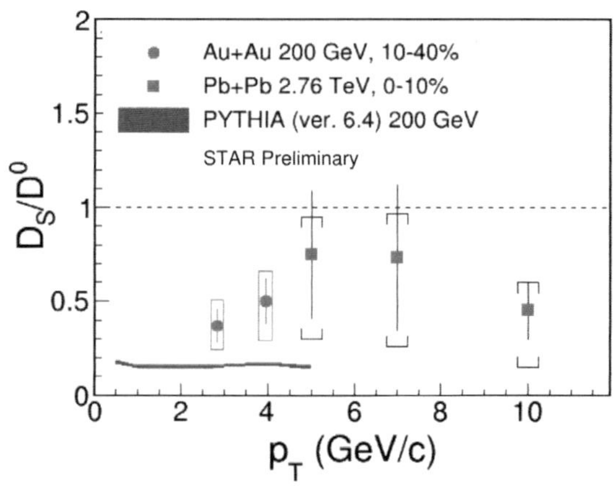

Figure 3. (Color online) The D_s/D^0 ratio in 10-40% Au+Au collisions at $\sqrt{s_{\rm NN}} = 200$ GeV(circle) in the range of $1.5 < p_T < 5.0$ GeV/c. A similar measurement in 0-10% central Pb+Pb collisions at $\sqrt{s_{\rm NN}} = 2.76$ TeV (squares) from the ALICE collaboration is shown for comparison.

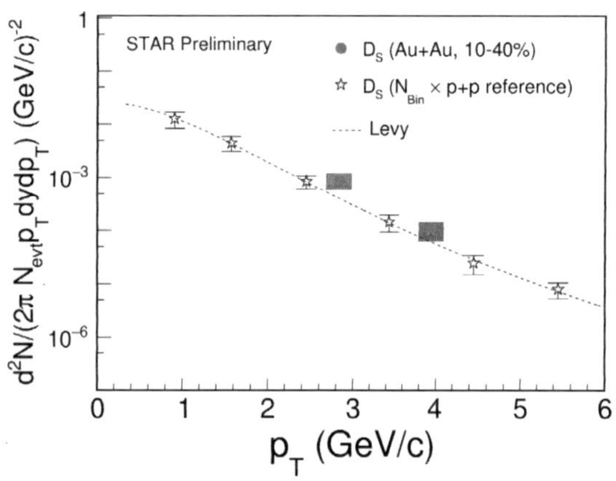

Figure 4. (Color online) The invariant yields of D_s in 10-40% central Au+Au collisions, compared to the scaled p+p reference

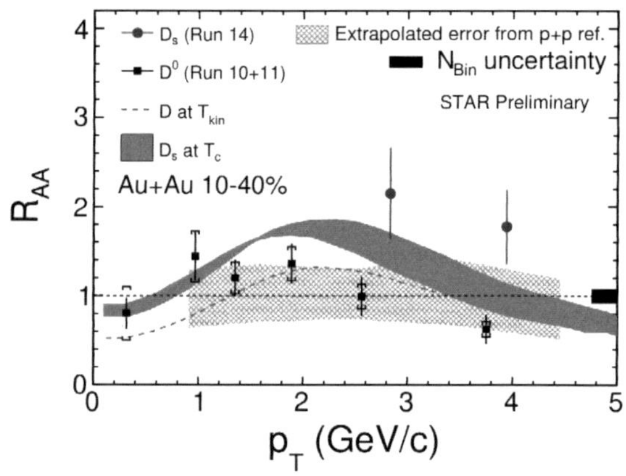

Figure 5. (Color online) The R_{AA} of D_s as a function of p_T in 10-40% central Au+Au collisions. Green band represents the extrapolated error from reference D_s spectra in p + p collisions.

4. Summary

We present the first measurement of D_s production in Au+Au collisions at $\sqrt{s_{\rm NN}} = 200$ GeV using the newly installed HFT at the STAR experiment. The D_s/D^0 ratio seems to be higher than that in p+p collisions predicted by PYTHIA at the intermediate p_T range. The nuclear modification factor of D_s are consistent with model calculations, and systematically higher than D^0 R_{AA}, indicating an enhancement of D_s production in Au+Au collisions. The R_{AA} of D_s is equal to 2.1 $\pm$ 0.5 (stat.) $\pm^{0.7}_{0.7}$ (sys.) and 1.7 $\pm$ 0.4 (stat.) $\pm^{0.5}_{0.7}$ (sys.) at p_T = 2.8 and 3.9

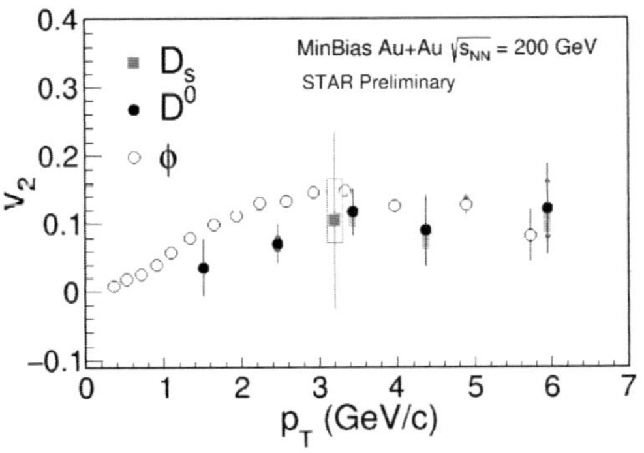

Figure 6. (Color online) Elliptic flow for D_s, D^0 [13] and ϕ [14] for 0-80% centrality bin in Au + Au collisions at $\sqrt{s_{NN}} = 200$ GeV. Open boxes and cap symbols are the systematic uncertainties on corresponding data points. Shaded grey band on D^0 v_2 is for non-flow estimation. Vertical lines are statistical errors.

GeV/c respectively. We have also presented the first measurement of D_s meson v_2 in heavy-ion collisions.

References

[1] He M, Fries R J, and Rapp R 2013 *Phys. Rev. Lett.* **110**, 112301
[2] Cacciari M, Nason P and Vogt R 2005 *Phys. Rev. Lett.* **95**, 122001
[3] Contin G(for the STAR Collaboration) QM2015 proceedings
[4] Adamczyk L et al. (STAR Collaboration) 2014 *Phys. Rev. Lett.* **113**, 142301
[5] Sjostrand T, Mrenna S and Skands P 2006 *JHEP* **0605**, 026
[6] Adam J et al. (ALICE Collaboration) 2016 *JHEP* **03**, 082
[7] Adamczyk L et al. (STAR Collaboration) 2012 *Phys. Rev. D* **86**, 072013
[8] Chekano S et al. (ZEUS Collaboration) 2005 *Eur. Phys. J. C* **44**, 351
[9] Aktas A et al. (H1 Collaboration) 2005 *Eur. Phys. J. C* **38**, 447
[10] Abramowicz H, Abt I, Adamczyk L et al. 2013 *JHEP* **09**, 058
[11] Kolb P F et al. 2003 *Nucl. Phys. A* **715**, 653
[12] Poskanzer A M and Voloshin S. A. 1998 *Phys. Rev. C* **58**, 1671
[13] Lomnitz M (for the STAR Collaboration) 2016 **arXiv 1601.00743**
[14] Adamczyk L et al. (STAR Collaboration) 2015 **arXiv 1507.05247**
[15] Adler C et al. 2002 *Phys. Rev. Lett.* **89** 202301

IOP Conf. Series: Journal of Physics: Conf. Series **779** (2017) 012029 doi:10.1088/1742-6596/779/1/012029

Heavy flavor production in heavy-ion collisions from soft collinear effective theory

Zhong-Bo Kang, Felix Ringer, Ivan Vitev

Theoretical Division, Los Alamos National Laboratory, Los Alamos, NM 87545, USA

E-mail: zkang@lanl.gov, f.ringer@lanl.gov, ivitev@lanl.gov

Abstract. We review the approach to calculate open heavy flavor production in heavy-ion collisions based on Soft Collinear Effective Theory (SCET). We include both finite heavy quark masses in the SCET Lagrangian as well as Glauber gluons that describe the interaction of collinear partons with the hot and dense QCD medium. From the new effective field theory, we derive massive in-medium splitting kernels and we propose a new framework for including in-medium interactions consistent with next-to-leading order calculations in QCD. We present numerical results for the suppression of both D- and B-mesons and compare to results obtained within the traditional approach to parton energy loss. We find good agreement when comparing to existing data from the LHC at $\sqrt{s_{\rm NN}} = 5.02$ TeV and 2.76 TeV.

1. Introduction

The quark-gluon plasma (QGP) predicted to have existed in the early universe can be reproduced in heavy-ion collisions at RHIC and the LHC. Highly energetic particles and jets that traverse the hot and dense QCD medium provide an ideal probe for this new state of matter. In particular, open heavy flavor production plays a crucial role in probing and understanding the QGP and has received a growing attention by both the experimental and theoretical communities in the past years. The LHC experiments CMS and ALICE have provided high precision data, see e.g. [1, 2, 3], for the nuclear modification factor R_{AA} most commonly used to study the quenching of particle yields in heavy-ion collisions. Interestingly, preliminary results from the CMS collaboration [3] for Pb+Pb collisions at $\sqrt{s_{\rm NN}} = 5.02$ TeV show that the suppression rate for D^0 mesons is the same as for light charged hadrons within the experimental uncertainty.

We present new theoretical calculations beyond the traditional framework of parton energy loss [4, 5, 6] based on recently developed techniques using Soft Collinear Effective Theory (SCET) [7, 8, 9, 10]. In [11, 15], an effective field theory based on SCET was developed in order to describe highly energetic massless partons traversing the QCD medium. The main idea is to include a Glauber mode that describes the interaction of energetic partons with the QGP. In this work, we extend this calculation by including finite masses for heavy quarks, see [13] for more details. We label the resulting new effective field as $\mathrm{SCET}_{\rm M,G}$. See [14] for Glauber modes in a more general context. Following the massless calculations in [15, 16, 17, 18], we derive massive in-medium splitting functions to first order in opacity. With the new in-medium splitting functions, we are able to go beyond the traditional approach to parton energy loss. See also [19, 20, 21, 22]. In addition, we introduce a new way for implementing the in-medium corrections consistent with calculations at next-to-leading order (NLO) in QCD [23] for inclusive

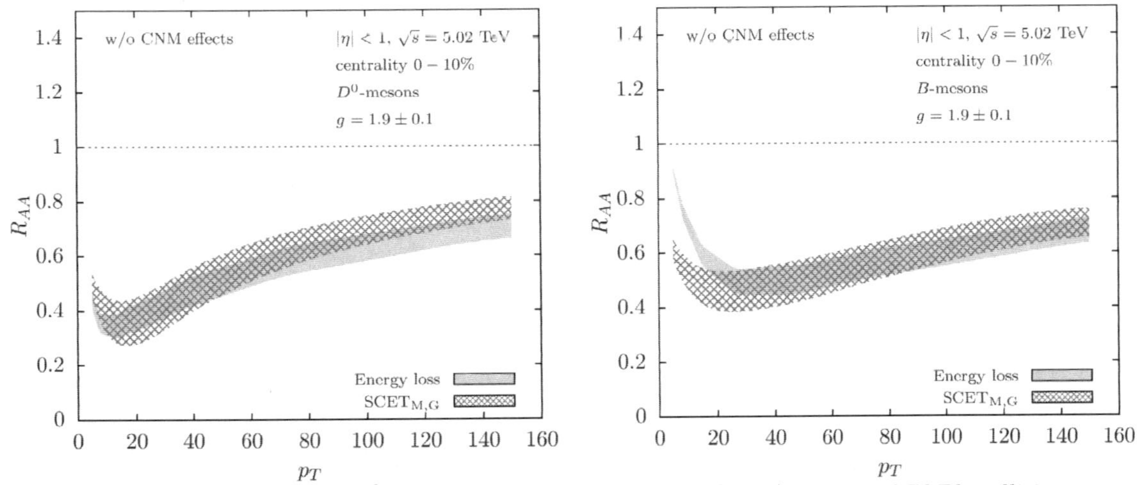

Figure 1. Suppression of D^0-mesons (left) and B-mesons (right) in central PbPb collisions at $\sqrt{s_{\rm NN}} = 5.02$ TeV. We show results obtained within the traditional approach to parton energy loss (green) and the new results using SCET$_{\rm M,G}$ (red).

hadron production. This can be achieved by formally defining medium modified fragmentation functions derived from the SCET$_{\rm M,G}$ splitting functions. As it turns out, the description of the underlying proton-proton baseline plays an important role [24, 25, 26, 27, 28].

The remainder of this paper is organized as follows. In section 2, we introduce the new formalism to calculate the in-medium corrections consistently at next-to-leading order using the new massive in-medium splitting functions based on SCET$_{\rm M,G}$. In section 3, we present numerical results for the suppression rates for both D- and B-mesons and compare to data from the LHC. In section 4, we conclude and give an outlook.

2. Massive in-medium splitting functions and their application to PbPb $\to HX$

The SCET$_{\rm M,G}$ Lagrangian is given by a sum of the massive SCET Lagrangian in the vacuum [29] and the Glauber gluon interaction terms as in the massless case [15]. The medium interaction terms are not modified due to the finite mass effects, which can be obtained from power counting. See [13] for more details. From the resulting Lagrangian, we can directly derive the massive vacuum splitting functions $Q \to Qg$, $Q \to gQ$ and $g \to Q\bar{Q}$ which were first obtained using traditional perturbative methods in QCD [30]. The in-medium results are obtained by taking into account single- and double-Born diagrams and by summing over the number of scattering centers. As it turns out, the in-medium splitting functions have a similar structure as the massless results in [15, 16]. Since the final expressions are quite lengthy, we refer the interested reader to [13]. In the soft emission limit, it is possible to reproduce the splitting functions obtained in traditional energy loss calculations for heavy quarks [6].

The interactions with the QGP affect only the partons after the hard-scattering event. Following the methods developed in [31], we consider both real and virtual corrections. The relevant part in the cross section for a fragmenting quark or gluon i can be schematically written as

$$\sum_j \sigma_i^{(0)} \otimes \mathcal{P}_{i \to jk} \otimes D_j^H \,, \tag{1}$$

where $\sigma_i^{(0)}$ is the leading-order production cross section for quarks or gluons. The $\mathcal{P}_{i \to jk}$ describe the splitting process $i \to jk$ and are given by the sum of vacuum and in-medium splitting

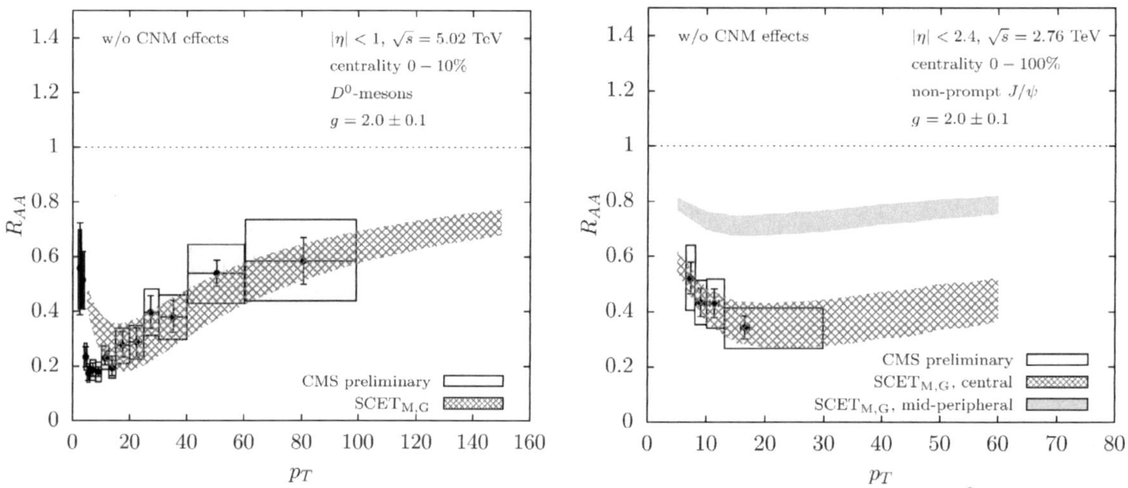

Figure 2. Comparison of SCET$_{\mathrm{M,G}}$ based results with preliminary CMS data for D^0-mesons [3] (left) and non-prompt J/ψ which originate from B-mesons [1] (right).

functions

$$\mathcal{P}_{i\ jk}(z,\mu) = \mathcal{P}^{\mathrm{vac}}_{i\ jk}(z,\mu) + \mathcal{P}^{\mathrm{med}}_{i\ jk}(z,\mu)\,. \tag{2}$$

The first terms in Eqs. (1) and (2) are simply part of the NLO calculation in the vacuum [23]. The second term in Eq. (2) gives the fixed order in-medium correction and can be formally considered as a medium modified fragmentation function. See [13] for a more detailed derivation.

3. Numerical results and comparison to data

In this section, we present results for the nuclear modification factor R_{AA} which is defined as the ratio of the PbPb and the proton-proton cross sections

$$R_{AA} = \frac{d\sigma^H_{\mathrm{PbPb}}/d\eta dp_T}{\langle N_{\mathrm{coll}}\rangle\, d\sigma^H_{pp}/d\eta dp_T}\,, \tag{3}$$

where $\langle N_{\mathrm{coll}}\rangle$ is the average number of binary nucleon-nucleon collisions for a given centrality. Note that for the proton-proton baseline, we choose to use the Zero Mass Variable Number Scheme (ZMVFNS) at NLO and the fragmentation functions of [26, 27]. We find that this scheme can be applied even for relatively low p_T of the observed hadron H. For our numerical results, we do not take into account Cold Nuclear Matter (CNM) effects. Firstly, we present a comparison to results obtained within the traditional approach to parton energy loss in Fig. 1. Secondly, in Fig. 2 we compare our SCET$_{\mathrm{M,G}}$ results to preliminary CMS data for D^0 mesons [1] (left) at $\sqrt{s_{\mathrm{NN}}} = 5.02$ TeV and for non-prompt J/ψ originating from B-hadrons [3] (right) at $\sqrt{s_{\mathrm{NN}}} = 2.76$ TeV. Note that the non-prompt J/ψ data is in fact only for minimum bias events instead of fixed centrality. Therefore, we present our calculation for both central as well as mid-peripheral collisions. We find that our central result (0-10% centrality) agrees very well with the data. The minimum bias results are dominated by central collisions as they are weighted with $\langle N_{\mathrm{coll}}\rangle$. We note that the heavy-ion results crucially depend on whether the heavy meson is produced by a fragmenting heavy quark or a gluon. In general, gluons lose more energy than (heavy) quarks when undergoing multiple scatterings in the medium before they eventually fragment into the observed heavy meson. See [13] for more details.

4. Conclusions

We derived a new version of Soft Collinear Effective Theory including both finite heavy quark masses and the interaction of collinear partons with the medium that are mediated by Glauber gluon exchange [13]. From the new effective field theory, we are able to deduce the in-medium massive splitting functions for $Q \to Qg$, $Q \to gQ$ and $g \to Q\bar{Q}$ to first order in opacity. In addition, we introduced a new formalism to include in-medium effects consistent with calculations at next-to-leading order in QCD. We presented numerical results for the suppression of D- and B-mesons in heavy-ion collisions. We found good agreement in comparison to existing data from the LHC at $\sqrt{s_{\rm NN}} = 5.02$ TeV and 2.76 TeV. In the future, we plan to extend the current framework of the in-medium modification to inclusive jet production [32].

Acknowledgments

We would like to thank Grigory Ovanesyan for his collaboration at the early stages of this work. In addition, we would like to thank Yen-Jie Lee, Emanuele Mereghetti, Rishi Sharma, Hongxi Xing and Zhiqing Zhang for helpful discussions. This work is supported by the U.S. Department of Energy under Contract No. DE-AC52-06NA25396, in part by the LDRD program at Los Alamos National Laboratory.

References

[1] CMS Collaboration [CMS Collaboration], CMS-PAS-HIN-12-014.
[2] A. Grelli [ALICE Collaboration], Nucl. Phys. A **904-905** (2013) 635c [arXiv:1210.7332 [hep-ex]].
[3] CMS Collaboration [CMS Collaboration], CMS-PAS-HIN-16-001.
[4] M. Gyulassy, P. Levai and I. Vitev, Phys. Rev. Lett. **85**, 5535 (2000) [nucl-th/0005032].
[5] M. Gyulassy, P. Levai and I. Vitev, Nucl. Phys. B **594**, 371 (2001) [nucl-th/0006010].
[6] M. Djordjevic and M. Gyulassy, Nucl. Phys. A **733**, 265 (2004) doi:10.1016/j.nuclphysa.2003.12.020 [nucl-th/0310076].
[7] C. W. Bauer, S. Fleming and M. E. Luke, Phys. Rev. D **63**, 014006 (2000), [hep-ph/0005275].
[8] C. W. Bauer, S. Fleming, D. Pirjol and I. W. Stewart, Phys. Rev. D **63**, 114020 (2001), [hep-ph/0011336].
[9] C. W. Bauer and I. W. Stewart, Phys. Lett. B **516**, 134 (2001), [hep-ph/0107001].
[10] C. W. Bauer, D. Pirjol and I. W. Stewart, Phys. Rev. D **65**, 054022 (2002), [hep-ph/0109045].
[11] A. Idilbi and A. Majumder, Phys. Rev. D **80**, 054022 (2009) [arXiv:0808.1087 [hep-ph]].
[12] G. Ovanesyan and I. Vitev, JHEP **1106**, 080 (2011) [arXiv:1103.1074 [hep-ph]].
[13] Z. B. Kang, F. Ringer and I. Vitev, [arXiv:1610.02043 [hep-ph]].
[14] I. Z. Rothstein and I. W. Stewart, JHEP **1608**, 025 (2016) [arXiv:1601.04695 [hep-ph]].
[15] G. Ovanesyan and I. Vitev, JHEP **1106**, 080 (2011) [arXiv:1103.1074 [hep-ph]].
[16] G. Ovanesyan and I. Vitev, Phys. Lett. B **706**, 371 (2012) [arXiv:1109.5619 [hep-ph]].
[17] M. Fickinger, G. Ovanesyan and I. Vitev, JHEP **1307**, 059 (2013) [arXiv:1304.3497 [hep-ph]].
[18] G. Ovanesyan, F. Ringer and I. Vitev, Phys. Lett. B **760**, 706 (2016) [arXiv:1512.00006 [hep-ph]].
[19] Z. B. Kang, R. Lashof-Regas, G. Ovanesyan, P. Saad and I. Vitev, Phys. Rev. Lett. **114**, no. 9, 092002 (2015) [arXiv:1405.2612 [hep-ph]].
[20] Y. T. Chien, A. Emerman, Z. B. Kang, G. Ovanesyan and I. Vitev, Phys. Rev. D **93**, no. 7, 074030 (2016) [arXiv:1509.02936 [hep-ph]].
[21] Y. T. Chien and I. Vitev, JHEP **1605**, 023 (2016) [arXiv:1509.07257 [hep-ph]].
[22] Y. T. Chien and I. Vitev, arXiv:1608.07283 [hep-ph].
[23] B. Jager, A. Schafer, M. Stratmann and W. Vogelsang, Phys. Rev. D **67**, 054005 (2003) [hep-ph/0211007].
[24] P. Nason, S. Dawson and R. K. Ellis, Nucl. Phys. B **327**, 49 (1989) Erratum: [Nucl. Phys. B **335**, 260 (1990)].
[25] M. Cacciari, M. Greco and P. Nason, JHEP **9805**, 007 (1998) [hep-ph/9803400].
[26] T. Kneesch, B. A. Kniehl, G. Kramer and I. Schienbein, Nucl. Phys. B **799**, 34 (2008) [arXiv:0712.0481 [hep-ph]].
[27] B. A. Kniehl, G. Kramer, I. Schienbein and H. Spiesberger, Phys. Rev. D **77**, 014011 (2008) [arXiv:0705.4392 [hep-ph]].
[28] M. Fickinger, S. Fleming, C. Kim and E. Mereghetti, arXiv:1606.07737 [hep-ph].
[29] A. K. Leibovich, Z. Ligeti and M. B. Wise, Phys. Lett. B **564**, 231 (2003) [hep-ph/0303099].
[30] S. Catani, S. Dittmaier and Z. Trocsanyi, Phys. Lett. B **500**, 149 (2001) [hep-ph/0011222].
[31] J. C. Collins and J. w. Qiu, Phys. Rev. D **39**, 1398 (1989).
[32] Z. B. Kang, F. Ringer and I. Vitev, arXiv:1606.06732 [hep-ph].

IOP Conf. Series: Journal of Physics: Conf. Series **779** (2017) 012030 doi:10.1088/1742-6596/779/1/012030

Heavy-quark dynamics in heavy-ion collisions

T Song[1,2], H Berrehrah[1], E L Bratkovskaya[2,3], D Cabrera[1], W Cassing[4], L Tolos[1,5], J M Torres-Rincon[1]

[1] Frankfurt Institute for Advanced Studies, University of Frankfurt, Frankfurt, Germany
[2] Institute for Theoretical Physics, University of Frankfurt, Frankfurt, Germany
[3] GSI Helmholtzzentrum für Schwerionenforschung GmbH, Darmstadt, Germany
[4] Institute for Theoretical Physics, University of Gießen, Gießen, Germany
[5] Institut de Ciencies de l'Espai (IEEC/CSIC), Universitat Autonoma de Barcelona, Spain

E-mail: E.Bratkovskaya@gsi.de

Abstract. The dynamics of partons and hadrons in ultra-relativistic nucleus-nucleus collisions are analyzed within the Parton-Hadron-String Dynamics (PHSD) transport approach, which is based on a dynamical quasiparticle model (DQPM) for the partonic phase including a dynamical hadronization scheme while reproducing lattice QCD results in thermodynamic equilibrium for the equation-of-state as well as transport coefficients like shear and bulk viscosities, the electric conductivity or the charm diffusion coefficient of the hot QCD medium. In this contribution, we report on recent results on the charm dynamics and elliptic flow in Au+Au collisions at RHIC and Pb+Pb collisions at the LHC.

1. Introduction

The dynamics of the early universe in terms of the 'Big Bang' may be studied experimentally by ultra-relativistic nucleus-nucleus collisions at Relativistic-Heavy-Ion-Collider (RHIC) or Large-Hadron-Collider (LHC) energies in terms of 'tiny bangs' in the laboratory. With sufficiently strong parton interactions, the medium in the collision zone can be expected to achieve local equilibrium after some initial delay and exhibit approximately hydrodynamic flow [1, 2]. In these collisions a new state of strongly interacting matter is created, being characterized by a very low shear viscosity η to entropy density s ratio, η/s, close to a nearly perfect fluid [3]. Lattice QCD (lQCD) calculations [4] indicate that a crossover region between hadron and quark-gluon matter should have been reached in these experiments.

Since the hot and dense matter produced in relativistic heavy-ion collisions appears only for a couple of fm/c, it is a big challenge for experiments to investigate its properties. Heavy-flavor mesons are considered to be promising probes in this respect since the production of heavy flavor requires a large energy-momentum transfer. Thus it takes place early in the heavy-ion collisions, and - due to the large energy-momentum transfer - should be described by perturbative quantum chromodynamics (pQCD). The produced heavy flavor then interacts with the hot dense matter (of partonic or hadronic nature) by exchanging energy and momentum. As a result, the ratio of the measured number of heavy flavors in heavy-ion collisions to the expected number in the absence of nuclear or partonic matter, which is the definition of $R_{\rm AA}$, is suppressed at high transverse momentum, and the elliptic flow of heavy flavor is generated by the interactions in noncentral heavy-ion collisions [5, 6]. Although it had been expected that the $R_{\rm AA}$ of heavy flavor is less suppressed and its elliptic flow is smaller as compared to the corresponding quantities for light hadrons, the experimental data show that the suppression of heavy flavor hadrons at high transverse momentum and its elliptic flow v_2 are comparable to those of light hadrons [5, 6]. This is a puzzle for heavy-flavor production and dynamics in relativistic heavy-ion collisions [7].

Since the heavy-flavor interactions are closely related to the dynamics of the partonic or hadronic degrees-of-freedom due to their mutual interactions, a proper description of the relativistic heavy-ion collisions and their bulk dynamics is necessary. In this contribution, we report about results from the parton-hadron-string dynamics (PHSD) approach [8], which has been successfully applied to p+A and Au+Au (Pb+Pb) collisions from SPS to LHC energies with respect to single-particle spectra, collective flows and electromagnetic observables [9].

2. Charm and bottom dynamics in PHSD

In the PHSD the charm and bottom quark pairs are produced through initial hard nucleon-nucleon scattering in relativistic heavy-ion collisions. We employ the PYTHIA event generator [10] to produce the heavy-quark pairs and modify slightly their transverse momentum and rapidity such that they are similar to those from the FONLL calculations [11]. The corrections employed at RHIC and LHC energies can be found in Refs. [12, 13, 14]. Accordingly, our tuned PYTHIA generator gives very similar charm and bottom distributions as those from FONLL calculations [11, 15], which fixes the input from pQCD in our approach.

The produced charm and bottom quarks in hard nucleon-nucleon interactions are hadronized in p+p collisions by emitting soft gluons, which is denoted by 'fragmentation' (cf. Ref. [12] for details). The excited $D^*(B^*)$ mesons first decay into $D(B)+\pi$ or $D(B)+\gamma$, and finally the $D-$ and $B-$mesons produce single electrons through the semileptonic decay [16]. In case of heavy-ion collisions, the shadowing effect is incorporated in the PHSD by employing the EPS09 package from Ref. [17] as in most of the related approaches [7]. The details of the implementation are given in Ref. [13].

In PHSD the baryon-baryon and baryon-meson collisions at high-energy produce strings, which above the critical energy density ($\sim$ 0.5 GeV/fm^3) melt into quarks and antiquarks with masses determined by the temperature-dependent spectral functions from the dynamical quasiparticle model (DQPM) [18], which has been fitted to thermodynamical quantities from lattice QCD. Massive gluons are formed through flavor-neutral quark and antiquark fusion in line with the DQPM. The heavy quarks and antiquarks produced in early hard collisions interact with the dressed lighter off-shell partons in the partonic phase. The cross sections for the heavy-quark scattering with massive off-shell partons have been calculated by considering explicitly the mass spectra of the final state particles in Ref. [19, 20]. The elastic scattering of heavy quarks in the QGP is treated in the PHSD by including the non-perturbative effects of the strongly interacting quark-gluon plasma (sQGP) constituents, i.e. the temperature-dependent coupling $g(T/T_c)$ as well as the effective propagators with broad spectral functions (imaginary parts) from the DQPM [18]. We note that heavy flavor interactions in the QGP – as described by the DQPM charm scattering cross sections – differ substantially from the pQCD scenario, however, the spacial diffusion constant for charm quarks $D_s(T)$ is consistent with the lQCD data [13, 21].

The heavy-quark hadronization in heavy-ion collisions is realized via 'dynamical coalescence' in competition to fragmentation. Here 'dynamical coalescence' means that the probability to find a coalescence partner is defined by Monte Carlo in the vicinity of the critical energy density $0.4 \leq \epsilon \leq 0.75$ GeV/fm^3 as explained in Ref. [13].

After the hadronization of heavy quarks and their subsequent decay into D, D^*, B and B^* mesons, the final stage of the evolution concerns the interaction of these states with the hadrons conforming the expanding bulk medium. A realistic description of the hadron-hadron scattering —potentially affected by resonant interactions— includes collisions with the states $\pi, K, \bar{K}, \eta, N, \bar{N}, \Delta$, and $\bar{\Delta}$. Such a description of their interactions has been developed in

Refs. [22, 23, 24] using effective field theory. The resulting cross sections are implemented in the PHSD.

3. Results for heavy-ion reactions

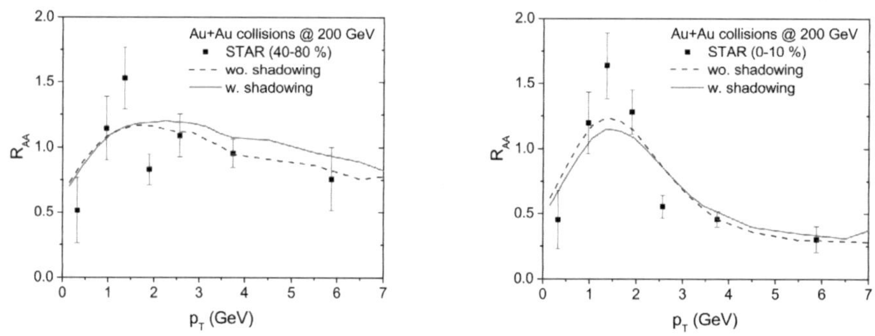

Figure 1. R_{AA} of $D-$mesons with (solid) and without (dashed) shadowing effect in Au+Au collisions at $\sqrt{s_{\mathrm{NN}}} = 200$ GeV for different centralities in comparison to the experimental data from the STAR collaboration [25].

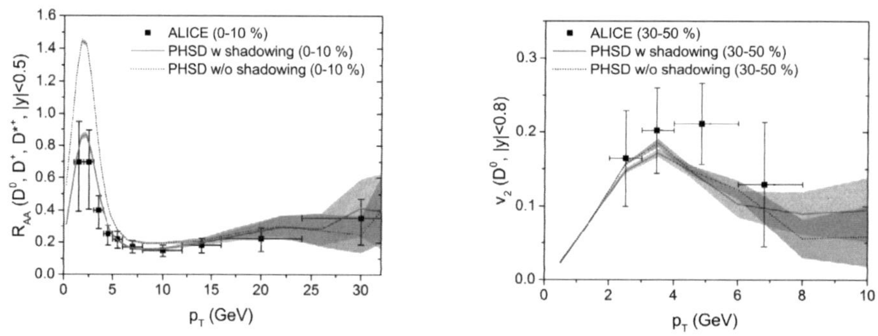

Figure 2. (left) The ratio R_{AA} of D^0, D^+, and D^{*+} mesons within $|y| < 0.5$ as a function of p_{T} in 0-10 % central Pb+Pb collisions at $\sqrt{s_{\mathrm{NN}}} = 2.76$ TeV compared with the experimental data from the ALICE collaboration [26]. The solid and dotted lines are, respectively, R_{AA} with and without (anti-)shadowing. (right) The elliptic flow v_2 of D^0 mesons within $|y| < 0.8$ in 30-50 % central Pb+Pb collisions at $\sqrt{s_{\mathrm{NN}}} = 2.76$ TeV compared with the experimental data from the ALICE collaboration [27].

Since the heavy flavor strongly interacts with the expanding matter in A+A reactions, it is also accelerated outwards. Such effects of the medium on the heavy-flavor dynamics are expressed in terms of the nuclear modification factor defined as [13]

$$R_{\mathrm{AA}}(p_{\mathrm{T}}) \equiv \frac{dN_{\mathrm{AA}}/dp_{\mathrm{T}}}{N_{\mathrm{binary}}^{\mathrm{AA}} \times dN_{\mathrm{pp}}/dp_{\mathrm{T}}}, \tag{1}$$

where N_{AA} and N_{pp} are, respectively, the number of particles produced in heavy-ion collisions and that in p+p collisions, and $N_{\mathrm{binary}}^{\mathrm{AA}}$ is the number of binary nucleon-nucleon collisions in the heavy-ion collision for the centrality class considered. Furthermore, in noncentral heavy-ion collisions the produced matter expands anisotropically due to the different pressure gradients in plane and out-of plane. If the heavy flavor interacts strongly with the nuclear matter, then it also follows this anisotropic motion to some extend. The dominant anisotropic flow is expressed in terms of the elliptic flow $v_2 = \langle (p_x^2 - p_y^2)/(p_x^2 + p_y^2) \rangle$ which is the dominant observable to quantify the collectivity of the reaction [13].

In Fig. 1 we show the R_{AA} of $D-$mesons in different centrality classes with (solid) and without (dashed) shadowing effect in Au+Au collisions at $\sqrt{s_{\mathrm{NN}}} = 200$ GeV in comparison to the experimental data from the STAR collaboration [25]. The shadowing effect decreases the charm production by ~8 % in central reactions. We note that the elliptic flow v_2 of D^0 mesons is enhanced by hadronic scattering by up to 30% at higher p_{T} [12].

Figure 2 (left) shows the ratio R_{AA} of D^0, D^+, and D^{*+} mesons within the rapidity range $|y| < 0.5$ as a function of p_{T} in 0-10 % central Pb+Pb collisions at $\sqrt{s_{\mathrm{NN}}} = 2.76$ TeV. The solid and dotted lines are, respectively, the R_{AA} of D mesons with and without (anti-)shadowing. We can see that the R_{AA} of D mesons decreases especially at small transverse momentum due to shadowing. When including shadowing our results are in a good agreement with the experimental data from the ALICE collaboration [26]. Figure 2 (right) shows the elliptic flow v_2 of D^0 mesons within the rapidity range $|y| < 0.8$ in 30-50 % central Pb+Pb collisions at $\sqrt{s_{\mathrm{NN}}} = 2.76$ TeV in comparison to the data from the ALICE collaboration [27]. The (anti-)shadowing effect slightly decreases the elliptic flow, because it reduces the production of low-p_{T} charm which more easily follows the bulk flow. We recall that so far it has been a challenge for theoretical models to reproduce the experimental data and to explain simultaneously the large energy loss of charm quarks (R_{AA}) and the strong collectivity (v_2) [7]. According to the studies performed above the microscopic PHSD approach appears to be quite in line with the experimental observations for charm quarks at top RHIC and LHC energies.

References

[1] Ollitrault J Y 1992 *Phys. Rev.* D **46** 229
[2] Heinz U and Kolb P 2002 *Nucl. Phys.* A **702** 269
[3] Shuryak E V 2005 *Nucl. Phys.* A **750** 64
[4] Aoki Y *et al.* 2009 *JHEP* **0906** 088
[5] Abelev B *et al.* [ALICE Collaboration] 2012 *JHEP* **1209** 112
[6] Abelev B *et al.* [ALICE Collaboration] 2013 *Phys. Rev. Lett.* **111** 102301
[7] Primo F Rapp R 2016 *J. Phys.* G **43** 093002
[8] Bratkovskaya E L *et al.* 2011 *Nucl. Phys.* A **856** 162
[9] Linnyk O Bratkovskaya E Cassing W 2016 *Prog. Part. Nucl. Phys.* **87** 50
[10] Sjostrand T Mrenna S Skands P Z 2006 *JHEP* **0605** 026
[11] Cacciari M Frixione S Houdeau N Mangano M L Nason P Ridolfi G 2012 *JHEP* **1210** 137
[12] Song T *et al.* 2015 Bratkovskaya E 2015 *Phys. Rev.* C **92** 014910
[13] Song T Berrehrah H Cabrera D Cassing W Bratkovskaya E 2016 *Phys. Rev.* C **93** 034906
[14] Song T Berrehrah H Cabrera D Torres-Rincon J M Tolos L Cassing W Bratkovskaya E 2016 *arXiv*:1605.07887
[15] Cacciari M Nason P Vogt R 2005 Phys. Rev. Lett. **95** 122001
[16] Olive K A *et al.* [Particle Data Group Collaboration] 2014 *Chin. Phys.* C **38** 090001
[17] Eskola K J Paukkunen H Salgado C A 2009 *JHEP* **0904** 065
[18] Cassing W 2009 *Eur. Phys. J. ST* **168** 3 Cassing W 2007 *Nucl. Phys.* A **795** 70
[19] Berrehrah H Gossiaux P B Aichelin J Cassing W Bratkovskaya E 2014 *Phys. Rev.* C **90** 064906
[20] Berrehrah H Bratkovskaya E Cassing W Gossiaux P B Aichelin J 2015 *Phys. Rev.* C **91** 054902
[21] Berrehrah H Bratkovskaya E Steinert T Cassing W 2016 *Int. Journal of Mod. Phys.* E **25** 1642003
[22] Tolos L Torres-Rincon J M 2013 *Phys. Rev.* D **88** 074019
[23] Torres-Rincon J M Tolos L Romanets O 2014 Phys. Rev. D **89** 074042
[24] Tolos L 2013 *Int. J. Mod. Phys.* E **22** 1330027
[25] Adamczyk L *et al.* [STAR Collaboration] 2014 *Phys. Rev. Lett.* **113** 142301
[26] Adam J *et al.* [ALICE Collaboration] 2015 *arXiv*:1509.06888 [nucl-ex]
[27] Abelev B B *et al.* [ALICE Collaboration] 2014 *Phys. Rev.* C **90** 034904

Heavy quark dynamics in QCD matter

S. K. Das[1], F. Scardina [1,2], S. Plumari[1,2] and V. Greco[1,2]

[1] Department of Physics and Astronomy, University of Catania, Via S. Sofia 64, I- 95125 Catania, Italy
[2] Laboratori Nazionali del Sud, INFN-LNS, Via S. Sofia 62, I-95123 Catania, Italy

E-mail: santosh@lns.infn.it

Abstract. Simultaneous description of heavy quark nuclear modification factor R_{AA} and the elliptic flow v_2 is a top challenge for all the existing models. We highlight how the temperature dependence of the energy loss/transport coefficients is responsible for addressing a large part of such a puzzle along with the full solution of the Boltzmann collision integral for the momentum evolution of heavy quarks in the medium. We consider four different models to evaluate the temperature dependence of drag coefficients of the heavy quark in the QGP. We have also highlighted the heavy quark dynamics in the presence of an external electromagnetic field which induces a sizable heavy quark directed flow, $v_1(y)$, that can be measurable at LHC.

1. Introduction

Ongoing experimental efforts at Relativistic Heavy Ion Collider (RHIC) and Large Hadron Collider (LHC) aim at creating and characterizing the properties of quark gluon plasma (QGP). The open heavy flavor mesons (mesons which contain one heavy quark, mainly c and b) constitute a novel probe of the QGP properties, because they are produced in the early stage of the heavy-ion collisions and they are the witness of the entire space-time evolution of the QGP.

The two main experimental observables related to the heavy quark dynamics in the QGP are: (i) the nuclear modification factor, R_{AA} [1, 2] which is the ratio of the p_T spectra of heavy flavored mesons (D and B) produced in nucleus + nucleus collisions to those produced in proton + proton collisions scaled with the number of binary collisions and (ii) the elliptic flow, $v_2 = \langle cos(2\phi_p)\rangle$ [2], which is a measure of the anisotropy in the azimuthal distribution of particle production. Several theoretical efforts have been made to study the R_{AA} and the v_2 measured in experiments within different models [3, 4, 5, 6, 7, 8, 9, 10, 11, 12, 13, 14, 15, 16] (see also for heavy baryons [17]). However all the approaches show some difficulties to describe both the R_{AA} and v_2 simultaneously.

2. Results

To achieve a simultaneous description of R_{AA} and v_2, we need to focus on the time evolution of R_{AA} and v_2 to understand how they develop during the expansion of the QGP. As shown in Ref. [18], the R_{AA} develops mostly at the very early stage of the QGP evolution within 3-4 fm/c, while the v_2 can be transferred to the heavy quarks only at later stages when the fireball has reached temperatures (T) close to the critical temperature. This indicates a more liquid-like behavior of the quark-gluon plasma and the T-dependence of the drag coefficient plays a significance role for a simultaneous description of R_{AA} and v_2 as they are sensitive to the two

IOP Conf. Series: Journal of Physics: Conf. Series **779** (2017) 012031 doi:10.1088/1742-6596/779/1/012031

different stages of the QGP evolution (T_i and T_c). To investigate the influence of the temperature dependence of the drag (and diffusion) coefficients on heavy quark observables, four different models have been used to calculate the drag and diffusion coefficients.

Model-I (pQCD): In this case, the elastic collisions of heavy quarks with the bulk consisting of light quarks, light anti-quarks and gluons have been considered within the framework of pQCD having temperature dependence of the coupling [19]:

$$g^{-2}(T) = 2\beta_0 ln\left(\frac{2\pi T}{a\,T_c}\right) + \frac{\beta_1}{\beta_0} ln\left[ln\left(\frac{2\pi T}{a\,T_c}\right)\right] \tag{1}$$

where $\beta_0 = (11 - 2N_f/3)/16\pi^2$, $\beta_1 = (102 - 38N_f/3)/(16\pi^2)^2$ and $a = 1.3$. N_f is the number of flavor and T_C is the transition temperature.

Model-II (AdS/CFT): In this second case, we consider the drag force from the gauge/string duality i.e. AdS/CFT [23], $\Gamma = C\frac{T^2_{QCD}}{M_{HQ}}$, where $C = 2.1 \pm 0.5$ and M_{HQ} the diffusion coefficient deduced from fluctuation-dissipation [24].

Model-III (QPM): In this case, we employ a quasi-particle model (QPM) [25, 26] with T-dependent quasi-particle masses, $m_q = 1/3g^2T^2$, $m_g = 3/4g^2T^2$, along with a T-dependent background field known as the bag constant which is tuned to match the thermodynamics of the lattice QCD. Such a fit leads to the coupling, $g^2(T) = \frac{48\pi^2}{[(11N_c-2N_f)ln[\lambda(\frac{T}{T_c}-\frac{T_s}{T_c})]^2}$, where λ=2.6 and T/T_s=0.57.

Model-IV ($\alpha_{QPM}(T), m_q = m_g = 0$): In this fourth case, we consider a model where the light quarks and gluons are massless but the coupling is taken from the QPM model discussed above. This case is merely a way to obtain a drag coefficient increasing toward T_c.

This fourth case has been considered to have a drag coefficient whose T dependence is similar to that in the T-matrix approach [4, 27].

In Fig 1, we have shown the variation of the drag coefficients obtained within the four different models discussed above. Our methodology is to reproduce the R_{AA} measured experimentally within all the four models, hence, these are the rescaled drag coefficients.

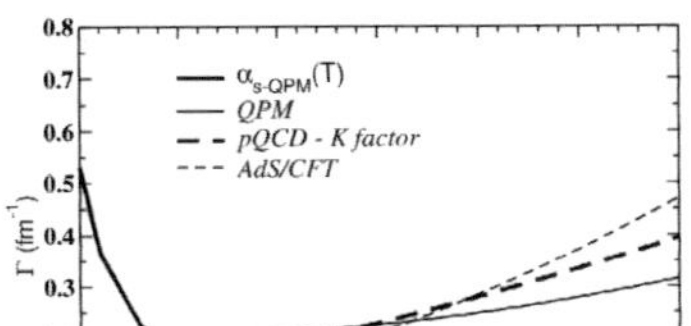

Figure 1. Drag coefficients as a function of temperature.

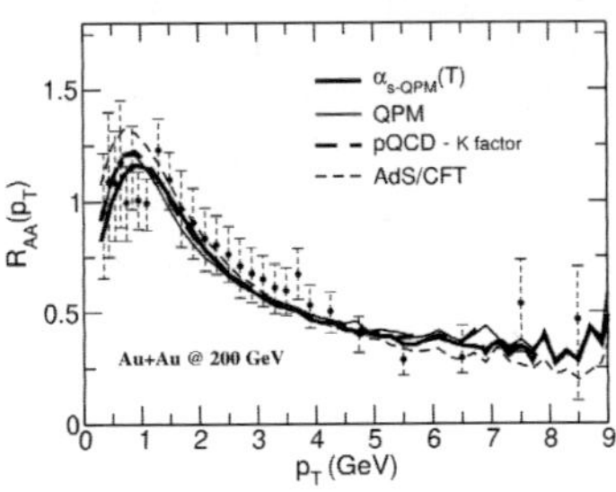

Figure 2. R_{AA} as a function of p_T for the minimum bias Au+Au collisions.

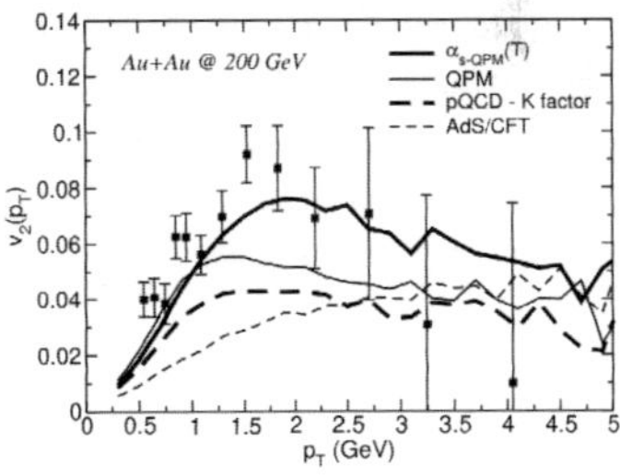

Figure 3. v_2 as a function of p_T for the minimum bias Au+Au collisions.

We have solved the Langevin dynamics to study the heavy quark momentum evolution in QGP starting from charm quark production in $p + p$ collision [22] as the initial charm quark distributions in the momentum space. To simulate the heavy quark dynamics in QGP, we need the bulk evolution. We are using the transport bulk which can reproduce the gross features of the bulk, i.e. the spectra and elliptic flow. For the $Au + Au$ collisions at the highest RHIC energy, we simulate the QGP evolution with the initial temperature at the center of the fireball

$T_i = 340$ MeV and the initial time $\tau_i = 0.6$ fm/c. For the details of the fireball evolution, we refer to our early works [20, 21].

In Fig 2, we have shown the variation of R_{AA} as a function of p_T for the four different models at RHIC energy. The v_2 for the same R_{AA} has been plotted in Fig 3 for all the four models. The experimental data on non-photonic electrons are taken from ref. [2]. It has been observed that for the same R_{AA}, the v_2 build-up can be quite different depending on the T-dependence of the drag coefficient in the model. The larger the drag coefficient at T_c, the larger the v_2 [8]. Similar effect has also been found in the light quark sector [28, 29]. This suggests that the correct temperature dependence of the drag coefficient has a vital role for a simultaneous reproduction of heavy quark R_{AA} and v_2. The heavy quark observables are also sensitive to the hadronic rescattering [18, 31], to the pre-equilibrium phase [32] as well as to the time evolution equation i.e. whether it follows the Langevin or Boltzmann equation [11]. In particular, the solution of the full two-body collision integral shows that the anisotropic flows are larger with respect to those predicted by a Langevin dynamics. For details we refer to our earlier works [11].

Recently it has been recognized that a very strong magnetic field [33, 34, 35] is created at early times in heavy-ion collisions. Since heavy quarks are produced at the very early stage of evolution due to their large masses, their dynamics can be affected by such a strong magnetic field [37, 38]. The $\vec{B}$ field generated in non-central heavy-ion collisions is dominated by the component along the $\vec{y}$ axis, so its main effect is the induction of a current in the xz plane. For the calculation of the electromagnetic field generated in ultra-relativistic heavy-ion collisions, we refer to the space-time solution developed in Ref. [36]. In Fig.4 we show the time dependence of the magnetic field B_y and electric field E_x at finite space rapidity $\eta = 1.5$ in a $Pb + Pb$ collision with $\sqrt{s} = 2.76$ ATeV at $b = 9.5$ fm for a medium of $\sigma_{el} = 0.023\,\text{fm}^{-1}$. The impact of the electromagnetic field has been taken into account through the Lorentz force as the external force in the Langevin equation. For details we refer to Ref. [39].

The Magnetic field introduces an anisotropy which eventually leads to a substantial directed flow, $v_1 = \langle p_x/p_T \rangle$, that is measurable experimentally. In Fig. 5 we show the resulting directed flow v_1 as a function of rapidity for charm (black solid line) and anti-charm quarks (dashed line). We find a substantial v_1 at finite rapidity with a peak at $y \simeq 1.75$. The flow is negative for positive charged particle (charm) at forward rapidity which means that the magnetic field dominates over the displacement induced by the Faraday current associated with the time dependence of the magnetic field. This is a non-trivial result and depends not only on the absolute magnitude of the magnetic field but also on the strength of heavy-light quark interaction. If we artificially increase the drag coefficient by a factor 5 to achieve a thermalization of the charm quarks similar to that of the light quarks, this will lead to a much smaller $v_1(y)$.

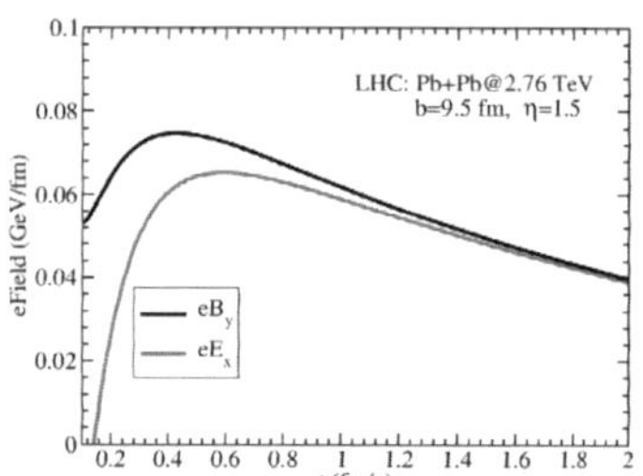

Figure 4. Time evolution of the magnetic field eB_y and the electric field eE_x in the forward rapidity region.

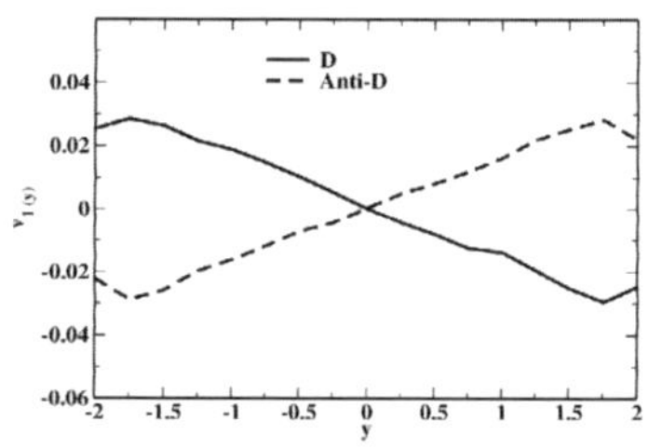

Figure 5. Directed flow v_1 as a function of the rapidity in $Pb + Pb$ collisions at $\sqrt{s} = 2.76$ ATeV with $b = 9.5$ fm for D meson $[c\bar{q}]$ (solid line) and anti-D meson $[\bar{c}q]$ (dashed line).

3. Summary and outlook

In summary, we have shown how the v_2 build-up for the same R_{AA} depends on the T-dependence of the drag coefficient which is the key for a simultaneous description of heavy quark R_{AA} and v_2. The larger the drag coefficient at T_c, the larger the v_2. We have also studied the heavy quark dynamics in the presence of an external electromagnetic field. Heavy quarks develop a sizable directed flow (v_1) which could be measurable experimentally. The present study suggests that the heavy quark v_1 can be considered a sensitive probe to the creation and characterization of the magnetic field in the heavy-ion collisions

Acknowledgements

We acknowledge the support by the ERC StG under the QGPDyn Grant n. 259684. VG acknowledges the hospitality at UCAS-Beijing where the work was finalized also the support from the National Natural Science Foundation of China (Grant No. 11135011 and 11575190) and the CAS Presidents International Fellowship Initiative (Grant No. 2016VBC002).

References

[1] Abelev B I *et al.* (STAR Collaboration) 2007 *Phys. Rev.* Lett. **98**, 192301.
[2] Adare A *et al.*(PHENIX Collaboration) 2011 *Phys. Rev.* C **84** 044905
[3] Hees H van, Greco V and Rapp R 2006 *Phys. Rev.* C **73**, 034913
[4] Hees H van, Mannearelli M, Greco V and Rapp R 2008 *Phys. Rev. Lett.* **100** 192301
[5] Das S K, Alam J, Mohanty 2010 *Phys. Rev.* C **82** 014908 ; Majumdar S, Bhattacharyya T, Alam J, Das S K 2011 *Phys. Rev.* C **84** 044901
[6] Alberico W M *et al.* 2011 *Eur. Phys. J.* C **71** 1666
[7] Cao S., Qin G.-Y., Bass S A 2013 *Phys. Rev.* C **88**,4, 044907: 2015 *Phys.Rev.* C **92** 2, 024907
[8] Das S K, Scardina F, Plumari S and Greco V (2015) *Phys. Lett.* B **747**, 260
[9] Gossiaux P B, Aichelin J 2008 *Phys. Rev.* C **78** 014904
[10] Uphoff J, Fochler O, Xu Z and Greiner C 2011 *Phys. Rev.* C **84** 024908; Phys. Lett. B **717** (2012) 430
[11] Das S K, Scadina F, Plumari S, Greco V 2014 *Phys. Rev.* C **90** 044901.
[12] Song T *et al.* 2015 *Phys Rev.* C **92** 014910
[13] Scadina F, Das S, Plumari S, Greco V 2015 J.Phys.Conf.Ser. *636* 1, 012017; 2014 J.Phys.Conf.Ser *535* 012019
[14] Cao S, Luo T, Qin G Y, Wang X N 2016, *Phys. Rev.* C **94**, no. 1, 014909
[15] Younus M, Coleman-Smith C E, Bass S A and Srivastava D K, arXiv:1309.1276 [nucl-th].
[16] Tripathy S, Bhattacharyya T, Garg P, Kumar P, Sahoo R and Cleymans J, arXiv:1606.06898 [nucl-th].
[17] Das S K, Torres-Rincon J M, Tolos L, Scardina F and Greco V, arXiv:1604.05666 [nucl-th].
[18] Das S K, Scardina F, Plumari P, Greco V 2016 J. Phys. Conf. Ser. **668**, 1, 012051
[19] Kaczmarek O, Zantow F 2005 it Phys. Rev. D, **71**, 114510
[20] Ruggieri M, Scardina F, Plumari S, Greco V 2013 *Phys. Lett.* B **727** 177
[21] Scardina F, Colonna M, Plumari S and Greco V 2013 *Phys. Lett. B* **724** 296
[22] Cacciari M,Nason P, Vogt R 2005 *Phys. Rev. Lett.* **95**, 122001
[23] Maldacena J M 1998 Adv. Theor. Math. Phys. **2**, 231
[24] Das S K, Davody A 2014 *Phys. Rev.* C **89**, 054912.
[25] Plumari S, Alberico W M, Greco V, Ratti C 2011 *Phys. Rev.* D **84**, 094004
[26] Das S K, Chandra V, Alam J 2014 *J. Phys.* G **41** 015102.
[27] Riek F, Rapp R 2010 *Phys. Rev.* C **82**, 035201
[28] Liao J, Shuryak E 2009 *Phys. Rev. Lett.* **102** 202302
[29] Scardina F, Toro M Di, Greco V 2010 *Phys. Rev.* C **82** 054901.
[30] Greco V, Ko C M, Rapp R 2004 *Phys. Lett.* B **595** 202
[31] Das S K, Ghosh S, Sarkar S, Alam J 2012 *Phys Rev* D **85** 074017; 2014 *Phys Rev* D **88**, 017501
[32] Das S K, Ruggieri M, Mazumder S, Greco V, Alam J 2015 *J. Phys.* G **42** 9, 095108
[33] Kharzeev D E, McLerran L D, Warringa H J 2008 *Nucl. Phys.* A **803** 227
[34] Fukushima K, Kharzeev D E, Warringa H J 2008 *Phys. Rev.* D **78** 074033
[35] Tuchin K 2010 *Phys. Rev.* C **82** 034904; Erratum: [Phys. Rev. C **83** (2011) 039903]
[36] Gursoy U, Kharzeev D, Rajagopal K 2014 *Phys. Rev.* C **89** 5, 054905
[37] Fukushima K, Hattori K, Yee H U, Yin Y 2016 *Phys. Rev.* D **93**, 7, 074028
[38] Finazzo S I, Critelli R, Rougemont R, Noronha J, arXiv:1605.06061 [hep-ph].
[39] Das S K, Plumari P, Chatterjee S, Alam J, Scardina F, Greco V, arXiv:1608.02231 [nucl-th].

Constraining in-medium heavy-quark energy-loss mechanisms via angular correlations between heavy and light mesons

M Rohrmoser, P-B Gossiaux, T Gousset, and J Aichelin
Subatech, 4 rue Alfred Kastler, 44307 Nantes, F

E-mail: martin.rohrmoser@subatech.in2p3.fr, gossiaux@subatech.in2p3.fr, gousset@subatech.in2p3.fr, aichelin@subatech.in2p3.fr .

Abstract. Two-particle correlations obtained from parton showers that pass the hot and dense medium of the quark gluon plasma (QGP) can be used as an alternative observable, in addition to the combination of the nuclear modification factor R_{AA} and the elliptic flow v_2, to study the mechanisms of in-medium heavy quark energy-loss. In particular, angular correlations represent a promising tool to distinguish between energy loss due to collisional and radiative interactions of jet and medium particles. To this end, parton cascades were created in Monte-Carlo simulations, where individual particles can undergo both parton splitting as well as an effective jet-medium interaction. A first model simulates the effects of induced radiations on parton cascades. Its consequences on angular correlations of partons within jets were studied in detail, with particular focus on angular broadening. The results can be compared to a second model that effectively describes elastic scatterings of jet and medium particles.

1. Introduction

Energetic heavy quarks passing through the QGP, observed by the resulting mesons, are viewed as a suitable probe for the interactions inside the QGP, in particular for the mechanisms of energy loss, as they are less likely to thermalize within the medium and are mostly created at early stages of the medium evolution. However, models of both purely collisional energy loss and combinations of collisional and radiative energy loss are equally successful for reproducing the nuclear modification factor R_{AA} and the elliptic flow v_2 [1].

Angular correlations between two mesons represent alternative observables to make progress for discriminating between the two different interaction mechanisms. Previous studies, e.g. Ref. [2], used azimuthal correlations between pairs of heavy mesons such as D-$\bar{D}$ pairs to distinguish between the energy-loss scenarios described. As an alternative, one can investigate the angular correlations between pairs of heavy and light mesons (D and π) in the forward direction. This second approach is more sensitive to medium effects on the evolution of an individual jet, e.g. angular jet-broadening.

In order to perform such a study a Monte-Carlo code that simulates parton cascades, based on a similar approach as in Ref. [3], was created. This algorithm allows for parton branchings and, furthermore, includes interactions of the cascade-particles with the medium in an effective model. In case of parton branchings, the outgoing partons and their four-momenta are selected from the distributions of splitting functions for collinear splitting, of leading order of α_s, cf.

IOP Conf. Series: Journal of Physics: Conf. Series **779** (2017) 012032 doi:10.1088/1742-6596/779/1/012032

Ref. [4, 5]. The parton cascades are initiated by a single heavy quark with a certain energy E_{ini}, and a maximal virtuality $Q_\uparrow$ and follow an evolution down to a virtuality threshold $Q_\downarrow$

The angular two-particle correlations considered are distributions over the angles $\Delta\theta$ between the direction of the three-momenta of the heavy quark and that of every other cascade particle in its final state. Studies involving heavy quarks have the crucial advantage that the heavy particle can be tagged among the jet particles. From Ref. [6] it can be deduced that for sufficiently high virtualities Q, where $Q \gg m$, the radiation behavior of a heavy quark can be approximated by that of a massless quark. Therefore, the same splitting functions are used for, heavy as well as light quarks, while the explicit inclusion of a quark mass remains to be done in further studies.

2. Model for inelastic jet-medium interaction

The jet-medium interaction is described by a continuous increase of squared virtuality $Q^2(t)$ over time, according to the law $\dot{Q}^2(t) = \hat{q}(t)$, during the propagation of the partons in the medium, cf. Ref. [3]. This kind of approach serves as an effective model for medium-induced radiation, since partons with these additional amounts of virtuality will, on average, branch more often than their vacuum counterparts. In the Monte-Carlo implementation of the in-medium propagation, the continuous interactions with the cascade particles were discretized over small time-steps Δt, which leads to the following changes in parton virtuality over time: $Q(t + \Delta t) \mapsto \sqrt{Q^2(t) + \Delta t\, \hat{q}(t)}$. In order to avoid confusion of the medium effects present in cascades of this kind of effective model with consequences of three-momentum transfers from medium to jet particles, the virtuality increase was attributed solely to an increase in parton energy, i.e.: $E(t + \Delta t) \mapsto \sqrt{E^2(t) + \Delta t\, \hat{q}(t)}$, $\vec{p}(t + \Delta t) \mapsto \vec{p}(t)$. The evolution of the transfer $\hat{q}(t)$ within the medium was fitted to a hydrodynamical description in Ref. [3]. In particular, it is assumed that the initial quarks of the cascades are created at the center of a medium that extends up to $L = 10$ fm/c. In order to quantify the strength of the jet-medium interaction, the quantity ΔQ^2 was used, defined as $\Delta Q^2 = \int_0^L \hat{q}(t)dt$.

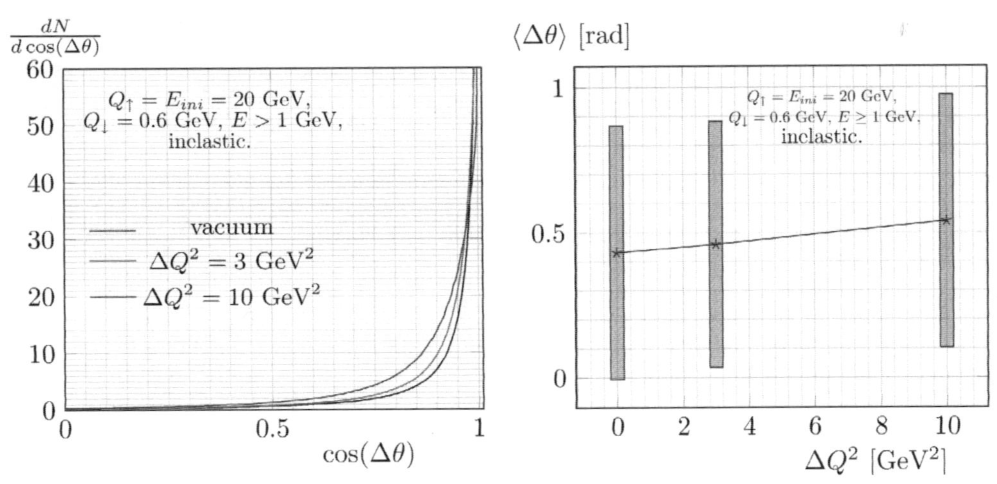

Figure 1. left panel: angular correlations in the form of $\frac{dN}{d\cos(\Delta\theta)}$, normalized to the average particle number per cascade. right panel: Average values of $\Delta\theta$ (stars) and corresponding standard deviation (boxes) for the three different angular correlations shown in the left panel.

3. Results for angular correlations

The investigation of medium effects on the evolution of parton cascades by means of two-particle correlations focused initially on a verification of angular broadening, which is later on studied via more refined observables. The left panel of Fig. 1 shows angular correlations in the form of $\frac{dN}{d\cos(\Delta\theta)}$ for different values of ΔQ^2. A broadening of the distribution with increasing ΔQ^2 can be observed. In order to quantify this medium effect, the mean values of $\Delta\theta$ are depicted in the right panel of Fig. 1 as functions of ΔQ^2.

To understand how angular correlations are built up during jet evolution, the sets of parton cascades obtained by Monte-Carlo simulations were sorted with regard to their topologies. A simple interesting case consists of a quark that splits two times successively into a quark and a gluon. The average values of the angles between the outgoing quark and gluon in the first and second splitting, ϑ_1 and ϑ_2, respectively, are shown in Fig 2. In contrast to the first branching angle the second branching angle increases strongly with ΔQ^2. This behavior can be explained with the much smaller virtualities of the quark after the first splitting than in the initial state. Thus, the time between the first and second splitting is longer than the time before the first splitting, which results in higher amounts of accumulated virtuality from jet-medium interactions.

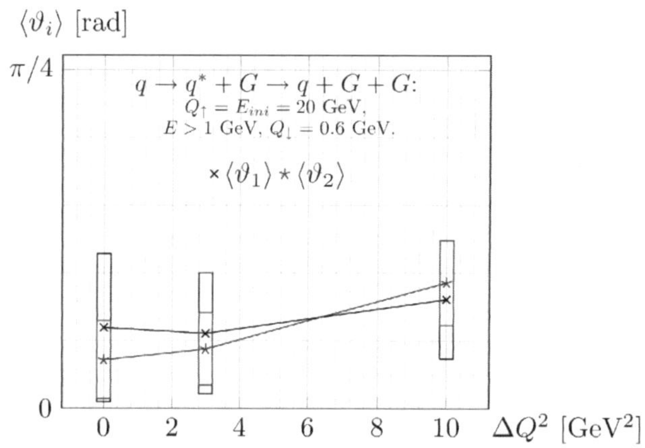

Figure 2. Average values of the first and second branching angles, ϑ_1 (crosses) and ϑ_2 (stars), respectively, as a function of ΔQ^2 for cascades with two consecutive gluon radiations from a single quark line together with the standard deviations of the distributions over ϑ_1 and ϑ_2 (boxes). The cascades were extracted from the set used for the results in Fig. 1.

Furthermore, in Fig. 2, one observes angular ordering in the form $\vartheta_1 > \vartheta_2$ in the vacuum. In the medium, this behavior is weakened, and for $\Delta Q^2 = 10$ GeV^2 even inversed. In order to find realistic observables that allow to look for such a kind of effect, and, furthermore, to discriminate between elastic and inelastic jet-medium interactions, the distribution $\frac{d^2N}{d\Delta\theta_{jQ}d\Delta\theta}$ was considered. There, $\Delta\theta_{jQ}$ is the angle between the directions of the three-momenta of the entire jet and of the tagged heavy quark. For the elastic jet-medium interactions, an effective model analogous to the one for inelastic scattering was used: In this model, however, the virtualities Q are constant during in-medium propagation, while the partons experience a longitudinal drag force and transverse momentum transfers following the description in Ref. [7]. In the elastic model, the transport coefficient is $\hat{q} = \frac{d\langle p_T^2\rangle}{dt}$. Fig. 3 compares the distributions $\frac{d^2N}{d\Delta\theta_{jQ}d\Delta\theta}$ for both models with the same value $\int_0^L \hat{q}dt = 10$ GeV^2. One can observe a strong broadening in $\Delta\theta$ for the elastic model, while this effect is less apparent for the inelastic model. For the latter,

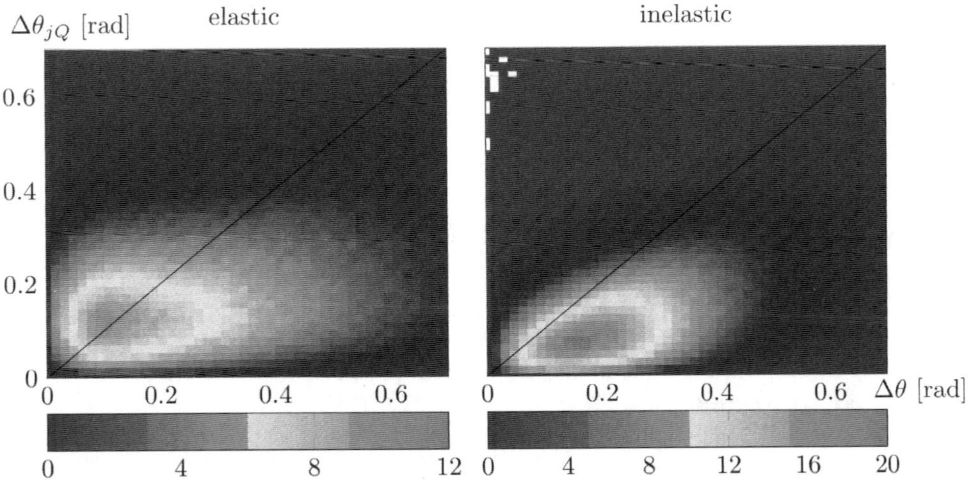

Figure 3. Distributions $\frac{d^2N}{d\Delta\theta_{jQ}d\Delta\theta}$ for parton cascades with $Q_\uparrow = E_{ini} = 20$ GeV and $Q_\downarrow = 0.6$ GeV that are subjected to elastic (left panel) and inelastic (right panel) jet-medium interactions, both for $\int_0^L \hat{q}dt = 10$ GeV^2. Only particles with a three-momentum that fulfills $\|\vec{p}\| \geq 2$ GeV were considered. The distributions are normalized to the average number of the corresponding parton pairs per cascade.

the peak of the distribution can be found at larger values of $\Delta\theta$ than for the elastic model, where the peak of the distribution occurs at higher $\Delta\theta_{jQ}$ values.

4. Conclusions and outlook

It was verified for parton cascades produced in an effective model for medium-induced radiation that angular two-particle correlations are sensitive observables to an angular broadening in the medium. A further investigation of cascades of particular topologies showed that later stages of jet evolution are more sensitive to this medium effect than earlier ones, which ultimately results in an inversion of angular ordering.

Currently, angular two-particle correlations are investigated in another effective model that is based on elastic collisions in an approach presented in Ref. [7]. It was shown that more sophisticated correlations could be useful to discriminate between the models for HQ-jet modifications by the medium, although they are more difficult to extract from the background.

Acknowledgments

M R thanks the organizers of the Strangeness in Quark Matter 2016 conference in Berkeley, CA, USA, and the project TOGETHER of the Region Pays de la Loire, France for their financial support that allowed for the participation in the conference.

References

[1] Gossiaux P-B, Aichelin J, Gousset T and Guiho V 2010 *J. Phys.* G **37** 094019
[2] Nahrgang M, Aichelin J, Gossiaux P-B and Werner K 2014 *J. Phys. Conf. Ser.* **509** 012047
[3] Renk T 2008 *Phys. Rev.* C **78** 034908
[4] Altarelli G and Parisi G 1977 *Nuclear Physics* B **126** 298 - 318
[5] Dokshitzer Y L 1977 *Sov. Phys. JETP* **46** 641-53
[6] Mele B and Nason P 1991 *Nucl. Phys.* B **361** 626-44
[7] Berrehrah H, Gossiaux P-B, Aichelin J, Cassing W and Bratkovskaya E (2014) *Phys. Rev.* C **90** 064906

IOP Conf. Series: Journal of Physics: Conf. Series **779** (2017) 012033 doi:10.1088/1742-6596/779/1/012033

D-meson observables in Pb-Pb and p-Pb collisions at LHC with EPOSHQ model

V Ozvenchuk[1], J Aichelin[2], P B Gossiaux[2], B Guiot[3], M Nahrgang[2] and K Werner[2]

[1] Institute of Nuclear Physics, Polish Academy of Science, Radzikowskiego 152, 31-342 Krakow, Poland

[2] Subatech, UMR 6457, IN2P3/CNRS, Universite de Nantes, Ecole des Mines de Nantes, 4 rue Alfred Kastler, 44307 Nantes cedex 3, France

[3] Universidad Tecnica Federico Santa Maria, Valparaiso, Chile

E-mail: `Vitalii.Ozvenchuk@ifj.edu.pl`

Abstract. We study the propagation of charm quarks in the quark-gluon plasma (QGP) created in ultrarelativistic heavy-ion and proton-nucleus collisions at LHC within EPOSHQ model. The interactions of heavy quarks with the light partons in ultrarelativistic heavy-ion collisions through the collisional and radiative processes lead to a large suppression of final D-meson spectra at high transverse momentum and a finite D-meson elliptic flow, v_2, whereas in proton-nucleus collisions the D-meson nuclear modification factor, R_{pA}, at high transverse momentum is compatible with unity. Our results are in good agreement with the available experimental data.

1. Introduction

The nuclear modification factor, R_{AA}, and the elliptic flow, $v2$, of D mesons are measured in ultrarelativistic heavy-ion ($PbPb$) collisions [1, 2, 3] and in proton-nucleus (pPb) collisions [4] at LHC. The R_{AA} of intermediate- and high-p_T D mesons measured in $PbPb$ collisions [1, 2] is significantly below unity. It demonstrates a large energy loss of high-p_T heavy quarks in the QGP produced in heavy-ion collisions. On the other hand, the R_{AA} of D mesons with intermediate- and high-p_T measured in pPb collisions [4] is compatible with unity. Many theoretical models describing the D-meson observables either in $PbPb$ collisions [5, 6, 7, 8] or in pPb collisions [9, 10, 11], but it is still a big challenge to describe both systems with the same model input.

In this contribution we present the nuclear modification factor and the elliptic flow of D mesons in $PbPb$ collisions calculated in MC@sHQ+EPOS2 model. Then we compare the MC@sHQ+EPOS2 results with the results for the R_{AA} and v_2 of D mesons obtained in MC@sHQ+EPOS3 model. These models couple a Monte Carlo propagation of heavy quarks to the 3+1 dimensional fluid dynamical evolution of the QGP from EPOS2 and EPOS3 initial conditions, respectively. Finally we calculate the R_{AA} of D mesons in $PbPb$ and pPb collisions within MC@sHQ+EPOS3 model implementing, in addition, the heavy quarks from EPOS3 initial conditions (EPOSHQ model).

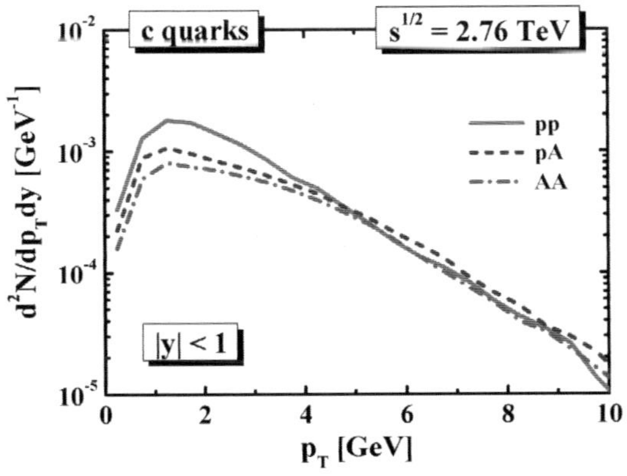

Figure 1. The initial spectrum of charm quarks obtained from EPOS3 for pp, pA and AA collisions.

2. The model

We calculate the D-meson observables using the different iterations of the EPOSHQ model, in particular, MC@sHQ+EPOS2, MC@sHQ+EPOS3 and EPOSHQ models. MC@sHQ+EPOS2 model couples a Monte Carlo propagation of heavy quarks [12, 13, 14], MC@sHQ, to the 3+1 dimensional ideal fluid dynamical evolution of the QGP from EPOS2 initial conditions [15, 16], which combine perturbative QCD (pQCD) calculations of the hard scattering with the Gribov-Regge theory. In MC@sHQ+EPOS3 model we couple MC@sHQ with the recent upgrade EPOS3 that includes a viscous fluid dynamical evolution, based on [17].

We initialize the heavy quarks at the original nucleon-nucleon scattering points according to the p_T-distribution from pQCD results in fixed order plus next-to-leading logarithm (FONLL) [18, 19, 20]. In EPOSHQ model we implement the heavy quarks directly from EPOS3 initial conditions with initial state shadowing included. The heavy quarks in EPOS3 can be produced during the spacelike cascade, the born process and the partonic shower. In Fig. 1 we show the initial spectrum of charm quarks from EPOS3 for pp, pA and AA collisions. Heavy quarks interact with plasma partons by either elastic or radiative collisions. The evolution of heavy quarks is described by the Boltzmann equation. The heavy quarks form hadrons via coalescence [12], predominantly for low-p_T heavy quarks, or fragmentation [21], predominantly for intermediate- and high-p_T quarks on the hypersurface of constant temperature $T = 155$ MeV.

3. The nuclear modification factor and elliptic flow

In Figs. 2 and 3 we present the D-meson nuclear modification factor and elliptic flow calculated within the MC@sHQ+EPOS2 model (blue lines) and the MC@sHQ+EPOS3 model (red lines) for the $PbPb$@2.76 TeV collisions at LHC. The MC@sHQ+EPOS2 results are taken from [22]. In each model we perform calculations for two scenarios, with purely collisional energy loss (dashed lines) or with collisional+radiative energy loss (solid lines). Both models show a reasonable agreement with the experimental data for the R_{AA} of D mesons at $p_T > 5$ GeV. The larger suppression of D-meson yield is seen for the MC@sHQ+EPOS3 results at intermediate p_T. If the initial state shadowing is not included the results from both models are overestimated at low p_T. For the whole range of p_T MC@sHQ+EPOS2 and MC@sHQ+EPOS3 model well describe the elliptic flow of D mesons. MC@sHQ+EPOS3 results show the enhancement of v_2 at intermediate p_T in comparison to MC@sHQ+EPOS2 results. The final elliptic flow of D mesons can be sensitive to the D-meson rescattering in the hadronic phase [23], we need to include the hadronic contribution for further study.

In Figs. 4 and 5 we show the c-quark and D-meson R_{pA} obtained within EPOSHQ model for the pPb@5.02 TeV collisions at LHC. The suppression of c-quark yield at low p_T for the scenario

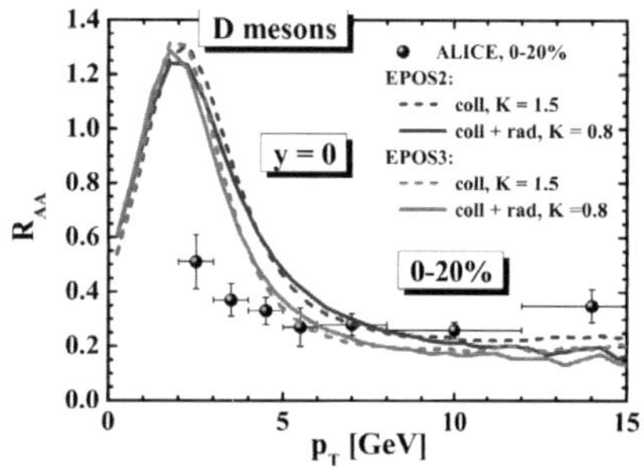

Figure 2. *D*-meson nuclear modification factor calculated in MC@sHQ+EPOS2 model (blue lines) in comparison to the MC@sHQ+EPOS3 results (red lines). Experimental data are taken from [1].

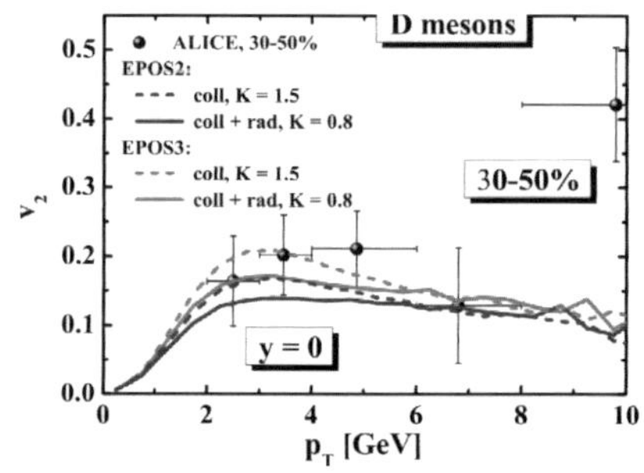

Figure 3. *D*-meson elliptic flow calculated in MC@sHQ+EPOS2 model (blue lines) in comparison to the MC@sHQ+EPOS3 results (red lines). Experimental data are taken from [3].

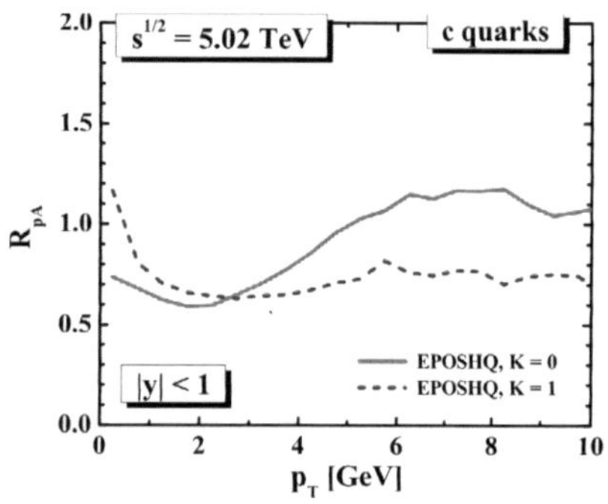

Figure 4. *c*-quark R_{pA} calculated in EPOSHQ model with (dashed line) and without (solid line) interactions with the bulk medium.

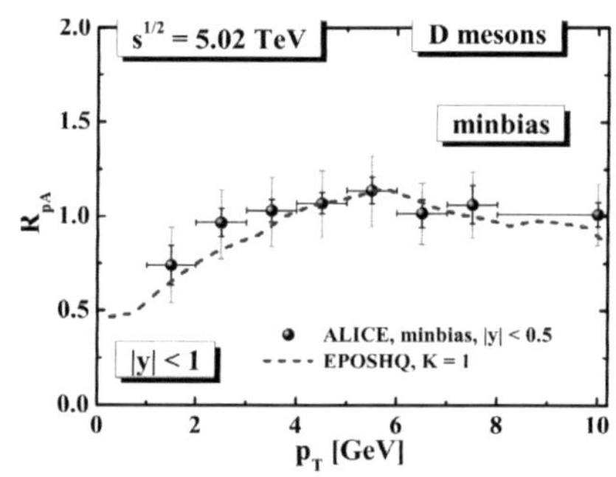

Figure 5. *D*-meson R_{pA} calculated in EPOSHQ model in comparison to the experimental data from [4].

with $K = 0$ (no interactions with the bulk medium) indicates the present of the initial state shadowing (see Fig. 1). The scenario with $K = 1$ corresponds to the collisional+radiative energy loss, where the suppression of c-quark yield at intermediate and high p_T is seen. The R_pA of D mesons is compatible with unity for $p_T > 4$ TeV, which is in agreement with experimental data.

Finally, in Fig. 6 we present the *D*-meson nuclear modification factor calculated in EPOSHQ model for the central $PbPb$@2.76 TeV collisions at LHC. The calculations show a reasonable agreement with the experimental data for whole range of p_T. For two scenarios (pure collisional and collisional+radiative) we obtained almost the same behavior. The presence of the initial state shadowing helped us to describe well the R_{AA} of D mesons at low p_T.

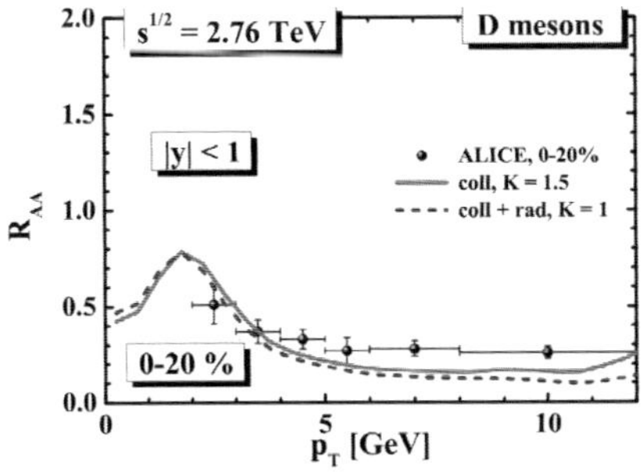

Figure 6. D-meson R_{AA} obtained within EPOSHQ model in comparison to the experimental data [1].

4. Conclusions

We have presented the results for D-meson observables in heavy-ion and proton-nucleus collisions at LHC within EPOSHQ model. EPOSHQ model is able to describe the nuclear modification factor and elliptic flow of D mesons both for the $PbPb$ and pPb collisions at LHC with the same initial setup. For the central $PbPb$ collisions the importance of including initial state shadowing was shown.

Acknowledgments

We thank J. Steinheimer and Iu. Karpenko for fruitful discussions. This work was supported by the Project TOGETHER (Pays de la Loire) and by the National Science Center, Poland under Grant No. 2014/14/E/ST2/00018.

References

[1] Abelev B *et al.* [ALICE Collaboration] 2012 *JHEP* **09** 112
[2] Sakai S *et al.* [ALICE Collaboration] 2013 *Nucl. Phys.* A **904-905** 661c
[3] Abelev B *et al.* [ALICE Collaboration] 2013 *Phys. Rev. Lett.* **111** 102301
[4] Abelev B *et al.* [ALICE Collaboration] 2014 *Phys. Rev. Lett.* **113** 232301
[5] Nahrgang M, Aichelin J, Bass S A, Gossiaux P B and Werner K 2014 *Nucl. Phys.* A **931** 575
[6] Song T, Berrehrah H, Cabrera D, Cassing W and Bratkovskaya E 2016 *Phys. Rev.* C **93** 034906
[7] Das S K, Scardina F, Plumari S and Greco V 2015 *Phys. Lett.* B **747** 260
[8] Cao S, Qin G Y and Bass S A 2013 *Phys. Rev.* C **88** 044907
[9] Mangano M, Nason P and Ridolfi G 1992 *Nucl. Phys.* B **373** 295
[10] Stump D, Huston J, Pumplin J, Tung W K, Lai H L, Kuhlmann S and Owens J F 2003 *JHEP* **0310** 046
[11] Xu Y, Cao S, Ke W, Nahrgang M, Auvinen J and Bass S A 2016 *Nucl. Part. Phys. Proc.* **276-278** 225
[12] Gossiaux P B, Bierkandt R and Aichelin J 2009 *Phys. Rev.* C **79** 044906
[13] Gossiaux P B and Aichelin J 2008 *Phys. Rev.* C **78** 014904
[14] Gossiaux P B, Aichelin J, Gousset T and Guiho V 2010 *J. Phys.* G **37** 094019
[15] Werner K, Karpenko I, Pierog T, Bleicher M and Mikhailov K 2010 *Phys. Rev.* C **82** 044904
[16] Werner K, Karpenko I, Bleicher M, Pierog T and Porteboeuf-Houssais S 2012 *Phys. Rev.* C **85** 064907
[17] Karpenko I, Huovinen P and Bleicher M 2014 *Comput. Phys. Commun.* **185** 3016
[18] Cacciari M, Greco M and Nason P 1998 *JHEP* **05** 007
[19] Cacciari M, Frixione S and Nason P 2001 *JHEP* **03** 006
[20] Cacciari M, Frixione S, Houdeau N, Mangano M L, Nason P and Ridolfi G 2012 *JHEP* **10** 137
[21] Cacciari M, Nason P, and Vogt R 2005 *Phys. Rev. Lett.* **95** 122001
[22] Nahrgang M, Aichelin J, Gossiaux P B and Werner K 2014 *Phys. Rev.* C **89** 014905
[23] Ozvenchuk V, Torres-Rincon J M, Gossiaux P B, Tolos L and Aichelin J 2014 *Phys. Rev.* C **90** 054909

T-Matrix Approach to Strongly Coupled QGP

Shuai Y.F. Liu, Ralf Rapp

Cyclotron Institute and Department of Physics&Astronomy, Texas A&M University, College Station, TX 77843-3366, USA

E-mail: lshphy@gmail.com, rapp@comp.tamu.edu

Abstract. Based on a thermodynamic T-matrix approach we extract the potential V between two static charges in the quark-gluon plasma (QGP) from fits to the pertinent lattice-QCD free energy. With suitable relativistic corrections we utilize this new potential to compute heavy-quark transport coefficients and compare the results to previous calculations using either F or U as potential. We then discuss a generalization of the T-matrix re-summation to a "matrix log" re-summation of t-channel diagrams for the grand partition function of the QGP in the Luttinger-Ward skeleton diagram formalism. With V as a non-perturbative driving kernel in the light-parton sector, we obtain the QGP equation of state from fits to lattice-QCD data. The resulting light-parton spectral functions are characterized by large thermal widths at small momenta, indicating the dissolution of quasi-particles in a strongly coupled QGP.

1. Introduction

The large suppression and elliptic flow of heavy-flavor (HF) spectra in heavy-ion collisions (HICs) at RHIC and LHC indicate that heavy quarks couple strongly to the quark-gluon plasma (QGP), see, *e.g.*, Ref. [1] for a recent review. To address this problem non-perturbatively, we have been developing a many-body T-matrix approach that includes fundamental features of the QCD force constrained by lattice QCD (lQCD) [2, 3]. One of the debated issues in such approaches is precisely which quantity is the appropriate heavy-quark (HQ) potential, V. This may depend on the definition of the potential in a given approach. In practical applications, one often employs the free (F) and internal (U) energies, which are directly available from lQCD, as limiting cases. The former gives rise to a rather weak in-medium force, while for U the in-medium force is substantially stronger, which is, in fact, preferred by phenomenological applications, *e.g.*, to HF diffusion in HICs [4]. However, a more rigorous and quantitative determination of the potential is desirable. Toward this end, in recent work [5], we have defined the potential as a driving kernel in a thermodynamic T-matrix equation. We have derived the pertinent free energy and, through fits to lQCD data for F, extracted the quantitative form of the underlying V.

In the present paper, we will briefly review this work and then discuss first applications to the HQ transport coefficients and, by extending it to the light sector, to the equation of state (EoS) of the QGP.

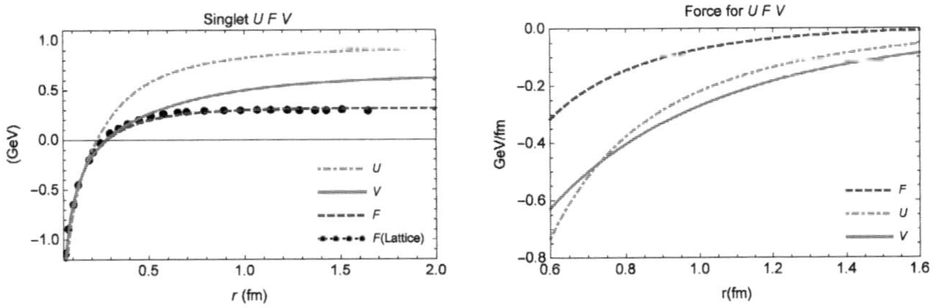

Figure 1: Color-singlet U, F and V (left) and corresponding forces (right) at T=0.24 GeV.

2. Heavy-Quark Potential and Transport Coefficients

In Ref. [5], the thermodynamic T-matrix approach has been used as a starting point for defining the HQ potential, V. The resulting expression for the static HQ free energy, F, was obtained as

$$F(r) = -\frac{1}{\beta}\ln\left[\int_{-\infty}^{\infty} d\omega \frac{-1}{\pi}\mathrm{Im}\left[\frac{1}{[\hat{G}(E+i\epsilon)]^{-1} - V(E+i\epsilon, r)}\right] e^{-\beta\omega}\right]$$

where $\hat{G}$ is the non-interacting 2-body Green's function including in-medium HQ self-energies. A 4-parameter ansatz for V was made consisting of screened Cornell and string terms. The resulting fit to lQCD data for F [6, 7, 8] is shown in Fig. 1 left, along with the underlying potential and the ensuing internal energy, U. While the magnitude of V lies between F and U, it turns out that V actually has the largest force at large distance, for $r \geq 0.7$ fm (Fig. 1 right). This will allow heavy quarks to interact with more medium partons at a time, with important consequences for the transport coefficients. Next, we use V in the T-matrix formalism [2, 3, 9] to calculate the thermal HQ relaxation rate within a Fokker-Planck framework [10] assuming either F, U or V as potential, see Fig. 2 left. At low momenta, the rate from V is the largest, due to the long-range force in the potential as mentioned above. Quite remarkably, the temperature ordering is reversed, due to large contributions from the string term. A schematic dimensional analysis suggests the cross section for the string term to be proportional to σ/T^4, where σ is the string tension. This low-T enhancement overcomes the decrease in density ($\sim T^3$) when T decreases, thus leading to an increased thermalization rate. With increasing momentum, the strong coupling transitions into a weakly coupled regime driven by short-range color-Coulomb interactions at high momentum, recovering a normal temperature hierarchy (see insert in Fig. 2 left). The resulting spatial diffusion coefficients, scaled by the thermal wavelength, are displayed in Fig. 2 right. The results for U and F slightly differ from previous results [2, 3] due to the inclusion of higher partial waves, interference effects and a different normalization factor.

3. Equation of State

The large drag coefficient obtained from V translates into large HQ self-energies, with pertinent widths of order 0.5-1 GeV. To investigate the impact of the strong coupling on the "light"-parton sector, we have developed a generalized T-matrix formalism to compute the EoS of QGP. Specifically, we re-sum all t-channel diagrams in the Luttinger-Ward skeleton expansion [11, 12] through a novel matrix "log" technique which allows us to include both single-parton and resonance (bound-state) contributions on the same footing, with self-consistent thermal widths. More explicitly, the Luttinger-Ward functional can be schematically written as

$$\Omega(T) = \sum_{s,c,f} \pm \int d\tilde{^4p}\left[\left[\ln\left(-G^{-1}\right) + \Sigma G\right] - \sum_{\nu}\frac{1}{2\nu}\Sigma_\nu G\right] . \quad (1)$$

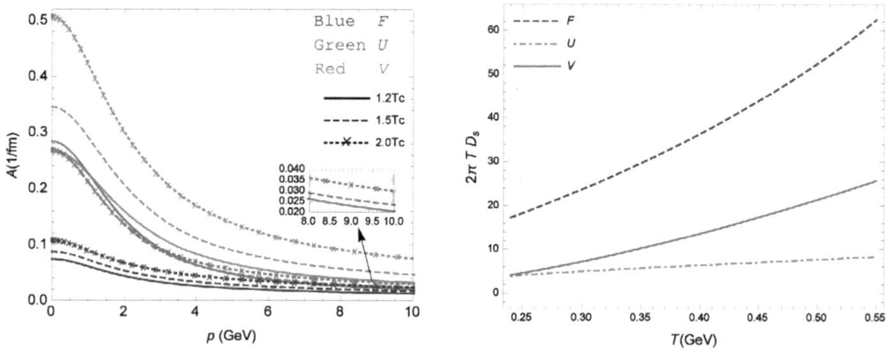

Figure 2: Drag (left) and spatial diffusion coefficients (right) calculated for F, U and V.

The single-particle self-energy in T-matrix approximation can be written as

$$\Sigma = \sum_{s,c,f} \int d\tilde{^4p} TG = \sum_{s,c,f} \int d\tilde{^4p} \left\{ V + V\hat{G}V + \ldots + V\hat{G}V\hat{G}\ldots.V \right\} G$$
$$= \sum_{s,c,f} \int d\tilde{^4p} V \left(1 - \hat{G}V\right)^{-1} G \,, \tag{2}$$

where the re-summation of the ladder series is carried out through matrix inversion. In analogy, we identify the series for the skeleton expansion via a natural-log function (ln) which can be resummed into a "matrix-log",

$$\sum_{\nu} \frac{1}{2\nu} \Sigma_\nu = \frac{1}{2} \sum_{s,c,f} \int d\tilde{^4p} \left\{ V + \frac{1}{2} V\hat{G}V + \ldots + \frac{1}{\nu} V\hat{G}V\hat{G}\ldots.V \right\} G$$
$$= \frac{1}{2} \sum_{s,c,f} \int d\tilde{^4p} \left\{ -\hat{G}^{-1} \ln\left(1 - \hat{G}V\right) \right\} G \,. \tag{3}$$

Applying this expression with the in-medium V as the potential, we calculate the EoS of the QGP *self-consistently*. Similar to quasi-particle approaches [13], we employ the quark and gluon masses as two fit parameters to the EoS of lQCD [14], see Fig. 3. The self-consistent formalism generates self-energies with non-trivial energy-momentum dependencies, in particular thermal widths. The latter turn out to be very large at low parton momenta, of order 1(0.5) GeV at T=1.2(2.0)T_c, see Fig. 4. This implies the loss of well-defined quasi-particle excitations as T_c is approached from above, while at higher T and/or 3-momenta (p), quasi-particle peaks are resurrected in the spectral functions. Our calculations thus suggest a rich structure of the QGP, with strong coupling for soft modes and a transition to a quasi-particle like, more weakly coupled QGP for harder modes. Again, we believe that these structures are critically driven by "cross sections" (although we do not use this notion) for the string term ($\sim\sigma/T^4$) which dominate over the reduction in density ($\sim T^3$) as T is lowered, rendering a more strongly coupled medium. This feature is intimately related to the T and p dependencies of the HQ transport coefficients discussed in the previous section. We note that we currently encounter instabilities of the self-consistent iteration procedure when computing gluon self-energies from the T-matrix close to T_c. These instabilities require further studies in the future, together with calculations of susceptibilities at finite chemical potential. This may help us to understand the phase structure generated by the T-matrix approach and its relation to QCD phase structure.

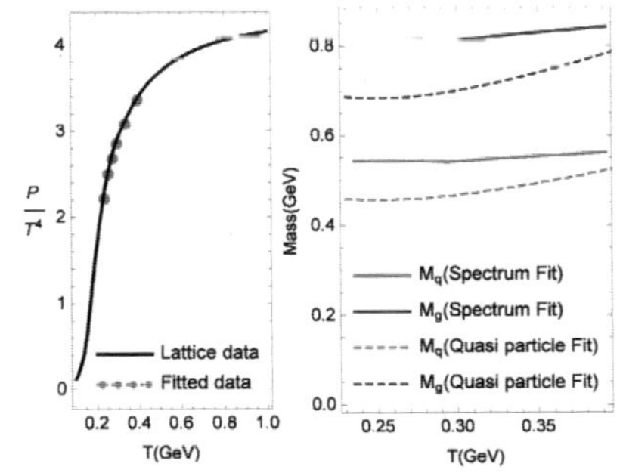

Figure 3: Fit to the lattice-QCD EoS (left) and the fitted quark and gluon masses (right).

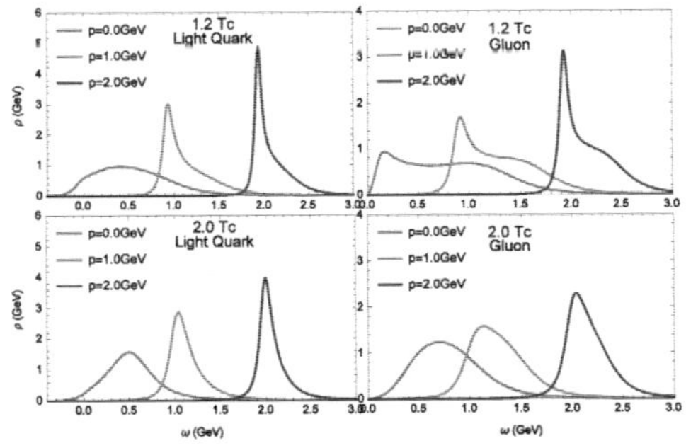

Figure 4: Light-parton spectral functions at different temperatures and momenta (T_c=196 MeV).

4. Conclusion and Perspective

We have employed a thermodynamic T-matrix approach with non-perturbative interactions to study HQ transport and the EoS of the QGP. We have extracted a HQ potential V by fitting the HQ free energy to lQCD data, and compared HQ transport coefficients with different driving kernels (F, U and V) for heavy-light interactions. We find that the low-momentum drag coefficient is largest for V, due to a long-range remnant of the confining force, with a nontrivial temperature dependence. We have extended the T-matrix to the light sector to study the EoS of the QGP. Within a self-consistent skeleton formalism, we find light-parton spectral functions with large widths at low momenta and temperatures, decreasing with temperature. This behavior directly reflects the properties of the HQ drag coefficients. Large values for parton widths and HQ drag coefficients suggest the QGP to be strongly coupled, where soft modes are no longer quasi-particles while heavy quarks may still serve as a suitable probe. In the future, we will work toward a unified non-perturbative T-matrix framework for heavy and light sectors that can simultaneously describe various lQCD data such as the EoS, susceptibilities, correlator ratios, and static free energies [5]. We believe that the variety of lQCD data can help to constrain the uncertainties of the framework so that we can perform controlled calculations of transport coefficients that directly link to experimental observables in HICs.

Acknowledgments

This work is supported by the U.S. NSF through grant no. PHY-1614484.

References

[1] Prino F and Rapp R 2016 *J. Phys. G* **43** 093002
[2] van Hees H, Mannarelli M, Greco V and Rapp R 2008 *Phys. Rev. Lett.* **100** 192301
[3] Riek F and Rapp R 2010 *Phys. Rev. C* **82** 035201
[4] He M, Fries R J and Rapp R 2014 *Phys. Lett. B* **735** 445–450
[5] Liu S Y F and Rapp R 2015 *Nucl. Phys. A* **941** 179–187
[6] Kaczmarek O and Zantow F 2005 *Phys. Rev. D* **71** 114510
[7] Kaczmarek O 2007 *PoS* **CPOD07** 043
[8] Petreczky P and Petrov K 2004 *Phys. Rev. D* **70** 054503
[9] Huggins K and Rapp R 2012 *Nucl. Phys. A* **896** 24–45
[10] Svetitsky B 1988 *Phys. Rev. D* **37** 2484–2491
[11] Luttinger J M and Ward J C 1960 *Phys. Rev.* **118**(5) 1417–1427
[12] Baym G 1962 *Phys. Rev.* **127** 1391–1401
[13] Plumari S, Alberico W M, Greco V and Ratti C 2011 *Phys. Rev. D* **84** 094004
[14] Borsanyi S, Endrodi G, Fodor Z, Jakovac A, Katz S D, Krieg S, Ratti C and Szabo K K 2010 *JHEP* **11** 077

IOP Conf. Series: Journal of Physics: Conf. Series **779** (2017) 012035 doi:10.1088/1742-6596/779/1/012035

Heavy flavor R_{AA} and v_n in event-by-event viscous relativistic hydrodynamics

Caio A. G. Prado[1], Jacquelyn Noronha-Hostler[2], Mauro R. Cosentino[3], Marcelo G. Munhoz[1], Jorge Noronha[1], Alexandre A. P. Suaide[1]

[1] Instituto de Física, Universidade de São Paulo, C.P. 66318, 05315-970 São Paulo, SP, Brazil
[2] Department of Physics, University of Houston, Houston TX 77204, USA
[3] Centro de Ciências Naturais e Humanas, Universidade Federal do ABC, Av. dos Estados 5001, Bairro Santa Terezina, 09210-508 Santo André, SP, Brazil

E-mail: cagprado@if.usp.br

Abstract. Recently it has been shown that a realistic description of the medium via event-by-event viscous hydrodynamics plays an important role in the long-standing R_{AA} vs. v_2 puzzle at high p_T. In this proceedings we begin to extend this approach to the heavy flavor sector by investigating the effects of full event-by-event fluctuating hydrodynamic backgrounds on the nuclear suppression factor and $v_2\{2\}$ of heavy flavor mesons and non-photonic electrons at intermediate to high p_T. We also show results for $v_3\{2\}$ of B^0 and D^0 for PbPb collisions at $\sqrt{s} = 2.76$ TeV.

1. Introduction

Heavy quarks, such as bottom and charm, are very useful probes of Quark-Gluon Plasma (QGP) dynamics. Because of their large mass, these quarks are produced at the very early stages of the collision via hard processes and their subsequent propagation through the hot and dense medium is sensitive to the whole hydrodynamic evolution of the system. The energy loss experienced by the heavy quarks during their path within the plasma has been studied using the nuclear modification factor (at mid-rapidity), defined as

$$R_{\mathrm{AA}}(p_T, \varphi) = \frac{\frac{\mathrm{d}N_{\mathrm{AA}}}{\mathrm{d}p_T\,\mathrm{d}\varphi}}{N_{\mathrm{coll}}\frac{\mathrm{d}N_{\mathrm{pp}}}{\mathrm{d}p_T}}, \tag{1}$$

where $\mathrm{d}N_{\mathrm{AA}}/\,\mathrm{d}p_T = \frac{1}{2\pi}\int_0^{2\pi} d\varphi \frac{\mathrm{d}N_{\mathrm{AA}}}{\mathrm{d}p_T\,\mathrm{d}\varphi}$ is the spectrum of heavy quarks in AA collisions while $\mathrm{d}N_{\mathrm{pp}}/\,\mathrm{d}p_T$ is the corresponding proton-proton yield, φ is the azimuthal angle in the plane transverse to the beam direction, and N_{coll} is the number of binary collisions (computed within the Glauber model).

The dependence of $R_{\mathrm{AA}}(p_T, \varphi)$ on the azimuthal angle φ can be used to study the energy loss and its path length dependence in the plasma. The degree of anisotropy in $R_{\mathrm{AA}}(p_T, \varphi)$ is determined via the v_n^{heavy} coefficients of its Fourier expansion

$$\frac{R_{\mathrm{AA}}(p_T, \varphi)}{R_{\mathrm{AA}}(p_T)} = 1 + 2\sum_{n=1}^{\infty} v_n^{\mathrm{heavy}}(p_T)\cos\left[n\varphi - n\psi_n^{\mathrm{heavy}}(p_T)\right] \tag{2}$$

where $R_{\mathrm{AA}}(p_T) = \frac{1}{2\pi}\int_0^{2\pi} d\varphi\, R_{\mathrm{AA}}(p_T,\varphi)$ is the azimuthal average and

$$v_n^{\mathrm{heavy}}(p_T) = \frac{\frac{1}{2\pi}\int_0^{2\pi} d\varphi \cos\left[n\varphi - n\psi_n^{\mathrm{heavy}}(p_T)\right] R_{\mathrm{AA}}(p_T,\varphi)}{R_{\mathrm{AA}}(p_T)} \tag{3}$$

and

$$\psi_n^{\mathrm{heavy}}(p_T) = \frac{1}{n}\tan^{-1}\left(\frac{\int_0^{2\pi} d\varphi \sin(n\varphi)\, R_{\mathrm{AA}}(p_T,\varphi)}{\int_0^{2\pi} d\varphi \cos(n\varphi)\, R_{\mathrm{AA}}(p_T,\varphi)}\right). \tag{4}$$

As pointed out in [1, 2], quantities such as $v_n^{\mathrm{heavy}}(p_T)$ do not actually correspond to what is measured. In fact, just as it is done in the soft sector, the harmonic flow coefficients either at high p_T or in the heavy flavor sector are defined via correlation functions between a soft particle (event plane) with a high p_T particle or a heavy flavor candidate. This intrinsic correlation necessarily requires [1, 2] the use of event-by-event viscous hydrodynamic simulations for the QGP to correctly describe the underlying flow, and its fluctuations, in the soft sector. Through such an approach one can obtain a nonzero triangular flow at high p_T, as shown in [1, 2].

In this proceedings we initiate the investigation of event-by-event viscous hydrodynamic fluctuations on observables in the heavy flavor sector for PbPb collisions at $\sqrt{s} = 2.76$ TeV, as very briefly described below.

2. Simulation

We developed a new framework to describe the propagation of heavy quarks on top of energy density and hydrodynamic flow profiles obtained via event-by-event viscous hydrodynamics. The simulation follows a modular paradigm, which allows for the separate study of different aspects of the collision. We simulate the propagation of heavy quarks (bottom and charm) in an expanding (boost invariant) medium described by the v-USPhydro code [3, 4] on an event-by-event basis for PbPb collisions at $\sqrt{s} = 2.76$ TeV. Only shear viscosity effects are taken into account and we assume $\eta/s = 0.11$ [1]. MCKLN initial conditions [5] are used for the hydrodynamic evolution. The system is evolved separately from the heavy quark propagation, which are treated as probes, and thus we neglect any effect of the probes on the medium, unlike [6]. The hydrodynamic evolution takes place until the complete freeze-out of the system, with the freeze-out temperature set as $T_{\mathrm{FO}} = 140$ MeV, occurs. The hydrodynamic simulation gives evolving profiles on the transverse plane for the energy density ε (and temperature T) and the components of the transverse velocity v_x and v_y on an event-by-event basis.

Heavy quarks are sampled at the beginning of every hydrodynamic event (1000 events for each centrality class). The initial position of the quarks is defined by the number of binary collisions in the event. We set a random initial propagation direction φ_{quark} and the initial momentum distribution of the quarks is given by a pQCD calculation (FONLL) [7]. The heavy quarks lose energy to the evolving medium via the simple energy loss model [8]

$$\frac{\mathrm{d}E}{\mathrm{d}x} = \alpha\Gamma_{\mathrm{flow}} f(T,E,x), \tag{5}$$

where $\Gamma_{\mathrm{flow}} = \gamma\left[1 - v\cos(\varphi_{\mathrm{quark}} - \varphi_{\mathrm{flow}})\right]$ takes into account the local flow boost of the medium [1], $f(T,E,x)$ is a function that specifies the energy loss model dependence with the medium temperature, heavy quark energy E, and path length x. The coupling constant parameter α is found by comparison to data and here we use the D^0 meson R_{AA} spectrum for central collisions to obtain it for the charm quark. After fixing the value of α for charm, we take the electron contribution of both heavy quarks and fit the parameter for the bottom quark using electron R_{AA} data. The heavy quarks propagate in the QGP until they find a region where the local temperature is smaller than a certain value, which we call the jet-medium decoupling parameter, $T_d = 120$ MeV, below which energy loss stops and fragmentation is employed. We do not include coalescence effects in this paper [9] and, thus, we will restrict ourselves to the high p_T region where these effects are minimal. The final electron spectra is obtained from the meson decays, calculated using PYTHIA8 [10]. The calculation of the differential flow coefficients follows Ref. [11] where here the soft and heavy flow harmonics are correlated, which is only possible in event-by-event calculations.

3. Results

In Fig. 1 our results for the nuclear modification factor are compared to D^0 meson and heavy flavor electron data for central PbPb collisions at $\sqrt{s} = 2.76$ TeV. The plot presents results for two different energy loss models where $dE/dx \sim T^2$ and $dE/dx =$ constant. The 2-particle elliptic flow cumulant $v_2\{2\}$ for non-central collisions is presented in Fig. 2 for D^0 meson and heavy flavor electrons. At intermediate $p_T < 10$ GeV our results are comparable to previous calculations [9, 12] when coalescence effects are not present. The two energy loss models that yielded the same nuclear modification factor give slightly different results for $v_2\{2\}$, which shows the sensitivity of the azimuthal distribution with the energy loss mechanisms.

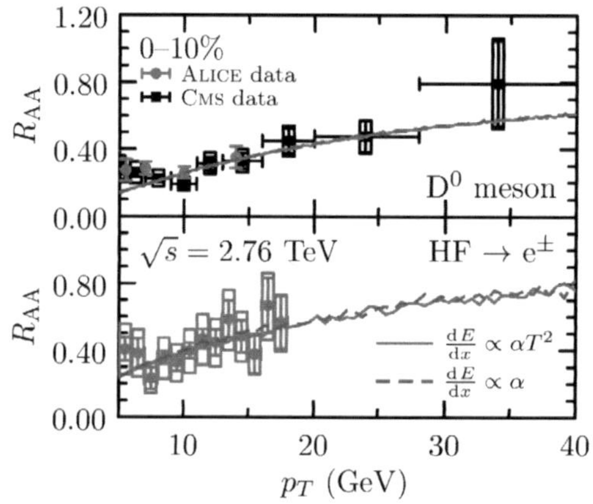

Figure 1. R_{AA} for D^0 meson (top) and heavy flavor electron (bottom) for $\sqrt{s} = 2.76$ TeV PbPb collisions and two different energy loss models. Data from ALICE and CMS [13–15] are presented with the simulation results.

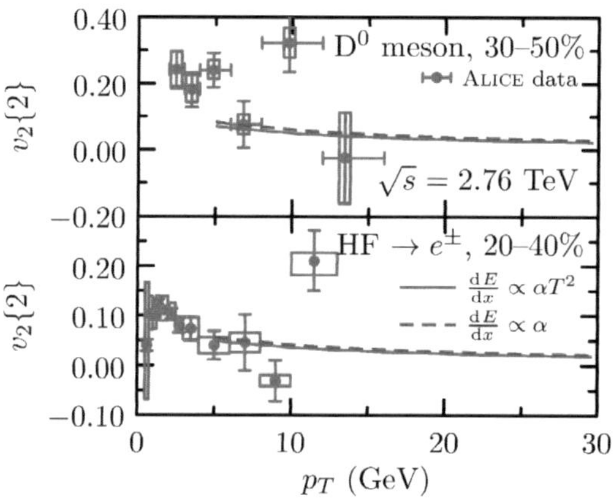

Figure 2. Differential $v_2\{2\}$ for D^0 meson (top) and heavy flavor electron (bottom) for $\sqrt{s} = 2.76$ TeV PbPb collisions and two different energy loss models. Data from the ALICE experiment [13, 16] is also presented.

Fig. 3 presents the first calculation of the 2-particle cumulant of triangular flow $v_3\{2\}$ for D^0 meson and heavy flavor electrons for non-central PbPb collisions at $\sqrt{s} = 2.76$ TeV for two energy loss models (we note that heavy flavor triangular flow, defined by the event plane method, has been computed in [17]). We see that the distinction between the energy loss models is even more significant than what is found in the case of $v_2\{2\}$ in Fig. 2 and, thus, $v_3\{2\}$ may be an even better tool than $v_2\{2\}$ to learn about jet-medium interactions.

4. Conclusions

We developed a new framework to study heavy probes in an event-by-event hydrodynamically expanding viscous QGP. Results for R_{AA} and $v_2\{2\}$ in the heavy flavor sector from this approach

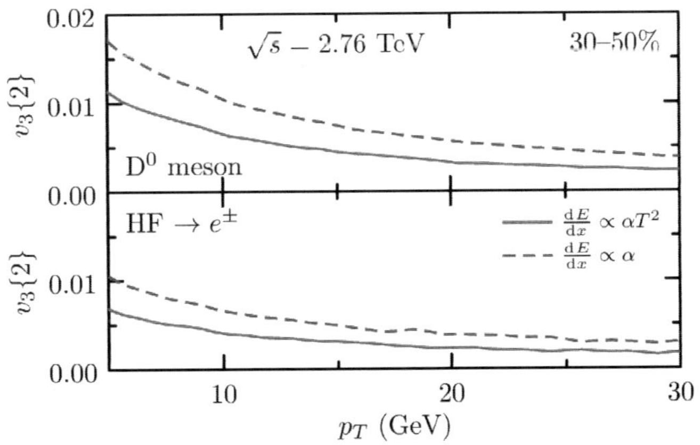

Figure 3. Differential $v_3\{2\}$ for D^0 meson (top) and heavy flavor electron (bottom) for $\sqrt{s} = 2.76$ TeV PbPb collisions and two different energy loss models.

were compared to available experimental data. We also presented the first calculation of $v_3\{2\}$ for heavy flavor, which revealed to be more sensitive to the choice of energy loss than $v_2\{2\}$. Future work includes the calculation of multiple-particle cumulants of harmonic flow and the inclusion of coalescence effects to also describe the intermediate p_T sector.

Acknowledgments

J.N.H. was supported by the National Science Foundation under grant no. PHY-1513864. The authors thank CNPq and FAPESP for financial support.

References

[1] Noronha-Hostler J, Betz B, Noronha J and Gyulassy M 2016 *Phys. Rev. Lett.* **116**(25) 252301 (*Preprint* 1602.03788)

[2] Betz B, Gyulassy M, Luzum M, Noronha J, Noronha-Hostler J, Portillo I and Ratti C 2016 (*Preprint* 1609.05171)

[3] Noronha-Hostler J, Denicol G S, Noronha J, Andrade R P G and Grassi F 2013 *Phys. Rev.* C **88**(4) 044916

[4] Noronha-Hostler J, Noronha J and Grassi F 2014 *Phys. Rev.* C **90**(3) 034907 (*Preprint* 1406.3333)

[5] Drescher H J and Nara Y 2007 *Phys. Rev.* C **75** 1–8 (*Preprint* nucl-th/0611017)

[6] Andrade R P G, Noronha J and Denicol G S 2014 *Phys. Rev.* C **90** 1–11

[7] Cacciari M, Greco M and Nason P 1998 *J. High Energy Phys.* **1998**(05) 007–007

[8] Betz B and Gyulassy M 2014 *J. High Energy Phys.* **2014**(8) 90 (*Preprint* 1404.6378)

[9] Cao S, Luo T, Qin G Y and Wang X N 2016 *Phys. Rev.* C **94**(1) 014909 (*Preprint* 1605.06447)

[10] Sjostrand T, Mrenna S and Skands P 2008 *Comput. Phys. Commun.* **178**(11) 852–867 (*Preprint* 0710.3820)

[11] Bilandzic A, Christensen C H, Gulbrandsen K, Hansen A and Zhou Y 2014 *Phys. Rev.* C **89**(6) 1–25 (*Preprint* 1312.3572)

[12] Das S K, Scardina F, Plumari S and Greco V 2015 *Phys. Lett.* B **747** 260–264 (*Preprint* 1502.03757)

[13] ALICE Collaboration 2014 *Phys. Rev.* C **90**(3) 034904

[14] CMS Collaboration 2015 Nuclear Modification Factor of prompt D^0 in PbPb Collisions at $\sqrt{s_{\rm NN}} = 2.76$ TeV Tech. rep. CERN CMS-PAS-HIN-15-005 Geneva

[15] ALICE Collaboration 2014 *AIP Conf. Proc.* vol 1625 pp 226–229

[16] ALICE Collaboration 2016 *Preprint* 1606.00321

[17] Nahrgang M, Aichelin J, Bass S, Gossiaux P B and Werner K 2015 *Phys. Rev.* C **91**(1) 014904 (*Preprint* 1410.5396)

IOP Conf. Series: Journal of Physics: Conf. Series **779** (2017) 012036 doi:10.1088/1742-6596/779/1/012036

Charmonium production in pp collisions with ALICE at the LHC

Hugo Pereira Da Costa, for the ALICE Collaboration
Commissariat à l'Energie Atomique, IRFU, Saclay, France
E-mail: pereira@hep.saclay.cea.fr

Abstract. We report on forward-rapidity charmonium production in pp collisions at a center-of-mass energy $\sqrt{s} = 13$ TeV, as measured by ALICE at the LHC. Differential cross sections for both J/ψ and ψ(2S) are presented as a function of the charmonium transverse momentum and rapidity. Results are compared to similar measurements performed by LHCb, to measurements at lower energies and to state-of-the-art model calculations.

Charmonia (e.g. J/ψ and ψ(2S)) are mesons formed of a charm and anti-charm quark pair. In high-energy hadronic collisions, their productions result mainly from the hard scattering of two gluons in a process characterized by a timescale $\sim 1/2m_c = 0.08$ fm/c (with m_c the mass of the charm quark), followed by the hadronization of the charm quark pair into a charmonium state with a timescale $\sim 1/\alpha_s m_c = 0.6$ fm/c (with α_s the strong coupling constant). While the production of the charm quark pair is relatively well described by perturbative QCD calculations [1], their binding into a charmonium state is inherently non-perturbative and poses a challenge to models [2]. It is therefore important to measure the charmonium cross sections in proton-proton (pp) collisions over large transverse momentum (p_T) and rapidity (y) ranges, and for as many collision energies as possible, in order to understand their production mechanism. Moreover these measurements also provide a baseline for similar measurements in proton-nucleus and nucleus-nucleus collisions, which in turn are used to quantify the properties of the Quark-Gluon Plasma [3].

Figure 1 shows the inclusive J/ψ (top) and ψ(2S) (bottom) production cross sections as a function of p_T (left) and y (right), measured by ALICE at forward rapidity ($2.5 < y < 4$) in pp collisions at a center-of-mass energy $\sqrt{s} = 13$ TeV. The analysis technique used for these measurements is very similar to that of charmonium measurements performed by ALICE at lower energies, as described for instance in [4]. In particular, the charmonium yield is measured using fits to the invariant mass distribution of opposite sign muon pairs detected in the ALICE Muon Spectrometer [5, 6]. The data sample used for this analysis corresponds to an integrated luminosity $L = 3.2$ pb^{-1}. Systematic uncertainties on the cross sections include contributions from signal extraction, acceptance and efficiency corrections as well as the integrated luminosity. They amount to about 7% for J/ψ and 10% for ψ(2S). For J/ψ, cross sections are compared to measurements done by LHCb in the same rapidity range [7]. An excellent agreement is observed between the two experiments. The ALICE J/ψ measurement extends the p_T reach from 14 to 30 GeV/c. For ψ(2S) this is the only measurement available at this energy and in this rapidity

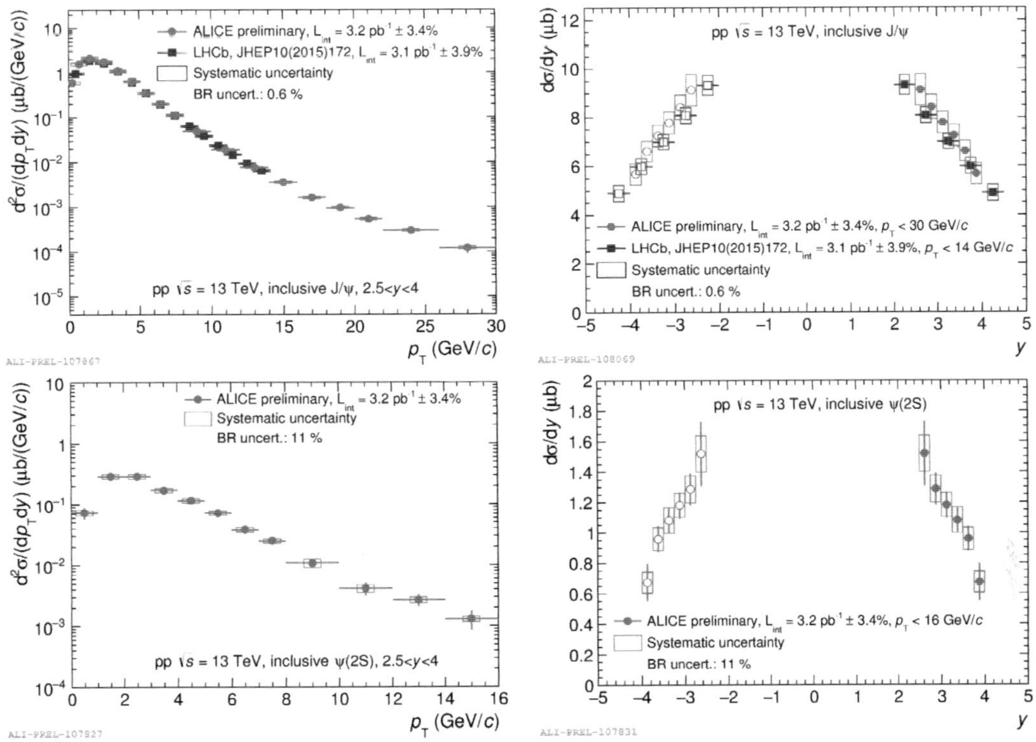

Figure 1. Inclusive forward-rapidity J/ψ (top) and $\psi(2S)$ (bottom) cross sections as a function of p_T (left) and y (right) in pp collision at $\sqrt{s} = 13$ TeV. For the figures on the right, open symbols are the reflection of the positive-y measurements with respect to $y = 0$.

range.

All cross sections shown in figure 1 are inclusive. They consist of a prompt contribution that includes direct production and for J/ψ, decays from $\psi(2S)$ and χ_c, and a non-prompt contribution from b-hadron decays. Both the prompt and non-prompt contributions must be properly accounted for when comparing the data to calculations.

One such comparison is shown in figure 2. The left panel shows the inclusive J/ψ cross section as a function of p_T, together with three calculations: (i) at high p_T in grey, a calculation of the prompt J/ψ production, in the framework of Non-Relativistic QCD (NRQCD) at Next-to-Leading Order (NLO) [8], (ii) at low p_T in blue, a Leading Order (LO) NRQCD calculation coupled to a Color Glass Condensate (CGC) description of the low-x gluons in the proton [9] and (iii) in red, a Fixed-Order Next-to-Leading Logarithm (FONLL) calculation of the non-prompt J/ψ production [1]. Based on these calculations, the non-prompt contribution to inclusive J/ψ production is of order 10% at low p_T, it increases steadily with increasing p_T and dominates for p_T larger than $\sim$ 20 GeV/c. At low p_T, where the non-prompt contribution is small, the NRQCD+CGC prompt J/ψ calculation reproduces the data reasonably well. For $p_T > 8$ GeV/c the NRQCD calculation is significantly below the data, which is expected due to the onset of the non-prompt contribution.

In the right panel of figure 2, the calculations for prompt and non-prompt J/ψ production are summed, assuming fully uncorrelated uncertainties. A good description of the data is obtained over the full p_T range, spanning more than 4 orders of magnitude in cross sections. A similar

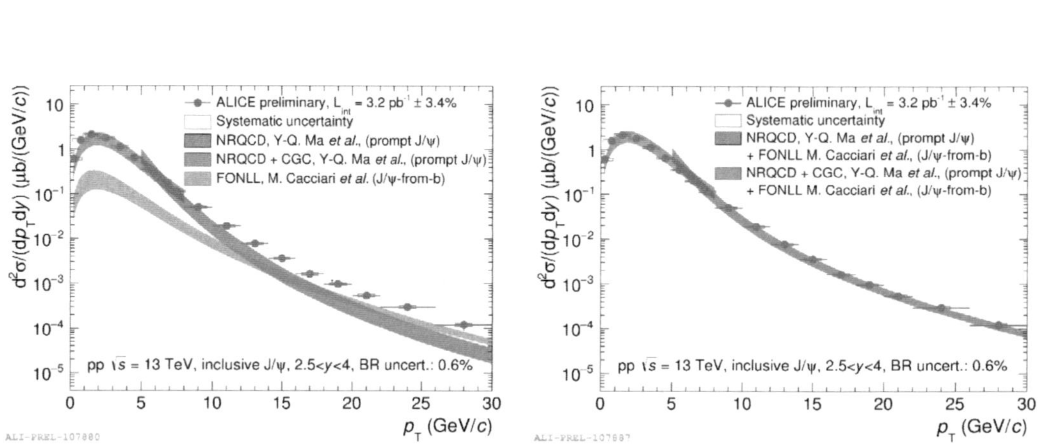

Figure 2. Inclusive forward-rapidity J/ψ cross section as a function of p_T compared to model calculations. See text for details about the models.

level of agreement is also observed when using the NRQCD calculation from [10] instead of the ones shown here or when considering the inclusive production of the $\psi(2S)$ meson instead of J/ψ.

Since the calculation from [9] extends down to $p_T = 0$ GeV/c, it can be integrated over p_T and compared to ALICE y-differential cross section measurements. A reasonable agreement is observed between model and data for both J/ψ and $\psi(2S)$, as shown in figure 3, although the theoretical uncertainties are rather large. This agreement would be further improved if one would add the non-prompt contribution to the model, which amounts to $10 - 15\%$ according to FONLL.

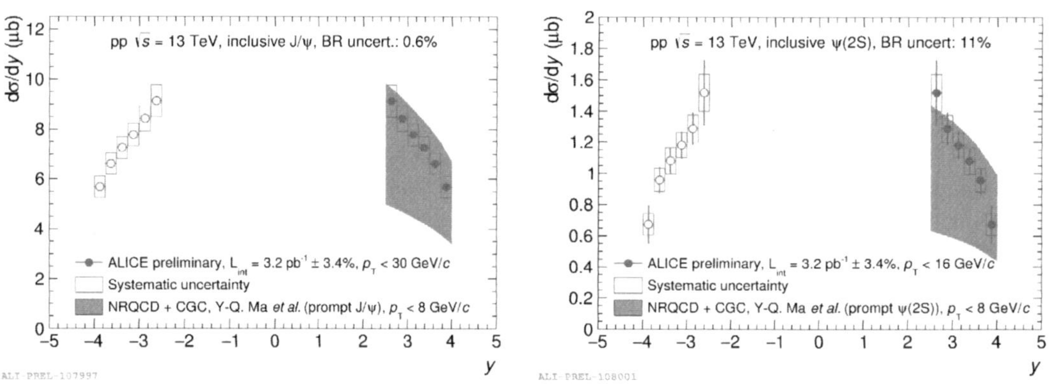

Figure 3. Inclusive forward-rapidity J/ψ (left) and $\psi(2S)$ (right) cross section as a function of y compared to a NRQCD + CGC model calculation.

ALICE has also measured the $\psi(2S)$-to-J/ψ inclusive cross section ratio as a function of p_T and y. In figure 4, this ratio is compared to a NLO NRQCD calculation performed by the same group as in figure 3 (left) and to measurements performed at lower center-of-mass energies, namely $\sqrt{s} = 7$ [11] and 8 TeV [4] (right). The motivation for measuring this quantity is that many sources of uncertainties cancel when forming the $\psi(2S)$-to-J/ψ ratio, for both data and model calculations. With these reduced uncertainties, it is observed that the NRQCD calculation slightly overestimates the ratio, whereas it was able to reasonably reproduce the $\psi(2S)$ and J/ψ

cross sections separately. In principle, the contribution from non-prompt $\psi(2S)$ and J/ψ should be added to the calculation before comparing to the data. It was checked however that the impact of doing so is small, even at high p_T, because the said contribution enters both the numerator and the denominator, and largely cancels.

Finally, considering the right panel of figure 4, no collision energy dependence is observed for the $\psi(2S)$-to-J/ψ cross section ratio, for energies ranging from $\sqrt{s} = 7$ TeV to 13 TeV. This is also consistent with models.

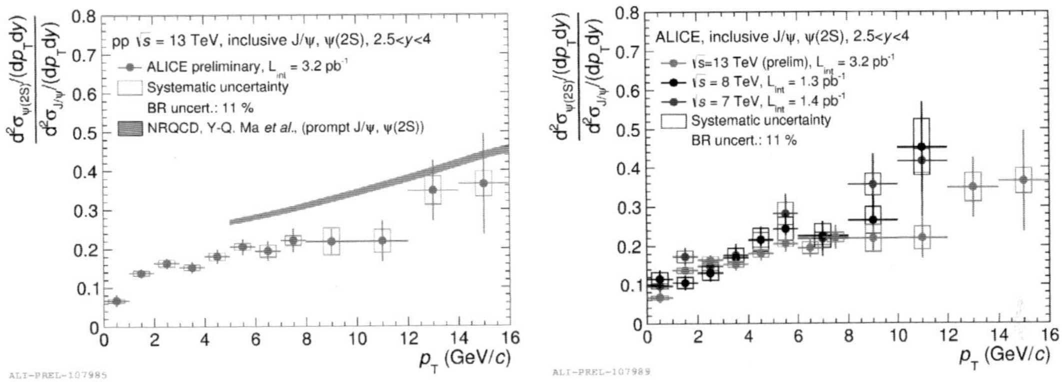

Figure 4. Forward-rapidity $\psi(2S)$-to-J/ψ inclusive cross section ratio as a function of p_T in pp collisions at $\sqrt{s} = 13$ TeV, compared to a NLO NRQCD calculation (left) and to lower energy measurements from ALICE.

To summarize, ALICE has measured the inclusive J/ψ and $\psi(2S)$ production cross sections at forward rapidity in pp collisions at $\sqrt{s} = 13$ TeV. These results complement the measurements performed by ALICE in the same rapidity range at $\sqrt{s} = 2.76$ [12], 5 [13], 7 [11] and 8 TeV [11]. The J/ψ cross sections are in excellent agreement with the ones measured by LHCb at the same energy in the same rapidity range and extend the p_T reach of the measurement from 14 to 30 GeV/c. For $\psi(2S)$, these cross sections are the first ones available at $\sqrt{s} = 13$ TeV in this rapidity range. Cross section measurements are well described by NRQCD calculations provided that the non-prompt contribution is properly accounted for, for instance using FONLL. The same NRQCD calculation however slightly overestimates the $\psi(2S)$-to-J/ψ ratio. When comparing this ratio to measurements at lower energies, no strong collision energy dependence is observed.

References

[1] Cacciari M *et al.* 2012 *J. High Energy Phys.* **1210** 137 (*Preprint* 1205.6344)
[2] Brambilla N *et al.* 2011 *Eur. Phys. J.* **C71** 1534 (*Preprint* 1010.5827)
[3] Shuryak E V 1978 *Phys. Lett.* **B78** 150 [Yad. Fiz.28,796(1978)]
[4] Adam J *et al.* (ALICE Collaboration) 2016 *Eur. Phys. J.* **C76** 184 (*Preprint* 1509.08258)
[5] Aamodt K *et al.* (ALICE Collaboration) 2008 *J. Instrum.* **3** S08002
[6] Aamodt K *et al.* (ALICE Collaboration) 2011 *Phys. Lett. B* **704** 442 (*Preprint* 1105.0380)
[7] Aaij R *et al.* (LHCb Collaboration) 2015 *JHEP* **10** 172 (*Preprint* 1509.00771)
[8] Ma Y Q, Wang K and Chao K T 2011 *Phys. Rev. Lett.* **106** 042002 (*Preprint* 1009.3655)
[9] Ma Y Q and Venugopalan R 2014 *Phys. Rev. Lett.* **113** 192301 (*Preprint* 1408.4075)
[10] Butenschoen M and Kniehl B A 2011 *Phys. Rev. Lett.* **106** 022003 (*Preprint* 1009.5662)
[11] Abelev B B *et al.* (ALICE Collaboration) 2014 *Eur. Phys. J.* **C74** 2974 (*Preprint* 1403.3648)
[12] Abelev B *et al.* (ALICE Collaboration) 2012 *Phys. Lett.* **B718** 295–306 [Erratum: Phys. Lett.B748,472(2015)] (*Preprint* 1203.3641)
[13] Adam J *et al.* (ALICE Collaboration) 2016 (*Preprint* 1606.08197)

Charmonium production in Pb-Pb collisions at $\sqrt{s_{\rm NN}} = 2.76$ and 5.02 TeV with ALICE

Biswarup Paul on behalf of the ALICE Collaboration
INFN Torino, Via Pietro Giuria 1, I-10125 Torino, Italy
E-mail: biswarup.paul@cern.ch

Abstract.
The production of charmonium states, J/ψ and ψ(2S), in heavy-ion collisions, is an important probe to investigate the formation of a plasma of quarks and gluons. In such a hot and deconfined medium, quarkonium production is, indeed, expected to be significantly modified, with respect to the pp yields, due to the interplay of color screening and charm-quark recombination mechanisms. The ALICE Collaboration at the LHC has measured charmonium production in Pb-Pb collisions at $\sqrt{s_{\rm NN}} = 2.76$ TeV. The nuclear modification factor of inclusive J/ψ, evaluated at forward ($2.5 < y < 4$) rapidity, is measured as a function of the event centrality and of the J/ψ kinematic variables such as transverse momentum and rapidity. In this article, we report on the new J/ψ results, obtained at forward rapidity, at $\sqrt{s_{\rm NN}} = 5.02$ TeV. These new results are compared to that obtained at $\sqrt{s_{\rm NN}} = 2.76$ TeV and to the available theoretical predictions.

1. Introduction

The suppression of quarkonia (bound states of a heavy quark and its anti-quark) in ultra-relativistic heavy-ion collisions is one of the most prominent probes used to investigate and quantify the properties of the Quark-Gluon Plasma (QGP). The in-medium dissociation probability of the different quarkonium states could provide an estimate of the temperature of the fireball since the dissociation is expected to take place at different temperatures depending on the binding energies of the quarkonium states when the medium reaches or exceeds the critical temperature of the phase transition ($T_{\rm c}$). For charmonium ($c\bar{c}$) states, the J/ψ is likely to survive significantly above $T_{\rm c}$ (1.5 – 2 $T_{\rm c}$) whereas χ_c and ψ(2S) melt near $T_{\rm c}$ (1.1 – 1.2 $T_{\rm c}$) [1,2]. At LHC energies, due to the large increase of the $c\bar{c}$ production cross-section with the collision energy, there is a possibility of J/ψ production via recombination of deconfined c and $\bar{c}$. Thus, the observation of J/ψ production in nucleus-nucleus collisions via recombination also constitutes an evidence of QGP formation. The two effects act in opposite directions and the comparison of the J/ψ production at the different collision energies can provide insights to the evolution of the relative contributions of the two processes.

The ALICE experiment has studied inclusive J/ψ production in Pb–Pb collisions at $\sqrt{s_{\rm NN}} = 2.76$ TeV and 5.02 TeV through its dimuon decay channel. Muons are identified and tracked in the Muon Spectrometer, which covers the pseudorapidity range $-4 < \eta < -2.5$ [3]. The pixel layers of the Inner Tracking System allow the vertex determination, while forward VZERO scintillators are used for triggering purposes. The VZERO is also used to determine the centrality of the collisions. In order to improve the purity of the muon tracks the following selection criteria were applied: (1) both muon tracks with $p_{\mathrm{T}} > 1$ GeV/c and matched with trigger, (2) both muon tracks in the pseudo-rapidity range $-4 < \eta < -2.5$, (3) transverse radius coordinate of the tracks at the end of the hadron absorber (the longitudinal position of the absorber from the interation point (IP) is $-5.0 < z < -0.9$ m) in the range $17.6 < R_{\mathrm{abs}} < 89.5$ cm and (4) dimuon rapidity in the range $2.5 < y < 4$. These cuts eliminate mainly tracks hitting the edges of the spectrometers or crossing the thicker part of the beam shield. The data samples used in this analysis correspond to an integrated luminosity $L_{\mathrm{int}}^{\mathrm{Pb-Pb}} \approx 225\ \mu\mathrm{b}^{-1}$ for Pb–Pb at $\sqrt{s_{\mathrm{NN}}} = 5.02$ TeV and $L_{\mathrm{int}}^{\mathrm{pp}} \approx 106\,\mathrm{nb}^{-1}$ for pp collisions at $\sqrt{s} = 5.02$ TeV. The analysis techniques are described in detail in [4].

2. Results

The in-medium modification of the J/ψ production is quantified through the nuclear modification factor (R_{AA}), defined as the ratio of the J/ψ yield measured in Pb–Pb collisions to the expected yield obtained by scaling the J/ψ production cross section in pp collisions with the number of binary nucleon-nucleon collisions. The R_{AA} of the inclusive J/ψ in Pb–Pb collisions at $\sqrt{s_{\mathrm{NN}}} = 5.02$ TeV, integrated over the centrality, rapidity and transverse momentum (p_{T}) ranges (0–90%, $2.5 < y < 4$ and $p_{\mathrm{T}} < 12$ GeV/c) is $R_{\mathrm{AA}} = 0.65 \pm 0.01(\mathrm{stat}) \pm 0.05(\mathrm{syst})$ [4]. This shows that there is a suppression of the J/ψ in Pb–Pb collisions at $\sqrt{s_{\mathrm{NN}}} = 5.02$ TeV with respect to pp collisions at the same energy.

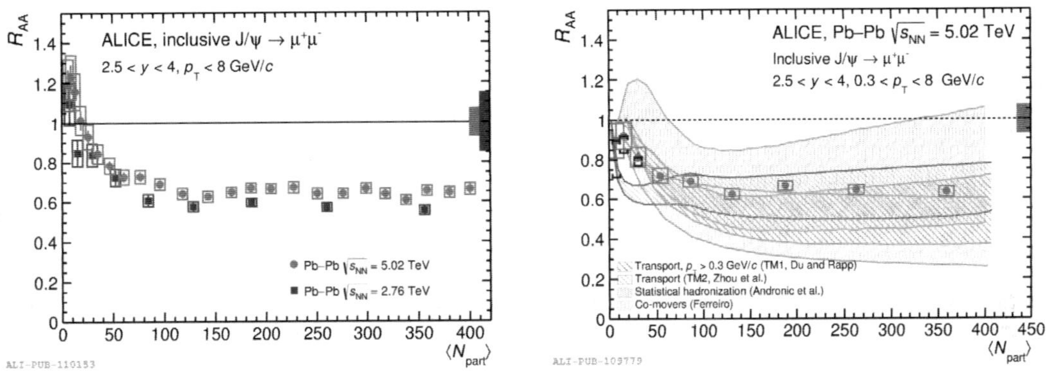

Figure 1. (Left) The inclusive J/ψ R_{AA} as a function of centrality at $\sqrt{s_{\mathrm{NN}}} = 5.02$ TeV compared to published results at $\sqrt{s_{\mathrm{NN}}} = 2.76$ TeV [5]. The statistical uncertainties are represented by the error bars. The boxes around the points represent the uncorrelated systematic uncertainties, while correlated global uncertainties are shown as filled boxes around $R_{\mathrm{AA}} = 1$. (Right) The inclusive J/ψ R_{AA} for $0.3 < p_{\mathrm{T}} < 8$ GeV/c compared to the theoretical models [6–11].

The centrality dependence of the inclusive J/ψ R_{AA} at $\sqrt{s_{\mathrm{NN}}} = 5.02$ TeV and the comparison to that obtained at $\sqrt{s_{\mathrm{NN}}} = 2.76$ TeV [5] are shown in Fig. 1. The large statistics collected at $\sqrt{s_{\mathrm{NN}}} = 5.02$ TeV allows to carry out the measurement in much narrower centrality bins w.r.t. 2.76 TeV. The figure shows that there is an increasing suppression with centrality up to $N_{\mathrm{part}} \sim 100$ followed by a roughly constant R_{AA} value at both energies. A systematic increase by about 15% is visible when comparing the two sets of results, although such an increase is within the overall uncertainty of the measurements. The R_{AA} of prompt J/ψ would be about 10% higher if $R_{\mathrm{AA}}^{\mathrm{non-prompt}} = 0$ (J/ψ's produced from b-hadron decays are called non-prompt J/ψ) and about 5% (1%) smaller if $R_{\mathrm{AA}}^{\mathrm{non-prompt}} = 1$ for central (peripheral) collisions.

An excess of J/ψ production at very-low p_{T}, which might originate from photo-production, was observed in peripheral Pb–Pb collisions at $\sqrt{s_{\mathrm{NN}}} = 2.76$ TeV [12]. According to the prediction from STARLIGHT MC simulation [13] a $p_{\rm T}$ cut at 0.3 GeV/c removes ~80% of the photo-produced J/ψ. The hadronic J/ψ $R_{\rm AA}$, for $0 < p_{\rm T} < 8$ GeV/c, is estimated to be about 34%, 17% and 9% smaller than the measured values in the 80–90%, 70–80% and 60–70% centrality classes, respectively [14]. The variation decreases to about 9%, 4% and 2%, respectively, when considering the $R_{\rm AA}$ for J/ψ with $0.3 < p_{\rm T} < 8$ GeV/c, due to the remaining small contribution from photo-produced J/ψ. After such a selection, a less pronounced increase of $R_{\rm AA}$ for peripheral events can indeed be seen by comparing the left and right plots in Fig. 1. The brackets shown in the right figure for the three most peripheral centrality intervals represent the range of variation for the hadronic J/ψ $R_{\rm AA}$ under extreme hypothesis [14] on the photo-production contamination.

The results have been compared to the theoretical calculations based on a statistical model approach [6], a transport model (TM1) [9,10], which is based on a thermal rate equation, a second transport model (TM2) [11], which implements a hydrodynamic description of the medium evolution and finally a 'co-mover' model [7, 8], where the J/ψ are dissociated via interactions with the partons/hadrons produced in the same rapidity range. The large uncertainties in the models are due to the choice of the corresponding input parameters and in particular of $d\sigma_{c\bar{c}}/dy$. Except for the statistical hadronization model, a better description is found when considering their upper limits. But for transport models this corresponds to the absence of nuclear shadowing, which can be clearly considered as an extreme assumption for J/ψ [15, 16].

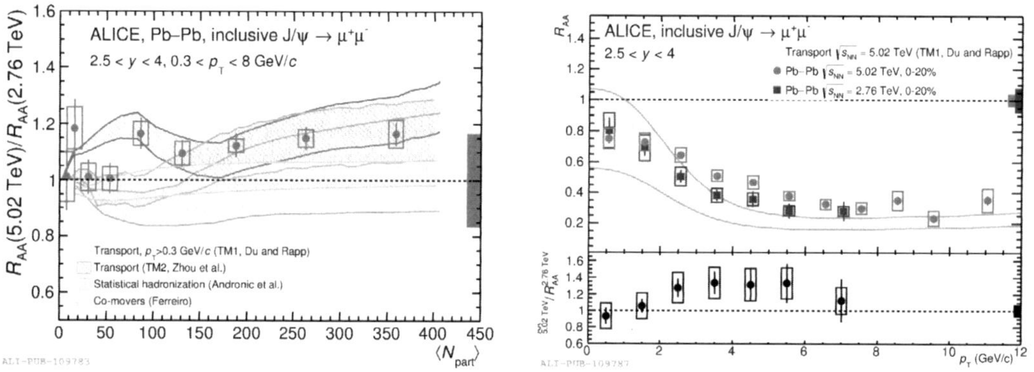

Figure 2. (Left) The ratio of the inclusive J/ψ $R_{\rm AA}$ for $0.3 < p_{\rm T} < 8$ GeV/c between $\sqrt{s_{\rm NN}} = 5.02$ and 2.76 TeV vs. centrality compared to theoretical models [6–11]. (Right) The inclusive J/ψ $R_{\rm AA}$ as a function of $p_{\rm T}$ at $\sqrt{s_{\rm NN}} = 5.02$ TeV in the centrality interval 0–20% compared to the corresponding result at $\sqrt{s_{\rm NN}} = 2.76$ TeV [5] and to the prediction of a transport model (TM1) [9, 10]. The $p_{\rm T}$ dependence of the ratio of $R_{\rm AA}$ between the two energies is also shown.

The ratio of the inclusive J/ψ $R_{\rm AA}$ between the two energies as a function of centrality compared with the theoretical models is shown in the left panel of Fig. 2. Many theoretical uncertainties cancel out due to the correlation between the parameters at the two energies. The uncertainties on the average nuclear overlap function ($\langle T_{\rm AA}\rangle$) cancel out in data. This ratio for the prompt J/ψ would be about 2% (1–2%) higher if beauty hadrons were fully (not) suppressed by the medium. The prediction from the transport model of Ref. [9, 10] (TM1) shows a decrease of this ratio with the increase of centrality due to the larger suppression effects at higher energy followed by an increase which is due to the regeneration that acts in the opposite direction and becomes dominant for central collisions. The other transport model (TM2) [11] also exhibits an increase for central collisions while for peripheral collisions the behaviour is different. No visible structure as a function of centrality is observed in the co-mover model [7, 8] and the calculation favours this ratio to be slightly below unity which implies that in this model the increase of the suppression effects with energy may be dominant over the regeneration effects for all centralities. Finally, the statistical model [6] predicts a continuous increase of this ratio with centrality dominated by the increase in the $c\bar{c}$ cross section with energy. The uncertainty bands in the theoretical models shown in left panel of Fig. 2 correspond to $\sim$5% variations of the $c\bar{c}$ cross section at $\sqrt{s_{\rm NN}} = 5.02$ TeV. Within the uncertainties the data show no clear centrality dependence and are compatible with the theoretical models. This ratio for central collisions and for $0.3 < p_{\rm T} < 8$ GeV/c is $1.17 \pm 0.04({\rm stat}) \pm 0.20({\rm syst})$.

The $R_{\rm AA}$ as a function of $p_{\rm T}$ for the centrality interval 0–20% compared to the corresponding results obtained at $\sqrt{s_{\rm NN}} = 2.76$ TeV and to theoretical models is shown in the right panel of Fig. 2. The $p_{\rm T}$ dependence of $R_{\rm AA}$ has an increasing trend at low $p_{\rm T}$, which is a sensitive test of the presence of the regeneration component in theoretical calculations. Since the contribution of J/ψ photo-production is negligible with respect to the hadronic one for central events [12], the region $p_{\rm T} < 0.3$ GeV/c is not excluded. The transport model of Ref. [9,10] (TM1) describes the data at low $p_{\rm T}$ but underestimates the data at intermediate $p_{\rm T}$ ($3 < p_{\rm T} < 7$ GeV/c). The ratio of $R_{\rm AA}$ vs. $p_{\rm T}$ between the two energies is also shown in the same figure. In the region $2 < p_{\rm T} < 6$ GeV/c, a hint of increase of $R_{\rm AA}$ with $\sqrt{s_{\rm NN}}$ is visible. The $R_{\rm AA}$ of prompt J/ψ is expected to be 7% larger (2% smaller) for $p_{\rm T} < 1$ GeV/c and 30% larger (55% smaller) for $10 < p_{\rm T} < 12$ GeV/c when the beauty contribution is fully (not) suppressed. The ratio of $R_{\rm AA}$ appears to be less sensitive to the non-prompt J/ψ contribution if we assume that $R_{\rm AA}^{\rm non-prompt}$ does not vary significantly between the two collision energies.

3. Conclusion

ALICE has measured the inclusive J/ψ production in pp and Pb–Pb collisions at $\sqrt{s_{\rm NN}} = 5.02$ TeV. Comparison with the previous measurement at $\sqrt{s_{\rm NN}} = 2.76$ TeV shows a systematic increase in $R_{\rm AA}$ by about 15%, although such an effect is within the total uncertainty of the measurements. A less pronounced increase of $R_{\rm AA}$ for peripheral events is observed upon removing the very-low $p_{\rm T}$ J/ψ ($p_{\rm T} < 0.3$ GeV/c). The rejection of photo-produced J/ψ [12] is a possible explanation for this decrease. The theoretical calculations describe $R_{\rm AA}$ and the ratio of $R_{\rm AA}$ between $\sqrt{s_{\rm NN}} = 5.02$ and 2.76 TeV closer to their upper limits. The $p_{\rm T}$ dependence of $R_{\rm AA}$ exhibits an increase at low $p_{\rm T}$ which is ascribed to the contribution of regenerated J/ψ. In the region $2 < p_{\rm T} < 6$ GeV/c, a hint of an increase of $R_{\rm AA}$ between $\sqrt{s_{\rm NN}} = 2.76$ and 5.02 TeV is observed. The results on J/ψ production support a picture where a combination of suppression and regeneration takes place in the QGP, with the two mechanisms being dominant at high and low $p_{\rm T}$, respectively.

References

[1] H. Satz, J. Phys. **G32**, (2006) R25.
[2] H. Satz, Int. J. Mod. Phys. **A28**, (2013) 1330043.
[3] **ALICE** Collaboration, K. Aamodt *et al.*, *JINST* 3, (2008) S08002.
[4] **ALICE** Collaboration, J. Adam *et al.*, `arXiv:1606.08197 [hep-ex]`
[5] **ALICE** Collaboration, B. Abelev *et al.*, *Phys. Lett.***B743** (2014) 314–327.
[6] A. Andronic, P. Braun-Munzinger, K. Redlich, and J. Stachel, *Nucl. Phys.***A904-905** (2013) 535c–538c.
[7] E. G. Ferreiro, *Phys. Lett.* **B731** (2014) 57–63.
[8] E. G. Ferreiro, *Phys. Lett.* **B749** (2015) 98–103.
[9] X. Zhao and R. Rapp, *Nucl. Phys.* **A859** (2011) 114–125.
[10] X. Du and R. Rapp, *Nucl. Phys.* **A943** (2015) 147–158.
[11] K. Zhou, N. Xu, Z. Xu, and P. Zhuang, *Phys. Rev.* **C89** no. 5, (2014) 054911.
[12] **ALICE** Collaboration, J. Adam *et al.*, *Phys. Rev. Lett.* **116** (2016) 222301.
[13] **ALICE** Collaboration, B. Abelev *et al.*, *Phys. Lett.***B718** (2013) 1273–1283.
[14] **ALICE** Collaboration, J. Adam *et al.*, *JHEP* **05** (2016) 179.
[15] **ALICE** Collaboration, J. Adam *et al.*, *JHEP* **06** (2015) 055.
[16] **ALICE** Collaboration, J. Adam *et al.*, *JHEP* **11** (2015) 127.

IOP Conf. Series: Journal of Physics: Conf. Series **779** (2017) 012038 doi:10.1088/1742-6596/779/1/012038

Charmonium production in pp, pPb and PbPb with CMS

Songkyo Lee on behalf of the CMS collaboration
Korea University, 145 Anam-ro, Seongbuk-gu, Seoul, Republic of Korea
E-mail: songkyo.lee@cern.ch

Abstract. We report on the analysis of charmonia in pp, pPb and PbPb collisions with the CMS experiment with various observables. The cross section for coherent J/ψ photoproduction is measured in ultra-peripheral PbPb collisions at 2.76 TeV. The ratio of prompt J/ψ yields at forward (proton-going direction) and backward (Pb-going direction) is studied as a function of p_T, rapidity, and the event activity in pPb collisions at 5.02 TeV. The nuclear modification factor of prompt J/ψ in PbPb collisions at 2.76 TeV is presented as a function of centrality, p_T, and rapidity. Finally, the production yields of $\psi(2S)$ and J/ψ in pp and PbPb collisions have been compared.

1. Introduction

In the early stage of high-energy heavy ion collisions, the transition from a hadronic phase into a deconfined system of quarks and gluons, the Quark-Gluon Plasma (QGP), is predicted. Due to the Debye screening of the heavy-quark potential in QGP, charmonium states are expected to dissociate sequentially according to their binding energies [1]. As an example, $\psi(2S)$ is less bound than J/ψ and hence, is expected to melt at a lower temperature. Given the abundance of charm quarks at LHC energies, charmonia can also be produced at the hadronization stage by the recombination of initially uncorrelated charm and anticharm quarks [2]. This type of process has as result an enhancement of charmonium yields in lead-lead (PbPb) collisions relative to those in proton-proton (pp) collisions. Aside from the formation of QGP, the charmonium production can also be affected by the so-called "cold nuclear matter" (CNM) effects, including modifications of the nuclear parton distribution functions (nPDFs), initial or final state parton energy loss, and nuclear absorption [3]. Such effects are best probed in proton-lead (pPb) collisions and ultra-peripheral collisions (UPC). In particular, the depletion of gluon PDFs in nuclei from those in a free proton can lead to the charmonium suppression at small values of the momentum fraction carried by a parton (x), a process known as shadowing. The values of x probed at the LHC correspond to $10^{-5} < x < 10^{-2}$. This conference proceeding reports the measurement of J/ψ in PbPb UPC, and in pPb collisions. The nuclear modification factor of prompt J/ψ, and the ratio of $\psi(2S)$ to J/ψ yields in pp and PbPb collisions are also presented.

2. Analysis methods and Results

2.1. J/ψ in Ultra Peripheral PbPb collisions

The CNM effects can be probed by the photon-induced reactions in UPC with impact parameters larger than the sum of the radii of the colliding nuclei [4]. If the photon interacts with the whole nucleus, the produced J/ψ is classified as coherent and dominantly distributed at $p_{\rm T} < 60$ MeV/c for PbPb collisions at 2.76 TeV. The UPC events are selected requiring a low energy deposition in the forward hadronic calorimeter, and only two muon tracks from J/ψ. In order to separate the coherent signal contribution from background (incoherent J/ψ and $\gamma\gamma \rightarrow \mu^+\mu^-$ interactions), a maximum likelihood fit is performed simultaneously to the dimuon invariant mass and $p_{\rm T}$ distributions determined from STARLIGHT simulation.

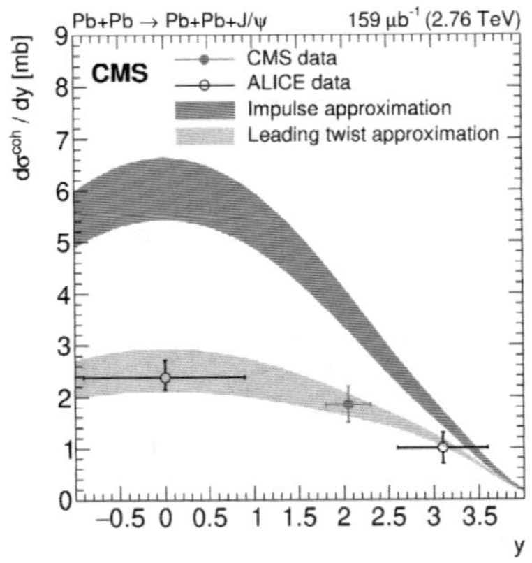

Figure 1. Differential cross sections of coherent J/ψ production as a function of rapidity in ultra-peripheral PbPb collisions at 2.76 TeV, measured by CMS (a filled circle) and ALICE (open circles) [5, 6]. The horizontal bars display the range of rapidity bins, and the vertical error bars represent the statistical and systematic uncertainties. The impulse approximation and the leading twist approximation calculations are also shown [7, 8].

In Fig. 1, the coherent J/ψ production cross sections measured by CMS using 159 μb^{-1} PbPb data (a filled circle) is compared to the ALICE measurement (open circles) [5, 6], and to theoretical calculations [7, 8]. The impulse approximation prediction uses data from $\gamma + p$ collisions to estimate the γ+Pb collisions, neglecting all nuclear effects. The leading twist approximation implements a gluon recombination using a diffractive proton PDF as an input, which results in an effective nuclear gluon shadowing. Experimental results from ALICE and CMS favor a model with shadowing effects and show a steady decrease with rapidity.

2.2. J/ψ in pPb collisions

The processes affecting prompt J/ψ production are also studied using 34.6 nb^{-1} pPb data at 5.02 TeV. The prompt J/ψ and those coming from b-hadron decays are separated using the pseudo-proper decay length, $l_{J/\psi} = L_{xy}\, m_{J/\psi}/p_{\rm T}$, where L_{xy} is the transverse decay length in the laboratory frame. The invariant-mass spectrum and the $l_{J/\psi}$ distribution of $\mu^+\mu^-$ pairs are fitted simultaneously in an extended unbinned maximum likelihood fit [9]. The rapidity dependence of nuclear effect is investigated using the forward-to-backward production ratio, $R_{\rm FB}(p_{\rm T}, y) = [d^2\sigma(p_{\rm T}, y > 0)/dp_{\rm T}dy]/[d^2\sigma(p_{\rm T}, y < 0)/dp_{\rm T}dy]$, where the $d^2\sigma/dp_{\rm T}dy$ is the J/ψ production cross section decayed into muon pairs in a given $p_{\rm T}$ and y bin, and forward regions ($y > 0$) are defined by the proton-going direction.

Figure 2 shows the $p_{\rm T}$ dependence of $R_{\rm FB}$ in three different rapidity ranges. The $R_{\rm FB}$ is closer to unity at high $p_{\rm T}$, and decreases at low $p_{\rm T}$. The EPS09 shadowing calculation with next-to-leading order (NLO), based on the Color Evaporation Model (CEM) for the J/ψ production, is also shown by bands [10]. Data shows lower $R_{\rm FB}$ values than the model prediction at $5 < p_{\rm T} < 6.5$ GeV/c, hinting at other effects beyond the single nPDF parameterization.

The $R_{\rm FB}$ of prompt J/ψ is further analyzed in terms of an event activity variable $E_{\rm T}^{{\rm HF}|\eta|>4}$, the transverse energy deposited in the forward hadronic calorimeter in $4 < |\eta| < 5.2$, as shown in Fig. 3. The data points are plotted at the bin-averaged values. The clear decreasing tendency of the $R_{\rm FB}$ with increasing $E_{\rm T}^{{\rm HF}|\eta|>4}$ has been observed over whole $p_{\rm T}$ and rapidity ranges.

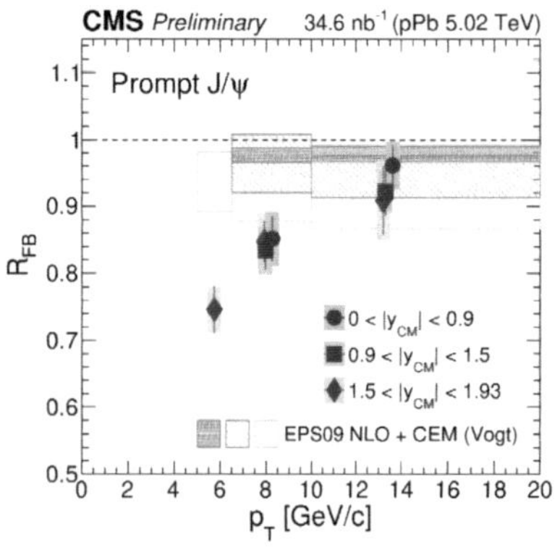

Figure 2. p_T dependence of R_{FB} for prompt J/ψ in three rapidity ranges. The EPS09(NLO) shadowing prediction is also shown. The error bars show the statistical uncertainties, and the shaded boxes represent the systematic uncertainties.

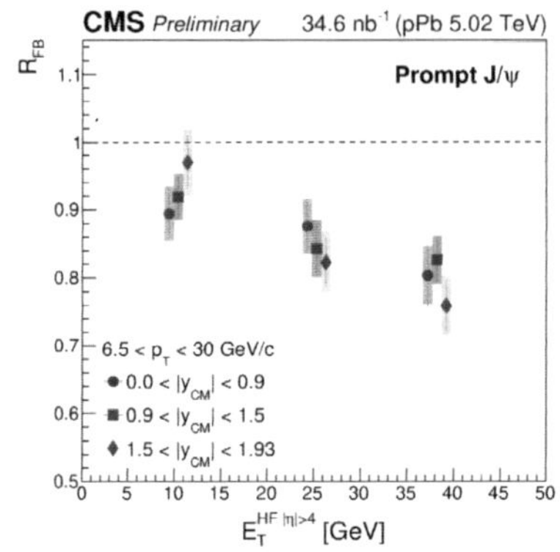

Figure 3. $E_T^{HF|\eta|>4}$ (right) dependence of R_{FB} for prompt J/ψ in three rapidity ranges. The data points are slightly shifted to avoid overlap. The error bars show the statistical uncertainties, and the shaded boxes represent the systematic uncertainties.

2.3. J/ψ and $\psi(2S)$ in PbPb collisions

The nuclear modification factor, R_{AA} of prompt J/ψ is measured using pp and PbPb data at 2.76 TeV with and integrated luminosity of 150 μb^{-1} and displayed in Fig. 4 [11]. The yields of J/ψ in PbPb are more suppressed compared to those in pp with increasing centrality, implying the signature of QGP. No strong p_T and rapidity dependence is observed.

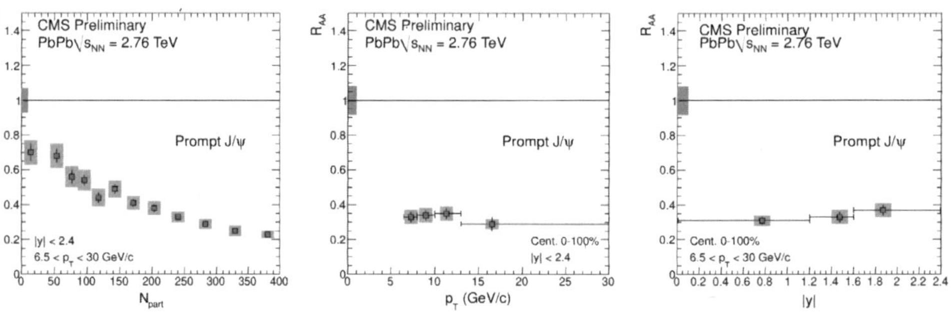

Figure 4. The nuclear modification factor of prompt J/ψ as a function of centrality (left), p_T (middle), and rapidity (right). The gray boxes plotted at unity show the scale of the global uncertainties. The bin boundaries are presented by small horizontal lines.

The results for J/ψ and $\psi(2S)$ are shown as a double ratio $(N_{\psi(2S)}/N_{J/\psi})_{PbPb}/(N_{\psi(2S)}/N_{J/\psi})_{pp}$ in Fig. 5, as a function of centrality [12]. At mid rapidity ($|y| < 1.6$ and $6.5 < p_T < 30$ GeV/c), the double ratio is always below unity without centrality dependence, which means more suppression of $\psi(2S)$ relative to J/ψ in PbPb. At forward rapidity extending down to lower p_T ($1.6 < |y| < 2.4$ and $3 < p_T < 30$ GeV/c), an increase of the double ratio with centrality is observed. The double ratio becomes $2.31 \pm 0.53(stat) \pm 0.37(syst) \pm 0.15(pp)$ in the most central bin, meaning $\psi(2S)$ is more produced than J/ψ in central PbPb collisions with respect to pp.

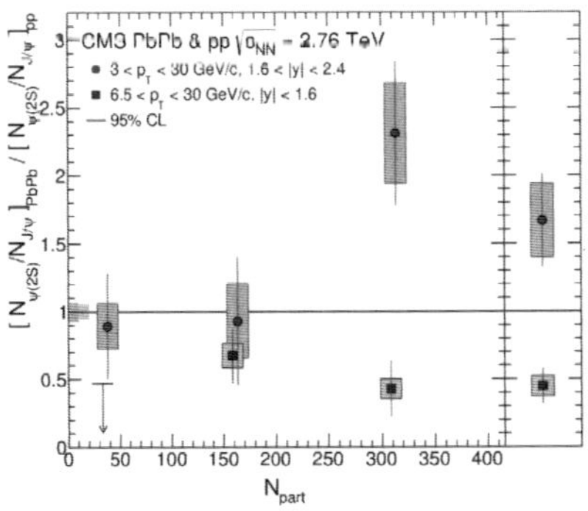

Figure 5. Double ratio of $\psi(2S)$ to J/ψ as a function of centrality for the mid rapidity (squares) and forward rapidity (circles, slightly shifted). Statistical uncertainties are shown as error bars, and systematics uncertainties are displayed as boxes. The boxes at unity represent global uncertainties. The right panel shows the centrality-integrated results.

3. Summary

The measurement of J/ψ in PbPb UPC has been reported. The result is in agreement with the model including nuclear gluon shadowing effects. The $R_{\rm FB}$ of J/ψ in pPb collisions is observed to be consistent with unity at high $p_{\rm T}$ and becomes lower than unity at low $p_{\rm T}$ indicating the presence of nuclear effects. The R_{AA} of prompt J/ψ is below unity and decreases with centrality providing the evidence of QGP. The double ratio results of $\psi(2S)$ to J/ψ are opposite to the expected sequential melting scenario in the forward rapidity bin, implying other physics processes such as recombination.

References

[1] T. Matsui, H. Satz, J/ψ suppression by quark-gluon plasma formation, Phys. Lett. B 178 (1986) 416. doi:10.1016/0370-2693(86)91404-8.

[2] X. Zhao, R. Rapp, Medium Modifications and Production of Charmonia at LHC, Nucl. Phys. A 859 (2011) 114. arXiv:1102.2194, doi:10.1016/j.nuclphysa.2011.05.001.

[3] A. Andronic, et al., Heavy-flavour and quarkonium production in the LHC era: from proton-proton to heavy-ion collisions, arXiv:1506.03981.

[4] V. Khachatryan, et al., Coherent J/ψ photoproduction in ultra-peripheral PbPb collisions at $\sqrt{s_{\rm NN}} = 2.76$ TeV with the CMS experiment, Submitted to: Phys. Lett. B. arXiv:1605.06966.

[5] B. Abelev, et al., Coherent J/ψ photoproduction in ultra-peripheral Pb-Pb collisions at $\sqrt{s_{\rm NN}} = 2.76$ TeV, Phys. Lett. B 718 (2013) 1273. arXiv:1209.3715, doi:10.1016/j.physletb.2012.11.059.

[6] E. Abbas, et al., Charmonium and e^+e^- pair photoproduction at mid-rapidity in ultra-peripheral Pb-Pb collisions at $\sqrt{s_{\rm NN}} = 2.76$ TeV, Eur. Phys. J. C 73 (11) (2013) 2617. arXiv:1305.1467, doi:10.1140/epjc/s10052-013-2617-1.

[7] V. Guzey, M. Strikman, M. Zhalov, Disentangling coherent and incoherent quasielastic J/ψ photoproduction on nuclei by neutron tagging in ultraperipheral ion collisions at the LHC, Eur. Phys. J. C 74 (7) (2014) 2942. arXiv:1312.6486, doi:10.1140/epjc/s10052-014-2942-z.

[8] V. Guzey, E. Kryshen, M. Strikman, M. Zhalov, Evidence for nuclear gluon shadowing from the ALICE measurements of PbPb ultraperipheral exclusive J/ψ production, Phys. Lett. B 726 (2013) 290. arXiv:1305.1724, doi:10.1016/j.physletb.2013.08.043.

[9] CMS Collaboration, Measurement of prompt and nonprompt J/ψ production in pPb collisions at $\sqrt{s_{\rm NN}} =$ 5.02 TeV, Tech. Rep. CMS-PAS-HIN-14-009, CERN, Geneva (2015).
URL `http://cds.cern.ch/record/2044699`

[10] R. Vogt, Shadowing effects on J/ψ and Υ production at energies available at the CERN Large Hadron Collider, Phys. Rev. C 92 (3) (2015) 034909. arXiv:1507.04418, doi:10.1103/PhysRevC.92.034909.

[11] CMS Collaboration, J/ψ results from CMS in PbPb collisions, with 150 μb^{-1} data, Tech. Rep. CMS-PAS-HIN-12-014, CERN, Geneva (2012).
URL `https://cds.cern.ch/record/1472735`

[12] V. Khachatryan, et al., Measurement of Prompt $\psi(2S) \rightarrow J/\psi$ Yield Ratios in Pb-Pb and $p-p$ Collisions at $\sqrt{s_{\rm NN}} = 2.76$ TeV, Phys. Rev. Lett. 113 (26) (2014) 262301. arXiv:1410.1804, doi:10.1103/PhysRevLett.113.262301.

Excess of J/ψ yield at very low transverse momenta in Au+Au collisions at $\sqrt{s_{NN}}$ = 200 GeV and U+U at $\sqrt{s_{NN}}$ = 193 GeV with STAR

Wangmei Zha (for the STAR collaboration)

University of Science and Technology of China, Hefei City, Anhui Province, China

wangmei@rcf.rhic.bnl.gov

Abstract. In this article, we report the STAR measurements of J/ψ production at very low transverse momenta (p_T) in hadronic Au+Au collisions at $\sqrt{s_{NN}}$ = 200 GeV and U+U collisions at $\sqrt{s_{NN}}$ = 193 GeV at mid-rapidity. Centrality dependence of J/ψ yields and nuclear modification factors at very low p_T are also presented.

1. Introduction

The Relativistic Heavy Ion Collider (RHIC) is built to search for the Quark-Gluon Plasma (QGP) and to study its properties in laboratory through high energy heavy-ion collisions [1]. J/ψ suppression in "hadronic" heavy-ion collisions, due to color screening of quark and antiquark potential in the deconfined medium, has been proposed as a signature of the QGP formation [2]. Other mechanisms, such as the cold nuclear matter effect [3,4] and charm quark recombination [5], are likely to contribute to the observed modification of J/ψ produced in "hadronic" heavy-ion collisions.

The J/ψ can also be produced via the strong electromagnetic fields generated by heavy ions, e.g. photon-nucleus coherent or incoherent interactions [6], in ultra-relativistic heavy-ion collisions. This has been studied in detail in Ultra-Peripheral Collisions (UPC), where the impact parameter can reach several tens of femtometers and no hadronic interactions occur. Can this electromagnetic process also exist in "hadronic" heavy-ion collisions?

Recently, a significant excess of J/ψ yield at very low transverse momenta (p_T < 0.3 GeV/c) has been observed by the ALICE collaboration in peripheral hadronic Pb+Pb collisions at $\sqrt{s_{NN}}$ = 2.76 TeV at forward-rapidity [7], which cannot be explained within the hadronic J/ψ production modified by the cold and hot medium effects. The excess observed may originate from the coherent photon-nucleus interactions, which would be very challenging for the existing models. Measurements of J/ψ production at very low p_T in different collision energies, collision systems, and collision geometries can shed new light on the origin of the excess.

In this article, we report the STAR measurements of J/ψ production at very low p_T in hadronic Au+Au collisions at $\sqrt{s_{NN}}$ = 200 GeV and U+U collisions at $\sqrt{s_{NN}}$ = 193 GeV at mid-rapidity. Centrality dependence of J/ψ yields and nuclear modification factors at very low p_T are also presented.

2. Experiment and Analysis

The STAR experiment is a large-acceptance multi-purpose detector which covers full azimuth and pseudorapidity of $|\eta| < 1$ [8]. The Au+Au and U+U data were obtained using a minimum-bias trigger which requires coincidence signals in the Vertex Position Detector (VPD) and the Zero Degree Calorimeter (ZDC). In this analysis, the J/ψs are reconstructed through their decay into electron-position pairs, $J/\psi \rightarrow e^+e^-$ (branching ratio Br_{e+e-}=5.97±0.03% [9]). The primary detectors used in this analysis are the Time Projection Chamber (TPC) [10], the Time-of-Flight (TOF) detector [11], and the Barrel Electromagnetic Calorimeter (BEMC) [12]. The electron identification and J/ψ reconstruction techniques are similar to those shown in [13,14].

3. Results

Figure 1 shows the J/ψ invariant yields for Au+Au collisions at $\sqrt{s_{NN}} = 200$ GeV and U+U collisions at $\sqrt{s_{NN}} = 193$ GeV as a function of p_T for different centralities at mid-rapidity ($|y| < 1$). The error bars depict the statistical uncertainties. The boxes represent the systematic uncertainties. The solid lines in the figure are the fits using Eq. 1, wherein a, b, and n are free parameters, to data points in the range $p_T > 0.2$ GeV/c. The dashed lines are the extrapolations of the fits. As shown in the figure, the fits describe the data points above 0.2 GeV/c very well, but significantly underestimate the yield bellow 0.1 GeV/c for all three centrality bins.

$$\frac{d^2N}{p_T dp_T} = a \times \frac{1}{(1+b^2p_T^2)^n} \quad (1)$$

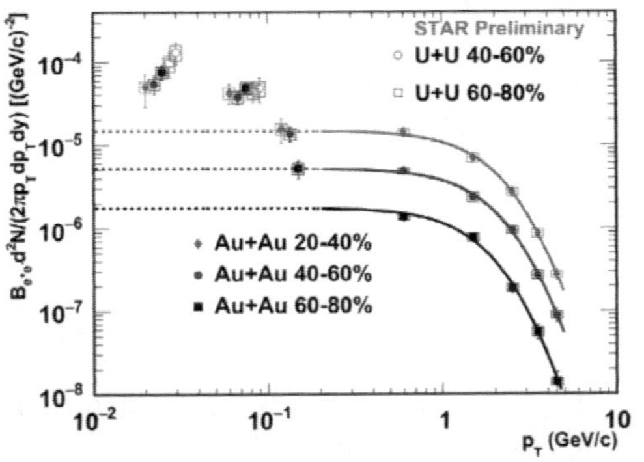

Figure 1. J/ψ invariant yields for Au+Au collisions at $\sqrt{s_{NN}} = 200$ GeV and U+U collisions at $\sqrt{s_{NN}} = 193$ GeV as a function of p_T for different centralities at mid-rapidity.

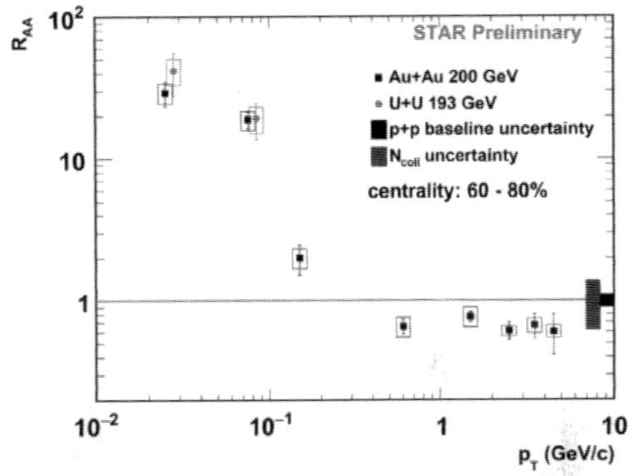

Figure 2. J/ψ R_{AA} as a function of p_T for centrality 60-80%.

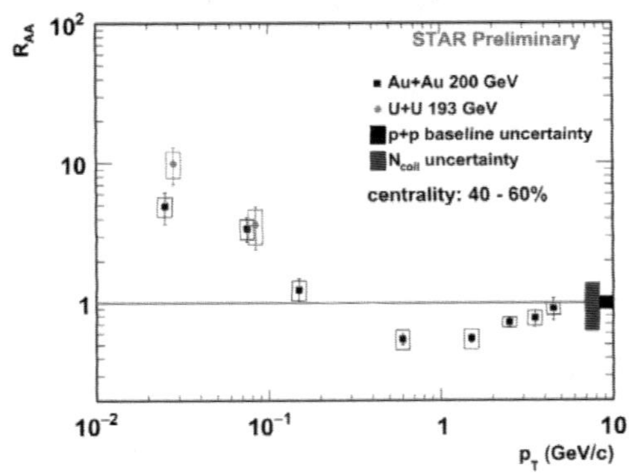

Figure 3. J/ψ R_{AA} as a function of p_T for 40-

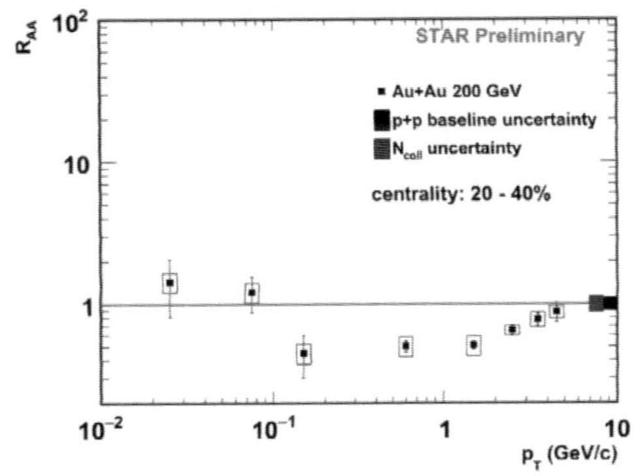

Figure 4. J/ψ R_{AA} as a function of p_T for 20-

60%. 40%.

The J/ψ R_{AA} as a function p_T for centralities 60-80%, 40-60%, and 20-40% are shown in Figures 2, 3, and 4, respectively. The pp baselines used for R_{AA} calculations are derived from [15]. Significant excess of J/ψ yield in the p_T range 0-0.1 GeV/c is observed for peripheral collisions (40-80%). The kinematic range of the enhancement is almost the same as that of coherent photon nucleus interactions, which indicates that the excess may originate from the coherent production.

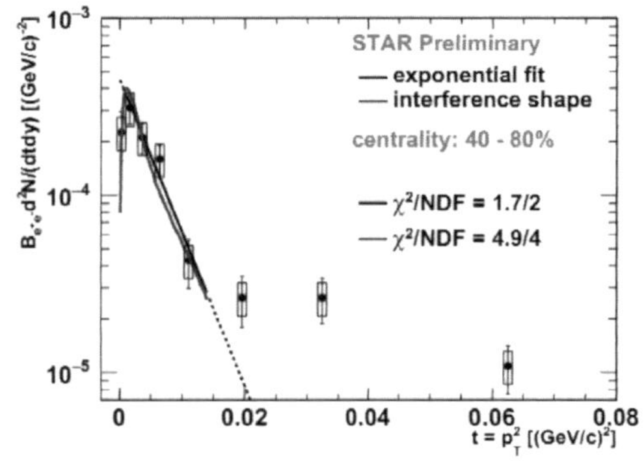

Figure 5. J/ψ invariant yield as a function of the momenta transfer squared t for 40-80% Au+Au collisions at $\sqrt{s_{NN}} = 200$ GeV.

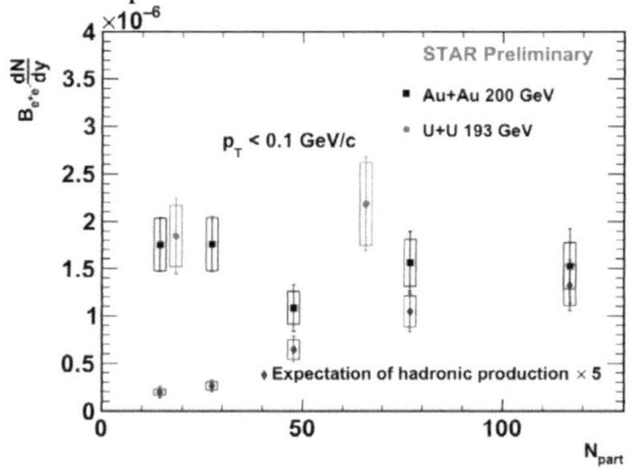

Figure 6. The p_T integrated J/ψ yields (p_T < 0.1 GeV/c) as a function of N_{part} for Au+Au collisions at $\sqrt{s_{NN}} = 200$ GeV. The expectation of hadronic production (scaled by a factor of 5) is also shown for comparison.

Assuming that the coherent photoproduction causes the excess at very low p_T, the J/ψ yield as a function of the momenta transfer squared t ($t \approx p_T^2$) for 40-80% Au+Au collisions at $\sqrt{s_{NN}} = 200$ GeV are shown in Figure 5. The structure of the distribution is very similar to that observed in UPC case [16]. An exponential fit has been applied to the distribution in t range 0.002-0.015 (GeV/c)^2. The extracted slope parameter is 196±31 (GeV/c)$^{-2}$, which is consistent with that of Au nucleus (199 [GeV/c]$^{-2}$). As shown in the figure, the first data point is significantly lower than the extrapolation of the exponential fit, which may be an indication of interference. The theoretical calculation with interference for the UPC case [17], shown as the blue curve in Figure 5, can describe the data very well ($\chi^2/NDF = 4.9/4$) for t < 0.015 (GeV/c)$^{-2}$.

Figure 6 shows the p_T integrated J/ψ yields (p_T < 0.1 GeV/c) as a function of N_{part} for 30-80% Au+Au collisions at $\sqrt{s_{NN}} = 200$ GeV. The integrated J/ψ yields show no significant centrality dependence. This is beyond the expectation of hadronic production, which, scaled by a factor of 5, is also plotted for comparison. As depicted in the figure, the contribution of hadronic production is not dominant in the excess range. The integrated J/ψ yields at p_T interval 0-0.1 GeV/c in U+U collisions are similar to those in Au+Au collisions.

4. Summary

In summary, we report the recent STAR measurements of J/ψ production at very low p_T in hadronic Au+Au collisions at $\sqrt{s_{NN}} = 200$ GeV and U+U collisions at $\sqrt{s_{NN}} = 193$ GeV at mid-rapidity. Centrality dependence of J/ψ production, nuclear modification factors, and dN/dt distributions at very low p_T are also reported. Significant excess of J/ψ yield at p_T interval 0-0.1 GeV/c is observed for peripheral collisions (40-80%). The excess for 30-80% centrality range shows no significant centrality dependence within uncertainties, which is beyond the expectation from hadronic production. The characteristics of the excess are consistent with the expectation of coherent photon-nucleus interactions.

References

[1] M. Harrison *et al.*, Nucl Instrum Meth A **499**, 235 (2003).
[2] T. Matsui and H. Satz, Phys Lett B **178**, 416 (1986).
[3] E. G. Ferreiro *et al.*, Phys Lett B **680**, 50 (2009).
[4] F. Arleo and S. Peigene, J High Energy Phys **10**, 073 (2014).
[5] L. Grandchamp *et al.*, Phys Rev Lett **92**, 212301 (2004).
[6] C. A. Bertulani *et al.*, Annu Rev Nucl Part S **55**, 271 (2005).
[7] J. Adam (ALICE Collaboration) *et al.*, Phys Rev Lett **116**, 222301 (2016).
[8] K. H. Ackermann *et al.*, Nucl Instrum Meth A **499**, 624 (2003).
[9] K. A. Olive *et al.*, Chinese Phys C **38**, 090001(2014).
[10] M. Anderson *et al.*, Nucl Instrum Meth A **499**, 659 (2003).
[11] W. J. Llope, Nucl Instrum Meth A **661**, S110 (2012).
[12] M. Beddo *et al.*, Nucl Instrum Meth A **499**, 725 (2003).
[13] L. Adamczyk (STAR Collaboration) *et al.*, Phys Rev C **90**, 024906 (2014).
[14] L. Adamczyk (STAR Collaboration) *et al.*, arXiv preprint arXiv:1607.07517 (2016).
[15] W. M. Zha *et al.*, Phys Rev C **93**, 024919 (2016).
[16] B. I. Abelev (STAR Collaboration) *et al.*, Phys Rev C **77**, 034910 (2008).
[17] S. R. Klein and J. Nystrand, Phys Rev Lett **84**, 2330 (2000).

IOP Conf. Series: Journal of Physics: Conf. Series **779** (2017) 012040 doi:10.1088/1742-6596/779/1/012040

Quarkonium measurements via the di-muon decay channel in p+p and Au+Au collisions with the STAR experiment

Takahito Todoroki (for the STAR collaboration)

Brookhaven National Laboratory, Upton, New York 11973, USA

E-mail: todoroki@bnl.gov

Abstract. We present the first J/ψ and Υ measurements in the di-muon decay channel at mid-rapidity at RHIC using the newly installed Muon Telescope Detector. In p+p collisions at $\sqrt{s} = 500$ GeV, inclusive J/ψ cross section can be described by CGC+NRQCD and NLO NRQCD model calculations for $0 < p_T < 20$ GeV/c. In Au+Au collisions at $\sqrt{s_{NN}} = 200$ GeV, we observe (i) clear J/ψ suppression indicating dissociation; (ii) J/ψ R_{AA} can be qualitatively described by transport models including dissociation and regeneration with a tension at high p_T; and (iii) hint of less melting of $\Upsilon(2S+3S)$ relative to $\Upsilon(1S)$ at RHIC compared to that at LHC.

1. Introduction

Quarkonia are an essential probe to study the properties of the Quark Gluon Plasma (QGP). The suppression of J/ψ due to color-screening effects in the medium was initially proposed as a direct evidence of the QGP formation [1]. However, the interpretation of the J/ψ suppression is still a challenge due to the contributions from the regenerated J/ψ by the recombination of $c\bar{c}$ pairs in the medium and the cold nuclear matter effects. Therefore it is important to have more precise J/ψ measurements over a broad kinematic range and even cleaner Υ state measurements. The latter do not suffer from the regeneration contribution due to the much smaller $b\bar{b}$ pair cross section, i.e. $\sigma_{b\bar{b}} \sim 2$ μb [2] while $\sigma_{c\bar{c}} \sim 800$ μb [3] at top RHIC energy. The newly installed Muon Telescope Detector (MTD), which provides both the di-muon trigger and the muon identification capability at mid-rapidity, opens the door to measuring quarkonia via the di-muon decay channel at STAR. Compared to the di-electron decay channel, the di-muon decay channel suffers much less from bremsstrahlung and thus provides much better invariant mass resolution to separate different Υ states. Using the MTD di-muon trigger, the STAR experiment recorded data corresponding to an integrated luminosity of 28.3 pb^{-1} in p+p collisions at $\sqrt{s} = 500$ GeV in the RHIC 2013 run, and 14.2 nb^{-1} in Au+Au collisions at $\sqrt{s_{NN}} = 200$ GeV in the RHIC 2014 run. In these proceedings, we report (1) the measurements of J/ψ production in p+p collisions at $\sqrt{s} = 500$ GeV; and (2) the measurements of the nuclear modification factor (R_{AA}) for J/ψ and the production of different Υ states in Au+Au collisions at $\sqrt{s_{NN}} = 200$ GeV.

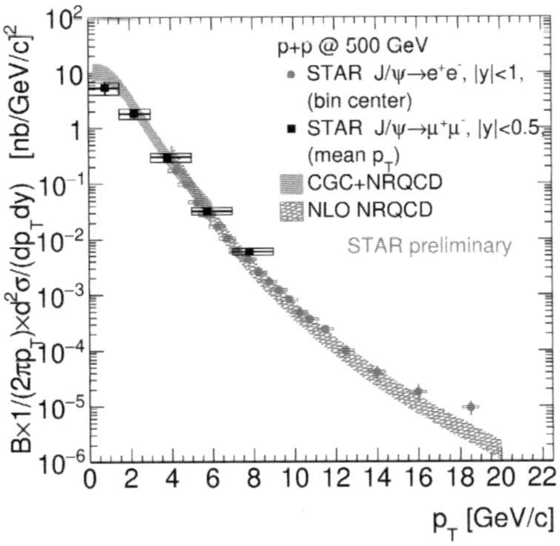

Figure 1. J/ψ cross section scaled by the branching ratio B as a function of p_T in the di-muon decay channel (black circle) and in the di-electron decay channel (red circle).

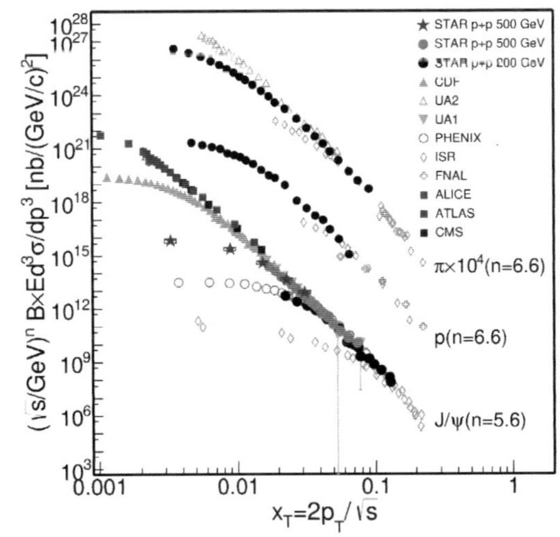

Figure 2. x_T scaling of J/ψ cross section scaled by the branching ratio B in the di-muon decay channel (blue star) and in the di-electron decay channel (red circle).

2. J/ψ measurements in p+p collisions at $\sqrt{s} = 500$ GeV

Figure 1 shows the cross section of J/ψ in p+p collisions at $\sqrt{s} = 500$ GeV in the di-electron and di-muon decay channels for $0 < p_T < 20$ GeV/c. The di-muon decay channel extends p_T reach down to 0 GeV/c. The results in these decay channels are consistent in the overlapping p_T range of $4 < p_T < 9$ GeV/c. The experimental results can be well described by CGC+NRQCD calculations at low p_T [4] and NLO NRQCD calculations at high p_T [5]. Figure 2 shows the $x_T = 2p_T/\sqrt{s}$ scaling of J/ψ cross section [6]. The J/ψ cross section in p+p collisions at $\sqrt{s} = 500$ GeV follows the common trend as a function of x_T at high p_T. The breaking of the x_T scaling at low p_T can be attributed to the soft processes.

3. J/ψ measurements in Au+Au collisions at $\sqrt{s_{NN}} = 200$ GeV

Figure 3 shows the invariant yield of J/ψ in Au+Au collisions at $\sqrt{s_{NN}} = 200$ GeV for different collision centralities. The new results in the di-muon decay channel are consistent with previous results in the di-electron decay channel [7, 8] within uncertainties.

The nuclear modification factor $R_{AA} = \frac{\sigma_{inel}}{\langle N_{coll}\rangle}\frac{d^2N_{AA}/dydp_T}{d^2\sigma_{pp}/dydp_T}$ of J/ψ in 0-40% central Au+Au collisions is shown in Fig. 4, compared with LHC results [13, 14]. The strong suppression at RHIC at low p_T indicates that dissociation plays a significant role in this p_T range. The hint of the increasing trend of R_{AA} at RHIC at high p_T can be explained by formation-time effects and feed-down of B hadrons. The less suppression of J/ψ at LHC at low p_T indicates larger regeneration contribution due to higher charm cross section, while more suppression of J/ψ at LHC at high p_T indicates larger dissociation rate due to higher temperature of the medium. Transport Models from Tsinghua [9, 10] and Texas A&M University (TAMU) [11, 12], including dissociation and regeneration effects, can qualitatively describe the p_T dependence of RHIC and LHC data.

Centrality dependence of J/ψ cross section is shown in Fig. 5 for integrated p_T and in Fig. 6 for $p_T > 5$ GeV/c. For integrated p_T, both models can describe centrality dependence at RHIC, but tend to overestimate suppression at LHC. For $p_T > 5$ GeV/c, there is tension among models

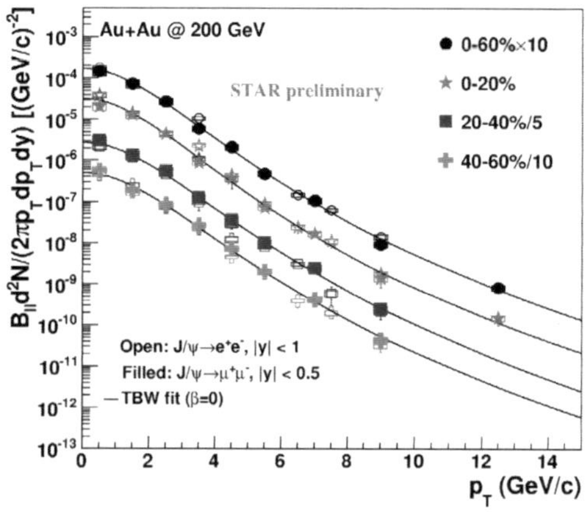

Figure 3. Invariant yield of J/ψ scaled by the branching ratio B_{ll} as a function of p_T in different centralities in the di-muon decay channel (filled) and in the di-electron decay channel (open) .

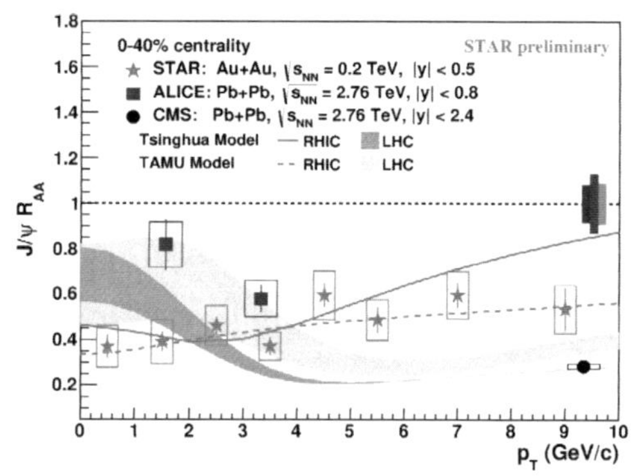

Figure 4. R_{AA} as a function of p_T at RHIC (red star) and at LHC (blue square, black circle). The lines and bands indicate Transport Model calculations for RHIC and LHC energies.

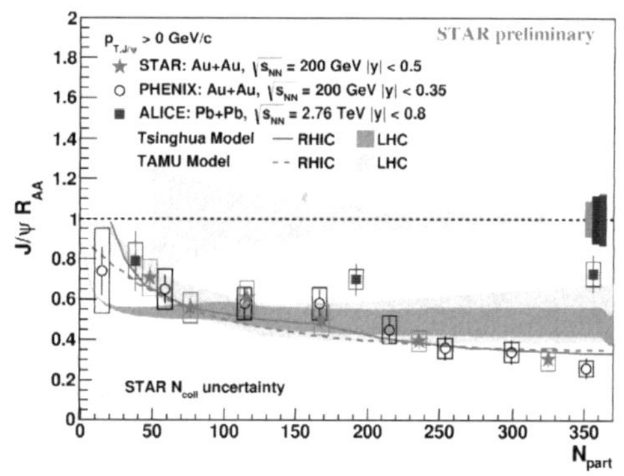

Figure 5. Nuclear modification factor R_{AA} for integrated p_T as a function of N_{part}.

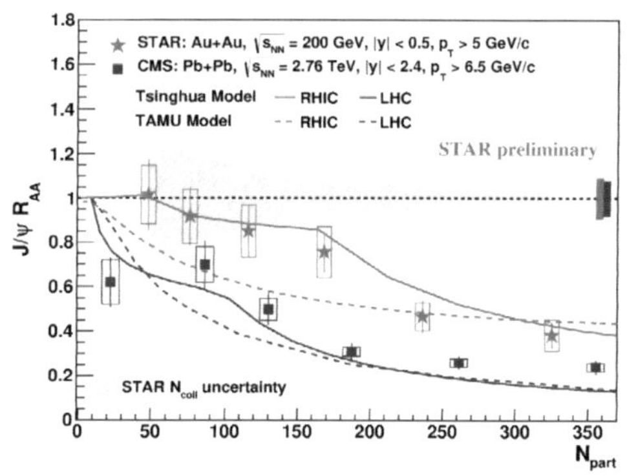

Figure 6. Nuclear modification factor R_{AA} for $p_T > 5$ GeV/c as a function of N_{part}.

and data. New measurements in the di-muon decay channel provide a distinguishing power for these transport models.

4. Υ measurements in Au+Au collisions at $\sqrt{s_{NN}} = 200$ GeV

Figure 7 shows the di-muon mass spectrum in Υ state mass range in Au+Au collisions at $\sqrt{s_{NN}} = 200$ GeV. We observe signs of an indication of $\Upsilon(2S+3S)$ signals in the di-muon decay channel. The raw yields of Υ states are obtained by a simultaneous fit to the like-sign and unlike-sign distributions. In this fit, (i) the Υ state masses are fixed to the PDG values and their widths are determined by simulation; (ii) the ratio of $\Upsilon(2S)/\Upsilon(3S)$ is fixed to the value in p+p collisions; and (iii) the shape of $b\bar{b}$ and Drell-Yan background is estimated using PYTHIA. Figure 8 shows the fitted $\Upsilon(2S+3S)/\Upsilon(1S)$ ratio compared with the world-wide p+p data [16] and CMS data [17, 18]. There is a hint of less melting of $\Upsilon(2S+3S)$ relative to $\Upsilon(1S)$ at RHIC

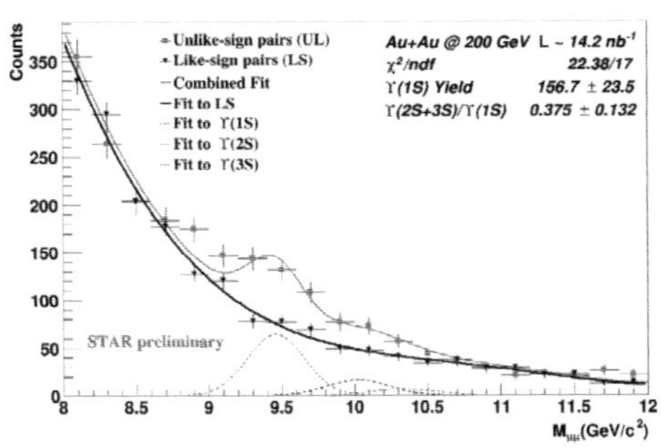

Figure 7. Di-muon mass spectrum in the Υ state mass range.

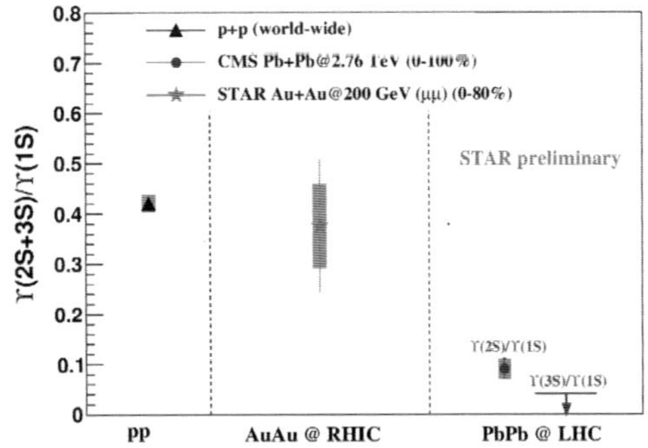

Figure 8. $\Upsilon(2S+3S)/\Upsilon(1S)$ ratio for world-wide p+p data, and for heavy-ion collisions at RHIC, and LHC energies.

than at LHC.

5. Summary and Outlook

We present the first J/ψ and Υ measurements in the di-muon decay channel at mid-rapidity at RHIC. In p+p collisions at $\sqrt{s} = 500$ GeV, inclusive J/ψ cross section can be described by CGC+NRQCD and NLO NRQCD model calculations for $0 < p_T < 20$ GeV/c. In Au+Au collisions at $\sqrt{s_{NN}} = 200$ GeV, we observe (i) clear J/ψ suppression indicating dissociation; (ii) J/ψ R_{AA} can be qualitatively described by transport models including dissociation and regeneration despite a tension at high p_T; and (iii) there is a hint of less melting of $\Upsilon(2S+3S)$ relative to $\Upsilon(1S)$ at RHIC compared to that at LHC. These measurements in Au+Au collisions will have better statistical precision by combining the similar amount of data recorded in the RHIC 2016 run.

References

[1] T. Matsui and H. Satz, *Phys. Lett. B* **178** 416-422 (1986)
[2] A. Adare *et. al.* (PHENIX Collaboration), *Phys. Rev. Lett.* **103** 082002 (2009)
[3] L. Adamczyk *et. al.* (STAR Collaboration), *Phys. Rev. D* **86** 072013 (2012)
[4] Yan-Qing Ma and Raju Venugopalan, *Phys. Rev. Lett.* **113** 192301 (2015)
[5] H. Shao *et. al.*, *JHEP* **05** 103 (2015)
[6] B. Abelev *et. al.* (STAR Collaboration), *Phys. Rev. C* **80** 041902 (2009)
[7] L. Adamczyk *et. al.* (STAR Collaboration), *Phys. Let. B* **722** 55-62 (2013)
[8] L. Adamczyk *et. al.* (STAR Collaboration), *Phys. Rev. C* **90** 024906 (2014)
[9] Y. Liu *et. al.*, *Phys. Let. B* **678** 72-76 (2009)
[10] K. Zhou *et. al.*, *Phys. Rev. C* **89** 054911 (2014)
[11] X. Zhao and R. Rapp, *Phys. Rev. C* **82** 064905 (2010)
[12] X. Zhao and R. Rapp, *Nucl. Phys. A* **859** 114-125 (2011)
[13] B. Abelev *et. al.* (ALICE Collaboration), *Phys. Let. B* **734** 314-327 (2014)
[14] S. Chatrchyan *et. al.* (CMS Collaboration), *JHEP* **05** 063 (2012)
[15] L. Adamczyk *et. al.* (STAR Collaboration), *Phys. Rev. Lett.* **111** 052301 (2013)
[16] W. Zha *et. al.*, *Phys. Rev. C* **88** 067901 (2013)
[17] S. Chatrchyan *et. al.* (CMS Collaboration), *Phys. Rev. Lett.* **109** 222301 (2012)
[18] S. Chatrchyan *et. al.* (CMS Collaboration), *JHEP* **04** 103 (2014)

Dynamical bottomonium-suppression in a realistic AA background

P-B Gossiaux and R Katz

Subatech, 4 rue Alfred Kastler, 44307 Nantes, France

E-mail: gossiaux@subatech.in2p3.fr, katz@subatech.in2p3.fr

Abstract. We refine our dynamical Schroedinger-Langevin scheme designed for dealing with the suppression of quarkonium states in URHIC in order to allow for feed-downs from higher states and present updated results for the R_{AA} as a function of the centrality.

1. Introduction

The suppression of upsilon $\Upsilon(1S)$ states in AA collisions, observed by the STAR collaboration at RHIC [1] and by the CMS [2] and ALICE [3] collaborations at LHC, is one of the most convincing evidence for the creation of the quark gluon plasma. The precise survival of excited $\Upsilon(2S)$ and $\Upsilon(3S)$ states vs ground state could even allow to measure the highest temperature reached in those collisions, according for instance to the sequential suppression scenario which is substantiated by calculations of the dissociation temperature based on lattice potentials evaluated at finite temperature. In a previous work [4], we have addressed the question of upsilon dissociation resorting to a dynamical approach, i.e. the non-linear Schroedinger-Langevin equation (SLE). In this scheme, the time-dependent real potential implements the Debye-screening while the stochastic forces express the (hard) interactions between the QGP and the $b\bar{b}$ state. The SLE allows to treat transitions between bound and open quantum states (the "suppression"), but also transitions between bound states, a feature usually lacking in other treatments. In a stationary QGP, SLE naturally leads to asymptotic distributions of Y(1S), Y(2S),... following correct statistical weights [5], which allows to make the link with models based on the hypothesis of statistical recombination. This sanity check is another appealing feature of our approach.

In this work, we iterate on our previous predictions [4] for upsilon suppression in URHIC at LHC energy. In order to achieve more reliable comparisons between our model and experimental measurements, two improvements were implemented, as described in sections 2 and 3.

2. The model

As detailed in [4], we resort to the so-called Schrödinger-Langevin equation (SLE) to deal with the dynamics of the $b-\bar{b}$ internal state:

$$i\hbar\frac{\partial\psi}{\partial t} = \left[\hat{H}_0+\hbar A\Big(S(\mathbf{x},t)-\int\psi^* S(\mathbf{x},t)\,\psi\,d\mathbf{x}\Big)-\mathbf{x}\cdot\mathbf{F}_R(t)\right]\psi \quad\text{with}\quad \hat{H}_0 = -\frac{\hbar^2\Delta}{m_b}+V_{\rm MF}(\mathbf{x})\,. \quad (1)$$

In this equation, the non linear term $\propto \hbar A$is a dissipative friction term based on the wave function phase S while the term $-\mathbf{x} \cdot \mathbf{F}_R(t)$ represents stochastic dipolar forces mocking the scatterings on QGP constituents. The Hamiltonian $\hat{H}_0$ encodes the dynamics in the absence of such stochastic interactions with the medium, resorting to the mean field potential $V_{\rm MF}$.

Solving (1) in 3+1D is still beyond our present implementation of the numerical scheme. Consequently, we adopt a 1+1D modeling of the $b\bar{b}$ pairs, with even (resp. odd) 1D-states mocking S (resp. P) 3D-states. In vacuum, the potential in $\hat{H}_0$ is taken as $V_{\rm MF}(T = 0, x) = V_{\rm vac}(x) = K|x|$, truncated to $V_{\rm max}$. Such an approximation was already adopted in [4], with (set I) $m_b = 4.575$ GeV, $K = 1.375$ GeV/fm and $V_{\rm max} = 1.2$ GeV, following lQCD calculations [6]. The masses of the various upsilon states were found in good agreement with their experimental values but the binding energies a bit too low (in particular for $\Upsilon(3S)$, whose binding energy was just 20 MeV in our model). In this work, we use slightly different values of the parameters, i.e. (set II) $m_b = 4.61$ GeV, $K = 1.246$ GeV/fm and $V_{\rm max} = 1.338$ GeV, which leads to very good values of the bottomonia masses ($E_{1S} = 9.46$ GeV, $E_{1P} = 9.77$ GeV, $E_{2S} = 9.99$ GeV, $E_{2P} = 10.18$ GeV, $E_{3S} = 10.35$ GeV) and even leads to a 3P state (not found for the set I), with $E_{3P} = 10.51$ GeV. With set II, the binding energy of the 3S state is 120 MeV, which is the correct value in the vacuum.

3. Production in pp

The main improvement of this contribution wrt our previous work consists in dealing with the feed-downs. For this purpose, we need a fair understanding of the direct production of the various bottomonium states within our scheme, starting from the pp case. This is not a trivial issue, as the debate between various descriptions (Color Evaporation Model, Color Singlet Model, Color Octet Model) is still not settled [8]. Our *bona fide* prescription is to consider some trial initial state $\psi_{b\bar{b}}(t = 0)$ in the singlet representation containing even ($\equiv$ S wave) and odd ($\equiv$ P wave) components:

$$\psi_{b\bar{b}}(t = 0, x) \propto \left(1 + a_{\rm odd}\frac{x}{\sigma}\right) e^{-\frac{x^2}{2\sigma^2}}, \tag{2}$$

where σ and $a_{\rm odd}$ are free parameters. In pp, the direct production of $\Upsilon(nS)$ or of $\chi_b(nP)$ is then modeled as $d\sigma^{\rm direct}_{\Upsilon(nS)/\chi_b(nP)} \propto |\langle nS/nP|\psi_{b\bar{b}}(t = 0)\rangle|^2$, where $|nS\rangle$ and $|nP\rangle$ are the eigenstates for $V_{\rm vac}$. The prompt production is then obtained from the branching ratios β_{ij} through $d\sigma_i^{\rm prompt} = d\sigma_i^{\rm direct} + \sum_{j>i} \beta_{ij} d\sigma_j^{\rm direct}$. This allows to deduce the various contributions from feed-downs and extract the optimal parameters $a_{\rm odd}$ and σ by comparison with experimental results [8]. Let us insist that this procedure is pretty much specific to quantum treatments while semi-classical modeling like transport equations "just" rely on the probabilities and can thus be initiated with the mere numbers extracted from measured $d\sigma^{\rm direct}$. The extraction of the optimal parameters was made considering the experimental values measured at small p_T which represents the bulk of the production. For the purpose of illustration we show in fig. 1 (left), the proportion of 1S prompt production stemming from 1P decay, as a function of both parameters. Rather unexpectedly, it appears to be possible to reproduce all low-p_T feed-downs ratios with the simple form (2) as initial state for $\sigma = 0.045$ fm and $a_{\rm odd} = 3.5$. With these parameters, the absolute (direct) weights in 1S, 1P, 2S, ..., 3P are 5.5%, 3.6%, 1.5%, 3.3%, 0.9% and 2.5% respectively. It should be noted, however, that rather strong p_T dependences of this "cocktail" has been reported in [8], f.i. an increase of the feed-down contribution from 1P→ 1S (LHCb) as well as the increase of the (prompt) $\Upsilon(nS)/\Upsilon(1S)$ ratios (ATLAS, CMS), reaching ~ 0.5 for $p_T \sim 30$ GeV/c which, to our knowledge has not yet been explained on theoretical grounds.

We have tried to accommodate these evolutions with the trial state (2), varying the parameters σ and $a_{\rm odd}$ with p_T but we have identified that the (prompt) ratios $\Upsilon(2S/3S)/\Upsilon(1S)$ cannot exceed $\approx 0.22/0.16$ (as illustrated on fig. 1 for the $\Upsilon(2S)/\Upsilon(1S)$ ratio) which are

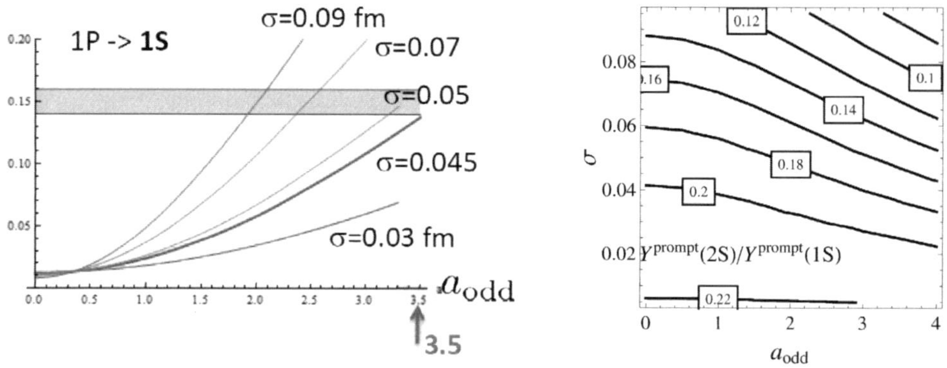

Figure 1. Left panel: proportion from the prompt 1S-bottomonia production stemming from 1P decay as a function of $a_{\rm odd}$ for various σ values. The band represent the experimental result at small p_T. Right panel: $\Upsilon(2S)/\Upsilon(1S)$ ratio as a function of σ and $a_{\rm odd}$.

significantly smaller than the ratios found experimentally at large p_T. As a consequence, more sophisticated production mechanisms should be investigated and understood in this domain, for which our predictions are thus less robust.

4. Production in PbPb and perspectives

In PbPb collisions, the initial state (2) is evolved as in [4]. The temperature and velocity profile is modeled by EPOS2 [9], which provides good predictions in the soft sector, despite resorting to an ideal fluid dynamics evolution (this being somehow compensated by broader initial flux tubes). In order to model the temperature dependence of the mean-field $V_{\rm MF}(T, x)$, we *bona fide* scale down $V_{\rm max}$ according to the asymptotic value of the so-called "weak" potential V_W introduced by Mocsy and Petreczky [6] (our privileged choice in these proceedings), characterized by a binding strength intermediate between the free energy F and the internal energy U [7].

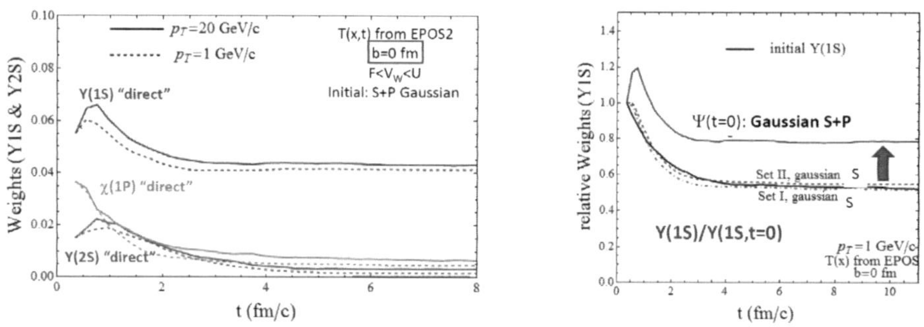

Figure 2. Left panel: average evolution of the 1S, 1P and 2S components of the quantum state in the EPOS2 temperature profile for a $b\bar{b}$ state produced according to an S+P initial Gaussian state with $p_T = 1$ GeV/c (dashed) and $p_T = 20$ GeV/c (solid). Right panel: "survival" of the 1S state for various models.

In fig. 2, we show the time evolution of the different components of the quantum state, after averaging on a) stochastic forces, b) production point and initial direction, and c) EPOS2 events, for a central PbPb collision at $\sqrt{s} = 2.76$ A GeV and 2 choices of p_T. Quite generally, one observes a strong suppression of excited states, whose weights evolve on longer time scales. We also notice that the S-states benefit from some extra population from the P states during the first fm/c, due to the dipolar nature of the stochastic forces, a feature lacking in [5] where an initial S-state corresponding to $a_{\rm odd} = 0$ was chosen. The p_T dependence is found to be mild. In the right panel of fig. 2, we show the "survival" of $\Upsilon(1S)$ – i.e. the instantaneous 1S-component of the state relative to the initial one – for various models. We see that this survival is not much affected by the new set of parameters, neither by the specific wave packet in the initial S wave (Gaussian of 1S-eigenstate), while starting from mixed S+P wave packet leads to a larger survival (of the order of 80% instead of 60%), mostly due to the initial P→ S transitions.

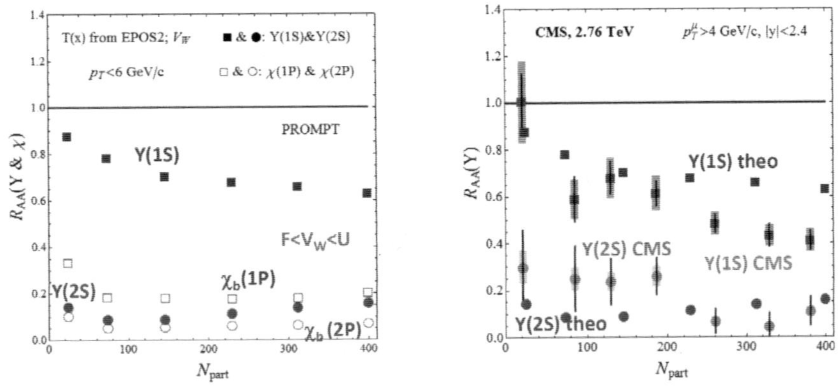

Figure 3. Left panel: R_{PbPb} of various bottomonium states as a function of the number of participants in the collision. Right panel: same for the upsilon states, compared with CMS data [2].

In fig. 3, we show the nuclear modification factor for the prompt production of all bottomonium-like states in our model, applying the suppression of the direct states before feed-downs. Left panel exhibits a pattern typical from a sequential-suppression "scenario" while, on the right panel, an agreement with experimental CMS data is obtained within error bars, except for the most central collisions where our model lacks some $\Upsilon(1S)$ suppression. Comparing with our previous results [4], this excess can be traced to be due to the larger 1S-survival – thanks to P→S transitions – not completely compensated by the suppression of the feed-downs. We fear we cannot conclude on this point without including cold nuclear matter effects and considering AuAu and UU collisions at RHIC as well, what will be done in a future work.

Acknowledgments
We gratefully acknowledge the support from the TOGETHER project, Région Pays de la Loire (France).

5. References

[1] STAR Collaboration, Phys.Lett. B**735** (2014) 127-137, Erratum: Phys.Lett. B**743** (2015) 537-541.
[2] CMS Collaboration, CMS PAS HIN-15-001 (2015).
[3] ALICE Collaboration, Phys. Lett. B**738** (2014) 361-372.
[4] P.B. Gossiaux and R. Katz, proceedings of Quark Matter 2015 conference [arXiv:1601.01443 [hep-ph]].
[5] R. Katz and P.B. Gossiaux, Annals Phys. 368 (2016) 267-295, [arXiv:1504.08087 [quant-ph]].
[6] A. Mocsy and P. Petreczky, Phys. Rev. D **77** (2008) 014501.
[7] O. Kaczmarek and F. Zantow, Eur. Phys. J. C **43** (2005) 63.
[8] A. Andronic et al., Eur.Phys.J. C**76** (2016) no.3, 107, [arXiv:1506.03981 [nucl-ex]].
[9] K. Werner, Iu. Karpenko, T. Pierog, M. Bleicher, K. Mikhailov, Phys.Rev. C **82** (2010) 044904.

ψ(2S) Production at the LHC

Xiaojian Du and Ralf Rapp

Cyclotron Institute and Department of Physics & Astronomy, Texas A&M University, College Station, TX 77843-3366, USA

E-mail: xjdu@physics.tamu.edu, rapp@comp.tamu.edu

Abstract. We calculate the production of $\psi(2S)$ and the pertinent double ratio of its nuclear modification factor ($R_{\rm AA}$) over that of the J/ψ in Pb-Pb collisions at the LHC. Based on a transport model with temperature dependent reaction rates, a sequential regeneration pattern emerges: the larger $\psi(2S)$ width, relative to the J/ψ, around and below the critical temperature, implies that most of the $\psi(2S)$ states are regenerated later in the evolution of the fireball. This has noticeable consequences for the transverse-momentum (p_T) spectra of the regenerated charmonia. While the total yield of $\psi(2S)$ meson remains smaller than those of J/ψ's, their harder p_T spectra can produce a double ratio above unity for a $p_T > 3\,$GeV cut, as applied by the CMS collaboration. A significant uncertainty in our calculations is associated with the values of the temperature where most of the $\psi(2S)$ regeneration occurs, *i.e.*, the quantitative temperature dependence of its inelastic width.

1. Introduction

Intense experimental efforts are ongoing to measure the production systematics of heavy quarkonia in heavy-ion collisions (HICs), to establish a robust benchmark for studying their modifications in hot and dense QCD matter [1, 2, 3, 4]. The sequence of bound states in the vacuum spectrum of heavy quarkonia provides a unique probe of how the basic QCD force evolves in the medium. In the charmonium sector, where extensive measurements of J/ψ production have revealed important information about its in-medium kinetics, the $\psi(2S)$ state is now becoming a rather hot topic. A strong $\psi(2S)$ suppression relative to the J/ψ has been observed both in proton-nucleus (pA) collisions at RHIC and the LHC [5, 6], and in nucleus-nucleus collisions [7] at the SPS, generally associated with final-state interactions together with a much smaller binding energy of the $\psi(2S)$. It was therefore rather intriguing when the CMS collaboration found an enhancement of the $\psi(2S)/J/\psi$ ratio in central Pb-Pb(2.76 TeV) collisions, relative to pp collisions [8]. More precisely, this enhancement was found in the kinematic range of forward-rapidities ($1.6<|y|<2.4$) and transverse momenta $p_T>3\,$GeV, while it turns out to be suppressed around mid-rapidity and $p_T>6.5\,$GeV. In Ref. [9] we suggested a sequential regeneration of J/ψ and $\psi(2S)$ states as a potential mechanism to understand this phenomenon within schematic model scenarios. In the present paper we expand on this work by revisiting the results within a more quantitative rate equation approach [10, 11, 12] (Sec. 2), discussing the double ratio and its main uncertainties in Pb-Pb(2.76 TeV) collisions (Sec. 3), and providing predictions at 5.02 TeV (Sec. 4). We conclude in Sec. 5.

2. Transport model for charmonia production

Our approach for quarkonium kinetics in HICs utilizes a rate equation,

$$\frac{\mathrm{d}N_\Psi}{\mathrm{d}\tau} = -\Gamma_\Psi(T)\left[N_\Psi - N_\Psi^{\mathrm{eq}}(T)\right] , \tag{1}$$

which evolves the charmonium yields, N_Ψ ($\Psi{=}J/\psi, \psi(2S), \chi_c(1P)$) through an expanding fireball. The pertinent transport coefficients are: (a) the inelastic reaction rate $\Gamma_\Psi(T)$ in the quark-gluon plasma (QGP) and in hadronic matter. "Quasifree" dissociation is adopted as the dominant mechanism to break up J/ψ and $\psi(2S)$ states in the QGP where both have relatively small binding energies, $E_B \leq T$, while hadronic dissociation rate is calculated from a meson exchange model with $SU_f(4)$ flavor symmetry; (b) the thermal equilibrium limit $N_\Psi^{\mathrm{eq}}(T)$ which controls the rate of regeneration. It is evaluated from the statistical model on either partonic or hadronic basis and includes non-equilibrium corrections due to a finite correlation volume and incomplete charm-quark thermalization [10].

To compute p_T spectra, we decompose the solution of the rate equation into two parts, one characterizing the primordial (direct) production and the other from regeneration of Ψ states. The Boltzmann equation is used to calculate the p_T-spectra of the primordial part via

$$\frac{\partial f(\vec{x},\vec{p},t)}{\partial t} + \vec{v}\cdot\frac{\partial f}{\partial \vec{x}} = -\Gamma_\Psi(\vec{p},T)f(\vec{x},\vec{p},t) \tag{2}$$

where f is the phase space distribution of charmonia and $\vec{v} = \vec{p}/E_p$ their velocity. Escape effects are included by setting the rate to zero if a state exits the fireball boundary. The p_T-spectra of the regeneration component are evaluated from a blastwave description assuming thermal equilibrium,

$$\frac{\mathrm{d}N}{p_T\mathrm{d}p_T} \simeq m_T\int_0^R r\mathrm{d}rK_1\left(\frac{m_T\cosh(\rho(r))}{T}\right)I_0\left(\frac{p_T\sinh(\rho(\mathrm{r}))}{T}\right) \tag{3}$$

where $m_T = \sqrt{p_T^2+m^2}$ is the transverse mass and $\rho(r) = \tanh^{-1}(v(r,t,b))$ the transverse-flow profile of the fireball. Later times in the evolution lead to a harder spectra, as lower temperatures are overcompensated by the blue-shift due to larger flow. The blastwave spectra are normalized to the regeneration yield obtained from the p_T-independent rate equation. The pertinent R_{AA}'s are computed in the usual way as a ratio of AA and pp spectra, $R_{\mathrm{AA}} = (N^{\mathrm{AA}})/(N_{\mathrm{coll}}N^{\mathrm{pp}})$, scaled by the binary collision number, N_{coll}, obtained from the optical Glauber model.

The space-time evolution of the temperature is constructed from a simple ansatz for an iso- and isentropically expanding fire cylinder volume, $V_{\mathrm{FB}}(t)$. Using conservation of total entropy,

$$S_{\mathrm{tot}} = s(T)V_{\mathrm{FB}}(t) , \tag{4}$$

together with a suitable equation of state for the entropy density, $s(T)$ (for which we use a QGP quasi-particle and hadron-resonance gas connected through a mixed phase at T_c=180 MeV), the time-dependent temperature is obtained for a given centrality by matching S_{tot} to the final-state hadron multiplicities.

3. Sequential regeneration in Pb-Pb(2.76 TeV)

Our results for the centrality dependence of the R_{AA} "double ratio", $R_{\mathrm{AA}}(\psi(2S))/R_{\mathrm{AA}}(J/\psi)$, as obtained from the kinetic rate equation are displayed in Fig. 1. They confirm the results of our previously published schematic-model study [9]; moderate quantitative deviations arise from the more restrictive model approach which intimately couples the suppression and regeneration yields. The basic trends of the CMS data in Pb-Pb(2.76 TeV) collisions for the two p_T cuts are still reproduced (left panel of Fig. 1), although the maximal enhancement in central collisions is

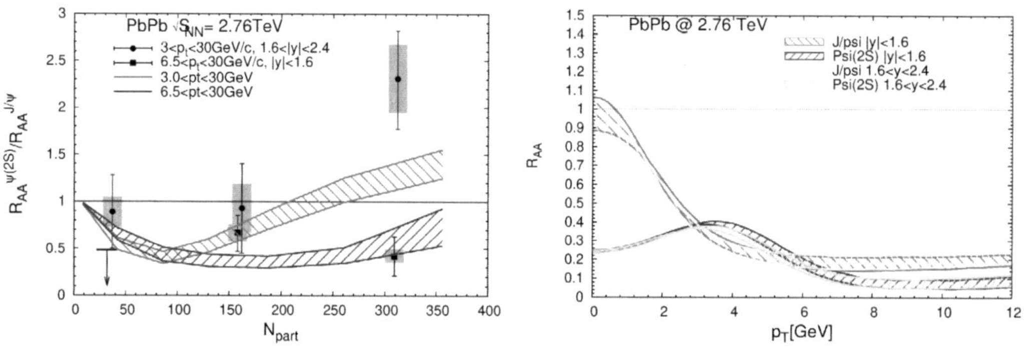

Figure 1. Charmonium production in Pb-Pb collisions at $\sqrt{s_{NN}}$=2.76 TeV within the kinetic rate equation approach. Left panel: centrality dependence of the double ratio $R_{\rm AA}(\psi(2S))/R_{\rm AA}(J/\psi)$ for p_T>6.5 GeV and $|y|$<1.6 (blue band) as well as p_T>3 GeV and 1.6<$|y|$<2.4 (red band), compared to CMS data (points with error bars) [8]. Right panel: p_T dependence of the individual J/ψ and $\psi(2S)$ $R_{\rm AA}$'s for central collisions. A 10 % shadowing is assumed.

not fully reproduced. Nevertheless, the marked increase of the double ratio beyond one remains a key signature of the *sequential regeneration* mechanism, whereby most of the $\psi(2S)$ are formed later in the evolution, thus being blue-shifted to p_T above 3 GeV (red band). The regenerated J/ψ are mostly concentrated at momenta p_T $\leq$3 GeV, consistent with ALICE data [13]. On the other hand, for p_T >6.5 GeV (blue band), the regeneration component has essentially faded away (being exponentially suppressed relative to the primordial power-law spectra), and the stronger suppression of the primordial $\psi(2S)$ relative to the J/ψ leads to a double ratio below one. The explicit p_T dependence of the $\psi(2S)$ and J/ψ $R_{\rm AA}$s for central Pb-Pb is depicted in the right panel of Fig. 1.

The largest contribution to the uncertainty bands in Fig. 1 is due to the choice of the average temperature, $\bar{T}_{\rm reg}$, at which the blast-wave expression is evaluated, representing the window where most of the regeneration occurs. From the time dependence of the regeneration yields we estimate $\bar{T}_{\rm reg}\simeq$160-165 MeV for the $\psi(2S)$ and $\bar{T}_{\rm reg}\simeq$180-200 MeV for the J/ψ [9] which can describe the ALICE data for $R_{\rm AA}^{J/\psi}(p_T)$ at 2.76 TeV [14]. We also varied the initial spectra by inclusion of a moderate Cronin effect with a broadening parameter of up to a_{gN}=0.2 GeV^2/fm.

4. Predictions for 5.02 TeV

For our predictions at $\sqrt{s_{NN}}$=5.02 TeV, we implement the following changes. The charged particle multiplicity, *i.e.*, $S_{\rm tot}$, is increased by 22.5% [15], yielding an increase of ~7% for the initial temperature. Using experimental [16] and theoretical [17] results as guidance, the charm cross section in pp, $\frac{d\sigma_{c\bar{c}}}{dy}$, is increased by ~40%, from 0.65(0.59) mb to 0.92(0.84) mb at mid-(forward) rapidity, and likewise for charmonia. Their initial p_T spectra in pp are updated (somewhat harder than at 2.76 TeV), and a 10% additional shadowing is assumed. Finally, for the J/ψ blast-wave spectra, we found that a somewhat lower temperature range of T=179-181 MeV (representing a time window around the mixed phase) better describes the preliminary ALICE dimuon data for the J/ψ $R_{AA}(p_T)$ at 5.02 TeV as presented at this meeting [18].

The resulting $R_{\rm AA}$ double ratios for the different p_T cuts show a trend of moving closer together (see left panel of Fig. 2). Due to the increase in transverse flow, more regenerated J/ψ are pushed beyond the p_T>3 GeV thus suppressing the red band, while more $\psi(2S)$ are pushed

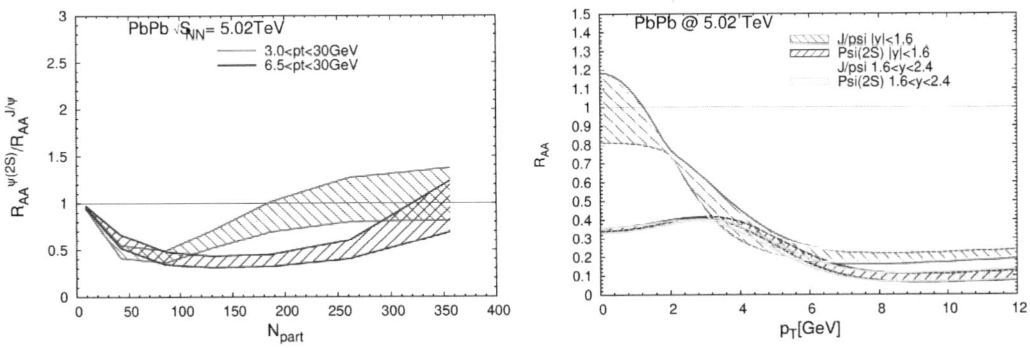

Figure 2. Same as Fig. 1 but for Pb-Pb collisions at $\sqrt{s_{NN}}$=5.02 TeV.

beyond p_T>6.5 GeV thus enhancing the blue band, cf. also the individual $R_{AA}(p_T)$'s in the right panel of Fig. 2.

5. Conclusions

In summary, the application of a kinetic rate equation approach to charmonium production in HICs suggests a scenario where $\psi(2S)$ states are regenerated significantly later in the fireball evolution than J/ψ mesons. This emerges from inelastic reaction rates which are significantly larger for the $\psi(2S)$ than the J/ψ in the later (hadronic) stages of the fireball. This "sequential regeneration" is a direct consequence of sequential suppression plus detailed balance. We have found that the phenomenological implications of this scenario can help to explain the puzzling observation of the CMS $\psi(2S)$-to-J/ψ R_{AA} double ratio.

Acknowledgments

This work is supported by the US National Science Foundation under grant no. PHY-1614484.

References

[1] L. Kluberg and H. Satz, arXiv:0901.3831 [hep-ph].
[2] P. Braun-Munzinger and J. Stachel, *Landolt-Bornstein* **23**, 424 (2010) arXiv:0901.2500 [nucl-th].
[3] R. Rapp, D. Blaschke and P. Crochet, *Prog. Part. Nucl. Phys.* **65**, 209 (2010)
[4] B. Chen, Y. Liu, K. Zhou and P. Zhuang, *Phys. Lett.* B **726**, 725 (2013)
[5] A. Adare *et al.* [PHENIX Collaboration], *Phys. Rev. Lett.* **111**, no. 20, 202301 (2013)
[6] J. Adam *et al.* [ALICE Collaboration], *JHEP* **1606**, 050 (2016)
[7] B. Alessandro *et al.* [NA50 Collaboration], *Eur. Phys. J.* C **49**, 559 (2007)
[8] V. Khachatryan *et al.* [CMS Collaboration], *Phys. Rev. Lett.* **113**, no. 26, 262301 (2014)
[9] X. Du and R. Rapp, *Nucl. Phys.* A **943**, 147 (2015)
[10] L. Grandchamp, R. Rapp and G. E. Brown, *Phys. Rev. Lett.* **92**, 212301 (2004)
[11] X. Zhao and R. Rapp, *Phys. Rev.* C **82**, 064905 (2010)
[12] X. Zhao and R. Rapp, *Nucl. Phys.* A **859**, 114 (2011)
[13] J. Adam *et al.* [ALICE Collaboration], *JHEP* **1605**, 179 (2016)
[14] B. B. Abelev *et al.* [ALICE Collaboration], *Phys. Lett.* B **734**, 314 (2014)
[15] H. Niemi, K. J. Eskola, R. Paatelainen and K. Tuominen, *Phys. Rev.* C **93**, 014912 (2016)
[16] A. Andronic *et al.*, *Eur. Phys. J.* C **76**, no. 3, 107 (2016)
[17] M. Cacciari, S. Frixione, N. Houdeau, M. L. Mangano, P. Nason and G. Ridolfi, *JHEP* **1210**, 137 (2012)
[18] J. Adam *et al.* [ALICE Collaboration], arXiv:1606.08197 [nucl-ex].

Strangeness production in p–Pb and Pb–Pb collisions with ALICE at LHC

Domenico Colella[1,2] for the ALICE Collaboration

[1] European Organization for Nuclear Research (CERN)
[2] Institute of Experimental Physics, Slovak Academy of Sciences (IEP SAS)

E-mail: domenico.colella@cern.ch

Abstract. The main goal of the ALICE experiment is to study the properties of the hot and dense medium created in ultra-relativistic heavy-ion collisions. The measurement of the (multi-)strange particles is an important tool to understand particle production mechanisms and the dynamics of the quark-gluon plasma (QGP). We report on the production of K^0_S, $\Lambda(\overline{\Lambda})$, $\Xi^-(\overline{\Xi}^+)$ and $\Omega^-(\overline{\Omega}^+)$ in proton-lead (p–Pb) collisions at $\sqrt{s_{NN}} = 5.02$ TeV and lead-lead (Pb–Pb) collisions at $\sqrt{s_{NN}} = 2.76$ TeV measured by ALICE at the LHC. The comparison of the hyperon-to-pion ratios in the two colliding systems may provide insight into strangeness production mechanisms, while the comparison of the nuclear modification factors helps to determine the contribution of initial state effects and the suppression from strange quark energy loss in nuclear matter.

1. Introduction

Strangeness production has been extensively studied in relativistic heavy-ion collisions. Its measurement represents an important tool to investigate the properties of the strongly interacting system created in the collision, as there is no net strangeness content in the initially colliding nuclei. In these proceedings we briefly discuss the technique used to identify K^0_S, $\Lambda(\overline{\Lambda})$, $\Xi^-(\overline{\Xi}^+)$ and $\Omega^-(\overline{\Omega}^+)$ and to measure their transverse momentum (p_T) spectra. Then we describe the main results about baryon anomaly, strangeness nuclear modification factor and strangeness enhancement.

2. Strange particle identification with the ALICE detector

The ALICE detector was designed to study heavy-ion physics at the LHC. At mid-rapidity, tracking and vertexing are performed using the Inner Tracking System (ITS), consisting of six layers of silicon detectors, and the Time Projection Chamber (TPC). The two innermost layers of the ITS and the V0 detector (scintillation hodoscopes covering the forward pseudo-rapidity region on either side of the interaction point) are used for triggering. The V0 is also used to estimate centrality in Pb-Pb collisions as well as multiplicity in p-Pb collisions. A complete description of the ALICE sub-detectors can be found in [1].

The single-strange (K^0_S and Λ) and multi-strange (Ξ and Ω) particles are reconstructed via their characteristic weak decay topologies into two and three particles, respectively. Tracks reconstructed by the TPC and the ITS are combined to select candidates satisfying a set of geometrical criteria. In addition, particle identification is performed by a selection on the

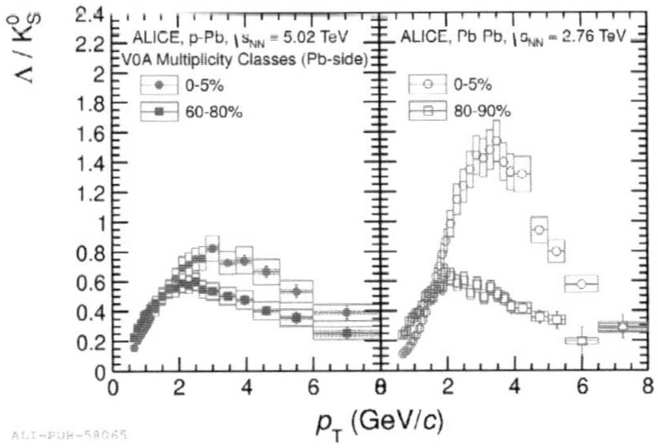

Figure 1. $(\Lambda+\overline{\Lambda})/K^0_S$ as a function of p_T in the highest and a low-multiplicity class in p–Pb (left) and Pb–Pb (right) collisions. Open boxes are total systematic uncertainties and shaded boxes represent the uncorrelated-over-multiplicity component.

specific energy loss in the TPC for the daughter tracks. Particle yields as a function of p_T are determined, in various multiplicity/centrality intervals, using an invariant mass analysis. Acceptance and efficiency corrections are calculated using Monte Carlo simulations. For more details, we refer to [2, 3, 4].

3. Results

Transverse momentum spectra for K^0_S, $\Lambda(\overline{\Lambda})$, $\Xi^-(\overline{\Xi}^+)$ and $\Omega^-(\overline{\Omega}^+)$ have been published in [6, 2] for Pb–Pb collisions and in [3, 4] for p–Pb collisions. In both systems the hardening of the spectral shape from low- to high-multiplicity collisions has been observed. Moreover, in the context of Blast Wave model, it has been shown that spectra for K^0_S and Λ are well predicted using parameters from a simultaneous fit to $\pi^\pm$, $K^\pm$ and p spectra in high multiplicity Pb–Pb and p–Pb collisions. This is also true for Ξ and Ω in p–Pb collisions, while in Pb–Pb collisions they cannot be described in a common freeze-out scenario, as they would require a lower mean transverse flow velocity and a higher kinetic freezeout temperature to be described properly [5].

The p_T-differential $(\Lambda+\overline{\Lambda})/K^0_S$ ratio is shown in Fig. 1 in the highest and a low-multiplicity class for p–Pb and Pb–Pb collisions. The presence of a peak in the 2 - 4 GeV/c p_T region and the multiplicity dependence of its characteristics are qualitatively similar in the two collision systems. In Pb–Pb collisions this phenomenon has been interpreted as a redistribution of baryons and mesons in momentum, when centrality increases, as consequence of the increased radial flow [6]. The flattening of the $(\overline{p}+p)/\phi$ ratio as a function of p_T in 0-10% central Pb–Pb collisions is further evidence supporting an hydrodynamical evolution of the system since due to the similar masses the boost from radial flow is almost the same. This data can also be described by parton recombination models [7].

The nuclear modification factor is defined as the ratio of the p_T spectra in Pb–Pb (R_{AA}) or p–Pb (R_{pPb}) and in pp collisions scaled by the number of nucleon-nucleon collisions. It has been shown that the strong suppression of hadron production at high p_T observed at the LHC in Pb–Pb collisions is not due to an initial-state effect but may be due to jet quenching in hot QCD matter [8]. In Fig. 2 the R_{AA} as a function of p_T for multi-strange baryons in central (0-10%) collisions is shown and compared with those for lighter hadrons (π, K and p). R_{AA} for Ξ is similar to the one for p, especially at high p_T (> 6 GeV/c), while π and K at low p_T (< 3 GeV/c) follow a clearly different trend, which can be interpreted as a consequence of having different masses in a radial flow scenario. R_{AA} for Ω is above unity, which might be the result of the larger strangeness enhancement compared to Ξ [2]. In the right canvas of the same figure, the R_{pPb} as a function of p_T in minimum-bias p–Pb collisions is shown for the same particles. In this case, there is no suppression for π, K and p at high p_T (> 6 GeV/c), as verified for unidentified particles [8]. In the so-called Cronin region ($2 < p_T < 6$ GeV/c) an increase of

IOP Conf. Series: Journal of Physics: Conf. Series **779** (2017) 012043 doi:10.1088/1742-6596/779/1/012043

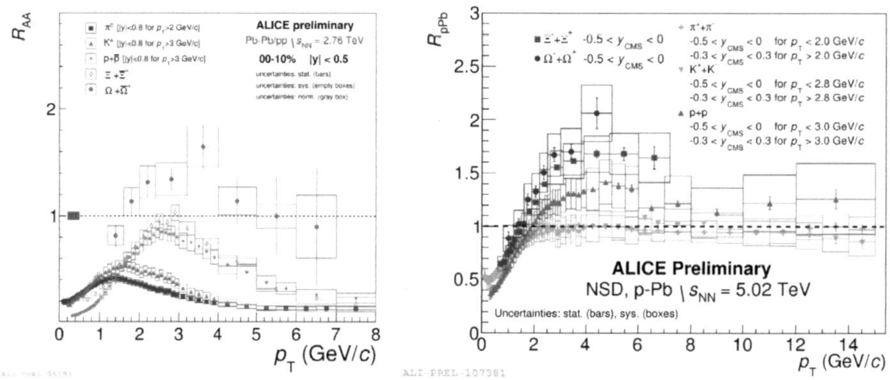

Figure 2. R_{pPb} in p–Pb collisions (left) and R_{AA} in 0-10% most central Pb–Pb collisions (right) as a function of p_{T} for π, K, p, Ξ and Ω particles. Open boxes are total systematic uncertainties.

the R_{pPb} with the mass of the particles is visible.

The strangeness enhancement is known as one of the proposed signatures for the QGP formation in relativistic heavy-ion collisions. Rafelski and Müller's expectation, first proposed in 1982 [9], is that QGP formation should lead to a higher abundance of strangeness per participating nucleon than in pp interactions. This phenomenon has been actually observed at SPS [10], RHIC [11] and the LHC [2], and has been found to increase with centrality and with the strangeness content of the particle, and decrease as the centre-of-mass energy increases. Statistical hadronization models (SHM) based on a grand-canonical approach have been demonstrated to be able to predict particle yield ratios in heavy-ion collisions over a large energy scale [12]. In this description, the energy dependence of strangeness enhancement has been understood as the consequence of a suppression of strangeness production due to the reduced phase-space volume in reference pp collisions (canonical suppression) [13]. At SPS and RHIC the strangeness enhancement has been studied looking at the ratio between the yield in nucleus–nucleus collisions and those in pp interactions at the same energy normalized to the mean number of participants ($\langle N_{part} \rangle$). It has been shown that this is not the ideal way to isolate the enhancement component due to strangeness content, since the production rates of charged particles do not scale linearly with N_{part} [14]. A better observable is the ratio to pion yields, shown in Fig. 3 for Λ, Ξ and Ω as a function of $\langle dN_{ch}/d\eta \rangle_{|\eta|<0.5}$ for Pb–Pb, p–Pb and pp collisions. In the case of multi-strange baryons, the ratio is shown to increase by up to a factor of about 2-3 going from pp collisions to Pb–Pb collisions and is more pronounced for the Ω; for the Λ a possible increase is not significant within the systematic uncertainty. It is shown that the predictions from statistical hadronization models using a chemical freezeout temperature of 156 MeV [15, 16] are comparable with the ratios measured in the most central Pb–Pb collisions for all the three particles. Looking at the p-Pb data, an increase of the (multi-)strange baryon-to-meson ratios with multiplicity is observed. The increase is seen to be more pronounced for particles with a larger strangeness quantum number. In particular $2\Lambda/(\pi^{-}+\pi^{+})$ and $(\Xi^{-}+\overline{\Xi}^{+})/(\pi^{-}+\pi^{+})$ ratios slightly exceed the saturation limit observed for Pb–Pb, while the $(\Omega^{-}+\overline{\Omega}^{+})/(\pi^{-}+\pi^{+})$ ratio is not higher than the one observed in peripheral Pb–Pb. Comparison of hyperon-to-pion ratios as a function of pion multiplicity to the trends observed in statistical hadronisation models, where the local conservation of the strangeness is required (as in a canonical ensemble), indicates that the behaviour is qualitatively consistent with the lifting of canonical suppression with increasing multiplicity [4].

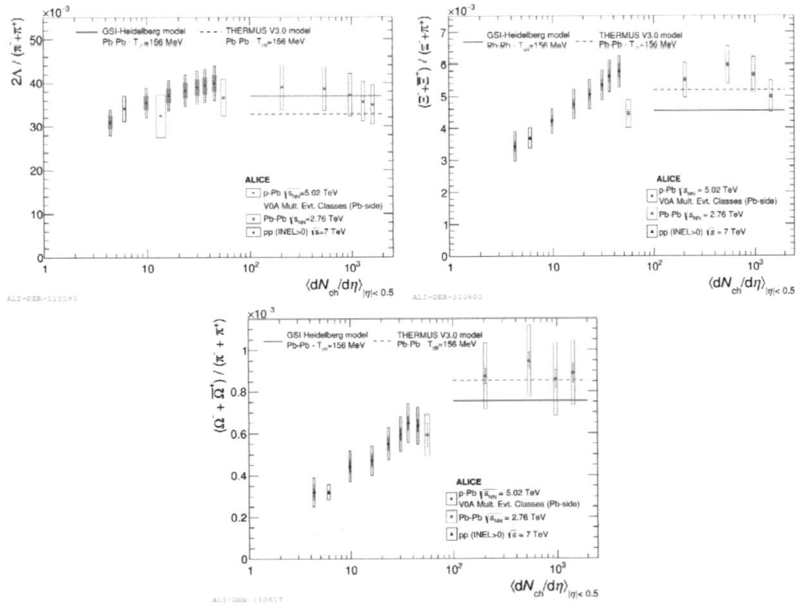

Figure 3. $2\Lambda/(\pi^{-}+\pi^{+})$, $(\Xi^{-}+\overline{\Xi}^{+})/(\pi^{-}+\pi^{+})$ and $(\Omega^{-}+\overline{\Omega}^{+})/(\pi^{-}+\pi^{+})$ ratios as a function of $\langle dN_{ch}/\mathrm{d}\eta\rangle_{|\eta|<0.5}$ for p–Pb (filled blue cross) and Pb–Pb (filled red square) collisions. Open boxes are total systematic uncertainties and shaded boxes show the uncorrelated-over-multiplicity component. Lines represent predictions from SHMs [15] (continuous) and [16] (dashed).

4. Conclusions

The production of (multi-)strange particles was measured in Pb–Pb and p–Pb collisions with the ALICE detector. The multiplicity dependence of the $(\Lambda+\overline{\Lambda})/\mathrm{K}^0_S$ ratio and the hardening of the spectra with increasing multiplicity point to similar collective behaviour. As observed for unidentified particles, R_{AA} for Ξ is clearly suppressed at high p_{T}, while R_{pPb} does not show any suppression. The fact that relative strangeness production in p–Pb not only increases with multiplicity but also bridges pp and Pb–Pb collision values is an important recent observation.

References

[1] Aamodt K *et al.* (ALICE Collaboration) 2008 *JINST* **3** S08002
[2] Abelev B *et al.* (ALICE Collaboration) 2014 *Phys. Lett. B* **728** 216-227
[3] Abelev B *et al.* (ALICE Collaboration) 2014 *Phys. Lett.* B **728** 25-38
[4] Adam J *et al.* (ALICE Collaboration) 2016 *Phys. Lett.* B **758** 389-401
[5] Melo I, Tomasik B 2015 *nucl-th*/1509.05383
[6] Abelev B *et al.* (ALICE Collaboration) 2013 *Phys. Rev. Lett.* **111** 222301
[7] Abelev B *et al.* (ALICE Collaboration) 2015 *Phys. Rev.* C **91** 024609
[8] Abelev B *et al.* (ALICE Collaboration) 2013 *Phys. Rev. Lett.* **110** 082302
[9] Rafelski J and Müller B 1982 *Phys. Rev. Lett.* **48** 1066
[10] Antinori F *et al.* (NA57 Collaboration) 2006 *J. Phys.* G **32** 427-441
[11] Abelev B *et al.* (STAR Collaboration) 2008 *Phys. Rev.* C **77** 044908
[12] Stachel J, Andronic A, Braun-Munzinger P and Redlich K 2014 *J. Phys. Conf. Ser.* **509** 012019
[13] Hamieh S, Redlich K and Tounsi A 2000 *Phys. Lett.* B **486** 61-66
[14] Aamodt K *et al.* (ALICE Collaboration) 2011 *Phys. Rev. Lett.* **106** 032301
[15] Andronic A, Braun-Munzinger P and Stachel J 2009 *Phys. Lett.* B **673** 142-145
[16] Cleymans J, Kraus I, Oeschler H, Redlich K and Wheaton S 2006 *Phys. Rev.* C **74** 034903

Studies of final state interactions via femtoscopy in ALICE

Lukasz Kamil Graczykowski (for the ALICE Collaboration)
Faculty of Physics, Warsaw University of Technology, ul. Koszykowa 75, 00-662, Warszawa, Poland
E-mail: lgraczyk@if.pw.edu.pl

Abstract. Femtoscopy is a technique enabling measurements of the space-time characteristics of particle-emitting sources. However, the femtoscopic analysis is also sensitive to the interaction cross-section. In this paper we show the first preliminary measurements of $K^0_SK^{\pm}$ correlation functions in Pb–Pb collisions at $\sqrt{s_{NN}} = 5.02$ TeV. These correlations originate from the final-state interactions which proceed through the $a_0(980)$ resonance only and can be employed to constrain its parameters. A similar approach can be applied to baryon pairs to extract the unknown interaction cross-sections for some (anti-)baryon–(anti-)baryon pairs. We show baryon–baryon and baryon–anti-baryon correlation functions of protons and lambdas, as well as discuss briefly the fitting method.

1. Introduction

The "traditional" method of femtoscopy (also known as "HBT" correlations), created to measure the space-time characteristics of the particle-emitting region at freeze-out, is based on measurements of correlations of two identical bosons (usually pions). These correlations are calculated as a function of relative momentum, expressed as $q = p_1 - p_2$ or $k^* = q/2$ [1, 2]. In general, the femtoscopic technique allows the width of the distribution of relative separation between the emission points of two particles, usually called the "HBT radius", to be extracted. The dependence of this parameter on event multiplicity (centrality) and transverse momentum of the pair $k_T = |p_1 + p_2|/2$ is frequently interpreted in terms of relativistic hydrodynamics as a signature of collective behavior of the created medium [2]. The mathematical formalism of femtoscopy is not limited only to identical bosons though and other pair combinations can be studied as well. The results of the femtoscopic analysis of two such not typical pair combinations are presented below.

2. Femtoscopy of $K^0_SK^{\pm}$ pairs

The results of charged and neutral femtoscopy of identical kaons have been published by ALICE in Pb–Pb and pp collisions [3–5]. In the case of identical kaons, the correlation effect at low relative momentum originates from the combination of the following physical effects: Bose-Einstein quantum statistics (for $K^{\pm}K^{\pm}$ and $K^0_SK^0_S$), Coulomb interaction (for $K^{\pm}K^{\pm}$), and final-state interaction (FSI) via the $f_0(980)/a_0(980)$ threshold resonances (for $K^0_SK^0_S$).

The non-identical correlations of $K^0_SK^{\pm}$ have never been studied before. In this case the FSI goes through one resonant state of $a_0(980)$[1]. This uniqueness of the $K^0_SK^{\pm}$ pair allows the properties of the a_0

[1] Both the $K^0_SK^0_S$ pair and the a_0 are in $I = 1$ isospin state. On the other hand, the f_0 is in $I = 0$ state, so this channel is not allowed since the isospin would not be conserved.

resonance itself to be constrained, which is discussed in many publications as a tetraquark (a 4-quark state), or a "$\overline{\rm K}$-K molecule". In this study we considered four a_0 parameterizations: "Achasov2" [6], "Achasov1" [7], "Antonelli" [8], and "Martin"[9].

Examples of $K^0_SK^+$ and $K^0_SK^-$ correlation functions measured by ALICE in two k_T ranges, plotted together with fits of the so-called Lednicky analytical model [10, 11] employing the "Achasov2" parameters of a_0, can be seen in Fig. 1. One can observe a suppression at low k^* which is a result of the strong FSI. By looking at the plots themselves one can judge that the fit with a_0 FSI gives a very good description of the experimental data.

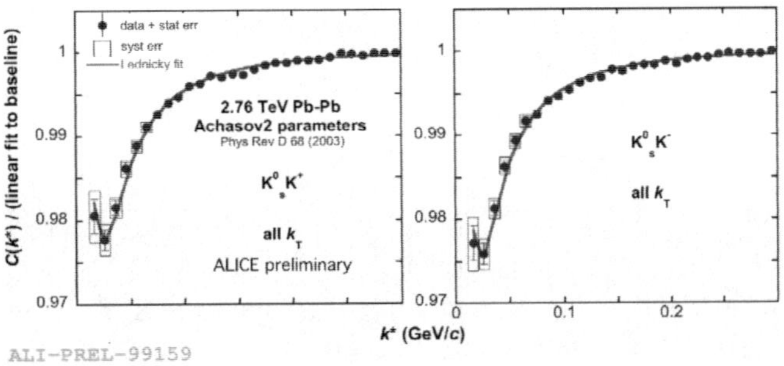

Figure 1. $K^0_SK^+$ (left) and $K^0_SK^-$ (right) correlation functions with Lednicky fits using "Achasov2" parameters of a_0.

The "Lednicky" fits allowed typical femtoscopic quantities, that is the "HBT radius" (the kaon source size) R and the correlation strength λ to be extracted for all a_0 parameters. It was found that for all parameterizations the fit results for $K^0_SK^+$ and $K^0_SK^-$ are consistent with each other, therefore both quantities were averaged. Both R and λ are presented in Fig. 2.

The results show a clear agreement between identical-kaon and $K^0_SK^\pm$ results for all a_0 parameters except "Martin" which are the lowest ones in comparison to other considered parameterizations. Since there is no physical reason to expect that femtoscopic quantities for $K^0_SK^\pm$ pairs should be different from those for $K^0_SK^0_S$ and $K^\pm K^\pm$ pairs, the experimental data seems to favor higher values of a_0.

3. Femtoscopy of baryon pairs

Femtoscopy of baryons also allows to study final-state interactions. For example, FSI effects can play a role in the rescattering phase for final-state hadrons and, going beyond the relativistic heavy-ion field, in understanding the properties of neutron stars [12, 13] or search for potential CPT symmetry breaking [14]. It is argued that baryon–anti-baryon annihilatation can be responsible for the p/π^+ ratio which falls short of thermal model expectations at the LHC [15]. Several theoretical calculations argue that rescattering in the final-state should be taken into account in the measurements of particle yields [16–18].

The ALICE experiment has already published results of pp and $\overline{\rm p}\overline{\rm p}$ femtoscopy in Pb–Pb collisions at $\sqrt{s_{\rm NN}} = 5.02$ TeV [5], which are complemented by still ongoing studies of other pairs with (anti-)protons and (anti-)lambdas. The ALICE measurements of $\Lambda\Lambda$ and $\overline{\Lambda}\overline{\Lambda}$ correlations will complement recently published STAR results in Au–Au collisions at $\sqrt{s_{\rm NN}} = 200$ GeV/c [19]. The upper panel of Fig. 3 shows baryon–baryon correlation functions for $\overline{\rm p}\overline{\rm p}$, $p\overline{\Lambda}+\overline{\rm p}\Lambda$, $\Lambda\Lambda+\overline{\Lambda}\overline{\Lambda}$ and $\Lambda\overline{\Lambda}$ pairs. A clear correlation signal is present for all systems. In the case of baryon–baryon correlations, the correlation effect at low relative momentum originates from the combination of the following phenomena: Fermi-Dirac quantum statistics (for $pp+\overline{\rm p}\overline{\rm p}$ and $\Lambda\Lambda+\overline{\Lambda}\overline{\Lambda}$), Coulomb interaction (for $pp+\overline{\rm p}\overline{\rm p}$), strong final-state interaction (for $pp+\overline{\rm p}\overline{\rm p}$, $\Lambda\Lambda+\overline{\Lambda}\overline{\Lambda}$, and $p\Lambda+\overline{\rm p}\overline{\Lambda}$).

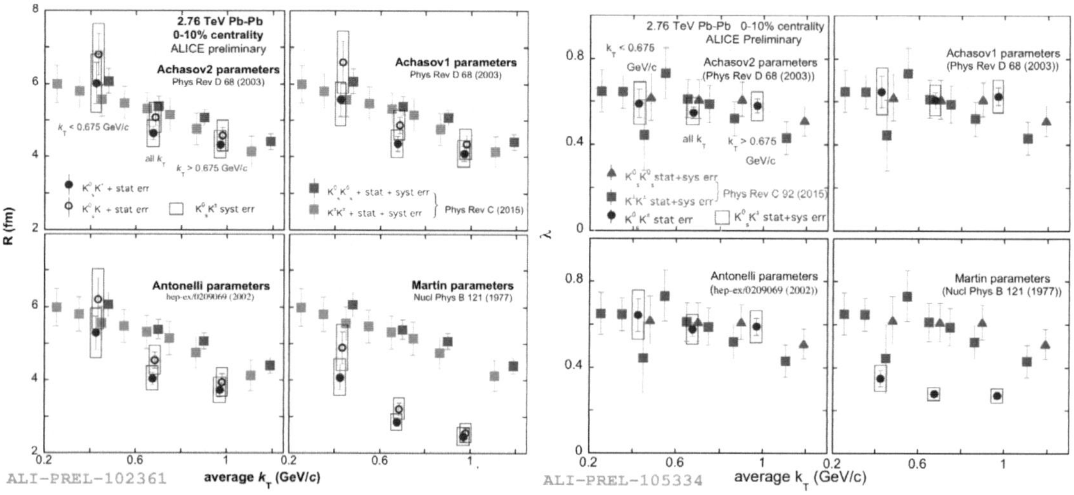

Figure 2. *R* (left) and λ (right) parameters of the fit from averaged $K^0_S K^{\pm}$ correlations compared to the results of identical-kaon femtoscopy from ALICE [5].

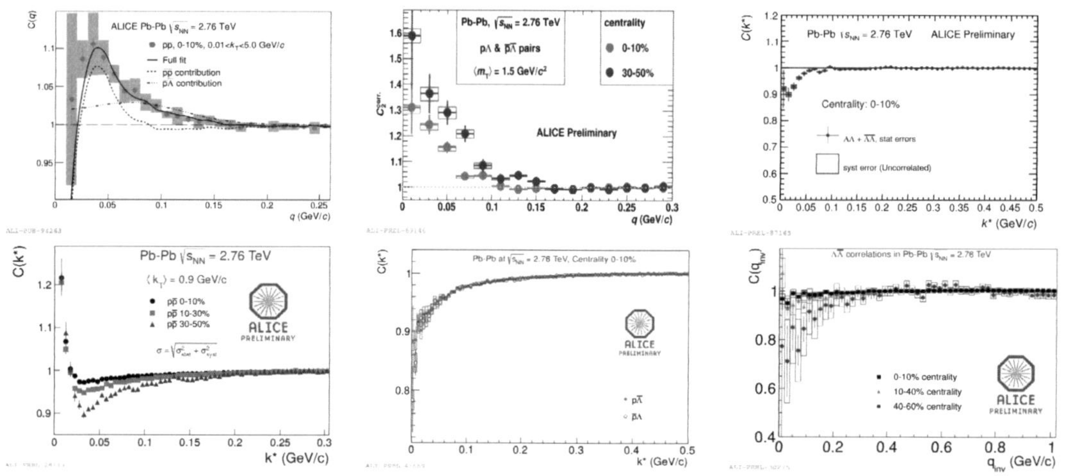

Figure 3. Upper panel: correlation functions for (anti-)baryon–(anti-)baryon pairs: $\overline{\rm p}\overline{\rm p}$ (left) [5], $\rm p\Lambda + \overline{\rm p}\overline{\Lambda}$ (middle), and $\Lambda\Lambda + \overline{\Lambda}\overline{\Lambda}$ (right). Lower panel: correlation functions for baryon–anti-baryon pairs: $\rm p\overline{p}$ (left), $\rm p\overline{\Lambda} + \overline{p}\Lambda$ (middle), and $\Lambda\overline{\Lambda}$ (right).

The $\overline{\rm p}\overline{\rm p}$ correlation function exhibits a maximum around $q = 40$ MeV/c which is caused by the strong interaction, since both Coulomb and Fermi-Dirac statistics produce a negative correlation. Similar behavior, originating from these three effects is visible also for other pairs. In the case of $\rm p\Lambda + \overline{p}\overline{\Lambda}$, where only strong FSI is present, we do not observe any negative correlation. For $\Lambda\Lambda + \overline{\Lambda}\overline{\Lambda}$ the strong interaction gives surprisingly weak contribution since the correlation function at low relative momentum is clearly below unity due to Fermi-Dirac quantum statistics (though a hint of enhancement from strong FSI is visible for for the first bin).

In the lower panel of Fig. 3 correlation functions for baryon–anti-baryon pairs are presented. In this case only two effects can be present: Coulomb interaction (for $p\overline{p}$) and strong FSI (for $p\overline{p}$, $p\overline{\Lambda}+\overline{p}\Lambda$, and $\Lambda\overline{\Lambda}$). The Coulomb interactions is present only in the case of $p\overline{p}$ pairs and, due to its attractive nature in case of opposite charges, produces a positive correlation. This effect is visible for the lowest values of k^*. Apart from this effect, all three pair combinations exhibit a wide suppression from strong FSI.

Fitting the baryon correlation functions is complicated by the presence of so-called "residual correlations". They arise from feed-down of weak decay products, which contaminate the experimental sample (e.g. an admixture of protons from decays of lambdas in the primary proton sample); see the $\overline{p}\overline{p}$ correlation function, top-left plot in Fig. 3. The fit with the $\overline{p}\overline{p}$ wave function only cannot describe the experimental data. The visible excess is however accounted for if a contribution from $\overline{p}\overline{\Lambda}$ pairs is considered. There are two methods incorporating "residual correlations" in the fitting procedure, "transformed residuals" [20] and "Gaussian residuals" [21]. The former method was used by both ALICE in [5] and STAR in [14] while the lattter one was used by STAR in [19].

4. Conclusions

Femtoscopy is a powerful tool capable not only to measure sizes of particle emitting sources but also gives access to final-state interactions. This feature allows to constrain strong interaction cross-sections for particle pairs where little or no other experimental data exist. The preliminary results from ALICE suggest that such detailed analysis for many meson and baryon pairs which has not been measured are possible at the LHC.

The author wishes to acknowledge the financial support of the Polish National Science Centre under decisions Nos. DEC-2013/08/M/ST2/00598, DEC-2014/13/B/ST2/04054, and DEC-2015/19/D/ST2/01600.

References

[1] Goldhaber G, Goldhaber S, Lee W Y and Pais A 1960 *Phys. Rev.* **120** 300–312
[2] Lisa M A, Pratt S, Soltz R and Wiedemann U 2005 *Ann. Rev. Nucl. Part. Sci.* **55** 357–402
[3] Abelev B *et al.* (ALICE) 2012 *Phys. Lett.* **B717** 151–161
[4] Abelev B *et al.* (ALICE) 2013 *Phys. Rev.* **D87** 052016
[5] Adam J *et al.* (ALICE) 2015 *Phys. Rev.* **C92** 054908
[6] Achasov N N and Kiselev A V 2003 *Phys. Rev.* **D68** 014006
[7] Achasov N N and Gubin V V 2001 *Phys. ReV. D* **63** 094007
[8] Antonelli A 2002 *eConf* **C020620** THAT06 [,65(2002)] (*Preprint* hep-ex/0209069)
[9] Martin A D, Ozmutlu E N and Squires E J 1977 *Nucl. Phys.* **B121** 514
[10] Lednicky R and Lyuboshits V L 1982 *Sov. J. Nucl. Phys.* **35** 770 [Yad. Fiz.35,1316(1981)]
[11] Bekele S and Lednicky R 2007 *Braz. J. Phys.* **37** 994–1001
[12] Schaffner-Bielich J 2008 *Nucl. Phys.* **A804** 309–321
[13] Wang Y N and Shen H 2010 *Phys. Rev.* **C81** 025801
[14] Adamczyk L *et al.* (STAR) 2015 *Nature* **527** 345–348
[15] Preghenella R (ALICE) 2012 *Primordial QCD matter in LHC era: Implications of QCD results on early universe Cairo, Egypt, December 4-8, 2011*
[16] Werner K, Karpenko I, Bleicher M, Pierog T and Porteboeuf-Houssais S 2012 *Phys. Rev.* **C85** 064907
[17] Karpenko I A, Sinyukov Yu M and Werner K 2013 *Phys. Rev.* **C87** 024914
[18] Steinheimer J, Aichelin J and Bleicher M 2013 *Phys. Rev. Lett.* **110** 042501
[19] Adamczyk L *et al.* (STAR) 2015 *Phys. Rev. Lett.* **114** 022301
[20] Kisiel A, Zbroszczyk H and Szymański M 2014 *Phys. Rev.* **C89** 054916
[21] Shapoval V M, Erazmus B, Lednicky R and Sinyukov Yu M 2015 *Phys. Rev.* **C92** 034910

IOP Conf. Series: Journal of Physics: Conf. Series **779** (2017) 012045 doi:10.1088/1742-6596/779/1/012045

Multi-strange hadrons and the precision extraction of QGP properties in the RHIC-BES domain

Jussi Auvinen

E-mail: jaa49@phy.duke.edu
Department of Physics, Duke University, Durham, NC 27708, USA

Krzysztof Redlich

Institute of Theoretical Physics, University of Wrocław, Pl-45204 Wrocław, Poland

Steffen A. Bass

Department of Physics, Duke University, Durham, NC 27708, USA

Abstract. We systematically compare an event-by-event transport+viscous hydrodynamics hybrid model to data from the RHIC beam energy scan using a general Bayesian method. We demonstrate how the inclusion of multistrange hadron observables affects the outcome of the Bayesian analysis and conduct an in depth analysis of the viability of ϕ and Ω as probes of the transition region between a deconfined quark-gluon plasma and hadronic phase in heavy ion collisions at higher-end RHIC collision energies. Utilizing UrQMD to model the final hadronic interactions, we examine the collision rates of ϕ and Ω and the modification to their transverse momentum spectra due to these interactions.

1. Introduction

Many modern models of relativistic heavy ion collisions employ a so called hybrid approach, where the hydrodynamical model of quark-gluon plasma evolution is coupled to a hadron transport "afterburner", allowing the chemical and kinetic freeze-outs to happen dynamically. However, this approach introduces an additional parameter to the model; namely, the switching condition between hydrodynamics and hadron transport. Typically this switching condition is either a particular value of temperature T_{SW}, or energy density ϵ_{SW}, which is close to, but below the transition temperature or energy density.

The justification of this approach comes from the assumption that both hydrodynamics and hadron transport describe the same system over a range of temperature / energy density values, and thus the exact value of T_{SW} or ϵ_{SW} should not matter. However, to make quantified statements about the actual size of the overlap between the two descriptions, one needs to identify the experimental observables which probe the transition region.

Multi-strange hadrons ϕ and Ω are potential probes of the transition stage, as they are produced at the phase boundary during hadronization of the quark-gluon plasma and exhibit small scattering cross section in the hadronic phase [1]. In this Article, we perform a detailed comparison of an event-by-event transport+viscous hydrodynamics hybrid model [2] to ϕ and Ω data from the RHIC beam energy scan [3, 4, 5].

2. Hybrid model

In the hybrid approach, the heavy ion collision is modeled in three separate phases.

The initial pre-equilibrium phase is simulated with UrQMD hadron+strings cascade [6, 7]. The hydrodynamical evolution starts after the two colliding nuclei have passed through each other: $\tau_0 \geq 2R_{\mathrm{nucleus}}/\sqrt{\gamma_{CM}^2 - 1}$. At this point, the particle properties such as energy and baryon number are converted to densities using 3D Gaussians with "smearing" parameters R_{trans}, R_{long} , each equal to $\sqrt{2}$ times the respective Gaussian width parameter σ_{trans}, σ_{long}.

In the local equilibrium phase, the system is evolved according to 3+1D viscous hydrodynamics [8], with viscosity parameter η/s kept constant during the full evolution. At the lower collision energies, the equation of state needs to include the effects of nonzero net-baryon density. For this purpose, a chiral model equation of state [9] is utilized in this investigation.

Finally, the transition from hydrodynamical evolution back to hadron transport ("particlization") happens when energy density ϵ drops below the chosen energy density value ϵ_{SW}. A hypersurface with constant energy density is constructed [10] and particles are sampled from this hypersurface according to the Cooper-Frye formula and propagated further using UrQMD. Both chemical and kinetic freeze-out thus happen dynamically.

As all ϕs will decay before the end of the simulation, we output the full interaction histories of the afterburner hadron cascades for each event. We search these histories for ϕs, and label the ones which have both their decay products surviving to the end of simulation as "detectable". It should be noted that the survivor decay products are allowed to have interacted via processes which preserve their identity. However, we have found that the fraction of survivor products which have interactions is less than 15% at investigated energies.

3. Statistical analysis and simulation setup

The optimal input parameter values are determined using Bayesian analysis similar to Refs. [11, 12]. The posterior distribution is sampled with Markov chain Monte Carlo (MCMC) method. We perform thousands of random walks in input parameter space, where step proposals are accepted or rejected based on a relative likelihood. However, it is not feasible to run the full hybrid model simulation for each evaluation of the likelihood function. To circumvent this problem, we use Gaussian processes to emulate the simulation output, based on ≈ 100 training points, which are samples of the input parameter space. The Latin hypercube method is used to achieve close to uniform distribution of training points on all 5 parameter dimensions.

4. Results

Figure 1 demonstrates the variance in transverse momentum distributions of ϕs and Ωs over the training points for $\sqrt{s_{NN}} = 39$ GeV. To eliminate some possible sources of discrepancies between simulation and experimental results, we follow the experimental method [3] and do also Levy fits on the transverse momentum spectra points which are within the p_T range reported by the experiments. We see that ϕ yield at low p_T is underpredicted by most of the input parameter combinations. This suggests that our criterion for ϕ meson detection is too stringent compared to the experimental reconstruction methods.

To verify the result of the statistical analysis, a full simulation was performed using the median values from the posterior distributions of the input parameters. The resulting Ω transverse momentum distributions for $\sqrt{s_{NN}} = 39$ GeV is shown in the right frame of Fig. 1. Although only the lowest transverse momentum bin was used for the calibration of model parameters, the experimental data is very well reproduced for the whole p_T spectrum.

IOP Conf. Series: Journal of Physics: Conf. Series **779** (2017) 012045 doi:10.1088/1742-6596/779/1/012045

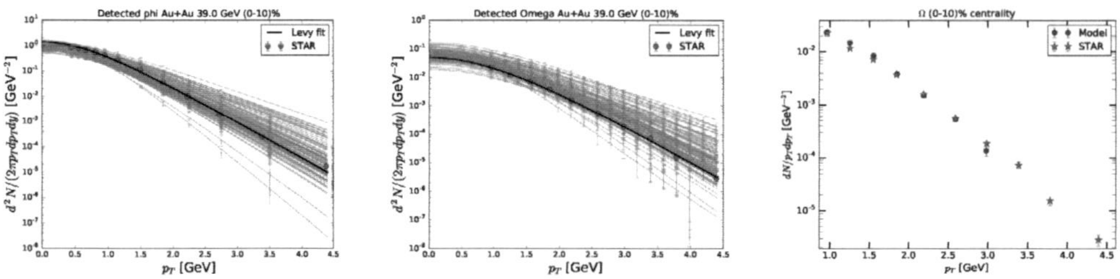

Figure 1. Transverse momentum spectra of ϕ (left) and Ω (middle) at $\sqrt{s_{NN}} = 39$ GeV. Curves represent Levy fits on model or STAR data points. Right: Transverse momentum spectrum of Ω at $\sqrt{s_{NN}} = 39$ GeV using the median values from statistical analysis. STAR data from [5].

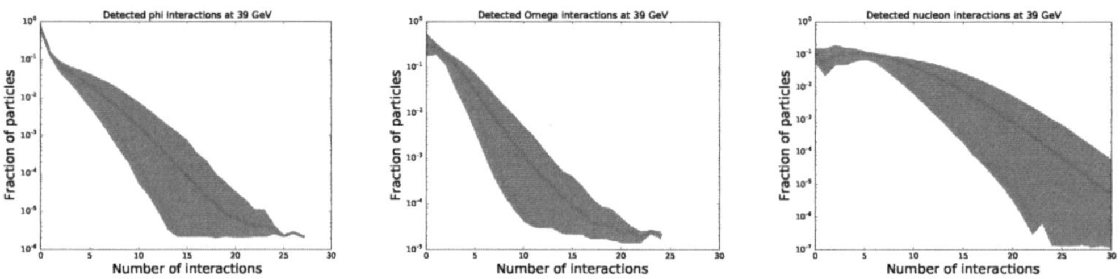

Figure 2. Distribution of average number of interactions for ϕ, Ω and N at $\sqrt{s_{NN}} = 39$ GeV. Red line represents the mean over all training points. Blue band represents the full range of values over all training points.

In addition of the yields, we can also calculate the mean p_T from the particle spectra. As we are interested in using ϕ and Ω as probes of the phase transition region, we need to quantify the change on $\langle p_T \rangle$ during the hadron gas phase. We find the detectable ϕ mean transverse momentum to be about 20% larger than the value at the hypersurface. The increase is slightly smaller, $\approx 15\%$, when including "undetectable" ϕs; this equals the change in $\langle p_T \rangle$ seen for Ω.

We also investigate the average number of interactions ϕ and Ω experience in the hadronic matter (Fig. 2). The analysis verifies that detectable ϕ mesons have hardly any interactions, and Ω baryons are likely to have only a few, compared to the 5-10 interactions of a typical nucleon.

To check the effect of multistrange hadron observables on the posterior distributions, the statistical analysis was performed with and without Ω yield data for $\sqrt{s_{NN}} = 19.6$, 39, and 62.4 GeV. As seen in Figs. 3 and 4, the analysis without Ω yields was not able to produce any constraints on switching energy density value, whereas the analysis including the Ω provide clear peak regions in probability distributions, revealing a visible dependence on the collision energy.

5. Summary

We have performed a detailed investigation of multistrange hadrons in relativistic heavy ion collisions, focusing on collision energies $\sqrt{s_{NN}} = 39$ and 62.4 GeV.

Based on their low number of interactions and the relatively small changes on their mean transverse momentum during the hadron gas phase, we find the assumption of ϕ and Ω as clear probes of the phase transition region justified. Performing a Bayesian analysis with and without Ω yield data has demonstrated the large effect the inclusion of multistrange hadron data can have on model-to-data comparisons when determining the "true" parameter values.

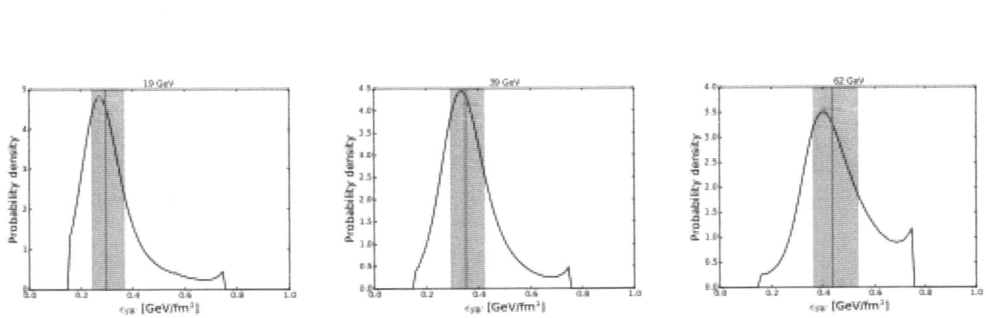

Figure 3. Posterior probability distributions of the switching energy density ϵ_{SW} for different collision energies: 19.6 GeV (left), 39 GeV (middle), and 62.4 GeV (right). Colored bands and median lines correspond to respective boxes and median lines in the right frame of Fig. 4.

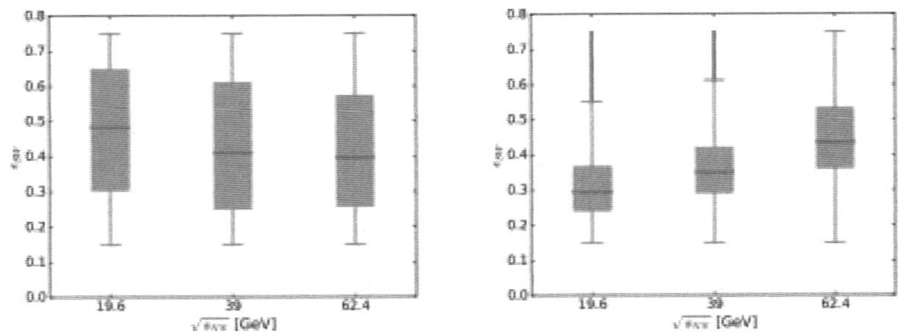

Figure 4. Box-whisker plots of the Bayesian posterior distribution of the switching energy density ϵ_{SW} vs. $\sqrt{s_{NN}}$. Left: without Ω yields. Right: with Ω yields. Boxes and median lines of the right-side figure correspond to respective colored bands and median lines in Fig. 3.

6. Acknowledgements

We thank Jonah E. Bernhard for useful discussions and providing a Python code package for the statistical analysis, and Iurii Karpenko for providing the 3+1D viscous hydrodynamics hybrid model code. The authors acknowledge support by the U.S. Department of Energy Grant no. DE-FG02-05ER41367. K. R. also acknowledges partial support of the Polish Science Center (NCN) under Maestro grant DEC-2013/10/A/ST2/00106.

References

[1] Shor A 1985 *Phys. Rev. Lett.* **54** 1122–1125
[2] Karpenko I A, Huovinen P, Petersen H and Bleicher M 2015 *Phys. Rev.* **C91** 064901
[3] Abelev B I *et al.* (STAR) 2009 *Phys. Rev.* **C79** 064903
[4] Aggarwal M M *et al.* (STAR) 2011 *Phys. Rev.* **C83** 024901
[5] Adamczyk L *et al.* (STAR) 2016 *Phys. Rev.* **C93** 021903
[6] Bass S A *et al.* 1998 *Prog. Part. Nucl. Phys.* **41** 255–369 [Prog. Part. Nucl. Phys.41,225(1998)]
[7] Bleicher M *et al.* 1999 *J. Phys.* **G25** 1859–1896
[8] Karpenko I, Huovinen P and Bleicher M 2014 *Comput. Phys. Commun.* **185** 3016–3027
[9] Steinheimer J, Schramm S and Stocker H 2011 *J. Phys.* **G38** 035001
[10] Huovinen P and Petersen H 2012 *Eur. Phys. J.* **A48** 171
[11] Novak J, Novak K, Pratt S, Vredevoogd J, Coleman-Smith C and Wolpert R 2014 *Phys. Rev.* **C89** 034917
[12] Bernhard J E, Marcy P W, Coleman-Smith C E, Huzurbazar S, Wolpert R L and Bass S A 2015 *Phys. Rev.* **C91** 054910

Signatures of Flux Tube Fragmentation and Strangeness Correlations in pp Collisions

Cheuk-Yin Wong

Physics Division, Oak Ridge National Laboratory, Oak Ridge, TN 37831, USA

E-mail: wongc@ornl.gov

Abstract.
In the fragmentation of a color flux tube in high-energy pp collisions or e^+-e^- annihilations, the production of q-$\bar{q}$ pairs along a color flux tube precedes the fragmentation of the tube. The local conservation laws in the production of these q-$\bar{q}$ pairs will lead to the correlations of adjacently produced hadrons. As a consequence, the fragmentation of a flux tube will yield a many-hadron correlation in the form of a chain of hadrons ordered in rapidity, with adjacent hadrons correlated in charges, flavor contents, and azimuthal angles. It will also lead to a two-hadron angular correlation between two hadrons with opposite charges or strangeness that is suppressed at $\Delta\phi \sim 0$ but enhanced at $\Delta\phi \sim \pi$, within a rapidity window $\Delta y{\sim}1/(dN/dy)$.

1. Introduction

The reaction mechanisms in pp collisions provide insights in AA collision at high energies. The importance of each reaction mechanism depends on the collision energy and the p_T domain. The flux-tube fragmentation process [1, 2] is expected to dominate at low p_T and low $\sqrt{s_{pp}}$, whereas the hard-scattering process [3, 4] at high p_T and high $\sqrt{s_{pp}}$. The two mechanisms cross-over at a certain transverse momentum $p_{Tb}(\sqrt{s_{pp}})$ that is a function of the collision energy, $\sqrt{s_{pp}}$. There are in addition other mechanisms such as the recombination of partons [5] and the production of resonances.

It is desirable to establish, even if approximately, the boundary function $p_{Tb}(\sqrt{s_{pp}})$ that separates the region of flux-tube fragmentation dominance from the region of hard-scattering dominance. We need signatures for these two mechanisms to facilitate such a separation.

The signature for the hard scattering process is well known. It is given as a two-hadron $\Delta\phi$–$\Delta\eta$ angular correlation which shows a peak at $(\Delta\phi, \Delta\eta){\sim}0$ and a ridge along $\Delta\eta$ at $\Delta\phi \sim \pi$. The peak at $\Delta\phi{\sim}0$ arises from the fragmentation of the jet associated with the trigger and the ridge at $\Delta\phi{\sim}\pi$ arises from the fragmentation of the other jet associated with colliding partons with unbalanced longitudinal momenta.

However, the signature for flux tube fragmentation has not been well studied. We would like to present here two signatures for flux-tube fragmentation [6, 7].

2. Many-hadron signatures of Flux Tube Fragmentation

In the semi-classical description of the flux-tube fragmentation process [1, 2] for hadron production, the production of quark-antiquark pairs along a color flux tube precedes the fragmentation of the tube. The production of these quark-antiquark pairs must however obey conservation laws at the local production points. As a consequence, the produced q-$\bar{q}$ pairs will lead to correlations of adjacently produced hadrons, and the hadrons are ordered according to their rapidities along the tube. The rapidity-space-time ordering and the local conservation laws will yield a many-body correlation of the hadrons in charge, flavor, and momentum, which may provide vital information on space-time dynamics of quarks and hadrons in the flux-tube fragmentation process.

As an example, we examine the fragmentation of a flux tube with an invariant mass of 8.65 GeV, corresponding to the average invariant mass of one of the two flux tubes in a *pp* collision at 17.3 GeV, with the flux tube formed by a quark of one proton with the diquark of the other proton. We carry out a Monte Carlo generation of hadrons in flux-tube fragmentation using the PYTHIA 6.4 program [8]. An example of the produced hadrons involving the production of a pair of strange hadrons is shown in Fig. 1, with the production of 5 hadrons listed in Table I.

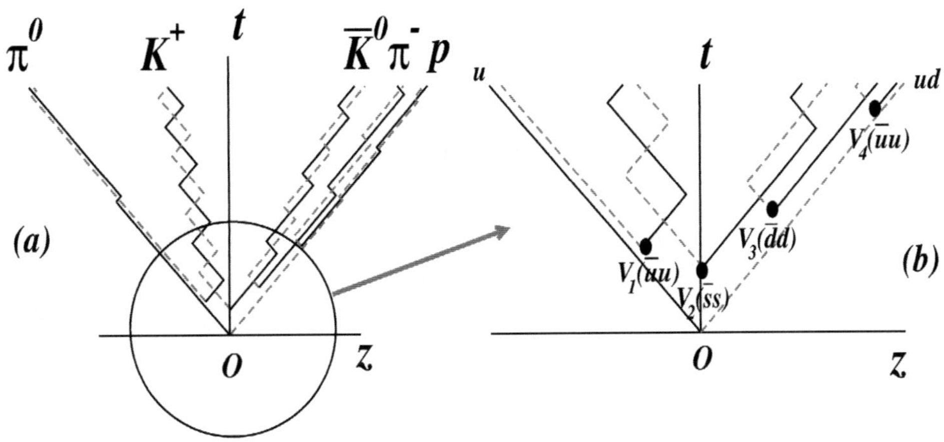

Figure 1. Fig. 1(a) shows an example of the space-time diagram in the fragmentation of a u-(ud) flux tube with an invariant mass of $\sqrt{s}$=8.65 GeV, obtained with PYTHIA 6.4 [8]. Quark lines are shown as solid lines, and antiquark (or diquark) lines as dashed lines. Fig. 1(b) is an expanded view of the pair-production vertices $V_i(\bar{q}_i q_i)$ of Fig. 1(a).

Table 1. An example of primary hadrons i, their rapidities y_i, their azimuthal angles ϕ_i and their constituents q_i-$\bar{q}_i$ produced in the fragmentation of the u-(ud) flux tube at an energy of $\sqrt{s}$=8.65 GeV obtained with PYTHIA 6.4 [8].

i	1	2	3	4	5
particle	π^0	K^+	$\bar{K}^0$	π^-	p
q_i-$\bar{q}_i$	u-$\bar{u}$	u-$\bar{s}$	s-$\bar{d}$	d -$\bar{u}$	u -(ud)
y_i	-1.55	- 1.15	- 0.75	0.27	1.78
ϕ_i	1.00	-2.01	1.44	-2.43	0.17
$\phi_i-\phi_{i+1}$	-3.01	3.46	-3.87	2.60	

In Table 1, the row of q_i-$\bar{q}_i$ shows that upon ordering the hadrons according to their repidities y_i as in a chain, the flavors of the constituent antiquark $\bar{q}_i$ and the flavors of the neighboring constituent quark q_{i+1} are correlated along the chain, on an event-by event basis. The row of ϕ_i-ϕ_{i+1} of neighboring hadrons in Table I indicates that neighboring pairs of hadrons are azimuthally correlated, approximately in a back-to-back manner. The many-hadron signature requires the identification of all hadrons detected in the events. Predicted signature of this kind is yet to be observed.

3. Two-hadron angular correlation signature of flux-tube fragmentation

Another signature of the flux tube fragmentation utilizes the two-hadron angular correlations arising from the productions of q-$\bar{q}$ pairs. Because of local conservation laws, the production of q-$\bar{q}$ pairs will lead to correlations of adjacently produced hadrons. Adjacently produced hadrons however can be signaled by their rapidity difference Δy falling within the window of $|\Delta y|{\sim}1/(dN/dy)$, on account of the space-time-rapidity ordering of produced mesons in a flux-tube fragmentation. Therefore, the local conservation laws of momentum, charge, and flavor will lead to a suppression of the angular correlation function $dN/(d\Delta\phi\, d\Delta y)$ for two hadrons with opposite charges or strangeness at $(\Delta\phi, \Delta y){\sim}0$, but an enhanced correlation on the back-to-back, away side at $\Delta\phi{\sim}\pi$, within the window of $|\Delta y|{\sim}1/(dN/dy)$. When we approximate the rapidity y as the pseudorapidity η, the two-hadron angular correlations can be used as signatures for the fragmentation of a color flux tube as shown in Fig 2.

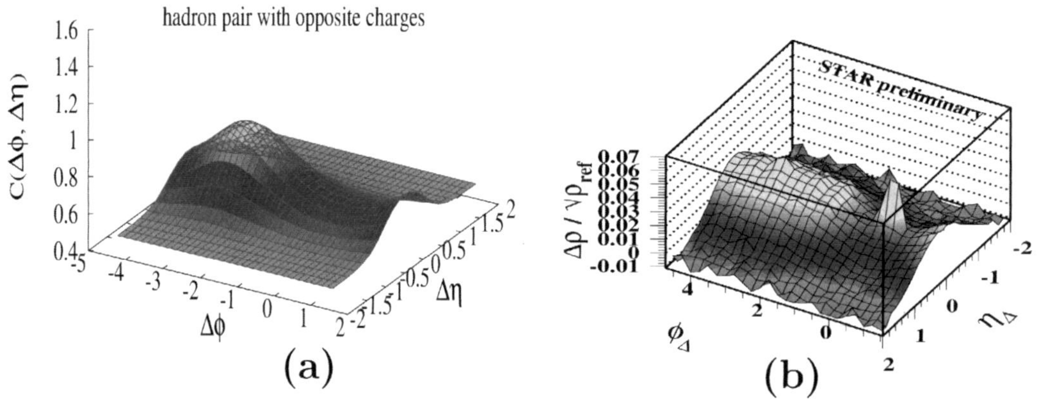

Figure 2. Fig. 2(a) gives the theoretical correlations for two two hadrons with unlike charges [6] and Fig 2(b) gives the experimental correlation data for two hadrons with unlike charges and p_T<0.5 GeV obtained by the STAR Collaboration for pp collisions at $\sqrt{s_{pp}}$=200 GeV [9]. The sharp peak at $(\Delta\phi, \Delta\eta) \sim 0$ in Fig. 2(b) is an experimental artifact.

Comparison of theoretical angular correlations for two hadrons with unlike charges in Fig. 2(a) [6] with experimental data from the STAR Collaboration in Fig. 2(b) [9] indicates that in high-energy pp collisions at $\sqrt{s_{pp}}$ =200 GeV, the production of unlike-charge hadron pairs in the region of p_T<0.5 GeV/c are qualitatively consistent with the flux-tube fragmentation mechanism. However, the correlations for two hadrons with unlike-charges in the region with p_T>0.5 GeV/c exhibit a completely different pattern. Namely, they shows a peak at $(\Delta\phi, \Delta\eta) \sim 0$, and a ridge along $\Delta\eta$ at $\Delta\phi \sim \pi$, which is a signature of the hard scattering process [9]. This indicates that for pp collisions energy at $\sqrt{s_{pp}}$ = 200 GeV, the boundary between the flux-tube fragmentation process and the hard-scattering process is p_{Tb} = 0.5 GeV/c. At the LHC energy of $\sqrt{s}_{pp}$ =5 TeV, the minimum bias data for the correlation of two hadrons with p_T>0.1 GeV/c show a pattern that appears to be a linear combination of the patterns for flux tube fragmentation and hard scattering [11]. Results in Fig. 2, together with the two hadron correlation data from lower energies [10] and higher energies [11] reveal that the p_{Tb} boundary of separation moves to lower p_T values as the collision energy increases.

Similarly, because of local conservation of strangeness in $q\bar{q}$ production, adjacently produced hadrons with opposite strangeness are correlated back-to-back in azimuthal angles.

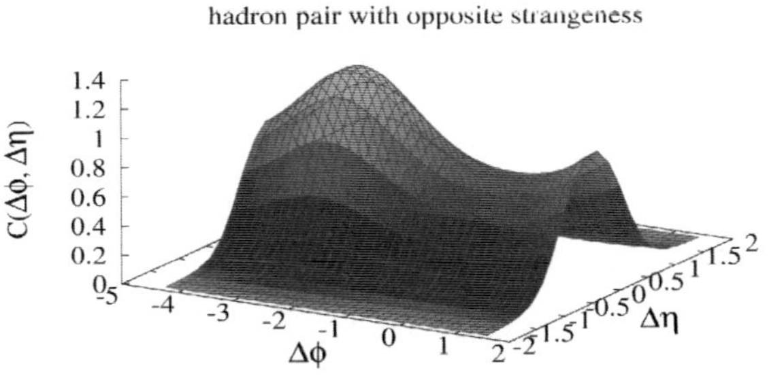

Fig.3 The correlation function for two hadrons with opposite strangeness in flux-tube fragmentation.

Adjacently produced mesons however can be signaled by their rapidity difference Δy falling within the window of $|\Delta y|{\sim}1/(dN/dy)$. The theoretical correlation function for two primary hadrons with opposite strangeness is shown in Fig. 3, with a suppression at $(\Delta\phi, \Delta\eta) \sim 0$ and an enhancement at $\Delta\phi{\sim}\pi$, within the window of $|\Delta y|{\sim}1/(dN/dy)$.

4. Conclusions and Discussions

In the fragmentation of a flux tube, the production of $q\bar{q}$ pairs obey local conservation laws and the hadrons follow space-time-rapidity ordering. As a consequence, a signature of the flux tube fragmentation consists of a chain of produced hadrons, correlated in rapidities, charges, flavors, and azimuthal angles. The observation of the chain of hadrons requires the identification of the produced hadrons which may be possible if the fraction of unobserved hadrons is small.

Another signature uses two-hadron angular correlations with opposite charges or strangeness, which exhibits a suppression at $(\Delta y, \Delta\phi){\sim}0$, but an enhancement at $\Delta\phi{\sim}\pi$, with the window of $|\Delta y|{\sim}1/(dN/dy)$. It should be kept in mind however that resonance production and resonance decay into hadrons will exhibit angular correlations similar to the pattern of the flux-tube fragmentation. The resonance fraction give rise to complications and the two-hadron signature will work well if the resonance fraction is not dominant. Various estimates give the resonance fractions to be of order 10 to 30% for *pp* collisions at $\sqrt{s_{pp}} = 17.3$ GeV [7].

In flux-tube fragmentation, the production of two adjacent hadrons with like charges or the same strangeness is prohibited [6]. Experimentally, two-hadron angular correlation of like charges with $\Delta\eta{\sim}0$ is not zero [9, 10] which indicates that there may be an additional mechanism for like charge production with $(\Delta\phi, \Delta\eta){\sim}$ 0. Further theoretical search for the origin of the source of like charge and strangeness correlations at $(\Delta\phi, \Delta\eta){\sim}$ 0 will be of interest.

Acknowledgments This work was supported in part by the Division of Nuclear Physics, U.S. Department of Energy, under Contract No. DE-AC05-00OR22725.

References

[1] X. Artru and G. Mennessier, Nucl. Phys. B70, 93 (1974)
[2] B. Andersson, G. Gustafson and C. Peterson, Z. Phys. C1, 105 (1979); B. Andersson, G. Gustafson and B. Söderberg, Z. Phys. C20, 317 (1983).
[3] R. Blankenbecler and S. J. Brodsky, Phys. Rev. D **10**, 2973 (1974); R. Blankenbecler, S. J. Brodsky and J. Gunion, Phys. Rev. D **12**, 3469 (1975).
[4] A.L.S. Angelis *et al.* (CCOR Collaboration), Phys. Lett. B **79**, 505 (1978).
[5] R. C. Hwa, Phys. Rev. D22, 1593 (1980).
[6] C. Y. Wong, Phys. Rev. D **92**, 074007 (2015).
[7] C. Y. Wong, Phys. arxiv:1510.01794 (2015).
[8] T. Sjöstrand *et at.*, *PYTHIA 6.4 Physics and Manual*, JHEP 05, 026 (2006), arXiv:hep-ph/0603175.
[9] R. J. Porter and T. A. Trainor, (STAR Collaboration), J. Phys. Conf. Ser. **27**, 98 (2005).
[10] M. Maksiak (NA61/SHINE Collaboration), arXiv:1503.02470; M. Gazdzicki (NA61/SHINE Collaboration), EPJ Web Conf. **95**, 01005 (2015), arxiv:1412.4243.
[11] CMS Collaboration, JHEP 1009, 091 (2010),[arXiv:1009.4122].

Flavour Oscillations in Dense Baryonic Matter

Peter Filip

Institute of Physics, Slovak Academy of Sciences, Dúbravská cesta 9, Bratislava 845 11, Slovakia

E-mail: peter.filip@savba.sk

Abstract. We suggest that fast neutral meson oscillations may occur in a dense baryonic matter, which can influence the balance of $s/\bar{s}$ quarks in the nucleus-nucleus and proton-nucleus interactions, if primordial multiplicities of neutral $K^0, \bar{K}^0$ mesons are sufficiently asymmetrical. The phenomenon can occur even if CP symmetry is fully conserved, and it may be responsible for the enhanced sub-threshold production of multi-strange hyperons observed in the low-energy A+A and p+A interactions.

1. Introduction

Strangeness conservation is usually assumed to take place in the strongly interacting matter created in nucleus-nucleus and proton-nucleus interactions [1]. This assumption is useful when baryonic density of the expanding QCD matter is negligible $\rho_B \approx 0$, or (naturally) in the vacuum, where net strangeness imbalance [2] develops at 10^{-3} level only due to CP symmetry violation.

In dense baryonic matter, however, the excess of baryons leads to the asymmetry in the population of $\bar{K}^0(\bar{d}s)/K^0(d\bar{s})$ mesons, since $\bar{s}$ antiquarks have little chance to become part of anti-baryons. Kaon multiplicity ratios $K^-(\bar{u}s)/K^+(u\bar{s}) \approx 10^{-2}$ are observed at low energy fixed-target experiments [3] and similar ratio $\bar{K}^0(\bar{d}s)/K^0(d\bar{s}) \approx 10^{-2}$ may be expected also for the primordial neutral kaons. When neutral kaons are reconstructed via $K^0_S \to \pi^+\pi^-$ decay channel, ratio $\bar{K}^0(ds)/K^0(d\bar{s})$ is unobservable, while the ratio $K^{0*}(d\bar{s})/\bar{K}^{0*}(ds)$ of vector mesons can be measured.

In this contribution we speculate about the possibility, that different in-medium potentials $V_K(\rho_B) \neq \bar{V}_{\bar{K}}(\rho_B)$ of neutral K^0, $\bar{K}^0$ mesons may result in the fast $(s\bar{d}) \rightsquigarrow (d\bar{s})$ and $(d\bar{s}) \rightsquigarrow (s\bar{d})$ transitions taking place in a dense baryonic matter. Due to the primordial excess of $(d\bar{s})$ mesons, the oscillations $(d\bar{s}) \leftrightarrow (s\bar{d})$ then result in the effective net strangeness non-conservation in A+A and p+A interactions, which can possibly explain the unexpected excess of s-quarks needed for the observed subthreshold production of multi-strange hyperons [4].

2. Neutral Kaon Oscillations in a Medium

Oscillations of neutral kaon states in the baryonic medium (regenerator) were studied in the pioneering experiment by Fitch et al. [5], in order to prove that interference of K^0_L and K^0_S meson states occurs, and CP violation is real. No modification of the mass difference $\Delta\tilde{m} = m(K^0_L) - m(K^0_S)$ inside the regenerator medium was expected [5], and this seems to be the assumption of all other experiments, where $K^0 \leftrightarrow \bar{K}^0$ oscillations were studied via strong interactions method [6].

However, in a dense baryonic matter, $K^0 \leftrightarrow \bar{K}^0$ oscillation length $\tau^K_{osc} = 2\pi\hbar/\Delta\tilde{m}$ can become shorter, due to larger mass difference $\Delta\tilde{m}$ of weak eigenstates $\tilde{K}^0_S$ and $\tilde{K}^0_L$ in the medium. Let us express Hamiltonian $\mathbb{H} = \mathbb{M} - \frac{i}{2}\mathbb{G}$ for the neutral kaon system [7] in medium as

$$\mathbb{H}' = \begin{bmatrix} M_{11} + V_{K^0}(\rho_B) & M_{12} \\ M_{21} & M_{22} + \bar{V}_{\bar{K}^0}(\rho_B) \end{bmatrix} - \frac{i}{2}\begin{pmatrix} \Gamma_{11} + A_{K^0}(\rho_B) & \Gamma_{12} \\ \Gamma_{21} & \Gamma_{22} + \bar{A}_{\bar{K}^0}(\rho_B) \end{pmatrix} \tag{1}$$

where potentials $V_{K^0}, V_{\bar{K}^0}$ and absorption coeficients $A_{K^0}, \bar{A}_{\bar{K}^0}$ are related to forward scattering amplitude difference $f_K(0) - \bar{f}_{\bar{K}}(0)$ of K mesons in medium. Let us assume linear dependence

$$V_{K^0}(\rho_B) = 20MeV[\,\rho_B\,/\rho_B^N] \quad ; \quad \bar{V}_{\bar{K}^0}(\rho_B) = -60MeV[\,\rho_B\,/\rho_B^N] \tag{2}$$

for V_{K^0} and $\bar{V}_{\bar{K}^0}$ potentials on density ρ_B up to the nuclear density value $\rho_B^N = 2.3{\cdot}10^{17}\mathrm{kg/m^3}$, in agreement with Eqs. 66, 67 in [8] (using values α_K=0.04 and $\alpha_{\bar{K}}$=−0.12 based on Figs. 3.1, 2.8, 2.5 in Ref.[8], with momentum dependence of $V_{K^+} \approx V_{K^0}$ and $\bar{V}_{K^-} \approx \bar{V}_{\bar{K}^0}$ potentials neglected). Diagonalization of Hamiltonian (1) with $M_{11} = M_{22} = 497\mathrm{MeV}$, and $\Gamma_{11} = \Gamma_{22} = 3.7 \cdot 10^{-12}$ (using $|\Gamma_{12}| = |\Gamma_{21}| = 3.48 \cdot 10^{-12}$ and $|M_{12}| = |M_{21}| = 1.74 \cdot 10^{-12}$) yields density-dependent mass difference $\Delta m(\rho)$ for $\tilde{K}^0_2, \tilde{K}^0_1$ eigenstates shown in Figure 1 (the absorption is neglected).

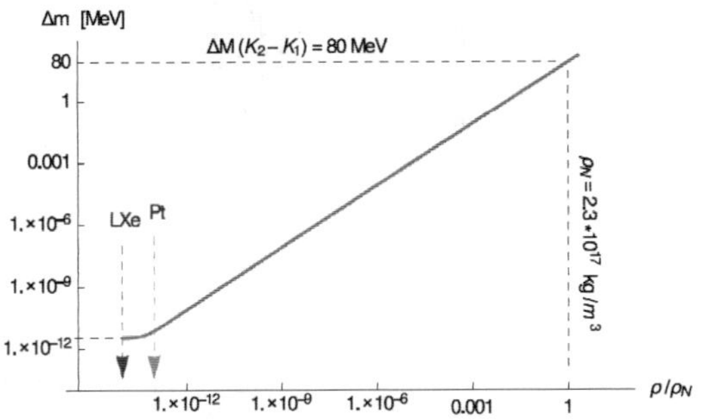

Figure 1. Mass difference of neutral kaon states in baryonic medium of density ρ/ρ_N.

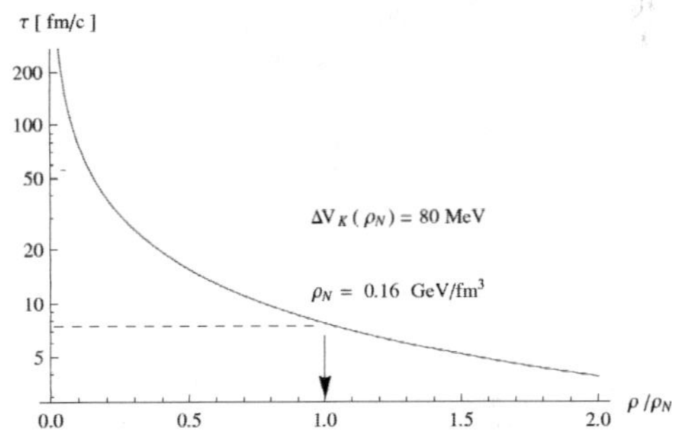

Figure 2. Oscillation time $\tau(\bar{s}d \to s\bar{d})$ in nuclear baryonic matter of density ρ/ρ_N.

Mass difference $\Delta\tilde{m}$ increases from its vacuum value $\Delta\tilde{m} = (3.48 \pm 0.01) \times 10^{-12}\mathrm{MeV}$ for densities larger than $\rho_{Fe} \approx 8\mathrm{g/cm^3}$. For W-Pt metals ($\rho \approx 21\mathrm{g/cm^3}$) the effect may be observable. However, measured forward scattering amplitudes $f_K(0), \bar{f}_{\bar{K}}(0)$ [9] should be used, instead of the rough linear approximation (2), to predict the magnitute of $\Delta\tilde{m}$ for dense metals.

3. Fast $d\bar{s} \leftrightarrow s\bar{d}$ Transitions in Dense Nuclear Medium

$K^0 \rightsquigarrow \bar{K}^0$ oscillation gives $\bar{s} \to s$ strangeness conversion length $\tau_{\bar{s}\to s} = \frac{1}{2}\tau_{osc} = \pi\hbar/\Delta\tilde{m}$, which for $\Delta\tilde{m} \approx 80\mathrm{MeV}$ (at nuclear density $\rho_B^N = 0.16\mathrm{GeV/fm^3}$) leads to $\tau_{\bar{s}\to s} \approx 7.5\mathrm{fm}/c$. This suggests, that neutral kaon oscillations (taking place in a dense baryonic medium) can potentially allow for $s \leftrightarrow \bar{s}$ conversion within time scales comparable to $p + A$ and $A + A$ interaction. Probability of $K^0 \leftrightarrow \bar{K}^0$ transition in the medium is [7] (CP symmetry means $|q_H/p_H|^2 = |p_L/q_L|^2$)

$$P[K^0 \rightsquigarrow \bar{K}^0] = \left|\frac{q_H}{p_H}\right|^2 |g_-(\tau)|^2 |(1-\theta)|^2 \quad ; \quad P[\bar{K}^0 \rightsquigarrow K^0] = \left|\frac{p_L}{q_L}\right|^2 |g_-(\tau)|^2 |(1-\theta)|^2 \tag{3}$$

where $|g_-(\tau)|^2 = \frac{1}{4}\left[e^{-\tau\Gamma_H} + e^{-\tau\Gamma_L} - 2\cos(\Delta\tilde{m}\tau)e^{-\tau(\Gamma_H+\Gamma_L)/2}\right]$, denoting $\tilde{K}^0_2 = p_H|K^0\rangle + q_H|\bar{K}^0\rangle$ and $\tilde{K}^0_1 = p_L|K^0\rangle - q_L|\bar{K}^0\rangle$ as eigenvectors of $\mathbb{H}'$. Since $|q_H/p_H| = 2|H_{21}|/|\Delta\mu(1-\theta)|$, where $\Delta\mu = \sqrt{4H_{12}H_{21} + (H_{22} - H_{11})^2}$ [7], parameter $(1-\theta)^2$ disappears in Eq.(3). For dense baryonic

IOP Conf. Series: Journal of Physics: Conf. Series **779** (2017) 012047 doi:10.1088/1742-6596/779/1/012047

matter $4|H_{12}H_{21}| \ll |H_{22} - H_{11}|^2$, since $|H_{22} - H_{11}| = |(\bar{V}_{\bar{K}^0} - V_{K^0}) - \frac{i}{2}(\bar{A}_{\bar{K}} - A_K)|$ with CPT symmetry conserved, and consequently

$$P[K^0 \rightsquigarrow \bar{K}^0] \;\approx\; \frac{4|M_{21} - \frac{i}{2}\Gamma_{21}|^2}{|H_{22} - H_{11}|^2}|g_-(\tau)|^2 = S(\rho_B)|g_-(\tau)|^2 \tag{4}$$

where suppression factor $S(\rho_B) = |2H_{21}|^2/|\Delta V_K - \frac{i}{2}\Delta A_K|^2$. If our approximations are correct, transitions $\bar{s}d \leftrightarrow s\bar{d}$ can happen within time scale 10fm/c in nuclear matter, though being suppressed by factor $S(\rho_B) \approx 4|H_{21}|^2/(80\mathrm{MeV})^2$.

For pseudoscalar $K^0(497)$ mesons (assuming dense medium does not influence M_{21}, Γ_{21} values originating from the weak interaction) one has $|H_{21}| \approx 2.5 \times 10^{-12}$MeV, and thus $K^0 \leftrightarrow \bar{K}^0$ transitions become irrelevant (negligible) with suppression factor $S(\rho_B) \approx 10^{-26}$.

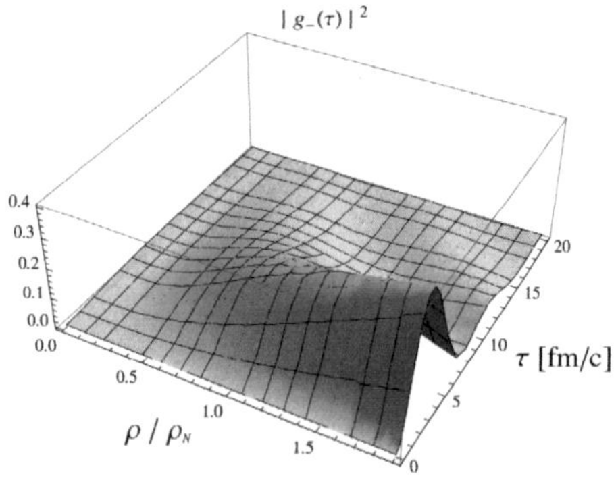

Figure 3. Interference term $|g_-(\tau)|^2$ for K^{0*}.

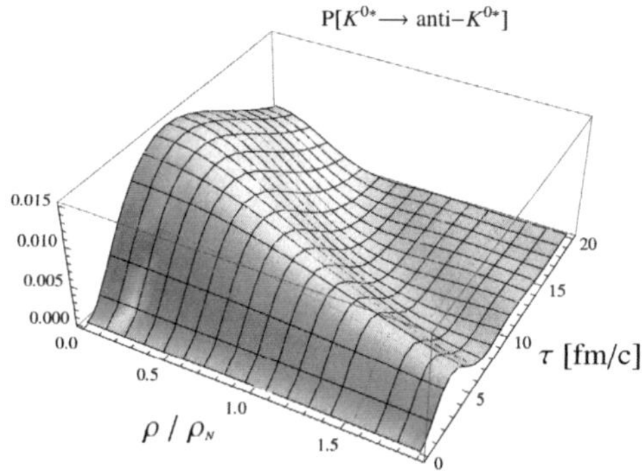

Figure 4. The probability $P[K^{0*} \to \bar{K}^{0*}]$.

However, the situation looks to be more promissing for $K^{0*}(896)$ vector mesons. Although $|M_{12}| \approx 10^{-12}$ remains to be small, value $\Gamma_{21}^* = \Gamma_{12} = \sum_n \rho_c^n \langle K^{0*}|H_w|f_n\rangle\langle f_n|H_w|\bar{K}^{0*}\rangle$ [7] may become much larger (in medium). For the upper estimate on Γ_{12} for K^{0*}, $\bar{K}^{0*}$ mesons, we write

$$\Gamma_{12} = \rho_c^S\langle K^{0*}|H_w|K_S^0\pi^0\rangle\langle K_S^0\pi^0|H_w|\bar{K}^{0*}\rangle + \rho_c^L\langle K^{0*}|H_w|K_L^0\pi^0\rangle\langle K_L^0\pi^0|H_w|\bar{K}^{0*}\rangle \tag{5}$$

Decay width of $K^{0*} \to K^0\pi^0 \to K_s^0\pi^0$ process is $\Gamma_{S\pi^o}^{K^{o*}} = (\frac{1}{2})48\mathrm{MeV}/3 = 8\mathrm{MeV}$, and the same applies to $K^{0*} \to K^0\pi^0 \to K_L^0\pi^0$ decay width: $\Gamma_{L\pi^o}^{K^{o*}} = 8\mathrm{MeV}$. For anti-$K^{0*}$ decays we assume similarly: $\Gamma_{S\pi^o}^{\bar{K}^{o*}} = \Gamma_{L\pi^o}^{\bar{K}^{o*}} = 8\mathrm{MeV}$ for $\bar{K}^{0*} \to \bar{K}^0\pi^0 \to K_{L,S}^0\pi^0$. Thus, we may expect $|\Gamma_{12}| \approx 16\mathrm{MeV}$ using Eq.(5), which makes the probability of $K^{0*} \leftrightarrow \bar{K}^{0*}$ oscillations significant. (Note, that quantum superposition of vector mesons $pK^{0*} + q\bar{K}^{0*}$ had been discussed by Littenberg [10].) Hamiltonian we have used for the system of K^{0*}, $\bar{K}^{0*}$ $(J = 1)$ mesons was

$$\mathbb{H}'_{K^{0*}} = \begin{bmatrix} 896 + V_{K^0}(\rho_B) & 1.7\cdot 10^{-12} \\ 1.7\cdot 10^{-12} & 896 + \bar{V}_{\bar{K}^0}(\rho_B) \end{bmatrix} - \frac{i}{2}\begin{pmatrix} 48 + A_{K^0} & 16\cdot e^{i\zeta} \\ 16\cdot e^{-i\zeta} & 48 + \bar{A}_{\bar{K}^0} \end{pmatrix}. \tag{6}$$

and $S(\rho_B) \approx 10^{-2}$ suppression factor was obtained for $\Delta V_{K^*} = 80\mathrm{MeV}$ at nuclear density $\rho_B = 0.16\mathrm{GeV/fm}^3$. If mass difference $\Delta\tilde{m}^* = \tilde{m}(K_2^{0*}) - \tilde{m}(K_1^{0*})$ in the baryonic medium approaches $\Delta\tilde{m}^* \approx 80\mathrm{MeV}$ at $\rho_B = \rho_B^N$, we obtain $\tau(\bar{s} \to s) \approx 7\mathrm{fm/c}$ again. Figures 3 and 4 show the interference term

$$|g_-(\tau)|^2 = \frac{1}{4}\left[e^{-\tau\Gamma_H^*(\rho_B)} + e^{-\tau\Gamma_L^*(\rho_B)} - 2\cos(\Delta\tilde{m}^*(\rho_B)\tau)e^{-\tau[\Gamma_H^*(\rho_B)+\Gamma_L^*(\rho_B)]/2}\right] \tag{7}$$

and transition probability $P[K^{0*} \rightsquigarrow \bar{K}^{0*}]$ obtained as a function of time and density for $K^*(896)$. The oscillation $D^{0*} \leftrightarrow \bar{D}^{0*}$ in medium appeared to be negligible: $P[D^{0*} \rightsquigarrow \bar{D}^{0*}] \leq 10^{-7}$ for $\tau < 20\mathrm{fm/c}$, due to upper bound $|\Gamma_{12}^{D^{o*}}| < 0.04\mathrm{MeV}$, originating from $\Gamma_{tot}(D^{0*}) \approx 40\mathrm{keV}$ [11].

4. Subthreshold $\Xi^-(ssd)$ Production

Main motivation for the considerations presented here was to explain the excessive subthreshold production of double-strange hyperons $\Xi^-(ssd)$ observed by HADES experiment in Ar+KCl and p+Nb interactions [4]. Our hypothesis is depicted in diagram (8): at subthreshold energy, only one $(s\bar{s})$ pair is created, and s-quark becomes part of Λ or Σ hyperon, while $\bar{s}$ antiquark enters $(d\bar{s})$ bound state. The oscillation $(d\bar{s} \rightsquigarrow s\bar{d})$ in dense baryonic matter creates the second s quark (e.g. with probability 1%) needed for double-strange Ξ hyperons.

$$\begin{bmatrix} p+A \\ A+A \end{bmatrix} \longrightarrow \left(\begin{smallmatrix} \bar{s} \\ s \end{smallmatrix}\right) \begin{array}{l} \nearrow \; (\bar{s}d)K^{0*} \quad \rightsquigarrow\rightsquigarrow \quad \bar{K}^{0*}(s\bar{d}) \; \searrow \\ \searrow \; (sdu)\Lambda,\Sigma \quad \longrightarrow \quad \Lambda,\Sigma(sdu) \; \nearrow \end{array} \longrightarrow (ssd)\Xi^- + (u\bar{d})\pi^+ \tag{8}$$

Strangeness recombination reaction $\Lambda + \bar{K} \to \Xi + \pi$ allows then excessive Ξ hyperons to be produced. Rescattering simulation of p+A / A+A collisions may confirm or deny our suggestion.

5. Summary

We suggest, that fast strangeness oscillation $(\bar{s}d) \rightsquigarrow (s\bar{d})$ may occur in dense baryonic medium with small $P(\bar{s} \rightsquigarrow s) \approx 10^{-2}$ probability. This phenomenon is potentially capable to increase the total yield of $\bar{K}$ mesons, if condition $K^0/\bar{K}^0 > 100$ for primordial neutral kaon multiplicities is fulfilled. Excessive double-strange Ξ hyperon production may then take place at subthreshold energies via strangeness recombination reactions (a quantitative estimate is needed).

$D^{0*} \leftrightarrow \bar{D}^{0*}$ oscillation probability appears to be negligible ($P \leq 10^{-7}$), if $|\Gamma_{12}^{D^*}| < 0.04$MeV restriction is kept for dense baryonic medium. We assume $|\Gamma_{12}^{K^*}| \approx 16$MeV also for small baryonic densities, and therefore, $P[K^{0*} \rightsquigarrow \bar{K}^{0*}] \approx 10^{-2}$ is obtained even for $\rho_B \to 0$ (Figure 4).

Fast $K^{0*} \rightsquigarrow \bar{K}^{0*}$ oscillations might be directly observable, if K^{0*} and $\bar{K}^{0*}$ spectra were carefully measured at AGS energies. Since experimental data on K^* production were not published by AGS experiments, a precise measurement of $\bar{K}^{0*}$ production in $p+p$, $p+A$ and $A+A$ interactions by NA61/SHINE at low SPS energies, by NICA/MPD experiment, and possibly also by STAR fixed target or BES II program at RHIC, could clarify the situation.

Experimental verification of the in-medium change of $\Delta m(K_2^0 - K_1^0)$ mass difference in dense metals (W, Pt) would affirm our understanding of the neutral kaon physics [12] to be complete.

The readers should be aware of the recent publication [13] which is also attempting to explain the excessive Ξ^- hyperon production, and of publication [14] on neutral kaons in dense medium.

Acknowledgements

This work was performed within project N. 2/0197/14 of the Slovak Grant Agency VEGA.

The author is grateful to organizers of SQM 2016 for supporting the presentation of these results at Berkeley. Kind hospitality experienced at CERN Library and support from the Committee for collaboration of Slovak Republic with CERN are acknowledged here.

[1] Tomášik B, Kolomeitsev E E 2016 *Eur. Phys. Jour.* A **52** 251
[2] Kabir P K 1970 *Phys. Rev.* D **2** 540; Dass G V, Sarma K V L, Kabir P K 1987 *Phys. Rev.* D **35** 1730
[3] Scheinast W, et al. 2006 *Phys. Rev. Lett.* **96** 072301; Agakishiev G, et al. 2009 *Phys. Rev.* C **80** 025209
[4] Agakishiev G, et al. 2009 *Phys. Rev. Lett.* **103** 132301; Agakishiev G, et al. 2015 *Phys. Rev. Lett.* **114** 212301
[5] Fitch V L, Roth R F, Russ J S, Vernon W 1965 *Phys. Rev. Lett.* **15** 73
[6] Camerini U, et al. 1962 *Phys. Rev.* **128** 362; Meisner G W, et al. 1966 *Phys. Rev. Lett.* **16** 278
[7] Branco G C, Lavoura L, Silva J P 1999 in book: *CP Violation*, Clarendon Press (Oxford)
[8] Hartnack C, et al. 2012 *Physics Reports* **510** 119
[9] Eberhard P H, Uchiyama F 1994 *Nucl. Inst. Meth. Phys. Res.* A **350** 144
[10] Littenberg L S 1980 *Phys. Rev.* D **21** 2027
[11] Nussinov S 1998 *Phys. Lett.* B **418** 383
[12] Angelopoulos A, et al. 2003 *Physics Reports* **374** 165
[13] Steinheimer J, Bleicher M 2016 *Jour. Phys. G: Nucl. Part. Phys.* **43** 015104
[14] Amelino-Camelia G, Kapusta J 1999 *Phys. Lett.* B **465** 291

IOP Conf. Series: Journal of Physics: Conf. Series **779** (2017) 012048 doi:10.1088/1742-6596/779/1/012048

An Application of Functional Renormalization Group Method for Superdense Nuclear Matter

G. G. Barnaföldi[1], **A. Jakovác**[2], **and P. Pósfay**[1,2]

[1] Wigner Research Centre for Physics of the H.A.S., P.O. Box, H-1525 Budapest, Hungary
[2] Institute of Physics, Eötvös University, 1/A Pázmány P. Sétány, H-1117 Budapest, Hungary

E-mail: `barnafoldi.gergely@wigner.mta.hu`

Abstract. We proposed a method, using the expansion of the effective potential in a base of harmonic functions, to study the Functional Renormalization Group (FRG) method at finite chemical potential. Within this theoretical framework we determined the equation of state and the phase diagram of a simple model of massless fermions coupled to scalars through Yukawa-couling at the zero-temperature limit. Here, we use our FRG-based equation of state to describe the superdense nuclear matter inside compact astrophysical objects. We calculated the mass-radius relation for a compact star using the TOV equation, which was compared to other results.

1. Introduction

Compact astrophysical objects, such as neutron-, quark-, or hybrid stars are the most extreme, high-density creatures of the present-day Universe. It is a challenging task to explore the inner structure of these *compact stars* due to the lack of direct probes or measurements of their interior. Recent spectroscopic radius measurements using X-ray data analysis [1] and even the gravitational-wave discoveries [2, 3] may provide such constraints, which led us to develop more reliable equation of state (EoS) of the superdense matter.

Modeling the high-density nuclear matter and providing its equation of state is still an research field. On the other hand the success of the above task is shaded by the *masquerade problem*, since different complex and sophisticated EoS result similar behavior and observables of the compact celestial bodies [4]. This motivates us not only to provide perfect EoS, but describe the phase structure of the cold and high-density nuclear matter [5].

The calculation of the equation of state in the high-density and zero temperature limit is usually considered in the mean-field or one-loop approximation. Functional Renormalization Group (FRG) method can extend this description in an exact way, taking into account the effect of quantum fluctuations in the effective action of the system.

In this work we use the Wetterich-equation to compute the EoS and Litim's regulator is applied regulating the scale dependence [6]. The Local Potential Approximation (LPA) is used to obtain the EoS for the *ansatz* containing a Yukawa-type interaction as described in Ref. [7]. We present there the calculated EoS which has a Maxwell-construction as an inner nature.

The calculated EoS is tested by solving the corresponding Tolman – Oppenheimer – Volkov (TOV) equations and investigating the properties by the mass-radius relation, $M(R)$ of compact stars. Comparison of the FRG-based equation of state to other high-density zero-temperature nuclear matter EoS and to the calculated $M(R)$ by various models are given.

2. The FRG Method for a Yukawa-type Model

The functional renormalization group method is a general way to find the effective action of a system. This formalism led us to calculate low-energy effective (observable) quantities by gradual momentum integration of a theory defined at some high-energy scale, k. Since low-scale effective quantities incorporate quantum fluctuations, using FRG at finite temperature one may calculate the equation of state of the system including the quantum fluctuations as well.

Within the FRG framework the quantum n-point correlation function is calculated by the gradual path integration. This can be achieved by introducing a regulator term, R_k in the generator functional, $Z_k[J]$, which acts as a mass term and suppresses modes below scale, k as explained in Refs. [8, 9]. Thanks to this regulator term, the effective action becomes scale-dependent, which scale dependence is given by the Wetterich-equation [10]

$$\partial_k \Gamma_k = \frac{1}{2} \int \mathrm{d}p^D \, \mathrm{STr} \left[(\partial_k R_k) \left(\Gamma_k^{(2)} + R_k \right)^{-1} \right], \tag{1}$$

where $\Gamma_k^{(2)}$ is the second derivative matrix of the effective action. The term 'STr' stands for the normal *trace* operation but includes a negative sign for fermionic fields and sums over all indices. The low-scale (observable) effective action is computed by integrating the Wetterich-equation (1), from the classical limit at some UV-scale $k = \Lambda$ to the IR-scale $k = 0$, where quantum effects are taken into account. The initial condition in this integration is the UV-scale (classical) action $\Gamma_{k=\Lambda}$, which has to be chosen in a way, that the low-scale effective action reproduces physical quantities, correctly.

Here, we use a simple Yukawa-type model with one bosonic and one fermionic degree of freedom described by the bare action. This is described by the bare action at scale Λ,

$$\Gamma_\Lambda[\varphi, \psi] = \int \mathrm{d}^4x \left[\bar{\psi}(i\partial\!\!\!/ - g_0\varphi)\psi + \frac{1}{2}(\partial_\mu \varphi)^2 - \frac{m_0^2}{2}\varphi^2 - \frac{\lambda_0}{24}\varphi^4 \right]. \tag{2}$$

As we described in Ref. [7] this model has two phases: (i) in the symmetric phase the fermion is massless, (ii) in the Spontaneous Symmetry Breaking (SSB) phase the fermion mass is $g\langle\varphi\rangle$.

To treat this model with the FRG method we need an *ansatz* for the effective action at scale k. As a consequence of LPA, we choose the simplest possible one, where only the bosonic effective potential depends on the scale:

$$\Gamma_k[\varphi, \psi] = \int \mathrm{d}^4x \left[\bar{\psi}(i\partial\!\!\!/ - g\varphi)\psi + \frac{1}{2}(\partial_\mu \varphi)^2 - U_k(\varphi) \right]. \tag{3}$$

Note, neither wave function renormalization, nor the running of the Higgs coupling are taken into account, however, both effects can be easily adapted into the present method.

The integrated Wetterich-equation (1) for this model can be rewritten after applying the three-dimensional Litim regulator at finite temperature T and at finite chemical potential μ,

$$\partial_k U_k = \frac{1}{2} \mathrm{STr} \ln \left[R_k + \Gamma_k^{(2)} \right] = \frac{k^4}{12\pi^2} \left[\frac{1 + 2n_B(\omega_B)}{\omega_B} + 4\frac{-1 + n_F(\omega_F - \mu) + n_F(\omega_F + \mu)}{\omega_F} \right], \tag{4}$$

where n_B and n_F are the Bose – Einstein and the Fermi – Dirac distributions, respectively

$$n_{B/F}(\omega) = \frac{1}{1 \mp e^{-\beta\omega_{B/F}}}, \quad \text{while} \quad \omega_B^2 = k^2 + \partial_\varphi^2 U, \quad \omega_F^2 = k^2 + g^2\varphi^2, \quad \text{and} \quad \beta = \frac{1}{T}. \tag{5}$$

In the case of $T = 0$ and $\mu > 0$ the Bose – Einstein distribution does not give contribution, but the Fermi – Dirac distribution reduces $n_F(\omega) \to \Theta(-\omega)$ and this simplifies equation (4) to:

$$\partial_k U_k = \frac{k^4}{12\pi^2} \left[\frac{1}{\omega_B} - 4\frac{\Theta(\omega_F - \mu)}{\omega_F} \right] \quad \text{with initial condition} \quad U_\Lambda(\varphi) = \frac{m_0^2}{2}\varphi^2 + \frac{\lambda_0}{24}\varphi^4. \tag{6}$$

The presence of the step function, $\Theta(\omega)$, generates two different domains, where two different differential equations evolve the potential in k. The boundary of these domains is called Fermi-surface S_F, which can be determined from requiring $\omega_F(k,\varphi)|_{S_F} = \mu$. The surface can be characterized either by $k = k_F(\varphi)$ or by $\varphi = \varphi_F(k)$ conditions. In our case these read as

$$k_F = \sqrt{\mu^2 - g^2\varphi^2} \qquad \text{and} \qquad \varphi_F = g^{-1}\sqrt{\mu^2 - k^2}. \tag{7}$$

The surface S_F, in terms of k and $g\varphi$, is a circle with radius μ, and for $\mu = 0$ it disappears. The Fermi-surface divides the coordinate space into two parts; we will denote the high energy regime by $\mathcal{D}_>$, the low energy regime by $\mathcal{D}_<$, where the following differential equations hold:

$$\partial_k U_k = \frac{k^4}{12\pi^2}\left[\frac{1}{\omega_B} - \frac{4}{\omega_F}\right], \qquad \text{if } (k,\varphi) \in \mathcal{D}_> := \left\{(k,\varphi)\,|\,k^2 + g^2\varphi^2 > \mu^2\right\}, \tag{8a}$$

$$\partial_k U_k = \frac{k^4}{12\pi^2}\frac{1}{\omega_B}, \qquad \text{if } (k,\varphi) \in \mathcal{D}_< := \left\{(k,\varphi)\,|\,k^2 + g^2\varphi^2 < \mu^2\right\}. \tag{8b}$$

and the solution is continuous at $k = k_F$.

For the solution we need to use standard FRG techniques, e.g. with discretization or with polynom expansion. The value of the potential at the boundary can be determined by cutting out the Fermi-surface from the zero chemical potential solution. In Ref. [7] we introduced a coordinate transformation which maps the circle-like Fermi-surface to a rectangular one, while it keeps the symmetries of the differential equations. Applying this circle-to-rectangle transformation and a harmonic expansion the Wetterich-equation can be solved numerically.

Since exact solution can be given in the mean field approximation, we took $v = f_\pi$, $g = m_N/v$, and $\lambda = 3m_\sigma^2/v^2$ with the values $m_N = m_\sigma = 0.938$ GeV and $f_\pi = 0.093$ GeV. We choose the chemical potential, μ_{MF}, close to the value of the first-order phase transition $\mu_{MF} \approx 0.6177 m_N$.

3. Results: Equation of State and Mass-radius Relation of Compact Stars

The potential solution provided by the FRG-method was found to converge quickly, where the potential is convex. On contrary it converges slowly at the concave part and contains a native Maxwell-construction, too. Comparison of the FRG-based EoS results to the mean-field and one-loop approximation is plotted on Fig. 1 as $p(\mu)$ in the zero temperature limit and at finite chemical potential. One can see, that FRG-based EoS is the softest, followed by the stiffer one-loop and finally the mean field approximation result.

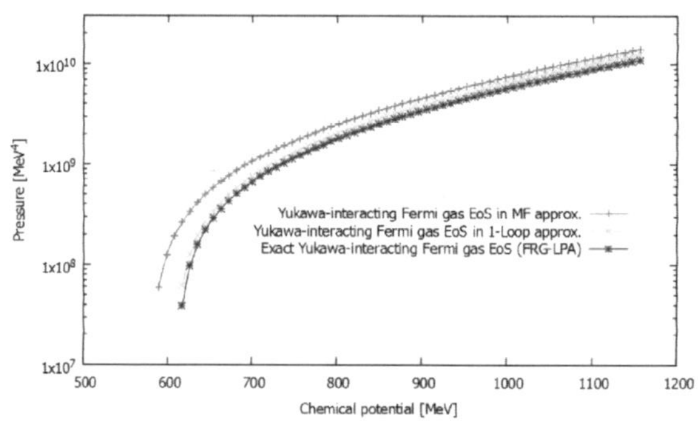

Figure 1. Equation of state $p(\mu)$, calculated from the functional renormalization group method.

IOP Conf. Series: Journal of Physics: Conf. Series **779** (2017) 012048 doi:10.1088/1742-6596/779/1/012048

We compared our FRG-based EoS results to some other EoS taken from Ref. [1], which are typically used in compact star models. On the *left panel* of Fig. 2, we found, our EoS calculations is the closest to the 'SQM3' model predictions at the highest, but differs at lower energy-density values. *Right panel* of Fig. 2 presents the calculated mass-radius diagram, $M(R)$ based on the Tolman – Oppenheimer – Volkov (TOV) equation combined together with the calculated FRG-based EoS. All these were drawn together with other model predictions from Ref. [1]. However these simple FRG-based EoS predict smaller but consistent compact stars with $M < 1.5M_{\odot}$ and $R < 8$ km, by this way we could successfully validate our concept described in Ref. [7].

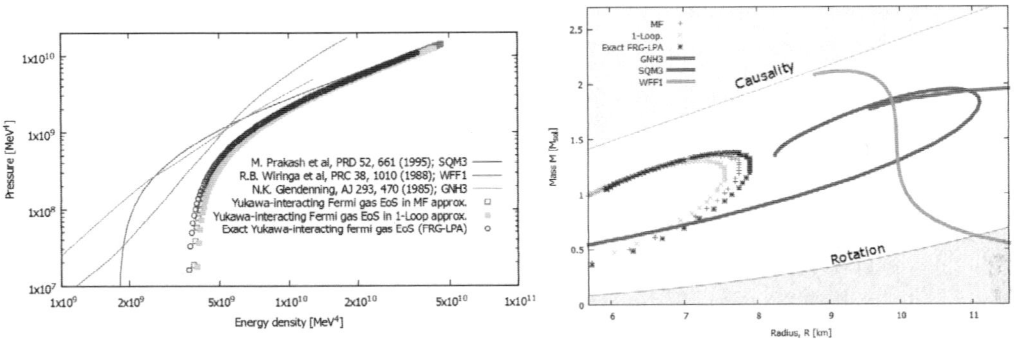

Figure 2. Comparison of different Equation of States and the corresponding $M(R)$ diagram for compact stars. Points were calculated by our method [7] and lines are from Ref. [1].

4. Summary

Using the Wetterich-equation the zero-temperature and high-density cold nuclear matter EoS was calculated in the FRG-framework based on the method and concept described in Ref. [7]. The calculated equation of state and the $M(R)$ diagram of compact stars were found to be consistent with the ones listed in Ref. [1], however with smaller stellar mass $M < 1.5M_{\odot}$ and radius $R < 8$ km.

Acknowledgments

This work is supported by the Hungarian Research Fund (OTKA) under contracts No. K104292, K104260, NK106119, K_16 120660, NKM-81/2016 MTA-UA bilateral mobility program, NIH TET_12_CN-1-2012-0016, and NewCompStar COST Action MP1304. Author GGB also thanks the János Bolyai Research Scholarship of the Hungarian Academy of Sciences.

References

[1] F. Ozel, D. Psaltis, T. Guver, G. Baym, C. Heinke and S. Guillot, Astrophys. J. **820**, no. 1, 28 (2016)
[2] B. P. Abbott *et al.* [LIGO Scientific and Virgo Collaborations], Phys. Rev. Lett. **116**, no. 06, 061102 (2016)
[3] B. P. Abbott *et al.* [LIGO Scientific and Virgo Collaborations], Phys. Rev. Lett. **116**, no. 24, 241103 (2016)
[4] M. Alford, M. Braby, M. W. Paris and S. Reddy, Astrophys. J. **629**, 969 (2005)
[5] P. Pósfay, G. G. Barnaföldi and A. Jakovác, PoS EPS-HEP2015 (2015) 369
[6] D. F. Litim, Phys. Rev. D **64**, 105007 (2001)
[7] G. G. Barnaföldi, A. Jakovác and P. Pósfay, arXiv:1604.01717 [hep-th].
[8] C. Wetterich, Nucl. Phys. B **352**, 529 (1991).
[9] H. Gies, Lect. Notes Phys. **852**, 287 (2012) [hep-ph/0611146].
[10] C. Wetterich, Phys. Lett. B **301**, 90 (1993).
[11] T. K. Herbst, J. M. Pawlowski and B. J. Schaefer, Phys. Rev. D **88**, no. 1, 014007 (2013)

Phenomenological Implications of the p_T Spectra of ϕ and Ω produced at LHC and RHIC

Rudolph C. Hwa[1] **and Lilin Zhu**[2]

[1]Institute of Theoretical Science and Department of Physics University of Oregon, Eugene, OR 97403-5203, USA

[2] Department of Physics, Sichuan University, Chengdu 610064, P. R. China

E-mail: hwa@uoregon.edu

Abstract. The data on the p_T spectra of ϕ and Ω at LHC can be presented in a format that shows exponential behavior up to $p_T \approx 6$ GeV/c with the same slope for both particles and for nearly all centralities. Those empirical properties are found at lower energies also with the inverse slope showing a power-law dependence on $\sqrt{s_{NN}}$. The shared properties of the spectra are shown to emerge naturally from the recombination model. No radial flow is needed. We find experimental hints for the possibility that ϕ and Ω are mostly produced in the ridge region, generated by minijets. Appropriate experimental test is suggested.

The production of ϕ and Ω at LHC with p_T reaching as high as 5 GeV/c (for ϕ) [1] and 7 GeV/c (for Ω) [2] has not been discussed in any theoretical framework ranging from hydrodynamics to QCD. Both the extent of the p_T range and the constituents being only s quarks present challenges to the usual theories. Those difficulties provide a good opportunity for us to investigate the subject in a novel phenomenological way with the hope of discovering clues to what may be flaws in the conventional interpretation of what happens in heavy-ion collisions at very high energies.

The p_T spectra of ϕ have been fitted in [1] by blast wave in the hydro model [3] using 3 free parameters for each centrality bin. Undoubtedly, the same can be done for the Ω spectra. So what does one learn from the 6 parameters for each centrality? We propose a different way to examine the p_T spectra. The usual presentation of data is in terms of $dN_h/p_T dp_T$. Let us define for meson and baryon data, respectively, the two functions

$$M_h(p_T) = \frac{dN_h}{p_T dp_T} \cdot \frac{m_T^h}{p_T}, \tag{1}$$

$$B_h(p_T) = \frac{dN_h}{p_T dp_T} \cdot \frac{m_T^h}{p_T^2}, \tag{2}$$

which are just different ways of presenting of the same data without theoretical adjustments. Note that they have different dimensions. In Figs. 1 and 2 we show the distributions $M_\phi(p_T)$ and $B_\Omega(p_T)$ using only the data points in [1, 2]. They show the strikingly simple exponential behaviors in p_T for nearly all centralities. The straight-line fits of both distributions amazingly have the same inverse slope

$$T_\phi = T_\Omega = 0.51 \text{ GeV/c}, \quad \text{for } \sqrt{s_{NN}} = 2.76 \text{ TeV} \tag{3}$$

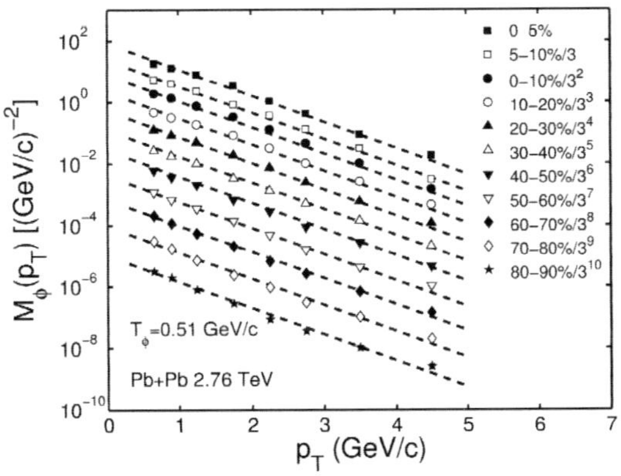

Figure 1. Plot of $M_\phi(p_T)$ from data in [1].

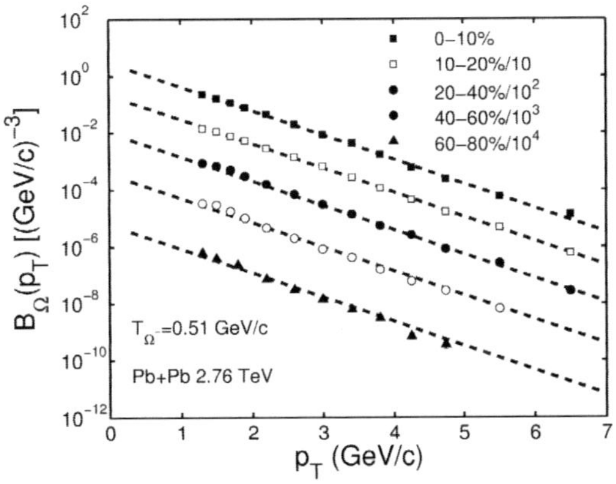

Figure 2. Plot of $B_\Omega(p_T)$ from data in [2].

a property that is totally hidden in the original data.

Seeing two different functions of different dimensions having the same inverse slope in semi-log plots stimulates curiosity. Exponential behavior in p_T usually means thermal distribution, but being valid out to $p_T \approx 6$ GeV/c is not what QCD thermodynamics can offer to explain. The fits in Figs. 1 and 2 can be expressed in the factorizable form

$$M_\phi(p_T, N_{\rm part}) = A_\phi(N_{\rm part})\exp(-p_T/T_\phi), \tag{4}$$
$$B_\Omega(p_T, N_{\rm part}) = A_\Omega(N_{\rm part})\exp(-p_T/T_\Omega), \tag{5}$$

where the centrality dependences for $N_{\rm part} > 70$ can be shown to satisfy

$$A_{\phi,\Omega}(N_{\rm part}) = A^0_{\phi,\Omega} N_{\rm part}^{a_{\phi,\Omega}}, \quad a_\phi = 0.9, \quad a_\Omega = 1.35. \tag{6}$$

What is found above for $\sqrt{s_{NN}} = 2.76$ TeV at LHC turns out to be also true at lower energies with different inverse slopes down to 7.7 GeV at RHIC [4, 5, 6, 7]. In Fig. 3 we show how $T_{\phi,\Omega}$ depends on $\sqrt{s_{NN}}$ empirically. The extension of the power-law behavior to 5.02 TeV yields $T_{\phi,\Omega} = 0.54$ GeV/c, which is a prediction without deeper theoretical explanation. We note that the values of $T_{\phi,\Omega}$ shown in Fig. 3 are not the temperatures discussed in hydro models.

Now, we summarize briefly how the observed behaviors of M_ϕ and B_Ω can be understood in the recombination model, a review of which can be found in the first few sections of [9]. The invariant distribution of ϕ at mid-rapidity is

$$E\frac{dN_\phi}{dp_T} = \int \frac{dp_1}{p_1}\frac{dp_2}{p_2}\mathcal{T}_s(p_1)\mathcal{T}_{\bar{s}}(p_2)R_\phi(p_1, p_2, p_T), \tag{7}$$

where only the thermal-parton distributions, $\mathcal{T}(p_i)$, are retained, since the production of strange shower partons is suppressed. The recombination function $R_\phi(p_1, p_2, p_T)$ contains a momentum-conservation factor $\delta((p_1 + p_2)/p_T - 1)$. Inserting into (7) $\mathcal{T}_s(p_1) = p_1^0 dN_s/dp_1 = C_s p_1 e^{-p_1/T_s}$ where T_s is the inverse slope for s quark distribution, and a similar form for $\mathcal{T}_{\bar{s}}(p_2)$, yields (with the assumption $T_s = T_{\bar{s}}$)

$$E\frac{dN_\phi}{dp_T} = C_s C_{\bar{s}} Z_\phi p_T^2 e^{-p_T/T_s}, \tag{8}$$

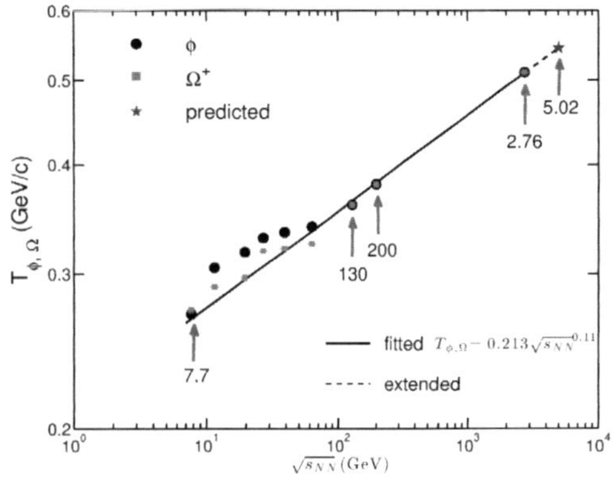

Figure 3. $T_{\phi,\Omega}$ derived from [1, 2, 4, 5, 6, 7].

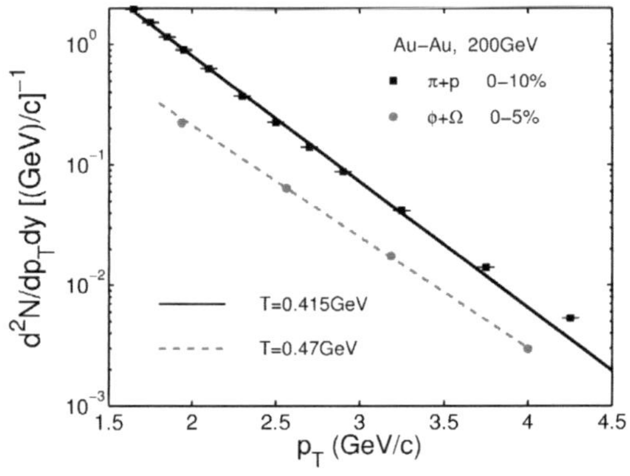

Figure 4. RHIC data on $\pi + p$ and $\phi + \Omega$ [4, 6, 8].

where Z_ϕ is a numerical factor arising from the integration. At $y \approx 0$ where $E \approx m_T$, Eq. (8) can be put into the form for M_ϕ given in (4) with $A_\phi(N_{\text{part}}) = C_s C_{\bar{s}} Z_\phi$ and $T_\phi = T_s$.

For Ω production three $\mathcal{T}_{\bar{s}}(p_i)$ are involved in EdN_Ω/dp_T, so one obtains (5) with $A_\Omega(N_{\text{part}}) = C_s^3 Z_\Omega$ and $T_\Omega = T_s$. Comparing M_ϕ with B_Ω, one gets $T_\phi = T_\Omega = T_s$ and that the ratio $A_\phi(N_{\text{part}})/A_\Omega(N_{\text{part}})$ is proportional to $C_s C_{\bar{s}}/C_s^3$, which agrees with the empirical N_{part} dependence in (6) at LHC, i.e., 2:3 in the exponents, in the reasonable assumption that $C_{\bar{s}}$ depends on N_{part} as C_s does.

What is important in our description above is that we have made no mention of radial flow. There is no need for radial velocity, T_{kin} or blast wave. The system must expand in some way, but it need not be prescribed by hydrodynamics without jets. We have for nearly all centralities used only one parameter T_s, which we have referred to as inverse slope, not temperature, although $\mathcal{T}_s(p_i)$ is regarded as thermal distribution. Since hadronization by recombination occurs at the end of the evolutionary process, we have not made use of any specific expansion dynamics. $\mathcal{T}_s(p_i)$ is the invariant s quark distribution at the time of hadronization without any contribution from the shower partons that arise from fragmentation of high-p_T jets. But at LHC minijets are copiously produced and must influence the bulk of the medium that hydro ignores. In our view $\mathcal{T}_s(p_i)$ must contain the effects of minijets.

The most notable feature about the above description is that the common value of the inverse slopes in Figs. 1 and 2 is $T_{\phi,\Omega} = 0.51$ GeV/c, as given in (3) for $\sqrt{s_{NN}} = 2.76$ TeV. Even at 7.7 GeV in the BES program the value is $T_{\phi,\Omega} = 0.27$ GeV/c, still much larger than the chemical freeze-out temperature of $T_{\text{ch}} < 0.165$ GeV and kinetic freeze-out temperature of $T_{\text{kin}} < 0.14$ GeV [10]. The fact that the exponential fits of $M_\phi(p_T)$ and $B_\Omega(p_T)$ are so good for such a wide range of p_T demands an explanation. An equation such as (8) relates T_s to the observed value of $T_{\phi,\Omega}$ but does not explain why T_s is so high. Our first step is to ask how high T_s is relative to the inverse slope for light quarks. It turns out that, when (1) is applied to pions, $M_\pi(p_T)$ cannot be fitted by a simple exponential because the fragmentation products and resonance decay contributions bend the distribution above a straight line at both high and low p_T regions. The situation is better for $B_p(p_T)$ for proton, but still not as good as in Fig. 2. Nevertheless, the common straight-line fits of $M_\pi(p_T)$ and $B_p(p_T)$ yields $T_{\pi,p} = 0.41$ GeV/c. That gives us an estimate of the difference between T_Ω and T_p: $\Delta T = T_\Omega - T_p = 0.1$ GeV/c at LHC. A similar comparison at RHIC-200 yields $\Delta T = 0.065$ GeV/c.

That leads us to recall the difference in inverse slopes when ridge was first discovered by STAR

[11]: $T_{\rm ridge} - T_{\rm incl}$=40-50 MeV. We use the STAR data for Au-Au collisions at 200 GeV and compare d^2N/dp_Tdy at mid-rapidity between $\pi + p$ (0-10%) and $\phi + \Omega$ (0-5%) as shown in Fig. 4. In the range between $p_T = 1.5$ and 4 GeV/c, which is the range on ridge in [11], the data can be well fitted by straight lines, showing the difference in inverse slopes to be $T_{\phi+\Omega} - T_{\pi+p} = 55$ MeV. That gives us a strong hint that ϕ and Ω may be a component in the ridge, but instead of being in the triggered events, they are identified.

So far there are no data from LHC on the $p_T^{\rm assoc}$ dependence of the particles in the ridge. However, CMS does have evidence that the ridge yield increases with decreasing $p_T^{\rm trig}$ [12]. Since a trigger particle comes from the fragmentation of a hard parton that emerges from the expanding medium, lower $p_T^{\rm trig}$ is more likely to be associated with a hard parton originated from the interior of the medium, thus losing a larger portion its energy traversing that medium. The energy lost enhances the thermal partons, so the ridge that comes from the enhanced thermal partons [9, 13] gets higher inverse slope and thus more yield. To confirm our interpretation that ϕ and Ω are in the ridge with higher $T_{\phi,\Omega}$ than $T_{\pi,p}$, we need data from LHC to identify ϕ and Ω in the triggered events and to compare the inverse slopes of their $p_T^{\rm assoc}$ distributions with (3).

In the usual QCD thermodynamics the chemical and kinetic freeze-out temperatures are less than 160 MeV. The larger $\langle p_T \rangle$, corresponding to our much larger value of inverse slope $T_{\phi,\Omega}$, is then achieved in hydro models by transverse flow, which is a description predicated on the assumption that the system evolves smoothly from early times. But we know from the parton distribution functions $F(x)$ that they increase steeply with decreasing momentum fraction x, so in an AA collision the probability is high for scattering among low-x partons resulting in a preponderance of minijets which can propagate inside the expanding medium and lose energy in ways that invalidate the rapid thermalization hypothesis. Minijets with $p_T \sim$ 2-3 GeV/c cannot be calculated reliably in pQCD, but their effects cannot be ignored. They are almost all absorbed by the environment, resulting in enhancement of the thermal partons that creates the ridge phenomenon [13, 14]. The minijets can produce more $s\bar{s}$ pairs than in equilibrium QCD thermodynamics. It is therefore natural for us to interpret the large value of $T_{\phi,\Omega}$ in (3) as a manifestation of s quarks in the ridge. If future LHC data can verify the correlation between (ϕ, Ω) and the ridge, then they not only give support to our view that is outside hydro flow, but also pose a strong challenge to any other approach for a satisfactory explanation.

Acknowledgments

This work was supported, in part, by the NSFC of China under Grant No. 11205106. We are grateful to Prof. Chunbin Yang for helpful discussions and to Dr. Xiaoping Zhang for providing us with the preliminary values of $T_{\phi,\Omega}$ from STAR's analysis of RHIC-BES data.

References

[1] Abelev B *et al* (ALICE Collaboration) 2015 *Phys. Rev.* C **91** 024609
[2] Abelev B *et al* (ALICE Collaboration) 2014 *Phys. Lett.* B **728** 216
[3] Schnedermann E, Sollfrank J and Heinz U 1993 *Phys. Rev.* C **45** 2462
[4] Abelev B I *et al* (STAR Collaboration) 2009 *Phys. Rev.* C **79** 064903
[5] Adams J *et al* (STAR Collaboration) 2004 *Phys. Rev. Lett.* **92** 182301
[6] Adams J *et al* (STAR Collaboration) 2007 *Phys. Rev. Lett.* **98** 062301
[7] Adamczyk L *et al* (STAR Collaboration) 2016 *Phys. Rev.* C **93** 021903
[8] Adare A *et al* (PHENIX Collaboration) 2013 *Phys. Rev.* C **88** 024906
[9] Hwa R C 2010 *Quark-Gluon Plasma 4*, ed Hwa R C and Wang Xin-Nian (Singapore: World Scientific), arXiv: 0904.2159.
[10] Mohanty B 2016 *Strangeness in Quark Matter 2016* (to be published in this volume)
[11] Abelev B I *et al* (STAR Collaboration) 2009 *Phys. Rev.* C **80** 064912
[12] Chatrchyan S *et al* (CMS Collaboration) 2011 *J. High. Energy Phys.* JHEP07(2011)076
[13] Hwa R C and Zhu L 2012 *Phys. Rev.* C **86** 024901
[14] Hwa R C 2013 *Int. J. Mod. Phys.* E **22** 1330003

IOP Conf. Series: Journal of Physics: Conf. Series **779** (2017) 012050 doi:10.1088/1742-6596/779/1/012050

Strangeness at finite temperature from Lattice QCD

J. Noronha-Hostler[a], R. Bellwied[a], J. Günther[b], P. Parotto[a], A. Pasztor[b], I. Portillo Vazquez[a], C. Ratti[a]

[a] *Department of Physics, University of Houston, Houston, TX 77204, USA*
[b] *Department of Physics, University of Wuppertal, Gaussstr. 20, D-42119 Wuppertal, Germany*

E-mail: jakinoronhahostler@gmail.com

Abstract. The precision reached by recent lattice QCD results allows for the first time to investigate whether the measured hadronic spectrum is missing some additional strange states, which are predicted by the Quark Model but have not yet been detected. This can be done by comparing some sensitive thermodynamic observables from lattice QCD to the predictions of the Hadron Resonance Gas model (with the inclusion of decays [3]). We propose a set of observables, defined as linear combinations of conserved charge fluctuations, where one can separate baryons by their strange quark content. These observables can isolate the multiplicity fluctuations of kaons from lattice QCD, which can then be compared with experimental results.

1. Introduction

In the 1960's Ralf Hagedorn proposed [1] that if there was a limiting temperature of the universe, now known as the Hagedorn Temperature, then the addition of increasingly more energy to a system would no longer increase the temperature but rather create more massive, highly degenerate resonances. The consequence of this idea was an exponentially increasing mass spectrum

$$N(M) = \sum_i d_i \Theta_i(M - M_i) \tag{1}$$

summed over the degeneracy, d_i, of the known hadrons. In 2004 [2] and 2015 [3] the experimentally measured hadrons from the Particle Data Group [4] confirmed the continually exponentially increasing mass spectrum, as Hagedorn originally suggested.

Meanwhile, high energy heavy-ion collisions at RHIC and the LHC probed temperatures surpassing Hagedorn's original limiting temperature, producing a deconfined state of matter known as the Quark Gluon Plasma. We now understand, thanks to first principle Lattice QCD calculations, that there is a cross-over phase transition [5], not a limiting temperature. In this framework we can understand the Hagedorn temperature as roughly equivalent to the critical temperature and then expect the effect of an exponentially increasing mass spectrum to appear close to the phase transition. Indeed, including missing resonances close to the phase transition can affect dynamical chemical equilibrium [6, 7, 8, 9], decrease the shear viscosity over entropy ratio [10, 11, 12, 13], affect the elliptical flow [14, 15], and improve thermal fits [16].

2. Missing States

Recent comparisons to Lattice Quantum Chromodynamic calculations [17] suggested that there may be missing strange hadrons as calculated from Quark Model states [18, 19] due to a mismatch in the strange chemical potential to baryon chemical potential in Lattice QCD vs. the Hadron Resonance gas model from the known PDG spectrum [4]. Further more, there were suggestions [17, 20, 21] that missing resonances could account for the p/π vs. strange hadron tension at LHC when it comes to the thermal fits [22]. However, in [17] the decays of the Quark Model states were not considered, which are necessary for thermal fits.

Fig. 1 shows the exponentially increasing mass spectra including Quark Model states, implying that these missing resonances are consistent with Hagedorn's original postulate. Using the known branching ratios from [4], we extrapolated up the branching ratios of the Quark Model states taking all quantum numbers into account. We analyzed the net-proton and net-charge fluctuations (χ_1/χ_2) as in [23], to extract the corresponding T and μ_B across energies in the Beam Energy Scan. The results in Fig. 2 show that the addition of the Quark Model states slightly decreases the freeze-out temperature but overall only has a small affect (similar to [16]).

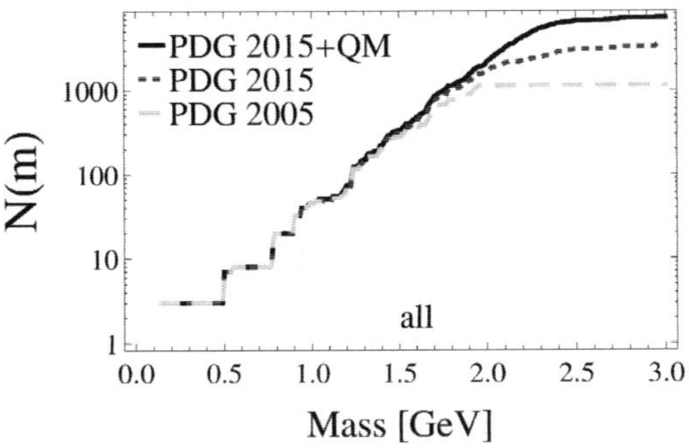

Figure 1. Mass spectrum, Eq. (1), of the strange mesons for the PDG05, PDG15, and PDG15+Quark Model states.

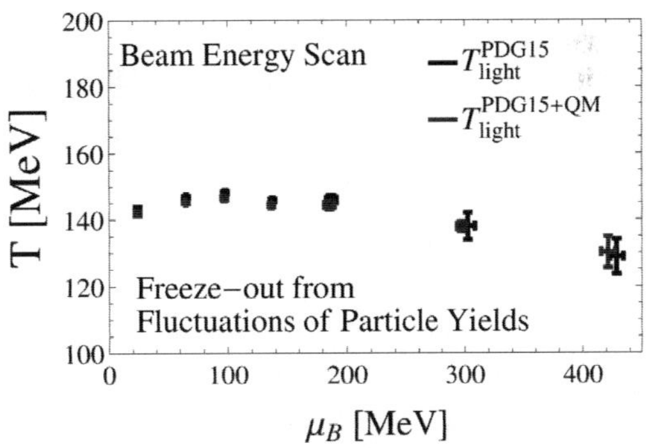

Figure 2. Freeze-out lines extracted from χ_1/χ_2 of the PDG vs. PDG+Quark Model states using net-p and net-Q fluctuations.

3. Strangeness chemical freeze-out from Lattice QCD

An alternative picture was also suggested to resolve the tension between the light and strange hadrons at LHC. In Lattice QCD, the inflection point of susceptibilities can provide clues of about the temperature of hadronization. Using the light susceptibility one finds an inflection point around $T \sim 150$ MeV whereas the inflection point for the strangeness susceptibility is around $T \sim 165$ MeV [24]. Considering there is ~ 15 MeV difference between light and strange hadrons, a logical consequence of this may be that strange hadrons reach chemical equilibrium at higher temperatures than light hadrons. If this is true, it would be consistent with the tension between the light and strange hadrons because it would increase the population of strange baryons, which are typically under-predicted using lower temperatures.

This idea is consistent with many dynamical models. In UrQMD there is no specified chemical freeze-out temperature such that each particle species reaches chemical equilibrium on a different time scale [25]. Similarly, using multi-body hadronic interactions via rate equations, one can also reach chemical equilibrium on different time scales depending on the species. However, if one can provide directly from first-principle Lattice calculations that different chemical equilibration temperatures are needed then it gives significantly more weight to these dynamical models.

Furthermore, it will then require the hadronization schemes to be updated uniformly to include different temperatures for light and strange hadrons.

In order to extract the chemical freeze-out temperature from the lattice we use ratios of susceptibilities as discussed in [23, 26, 27]. In the experiment, the only strange multiplicity fluctuations (and their corresponding moments of these distributions) that can be currently measured at the Beam Energy Scan are charged kaons [28]. However, on the lattice all particles exist as well as their corresponding interactions. In order to extract only the charged kaon contribution, we implement partial pressures for charged strange mesons as in [29] $\frac{\chi_2^K}{\chi_1^K} = \frac{\cosh(\hat{\mu}_S+\hat{\mu}_Q)}{\sinh(\hat{\mu}_S+\hat{\mu}_Q)}$ where $\hat{\mu}_S$ and $\hat{\mu}_Q$ are supplied from the Wuppertal Budapest collaboration.

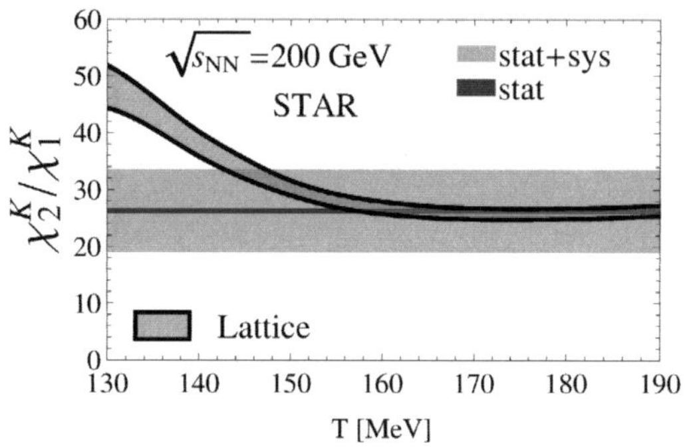

Figure 3. Net-K from Lattice QCD (Wuppertal-Budapest) vs. preliminary STAR data

In Fig 3 a comparison between the Lattice QCD kaon partial pressure is shown compared to the preliminary STAR data. Due to the large error bars, it is not yet possible to extract the strange chemical freeze-out temperature. Eventual smaller experimental error bars would give a decisive answer to the tension between light and strange hadrons.

4. Conclusions

We introduce two approaches to study possible differences in light vs. strange chemical equilibration temperatures. The first is to include states predicted from the Quark Model while modeling their decay channels to study the tension between light and strange hadrons. While the addition of extra resonances has only a small affect on the light freeze-out temperature, the affect on the strange freeze-out remains to be seen since the kaon fluctuation error bars are too large to extract a meaningful strangeness freeze-out temperature. The second approach is a new method to extract the charged, strange susceptibilities from Lattice QCD to determine the strange freeze-out temperature from first principles. Once the kaon error bars are decreased, we may be able to settle the tension between light and strange hadrons.

5. Acknowledgements

This work is supported by the National Science Foundation under grant no. PHY-1513864 and by the U.S. Department of Energy, Office of Science, Office of Nuclear Physics, within the framework of the Beam Energy Scan Theory (BEST) Topical Collaboration. This work contains lattice QCD data provided by the Wuppertal-Budapest Collaboration. An award of computer time was provided by the INCITE program. This research used resources of the Argonne Leadership Computing Facility, which is a DOE Office of Science User Facility supported under Contract DE-AC02-06CH11357. The work of J. G. and A. P. was supported by the DFG grant SFB/TR55. The authors gratefully acknowledge the Gauss Centre for Supercomputing

(GCS) for providing computing time for a GCS Large-Scale Project on the GCS share of the supercomputer JUQUEEN [30] at Jülich Supercomputing Centre (JSC).

References

[1] Hagedorn R 1965 *Nuovo Cim. Suppl.* **3** 147–186
[2] Broniowski W, Florkowski W and Glozman L Ya 2004 *Phys. Rev.* **D70** 117503 (*Preprint* hep-ph/0407290)
[3] Lo P M, Marczenko M, Redlich K and Sasaki C 2015 *Phys. Rev.* **C92** 055206 (*Preprint* 1507.06398)
[4] Olive K A *et al.* (Particle Data Group) 2014 *Chin. Phys.* **C38** 090001
[5] Aoki Y, Endrodi G, Fodor Z, Katz S D and Szabo K K 2006 *Nature* **443** 675–678 (*Preprint* hep-lat/0611014)
[6] Noronha-Hostler J, Greiner C and Shovkovy I A 2008 *Phys. Rev. Lett.* **100** 252301 (*Preprint* 0711.0930)
[7] Noronha-Hostler J, Beitel M, Greiner C and Shovkovy I 2010 *Phys. Rev.* **C81** 054909 (*Preprint* 0909.2908)
[8] Beitel M, Gallmeister K and Greiner C 2014 *Phys. Rev.* **C90** 045203 (*Preprint* 1402.1458)
[9] Beitel M, Greiner C and Stoecker H 2016 *Phys. Rev.* **C94** 021902 (*Preprint* 1601.02474)
[10] Noronha-Hostler J, Noronha J and Greiner C 2009 *Phys. Rev. Lett.* **103** 172302 (*Preprint* 0811.1571)
[11] Noronha-Hostler J, Noronha J and Greiner C 2012 *Phys. Rev.* **C86** 024913 (*Preprint* 1206.5138)
[12] Pal S 2010 *Phys. Lett.* **B684** 211–215 (*Preprint* 1001.1585)
[13] Kadam G P and Mishra H 2014 *Nucl. Phys.* **A934** 133–147 (*Preprint* 1408.6329)
[14] Noronha-Hostler J, Noronha J, Denicol G S, Andrade R P G, Grassi F and Greiner C 2014 *Phys. Rev.* **C89** 054904 (*Preprint* 1302.7038)
[15] Paquet J F, Shen C, Denicol G S, Luzum M, Schenke B, Jeon S and Gale C 2016 *Phys. Rev.* **C93** 044906 (*Preprint* 1509.06738)
[16] Noronha-Hostler J, Ahmad H, Noronha J and Greiner C 2010 *Phys. Rev.* **C82** 024913 (*Preprint* 0906.3960)
[17] Bazavov A *et al.* 2014 *Phys. Rev. Lett.* **113** 072001 (*Preprint* 1404.6511)
[18] Capstick S and Isgur N 1986 *Phys. Rev.* **D34** 2809 [AIP Conf. Proc.132,267(1985)]
[19] Ebert D, Faustov R N and Galkin V O 2009 *Phys. Rev.* **D79** 114029 (*Preprint* 0903.5183)
[20] Noronha-Hostler J and Greiner C 2014 (*Preprint* 1405.7298)
[21] Noronha-Hostler J and Greiner C 2014 *Nucl. Phys.* **A931** 1108–1113 (*Preprint* 1408.0761)
[22] Abelev B *et al.* (ALICE) 2012 *Phys. Rev. Lett.* **109** 252301 (*Preprint* 1208.1974)
[23] Alba P, Alberico W, Bellwied R, Bluhm M, Mantovani Sarti V, Nahrgang M and Ratti C 2014 *Phys. Lett.* **B738** 305–310 (*Preprint* 1403.4903)
[24] Bellwied R, Borsanyi S, Fodor Z, Katz S D and Ratti C 2013 *Phys. Rev. Lett.* **111** 202302 (*Preprint* 1305.6297)
[25] Bass S A, Dumitru A, Bleicher M, Bravina L, Zabrodin E, Stoecker H and Greiner W 1999 *Phys. Rev.* **C60** 021902 (*Preprint* nucl-th/9902062)
[26] Karsch F 2012 *Central Eur. J. Phys.* **10** 1234–1237 (*Preprint* 1202.4173)
[27] Borsanyi S, Fodor Z, Katz S D, Krieg S, Ratti C and Szabo K K 2014 *Phys. Rev. Lett.* **113** 052301 (*Preprint* 1403.4576)
[28] for [STAR] X J
[29] Noronha-Hostler J, Bellwied R, Gunther J, Parotto P, Pasztor A, Vazquez I P and Ratti C 2016 (*Preprint* 1607.02527)
[30] 2015 Juqueen: Ibm blue gene/q supercomputer system at the jülich supercomputing centre Tech. Rep. 1 A1 Jülich Supercomputing Centre http://dx.doi.org/10.17815/jlsrf-1-18

Production of π^0, $K^{\pm}$ and η mesons in Pb-Pb and pp collisions at $\sqrt{s_{\rm NN}}$ =2.76 TeV measured with the ALICE detector at the LHC

Astrid Morreale on behalf of the ALICE collaboration
Laboratoire de Physique Subatomique et des Technologies Associées SUBATECH.
4 Rue Alfred Kastler, 44300 Nantes, France

E-mail: astrid.morreale@cern.ch

Abstract. One of the key signatures of the Quark-Gluon Plasma (QGP) is the modification of hadron transverse momentum differential cross-sections in heavy-ion collisions (HIC) as compared to proton-proton (pp) collisions. Suppression of hadron production at high transverse momenta ($p_{\rm T}$) in HIC has been explained by the energy loss of the partons produced in the hard scattering processes which traverse the deconfined quantum chromodynamic (QCD) matter. The dependence of the observed suppression on the $p_{\rm T}$ of the measured hadron towards higher $p_{\rm T}$ is an important input for the theoretical understanding of jet quenching effects in the QGP and the nature of the energy loss, while suppression towards lower $p_{\rm T}$ gives information about collective behaviour. The ALICE experiment at the Large Hadron Collider (LHC) performs a variety of measurements including the spectra of neutral mesons and kaons at mid-rapidity in a wide $p_{\rm T}$ range in pp, p-Pb and Pb-Pb collisions. An overview of ALICE results in HIC and pp collisions at $\sqrt{s_{NN}}$ =2.76 TeV of neutral pions, kaons and eta mesons as a function of transverse momentum ($p_{\rm T}$) and centrality will be presented. Ratios η/π^0, $K^{\pm}/\pi^{\pm}$ as well as comparisons to model calculations will also be discussed.

1. Introduction

Light mesons offer an opportunity to study the QGP and quantify its properties. At low $p_{\rm T}$, mesons give insight about bulk properties and collective effects. Particles with different masses are expected to be affected differently by the collective motion of the medium; one would thus expect a mass ordering of the nuclear modification factor $R_{\rm AA}$ at low $p_{\rm T}$ due to radial flow. At high $p_{\rm T}$, hadrons result from the hadronization of partons created in initial hard scattering. The suppression of hadron production at high $p_{\rm T}$ in heavy-ion collisions is interpreted as energy loss of the scattered parton in the QGP. Gluons are expected to suffer a larger energy loss in the medium than quarks (Casimir factor). This effect may give rise to differences in suppression patterns between π^0 and η which can be quantified with $R_{\rm AA}$ measurements. While one would expect a larger energy loss of high $p_{\rm T}$ partons at higher $\sqrt{s}$, a smaller suppression may be observed at high $p_{\rm T}$ due to a less steeply falling spectrum for a given magnitude of partonic energy loss. A smaller suppression at higher $p_{\rm T}$ may additionally indicate a larger fractional quark to gluon partonic contribution to the production of mesons.

2. ALICE's neutral meson reconstruction and charged particle detection

The ALICE detector [1] measures photons with three fully complementary methods: photon conversion method (PCM) and electromagnetic shower calorimetry using two detectors: PHOS [2] which covers a pseudorapidity range of 0.26 units and a coverage in azimuth of 60° and the EMCal [3] with a pseudorapidity and azimuth coverage of 1.4 units and 100° respectively. The electromagnetic calorimeters are based on energy measurements via total absorption of particles.

The PCM method measures photons and meson yields by reconstructing e^+e^- pairs proceeding from photon conversions in the material of ALICE inner detectors. PCM measurements use the ALICE Inner Tracking System (ITS [4]) and the Time Projection Chamber (TPC [5]), measurements using this method have full azimuthal acceptance and 1.8 units of pseudo rapidity coverage.

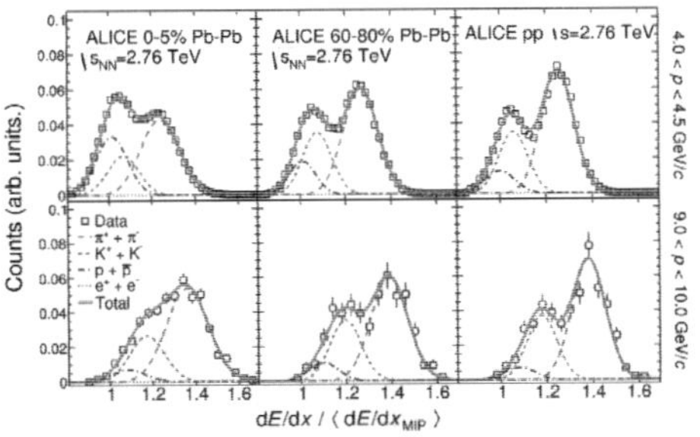

Figure 1: $\mathrm{d}E/\mathrm{d}x$ distributions of charged particle signals in Pb-Pb (two leftmost columns) and pp (right column). Signals are fitted with the sum of four Gaussian functions (solid line). Top row corresponds to particles with $4.0 < p < 4.5$ GeV/c while bottom row corresponds to $9.0 < p < 10$ GeV/c.

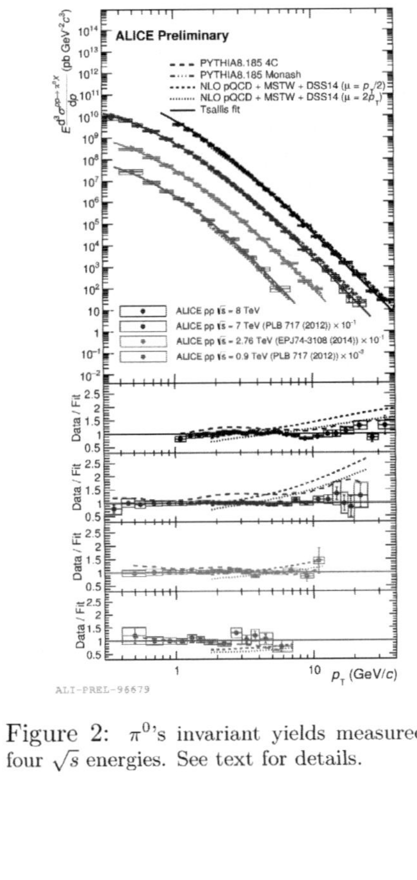

Figure 2: π^0's invariant yields measured at four $\sqrt{s}$ energies. See text for details.

Charged particle identification is done via the specific energy loss ($\mathrm{d}E/\mathrm{d}x$) of particles traversing the TPC gas. The candidate signals are fitted with the sum of four Gaussian functions (Fig. 1) where the remaining contribution from electrons remains below 1%. Particle identification is further improved at $p_T < 4$ GeV/c ($K^\pm$) with the use of the Time of Flight (TOF) and the High Momentum Particle Identification Detector (HMPID). Charged particles are detected up to $p_T = 20$ GeV/c and at mid-rapidity $|\eta| \lesssim 0.8$.

3. π^0 invariant yields in pp collisions at $\sqrt{s} = 0.9$, 2.76, 7 and 8 TeV

The ALICE experiment to this date has measured the invariant yields of π^0 at four $\sqrt{s}$. The measurements indicate a power law dependence at high p_T with a measured power value of $n = 6.0 \pm 0.1$ at $\sqrt{s} = 2.76$ TeV, a value lower than what has been observed with measurements performed at lower $\sqrt{s}$ energies [6]. The measurements are also compared to next-to-leading order pQCD (NLO pQCD) which use MSTW PDFs, and DSS14 fragmentation functions (FF) [7]. These comparisons as well as the ratio are illustrated in Fig. 2. The predictions with MSTW PDF and DSS14 FF describe the magnitude better than previous pQCD calculations [8], however an increasing discrepancy between pQCD and the measurements is observed with increasing $\sqrt{s}$ and p_T.

4. Invariant yields in Pb-Pb collisions compared to pp

Measurements of π^0, η, $\pi^\pm$ and $K^\pm$ mesons in HIC are presented in Fig. 3. The new π^0 results in Fig. 3 are from a data set offering ten times the integrated luminosity with respect to previous ALICE results[9]. This increase of statistics allowed for an improved measurement that probes the p_T region above 12 GeV/c and up to 20 GeV/c. As for η mesons (Fig. 3), this the first time this particle is measured at the LHC.

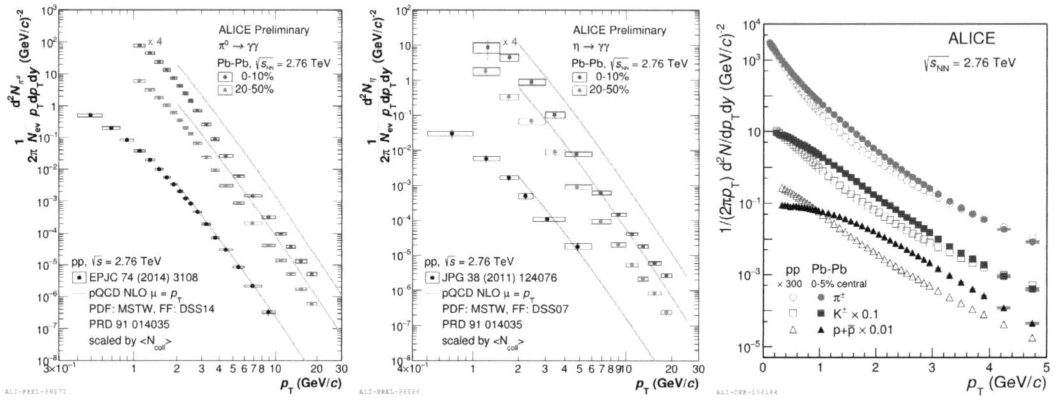

Figure 3: Invariant yields in Pb-Pb collisions of π^0 (left), η (middle) and $K^{\pm}$, $\pi^{\pm}$ (right)

The π^0 and η results are compared in two centrality classes to NLO pQCD predictions of the corresponding production in pp collisions [7] scaled by N_{coll}. These comparisons show a suppression in the Pb-Pb data with respect to scaled NLO pQCD. $K^{\pm}$ and $\pi^{\pm}$ in the 0-5% centrality class (Fig. 3) are compared to corresponding pp measurements (open markers), where a zoom to the p_T region below 5 GeV/c indicate differences in the spectra with respect to pp that is attributed to radial flow [10].

5. Nuclear modification factor R_{AA}

The nuclear modification factor R_{AA} is defined as: $R_{AA}(p_T) = \frac{1}{N_{coll}} \frac{dN_{AA}/dp_T}{dN_{pp}/dp_T}$ is measured to quantify nuclear effects in A-A collisions, where production in A-A is compared to production in scaled pp collisions. N_{coll} is the number of binary nucleon-nucleon collisions and is taken from Glauber Monte Carlo simulations [11]. Fig. 4 shows the measured R_{AA} for kaons and pions as well as other particles, where a large suppression in central Pb-Pb collisions is observed, 0-5%: R_{AA} ~0.1 for $p_T > 6$ GeV/c, and a moderate suppression at peripheral collisions, 60-80%: $R_{AA} \sim 0.6$ for $p_T > 6$ GeV/c. At high p_T (>8 GeV/c), all particle species exhibit a similar suppression trend which decreases towards large p_T. At low p_T the magnitude of the R_{AA} increases with the increasing particle's mass.

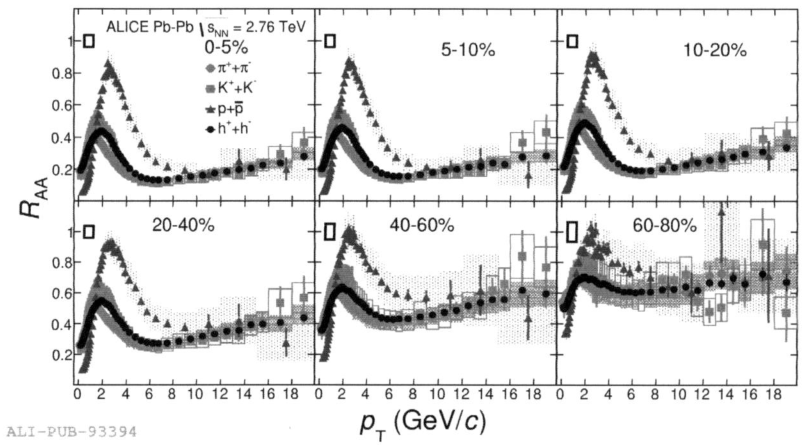

Figure 4: R_{AA} in six centrality classes. See text for details.

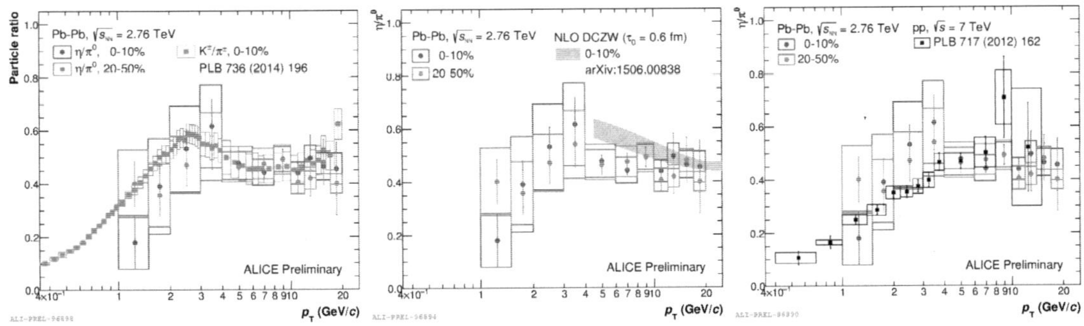

Figure 5: η/π^0 in two centrality classes compared to $K^{\pm}/\pi^{\pm}$(left), NLO pQCD (middle) and η/π^0 measured in pp collisions (right).

6. Particle ratios in Pb-Pb collisions

Fig. 5 shows the η/π^0 in two centrality classes compared to: $K^{\pm}/\pi^{\pm}$ (Fig. 5 left) at the same $\sqrt{s_{\mathrm{NN}}}$ and the η/π^0 (Fig. 5 right) measurement in pp collisions at $\sqrt{s}$ = 7 TeV. The trend and magnitude of the η/π^0 ratio is consistent with the $K^{\pm}/\pi^{\pm}$ ratio [10] as well as the corresponding ratio in pp collisions at $\sqrt{s}$ = 7 TeV. Further comparisons to jet quenching NLO pQCD predictions (Fig. 5 middle) [12] describe the ratio within the current uncertainties.

7. Summary

ALICE measures neutral and charged mesons in a wide p_{T} range thanks to complementary detectors. π^0, η and $K^{\pm}$ invariant yields have been measured both in pp as well as Pb-Pb collisions by ALICE and compared to NLO pQCD. π^0's magnitude is well described by NLO pQCD calculations, however, there is a growing discrepancy as a function of p_{T} and with increasing $\sqrt{s}$ (pp results).

A suppression of π^0 and η is observed via comparisons of the Pb-Pb invariant yields to scaled pp NLO pQCD. $K^{\pm}$ and $\pi^{\pm}$ suppression is observed via R_{AA} which is p_{T} and particle dependent ($p_T < 6$ GeV/c). At high p_{T} ($>$ 8GeV/c) the value and trend of the R_{AA} is particle independent pointing that the suppression may be a partonic effect alone. At low p_T a mass hierarchy is observed on the magnitude of the R_{AA} attributed to radial flow. Particle ratios measured in Pb-Pb collisions were presented and compared to jet quenching predictions as well as corresponding ratios measured in pp. $K^{\pm}/\pi^{\pm}$ reaches a constant value for $p_{\mathrm{T}} > 4$ GeV/c. For p_{T} below 4 GeV/c an enhancement is observed. η/π^0 reaches a constant value for $p_{\mathrm{T}} > 4$ GeV/c. For p_{T} below 4 GeV/c it is consistent with the pp and $K^{\pm}/\pi^{\pm}$ ratio. At high p_{T} no significant differences are seen in the particle ratios between Pb-Pb and pp.

References

[1] Aamodt K *et al.* (ALICE Collaboration) 2008 *J. Instrum.* **3** S08002
[2] *CERN/LHCC 99-4*
[3] Abelev B B *et al.* (ALICE) 2014 *Int. J. Mod. Phys.* **A29** 1430044
[4] Aamodt K *et al.* (ALICE Collaboration) *J. Instrum.*
[5] Alme J *et al.* (ALICE Collaboration) 2010 *Nucl. Instrum. Meth.* **A622** 316–367
[6] Adare A *et al.* (PHENIX) *Phys.Rev.* **C88** 024906
[7] de Florian D, Sassot R, Epele M, Hernandez-Pinto R J and Stratmann M 2015 *Phys. Rev.* **D91** 014035
[8] Abelev B *et al.* (ALICE) 2012 *Phys. Lett.* **B717** 162–172
[9] Abelev B B *et al.* (ALICE) 2014 *Eur. Phys. J.* **C74** 3108
[10] Abelev B B *et al.* (ALICE) 2014 *Phys. Lett.* **B736** 196–207
[11] Abelev B *et al.* (ALICE) 2013 *Phys. Lett.* **B720** 52–62
[12] Dai W, Chen X F, Zhang B W and Wang E 2015 *Phys. Lett.* **B750** 390–395

Measurement of dielectrons in pp, p–Pb and Pb–Pb collisions with ALICE at the LHC

Alberto Calivà on behalf of the ALICE Collaboration
NIKHEF Instite for subatomic physics and Utrecht University, Utrecht, NL
E-mail: alberto.caliva@cern.ch

Abstract.
Dielectrons are penetrating probes which carry unaffected information on the hot and dense medium created in ultra-relativistic heavy-ion collisions. The low-mass region of the dielectron spectrum ($m_{\pi^0} < m_{ee} < m_\rho$) is sensitive to virtual photon production and in-medium modifications of low-mass vector meson spectral functions. These modifications are connected to the restoration of chiral symmetry, which is expected at high temperature. Dielectrons are also sensitive to heavy-flavor production.

In these proceedings, we present the preliminary results of dielectron measurements in pp collisions at $\sqrt{s} = 7$ TeV, in p–Pb collisions at $\sqrt{s_{\rm NN}} = 5.02$ TeV and in central (0-10%) Pb–Pb collisions at $\sqrt{s_{\rm NN}} = 2.76$ TeV with the ALICE experiment at the LHC. The expected scenario and perspectives for the dielectron measurement after the ALICE upgrade are also discussed.

1. Introduction

Photons and dileptons are emitted continuously during the expansion of the hot and dense medium created in ultra-relativistic heavy-ion collisions [1]. The temperature of the system can be extracted from the direct photon $p_{\rm T}$-spectrum and from the dielectron invariant-mass distribution, with the advantage that in the latter the measurement is not affected by the Doppler shift caused by the rapid expansion of the emitting source.

Different regions of the dielectron spectrum are particularly sensitive to different stages of the collision. The low-mass region ($m_{ee} < m_\rho$) is interesting for studying the production of virtual direct photons and the modifications of the ρ-meson spectral properties in the hot hadron gas phase. These are connected to the chiral symmetry restoration, which is expected at high temperatures [2]. Dielectrons are also sensitive to heavy-flavor production.

2. Data analysis strategy

The ALICE experimental apparatus is described in detail in [3]. The main detectors used in the data analysis are the Inner Tracking System (ITS) (for vertexing, tracking and particle identification), the Time Projection Chamber (TPC) (for tracking and particle identification) and the Time-Of-Flight (TOF) detector (for particle identification). In Pb–Pb collisions the centrality selection has been done using the measured multiplicity in the V0 detector.

Track selection criteria have been optimized to ensure a good overall quality of the tracks and to reject secondary tracks, particularly those from photon conversion in the material. Electron identification has been based on the measurement of the average energy loss per unit path length in the TPC and ITS and on the time-of-flight measurement in the TOF detector.

The background has been estimated using the like-sign spectrum, corrected for acceptance effects using the event-mixing. The efficiency calculation has been based on MC simulations including a realistic description of the detector effects.

3. Results in pp and p–Pb collisions

The dielectron invariant-mass spectra measured in pp collisions at $\sqrt{s} = 7$ TeV and in p–Pb collisions at $\sqrt{s_{\rm NN}} = 5.02$ TeV have been compared to the expected contributions from hadronic sources, called *hadronic cocktail.*

The cocktail has been generated using the parametrizations of the $p_{\rm T}$-differential cross sections of π^0, η, ϕ and J/ψ measured by ALICE [4, 5, 6] in pp collisions, and those of charged pions for p–Pb collisions [7]. The remaining contributions from light-flavored mesons have been obtained using $m_{\rm T}$ scaling. The charm contribution in pp collisions has been obtained from PYTHIA 6, scaled to $\sigma_{c\bar{c}}$ measured by ALICE at $\sqrt{s} = 7$ TeV [8]. In p–Pb collisions, the charm and beauty contributions have been obtained using the interpolated values of $\sigma_{c\bar{c}}$ and $\sigma_{b\bar{b}}$ at $\sqrt{s_{\rm NN}} = 5.02$ TeV and scaling the spectra by the average number of binary collisions $\langle N_{\rm coll}\rangle$.

In both collision systems no enhanced dielectron yield is expected in the low-mass region. The measured invariant-mass spectra are in agreement with the cocktail within the statistical and systematic uncertainties (Fig. 1).

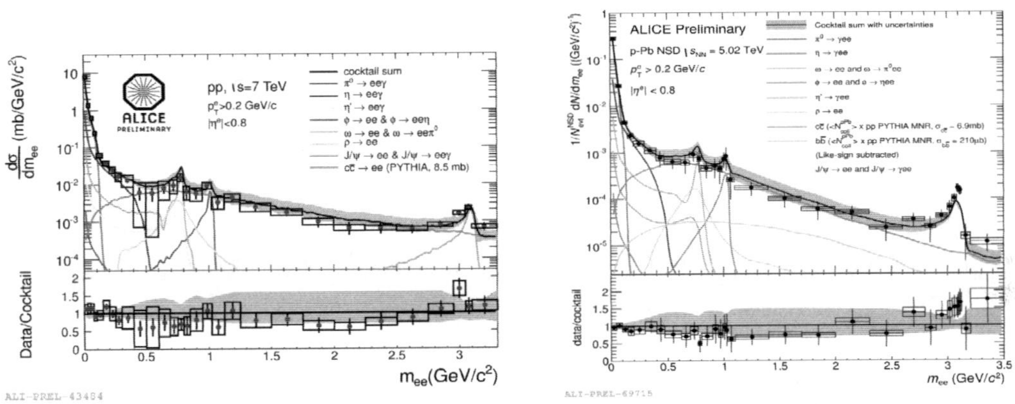

Figure 1. Dielectron invariant mass spectrum measured in pp collisions at $\sqrt{s} = 7$ TeV (left) and in p–Pb collisions at $\sqrt{s_{\rm NN}} = 5.02$ TeV (right) compared to the hadronic cocktail.

The virtual photon contribution has been extracted in pp collisions from the two-component fit in the kinematic range $p_{\rm T}^{\rm ee} \gg m_{\rm ee}$: $f(m_{\rm ee}) = r\cdot f_{dir}(m_{\rm ee}) + (1\text{-}r)\cdot f_c(m_{\rm ee})$, where f_{dir} is the mass distribution of virtual photons, given by the Kroll-Wada formula [10], and f_c is the cocktail shape. The fraction r of virtual direct photons extracted from the fit is consistent with the measurement of real direct photons using the photon conversion method (PCM) (Fig. 2) [11]. The virtual direct photon invariant cross section measured in pp collisions is consistent with NLO pQCD calculations [13].

4. Results in Pb–Pb collisions

The first measurement of the dielectron invariant-mass spectrum in central Pb–Pb collisions at $\sqrt{s_{\rm NN}} = 2.76$ TeV is shown in Fig. 3 in comparison with the hadronic cocktail.

The $p_{\rm T}$-differential cross section of π^0 [12] has been used as input parametrization, while $m_{\rm T}$ scaling was applied to obtain the remaining contributions of light-flavored mesons. The

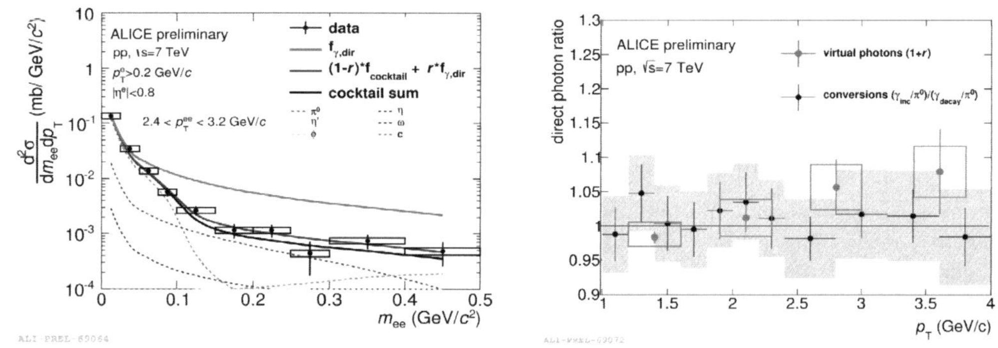

Figure 2. (Left) Example of the two-component fit to the dielectron spectrum in the low-mass region. (Right) Real direct photon ratio ($\gamma_{incl}/\gamma_{decay}$) measured in pp collisions at $\sqrt{s} = 7$ TeV with the Photon Conversion Method (black) [11] compared to virtual direct photon ratio (1+r) (red) from the dielectron measurement. The real photon measurement is consistent with zero direct photon yield within uncertainties, while the virtual photon measurement indicates a non-zero yield for $p_{\rm T} > 2.5$ GeV/c.

contributions from charm and beauty decays have been obtained from PYTHIA 6 for pp collisions at $\sqrt{s_{\rm NN}} = 2.76$ TeV (using $\sigma_{c\bar{c}} = 4.8$ mb [8] and $\sigma_{b\bar{b}} = 130$ μb [9]), and scaled by the average number of binary collisions. Energy losses of charm quarks in the medium are ignored.

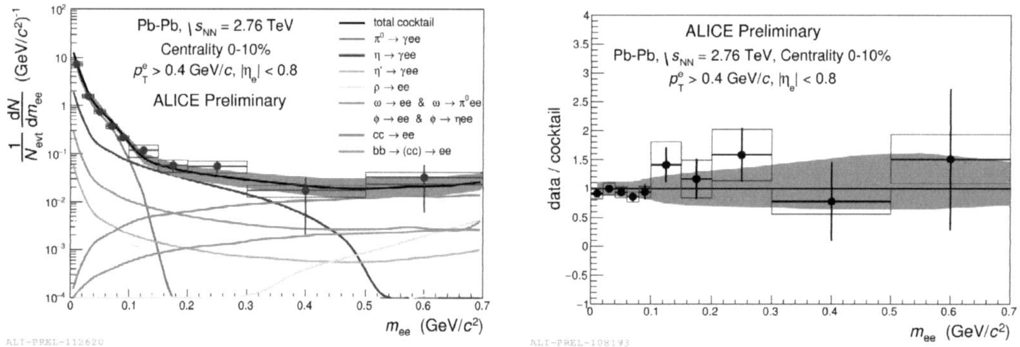

Figure 3. Dielectron invariant-mass spectrum measured in Pb–Pb collisions at $\sqrt{s_{\rm NN}} = 2.76$ TeV compared to the hadronic cocktail (left) and data-to-cocktail ratio (right).

While an enhanced dielectron production is expected in Pb–Pb collisions, originating from the QGP thermal radiation and the in-medium modification of the ρ-meson, the data-to-cocktail ratio is consistent with unity within the uncertainties.

An upper limit on the virtual photon production has been estimated at 90% C.L. (Fig. 4), from the analysis of the mass region $100 < m_{\rm ee} < 300$ MeV/c^2, which is consistent with the real photon measurement by ALICE [14] and previous dielectron measurements from PHENIX and STAR at $\sqrt{s_{\rm NN}} = 200$ GeV [15, 16].

5. Perspectives after the ALICE upgrade

The perspectives for the dielectron measurement after the ALICE upgrade for Run 3 at the LHC have been studied. The new ITS [17], with improved resolution in the determination of secondary vertices and reduced material budget, will allow a better suppression of the main background

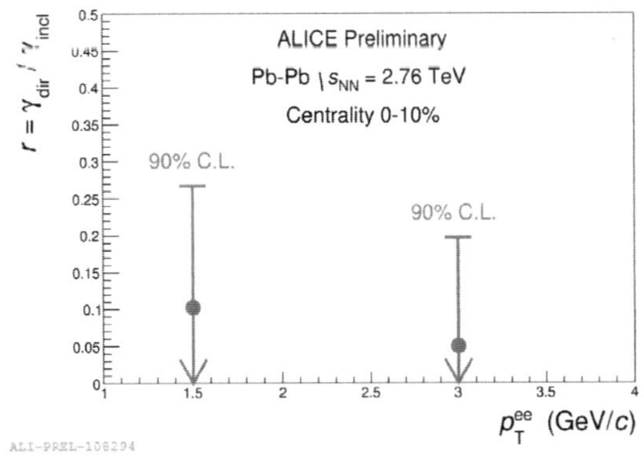

Figure 4. Upper limit at 90 % C.L. on virtual photon production in Pb–Pb collisions at $\sqrt{s_{\rm NN}}$ = 2.76 TeV.

contributions (Dalitz and charm decays, photon conversions), and the continuous TPC readout will increase the event rate by a factor $\sim$ 100 [18]. This will allow a precise measurement of the QGP temperature, from a fit to the intermediate mass region, and a detailed study of the in-medium modifications of the ρ-meson [19].

6. Conclusions

Dielectron invariant mass spectra have been measured in the three different collision systems at different energies. In pp collisions the measured contribution of virtual direct photons is consistent with the PCM measurement of real photons, and the direct photon invariant cross section is consistent with NLO pQCD calculations. In p–Pb collisions the dielectron spectrum is in agreement with the hadronic cocktail within the uncertainties. In Pb–Pb collisions, the extracted upper limit at 90% C.L. on virtual photon production is consistent with the real photon measurement from ALICE and previous dielectron measurements from PHENIX and STAR at RHIC. Regarding the expected scenario after the upgrade, the simulations indicate an improved significance of the temperature measurement and an enhanced sensitivity to in-medium effects of the ρ-meson with upgraded detectors and enhanced statistics in Run 3 at the LHC.

References

[1] J-e. Alam, S. Raha, and B. Sinha, Physics Reports, 273, pp. 243–362 (1996).
[2] R. Rapp and J. Wambach, Advances in Nuclear Physics, 25, pp. 1–205 (2000).
[3] K. Aamodt et al. (The ALICE Collaboration) JINST, 3, S08002 (2008).
[4] K. Koch (The ALICE Collaboration), Nucl. Phys. A, 855, pp. 281–284 (2011).
[5] B. Abelev et al. (The ALICE Collaboration), Eur. Phys. J. C, 72, p. 2183 (2012).
[6] K. Aamodt et al. (The ALICE Collaboration), Phys. Lett. B, 704, pp. 442–455 (2011).
[7] B. Abelev et al. (The ALICE Collaboration), Phys. Rev. Lett., 110, 082302 (2013).
[8] B. Abelev et al. (The ALICE Collaboration), J. High Energ. Phys. (2012).
[9] B. Abelev et al. (The ALICE Collaboration), Phys. Lett. B, 738, p. 97 (2014).
[10] N. Kroll, W. Wada, Phys. Rev., 98, p. 1355 (1955).
[11] J. Adam et al. (The ALICE Collaboration), Phys. Lett. B, 754, pp. 235–248 (2016).
[12] B. Abelev et al. (The ALICE Collaboration), Eur. Phys. J. C, 74, 10, p. 3108 (2014).
[13] Markus K. Kohler (The ALICE Collaboration), Nuclear Physics A 00 14 (2014).
[14] J. Adam et al. (The ALICE Collaboration), Phys. Lett. B, 754, pp. 235–248 (2016).
[15] A. Adare et al. (The PHENIX Collaboration), Phys. Rev. C, 93, 014904 (2016).
[16] L. Adamczyk et al. (The STAR Collaboration), Phys. Rev. Lett., 113, 022301 (2014).
[17] CERN-LHCC-2013-024, ALICE-TDR-017.
[18] CERN-LHCC-2013-020, ALICE-TDR-016.
[19] B. Abelev et al. (The ALICE Collaboration), J. Phys. G, 41, 087002 (2014).

Measurement of bottom contribution to the non-photonic electron production in p+p collisions at STAR

Wei Li (for the STAR Collaboration)

Shanghai Institute of Applied Physics, Chinese Academy of Sciences, Shanghai 201800, China

E-mail: liw2002@sinap.ac.cn

Abstract.
We present the STAR preliminary results of the azimuthal correlations between non-photonic electrons and charged hadrons at mid-rapidity in p+p collisions at $\sqrt{s}$ = 200 GeV and 500 GeV. The correlation distributions are fitted with PYTHIA templates to extract the relative contribution from bottom decays to non-photonic electrons. This could provide a precise p+p reference for the study of bottom production in heavy-ion collisions at RHIC.

1. Introduction

Models based on radiative energy loss mechanism can describe the large suppression of light hadrons in central Au+Au collisions at RHIC quite well [1, 2]. However the large suppression of non-photonic electron (NPE) yields from semi-leptonic decays of B and D mesons challenges such models which predicted that the energy loss of heavy quarks should be much smaller compared to that of light quarks due to their larger masses ("dead-cone effect") [3, 4]. Various models have been proposed to explain the large suppression of NPE yields and such calculations are crucially dependent on the relative contributions from B/D meson decays to NPE [5, 6]. Thus separating the B/D meson contributions to NPE yield experimentally is important for constraining models, and thus understanding the energy loss mechanism of heavy flavor quarks in the medium. Separation of B/D meson contributions to NPE in p+p collisions serves as a reference for the study in heavy-ion collisions. By comparing results from 200 GeV and 500 GeV p+p collisions, we can check the collision energy dependence of bottom contribution to NPE. In this study, we utilize an indirect method to disentangle B and D meson contibutions to NPE, which relies on the fact that electrons decayed from B and D mesons have different azimuthal correlations with charged hadrons on the near side [7, 8].

2. Analysis

The data used in this analysis were recorded in p+p collisions at $\sqrt{s} = 200$ GeV in 2012 and $\sqrt{s}$ = 500 GeV in 2011 by the STAR experiment. The main detectors used in this analysis are the Time Projection Chamber (TPC) [9], which is used for momentum determination and electron identification; Barrel Electromagnetic Calorimeter (BEMC) for electron identification as well as electron triggering at high transverse momenta (p_T) [10]; Barrel Shower Maximum Detector (BSMD) which is embedded in the BEMC and provides good shower size measurement for electron identification; Time Of Flight (TOF) which is used for low p_T electron identification [11]; and Vertex Position Detector (VPD) which provides the minimum bias trigger [12]. In order to obtain sufficient statistics at high p_T, we used BEMC triggered events, with single-tower energy thresholds of 2.6 and 4.3 GeV in p+p collisions at $\sqrt{s}$ = 200 GeV; and 2.6, 4.3 and 6.0 GeV in p+p collisions at $\sqrt{s}$ = 500 GeV.

We start with the semi-inclusive electron sample (Semi-Inc) to construct the correlation between non-photonic electrons and charged hadrons. The semi-inclusive electrons are obtained by removing the electrons that fall into the invariant mass cut after pairing with an opposite-sign partner (OppSign) from the inclusive electron sample. The invariant mass cut is set to be less than 0.24 GeV/c^2 to remove the background from photon conversions and π^0/η Dalitz decays. The correlation signal between non-photonic electrons and charged hadrons is calculated as following:

$$\Delta\phi_{Non\_Pho} = \Delta\phi_{Semi\_Inc} + \Delta\phi_{SameSign} - \Delta\phi_{Not\_Reco\_Pho} - \Delta\phi_{Hadron} \quad (1)$$

where $\Delta\phi_{SameSign}$ is an estimate of the combinatorial background using electrons that pass the invariant mass cut of 0.24 GeV/c^2 after pairing with a same-sign partner, this is a correction for the over subtraction of the true signal. $\Delta\phi_{Hadron}$ is the hadron-hadron correlation which is used to subtract the hadron contamination, and $\Delta\phi_{Not\_Reco\_Pho}$ is the remaining contribution from the correlation of photonic electrons with charged hadrons that was not captured by the invariant mass selection. It is calculated as:

$$\Delta\phi_{Not\_Reco\_Pho} = (\frac{1}{\varepsilon} - 1) \times \Delta\phi_{Reco\_Pho} \quad (2)$$

where $\Delta\phi_{Reco\_Pho}$ is the correlation between photonic electrons and hadrons, and ε is the efficiency for the photonic electron reconstruction which can be obtained from simulations. The detailed analysis procedure can be found in [7, 8].

3. Result

Panel (a) in Figure 1 shows the azimuthal correlation between NPE and charged hadrons in $p+p$ collisions at $\sqrt{s}$ = 200 GeV for NPE with $5.5 < p_T < 6.5$ GeV/c and $|\eta^{NPE}| < 0.7$. The associated hadrons are required to have $p_T > 0.3$ GeV/c and $|\eta^{asso}| < 1.0$. Clear azimuthal correlation signals can be seen on the near side. We use PYTHIA 8.1 combined with STAR-HF-Tune Version 1.1 to generate azimuthal correlations between electrons from B and D meson decay and charged hadrons in $p+p$ collision at $\sqrt{s}$ = 200 GeV [13]. STAR-HF Tune is a set of parameters that are tuned to describe the NPE and J/ψ measurements at RHIC. Clear correlation peak can also be seen on the near side for D and B meson decayed electrons, and the difference comes from the different decay kinematics. Fitting the experimental data with the NPE-h azimuthal correlations from PYTHIA, we can extract the B meson contribution to non-photonic electrons. The fit function is:

$$\Delta\phi_{exp} = (R \times \Delta\phi_B + (1-R) \times \Delta\phi_D) \times Norm \quad (3)$$

where R is the fit parameter which represents the relative B meson contribution to non-photonic electrons, $\Delta\phi_B$ ($\Delta\phi_D$) is the correlation between B (D) meson decayed electrons and charged hadrons from PYTHIA simulation, and Norm is the normalization parameter.

Panel (b) shows the extracted B meson contribution as a function of p_T (red circles), together with published results from Run 6 p+p collisions at $\sqrt{s}$ = 200 GeV (black stars) for comparison. The error bars represent the statistical errors and the systematic uncertainties are shown as boxes. The preliminary results agree with STAR Run 6 analysis for $p_T < 8.5$ GeV/c, and the systematic uncertainties are significantly reduced. We also compare the results with Fixed

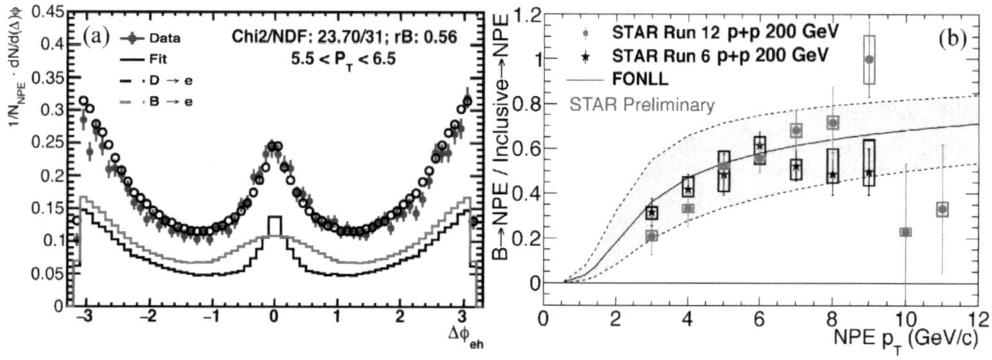

Figure 1. (Color online) (a) NPE-hadron azimuthal correlations from STAR fitted with PYTHIA templates in p+p collisions at $\sqrt{s} = 200$ GeV. (b) The extracted B meson contribution to non-photonic electrons as a function of p_T in $p + p$ collisions at $\sqrt{s} = 200$ GeV.

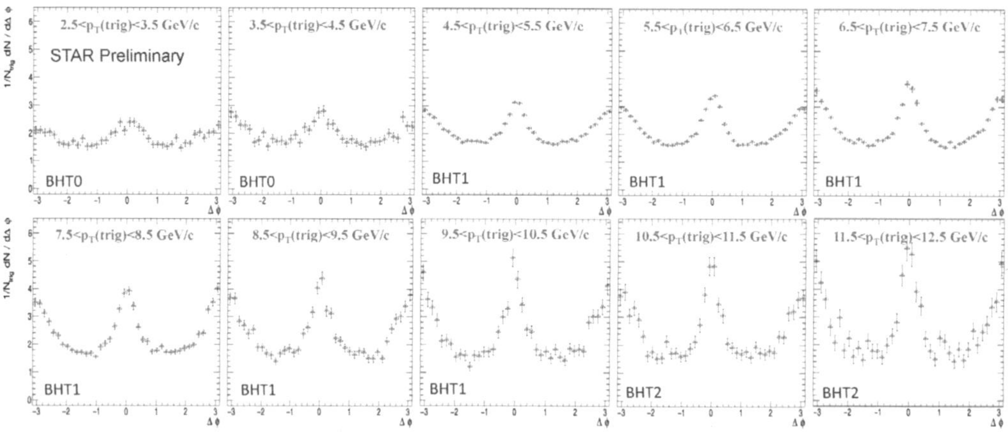

Figure 2. (Color online) NPE-hadron azimuthal correlations in $p + p$ collisions at $\sqrt{s} = 500$ GeV for different electron p_T bins from the STAR experiment.

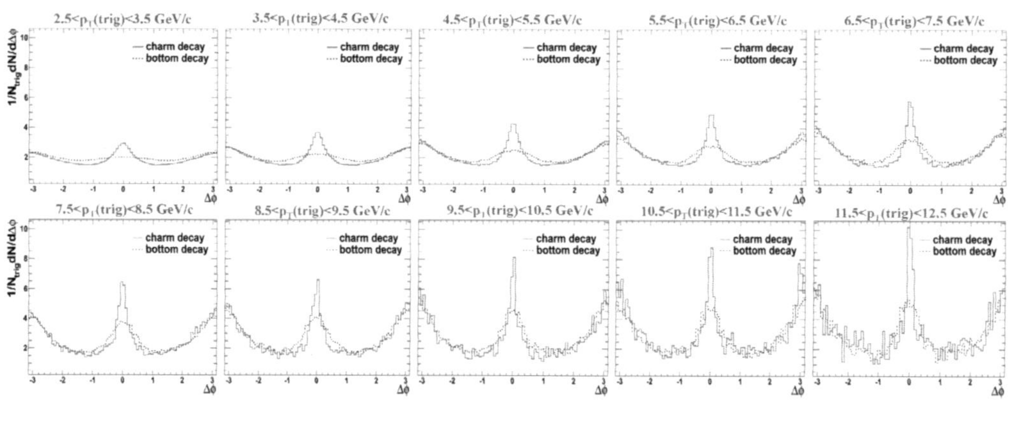

Figure 3. (Color online) NPE-hadron azimuthal correlation in $p + p$ collisions at $\sqrt{s} = 500$ GeV for different electron p_T bins from PYTHIA simulations.

Order Next to Leading Logarithm (FONLL) calculation, and find the results are consistent with FONLL calculation within uncertainties.

Figure 2 shows the raw azimuthal correlation between non-photonic electrons and charged hadrons in $p+p$ collisions at $\sqrt{s} = 500$ GeV for different trigger NPE p_T bins between $2.5 < p_T < 12.5$ GeV/c requiring $|\eta^{NPE}| < 0.7$. The associated hadrons are required to have $p_T > 0.3$ GeV/c and $|\eta^{asso}| < 1.0$. Clear azimuthal correlation can be seen on the near side and the correlation signal increases with increasing electron p_T. Figure 3 shows the correlations between B and D meson decayed electrons and charged hadrons from PYTHIA simulation for different electron p_T bins in $p+p$ collisions at $\sqrt{s} = 500$ GeV. A similar correlation structure is observed in $\sqrt{s} = 500$ GeV as in $\sqrt{s} = 200$ GeV. As the NPE p_T increases, the near side correlation signal increases for both B and D decayed electrons, but the difference between the correlation shapes reduces.

4. Summary

In this work, we study the azimuthal correlations bewteen non-photonic electrons and charged hadrons in $p+p$ collisions at $\sqrt{s} = 200$ GeV and $\sqrt{s} = 500$ GeV from the STAR experiment. By fitting the measured correlations with PYTHIA templates, we have extracted the B meson contribution to non-photonic electrons in the range of $2.5 < p_T < 11.5$ GeV/c for p+p collisions at $\sqrt{s} = 200$ GeV. The preliminary results agree with previous STAR Run 6 analysis for $p_T < 8.5$ GeV/c, and the systematic uncertainties are significantly reduced. The results are also consistent with FONLL calculation within uncertainties. For the analysis in p+p collisions at $\sqrt{s} = 500$ GeV, we have constructed the raw azimuthal correlation between non-photonic electrons and charged hadrons, and observed an increase to the near-side correlation signal as the electron p_T increases. We also used PYTHIA to simulate the azimuthal correlations between B and D meson decayed electrons and charged hadrons, and found that the correlation shapes are different on the near side for the two different species. This difference decreases as trigger electron p_T increases. Further study is under way to extract the B meson contribution to non-photonic electrons in p+p collisions at $\sqrt{s} = 500$ GeV.

Acknowledgments

This work was supported in part by the National Natural Science Foundation of China (Grant No.11205231), Major State Basic Research Development Program of China (Grant No.2014CB845403), and Youth Innovation Promotion Association CAS.

References

[1] M. Gyulassy and M. Plumer, Phys. Lett. B **243**, 432 (1990).
[2] A. Adil and M. Gyulassy, Phys. Lett. B **602**, 52 (2004).
[3] A. Adare et al., (PHENIX Collaboration), Phys. Rev. Lett. **98**, 172301 (2007).
[4] Y. L. Dokshitzer and D. E. Kharzeev, Phys. Lett. B **519**, 199 (2001).
[5] M. Djordjevic, Phys. Lett. B **632**, 81 (2006).
[6] A. Adil and I. Vitev, Phys. Lett. B **649**, 139 (2007).
[7] J. Adams et al. (STAR Collaboration), Phys. Rev. Lett. **105**, 202301 (2010).
[8] X. Lin, 2007. Ph.D Thesis, Central China Normal University, China.
[9] M. Anderson et al., Nuclear Instruments and Methods in Physics Research A **499**, 659 (2003).
[10] M. Beddo et al., Nucl. Instrum. Meth. A **499**, 725 (2003).
[11] B. Bonner et al., Nucl. Instr. Meth. A **508**, 181 (2003).
[12] W. J. Llope et al., Nucl. Instr. Meth. A **522**, 252 (2004).
[13] T. Sjöstrand, Comput. Phys. Commun. **135** 238 (2001).

Transverse-momentum spectra of strange particles produced in Pb+Pb collisions at $\sqrt{s_{\rm NN}} = 2.76$ TeV in the chemical non-equilibrium model

Wojciech Florkowski

Institute of Nuclear Physics Polish Academy of Sciences, PL-31342 Krakow, Poland

E-mail: `wojciech.florkowski@ifj.edu.pl`

Abstract. It is first shown that recent problems in heavy-ion collisions at the LHC energies, connected with thermal description of the proton yield and the pion spectra at low transverse-momenta, can be simultaneously explained within a chemical non-equilibrium statistical hadronization framework. Then the predictions of this approach for the production of strange particles are presented.

1. Introduction

Statistical hadronization model (SHM) of hadron production has become one of the cornerstones of our understanding of ultra relativistic heavy-ion collisions and also of more elementary hadronic processes including e^+e^- annihilation. In this situation, the data from the Large Hadron Collider (LHC) collected in Pb+Pb collisions at $\sqrt{s_{\rm NN}} = 2.76$ TeV came as a big surprise: the measured proton abundances do not agree with the most common version of the thermal model based on the grand-canonical ensemble [1, 2]. Possible explanations of this situation refer, for example, to hadronic rescattering present in the final stage [3] or to hadronization and subsequent freeze-out taking place off chemical equilibrium [4, 5]. The latter approach will be denoted below as NEQ SHM in contrast to the standard thermal framework to which we refer as to EQ SHM.

Besides the proton anomaly, the same LHC data exhibits another interesting feature, namely, the low-transverse-momentum pion spectra show enhancement by about 25%–50% with respect to the predictions of various thermal and hydrodynamic models. It turns out that the two problems may be solved simultaneously within the chemical non-equilibrium model. In this conference proceedings I review some of the findings which lead to this conclusion and present the NEQ SHM results for the strange particles [6, 7, 8].

2. Model assumptions

In this work I use the single-freeze out model [10, 11, 12] in the Monte-Carlo implementation given by the `THERMINATOR` code [13, 14]. The hadron spectra are calculated from the Cooper-Frye formula

$$\frac{dN}{dyd^2p_T} = \int d\Sigma_\mu p^\mu f(p\cdot u), \tag{1}$$

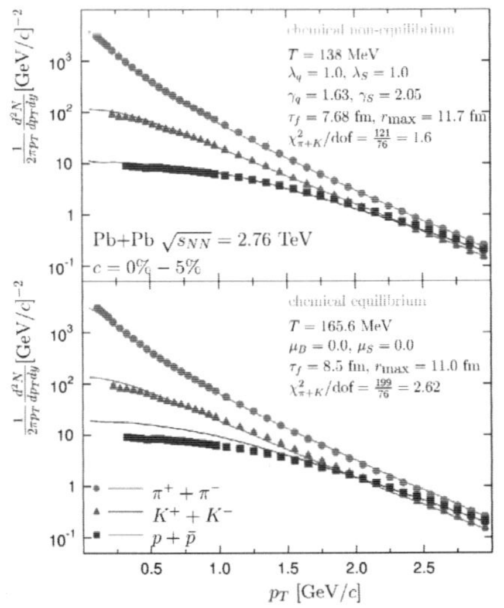

Figure 1. Comparison of the ALICE data [9] with the predictions of NEQ SHM and EQ SHM models.

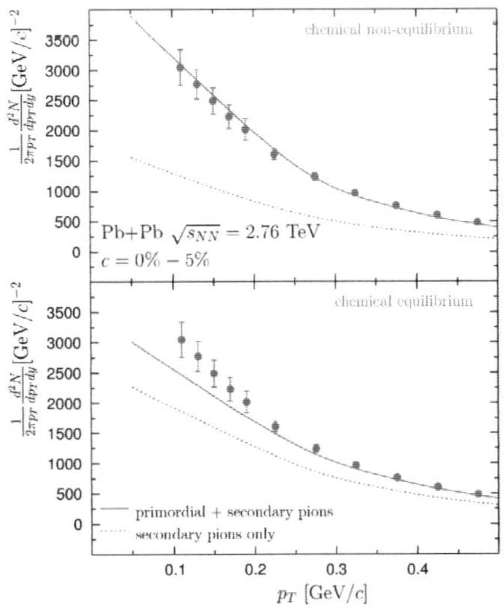

Figure 2. Same as Fig. 1 but with a linear vertical scale and low p_T region.

where the freeze-out hyper-surface is defined by the conditions

$$t^2 = \tau_f^2 + x^2 + y^2 + z^2, \quad x^2 + y^2 \le r_{\mathrm{max}}^2. \tag{2}$$

The hydrodynamic flow at freeze-out has the Hubble form $u^\mu = x^\mu/\tau_f$.

There are only two independent parameters in the model: the product $\pi\tau_f r_{\mathrm{max}}^2$ that defines the volume (per unit rapidity), and the ratio r_{max}/τ_f that determines the shape (slope) of the spectra for a given centrality class. The phase-space distributions include all well established resonances from PDG. The primordial distributions in the local rest frame have the form

$$f_i = g_i \int \frac{d^3p}{(2\pi)^3} \frac{1}{\Upsilon_i^{-1} \exp(\sqrt{m^2+p^2}/T) \pm 1}, \tag{3}$$

where the parameterization of the fugacity Υ_i is defined as in SHARE [15]

$$\Upsilon_i \;=\; (\lambda_q\gamma_q)^{N_q^i} (\lambda_s\gamma_s)^{N_s^i} (\lambda_{\bar{q}}\gamma_{\bar{q}})^{N_{\bar{q}}^i} (\lambda_{\bar{s}}\gamma_{\bar{s}})^{N_{\bar{s}}^i}. \tag{4}$$

Here $\lambda_q = \lambda_{\bar{q}}^{-1}, \lambda_s = \lambda_{\bar{s}}^{-1}$, $\gamma_q = \gamma_{\bar{q}}$, $\gamma_s = \gamma_{\bar{s}}$, and N_q^i and N_s^i are the numbers of light (u, d) and strange (s) quarks in the ith hadron, while $N_{\bar{q}}^i$ and $N_{\bar{s}}^i$ are the numbers of the antiquarks in the same hadron. Using the Gell-Mann–Nishijima formulas one finds $\lambda_q = \exp(\mu_B/3T)$ and $\lambda_s = \exp((-3\mu_S + \mu_B)/3T)$, which gives

$$\Upsilon_i = \gamma_q^{N_q^i + N_{\bar{q}}^i} \gamma_s^{N_s^i + N_{\bar{s}}^i} \exp\left(\frac{\mu_B B_i + \mu_S S_i}{T}\right), \tag{5}$$

where B_i and S_i are the baryon number and strangeness of the ith hadron. At the LHC, in the central rapidity region the baryon number density and strangeness are negligible, hence

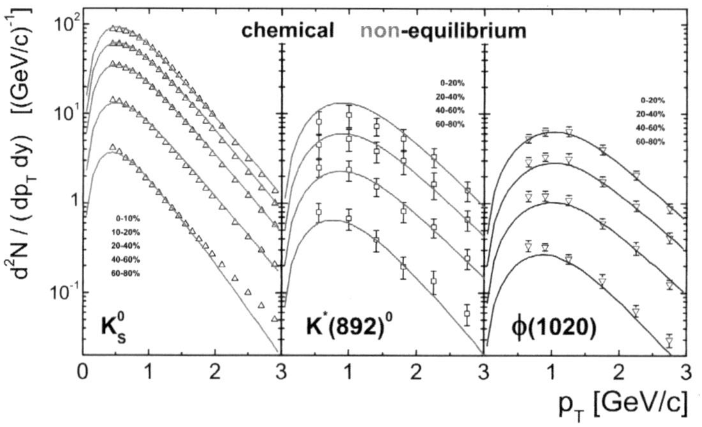

Figure 3. NEQ SHM predictions for K_S, $K^*(892)$, and ϕ, compared with the data from [16, 17].

the baryon and strangeness chemical potentials can be set equal to zero, $\mu_B \approx \mu_S \approx 0$. The parameters γ_q and γ_s are equivalent to the chemical potentials $\mu_q/T = \ln \gamma_q$ and $\mu_s/T = \ln \gamma_s$

$$\Upsilon_i \approx \quad = \exp\left(\frac{\mu_q\left(N_q^i + N_{\bar{q}}^i\right) + \mu_s\left(N_s^i + N_{\bar{s}}^i\right)}{T}\right)$$

The new potentials are connected with the conservation of the sum of the number of quarks and antiquarks during the hadronization process, similarly as μ_B and μ_S are connected with the conservation of the difference of the quark and antiquark numbers. The introduction of such chemical potentials make sense if the hadronization process is fast and there is no significant volume expansion during this process [4, 5]. There exist also hints for a non-zero value of μ_q found in the scenario presented in Ref. [18].

3. Results

In Fig. 1 we show our results for the spectra of pions, kaons, and protons for central Pb+Pb collisions at $\sqrt{s_{\rm NN}}$=2.76 TeV. In the upper panel we show the results for the chemical non-equilibrium model (NEQ SHM), whereas in the lower panel the spectra are shown for the chemical equilibrium scenario (EQ SHM). The thermodynamic parameters are taken from [5]: T=138 MeV, γ_q=1.63, γ_s=2.05. The fit of the spectra gives: τ_f=7.68 fm and $r_{\rm max}$=11.7 fm. Correspondingly, for EQ SHM we use: T =165.6 MeV, τ_f=8.5 fm and $r_{\rm max}$=11 fm. We observe that the chemical non-equilibrium model gives much better description of the ALICE data [9], especially for protons and in the low p_T region for pions. This is seen also more clearly in Fig. 2, where a linear scale is used.

Having fixed the model parameters with the spectra of pions, kaons and protons (for each accessible centrality bin), we can calculate the spectra of other hadrons. Fig. 3 shows the spectra of K_S, $K^*(892)$, and ϕ for various centrality selections. We find very good agreement between the model predictions and the data [16, 17]. This is quite interesting since $K^*(892)$ and ϕ have very different mean free paths in the medium. The fact that the spectra of these two particles are well reproduced by our framework supports the idea of a single freeze-out. Finally, in Fig. 4 we show the model spectra of hyperons and the data [16, 19]. The spectra of Λ are rather well reproduced for central collisions. We note that the model description can be further improved

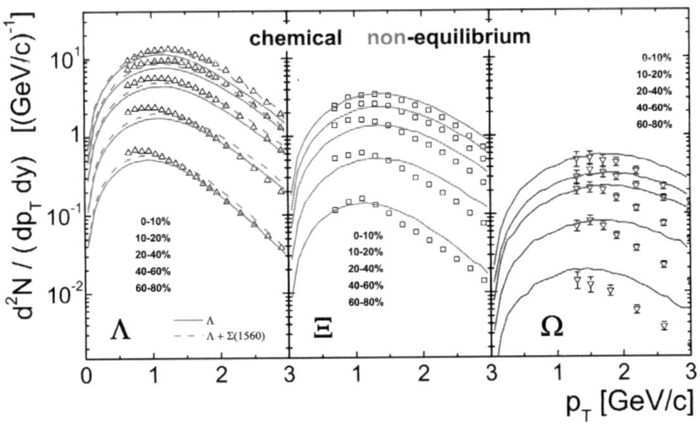

Figure 4. NEQ SHM model predictions for hyperons, compared with the data from [16, 19].

if the decay of $\Sigma(1560)$ is included. The spectra of Ξ and Ω are rather well described for the lowest measured values of p_T.

4. Conclusions

The non-equilibrium thermal model combined with the single-freeze-out scenario explains very well the spectra of pions, kaons, and protons. In particular, it explains the proton anomaly and the low-p_T enhancement of pions. With the parameters fixed by such fits, the NEQ SHM model gives also a rather good description of the strange hadron production (except for Ξ's and Ω's with large p_T). This work was supported by Polish National Science Center Grant DEC-2012/06/A/ST2/00390.

[1] Stachel J, Andronic A, Braun-Munzinger P and Redlich K 2014 *J. Phys. Conf. Ser.* **509** 012019 (*Preprint* 1311.4662)
[2] Floris M 2014 *Nucl. Phys.* **A931** 103–112 (*Preprint* 1408.6403)
[3] Becattini F, Bleicher M, Kollegger T, Schuster T, Steinheimer J and Stock R 2013 *Phys. Rev. Lett.* **111** 082302 (*Preprint* 1212.2431)
[4] Petran M and Rafelski J 2013 *Phys. Rev.* **C88** 021901 (*Preprint* 1303.0913)
[5] Petran M, Letessier J, Petracek V and Rafelski J 2013 *Phys. Rev.* **C88** 034907 (*Preprint* 1303.2098)
[6] Begun V, Florkowski W and Rybczynski M 2014 *Phys. Rev.* **C90** 014906 (*Preprint* 1312.1487)
[7] Begun V, Florkowski W and Rybczynski M 2014 *Phys. Rev.* **C90** 054912 (*Preprint* 1405.7252)
[8] Begun V and Florkowski W 2015 *Phys. Rev.* **C91** 054909 (*Preprint* 1503.04040)
[9] Abelev B *et al.* (ALICE) 2012 *Phys. Rev. Lett.* **109** 252301 (*Preprint* 1208.1974)
[10] Broniowski W and Florkowski W 2001 *Phys. Rev. Lett.* **87** 272302 (*Preprint* nucl-th/0106050)
[11] Broniowski W and Florkowski W 2002 *Phys. Rev.* **C65** 064905 (*Preprint* nucl-th/0112043)
[12] Rybczynski M, Florkowski W and Broniowski W 2012 *Phys. Rev.* **C85** 054907 (*Preprint* 1202.5639)
[13] Kisiel A, Taluc T, Broniowski W and Florkowski W 2006 *Comput. Phys. Commun.* **174** 669–687 (*Preprint* nucl-th/0504047)
[14] Chojnacki M, Kisiel A, Florkowski W and Broniowski W 2012 *Comput. Phys. Commun.* **183** 746–773 (*Preprint* 1102.0273)
[15] Torrieri G, Steinke S, Broniowski W, Florkowski W, Letessier J and Rafelski J 2005 *Comput. Phys. Commun.* **167** 229–251 (*Preprint* nucl-th/0404083)
[16] Abelev B B *et al.* (ALICE) 2013 *Phys. Rev. Lett.* **111** 222301 (*Preprint* 1307.5530)
[17] Abelev B B *et al.* (ALICE) 2015 *Phys. Rev.* **C91** 024609 (*Preprint* 1404.0495)
[18] Melo I and Tomasik B 2016 *J. Phys.* **G43** 015102 (*Preprint* 1502.01247)
[19] Abelev B B *et al.* (ALICE) 2014 *Phys. Lett.* **B728** 216–227 [Erratum: Phys. Lett.B734,409(2014)] (*Preprint* 1307.5543)

Anisotropic flow measurements in Pb–Pb collisions at $\sqrt{s_{\rm NN}} = 5.02$ TeV with ALICE

You Zhou (for the ALICE Collaboration)
Niels Bohr Institute, University of Copenhagen, Blegdamsvej 17, 2100 Copenhagen, Denmark
E-mail: you.zhou@cern.ch

Abstract.
Anisotropic flow is a sensitive probe of the initial conditions and the transport properties of the Quark Gluon Plasma (QGP) produced in heavy-ion collisions. In these proceedings, we present the first results of elliptic (v_2), triangular (v_3) and quadrangular flow (v_4) of charged particles in Pb–Pb collisions at $\sqrt{s_{\rm NN}} = 5.02$ TeV with the ALICE detector. In addition, comparisons of experimental measurements with various theoretical calculations will be discussed. This provides a unique opportunity to test the validity of the hydrodynamic picture and to discriminate between various possibilities for the temperature dependence of shear viscosity to entropy density ratio of the produced QGP.

1. Introduction

Collisions of high-energy heavy-ions, at the Brookhaven Relativistic Heavy Ion Collider (RHIC) and the CERN Large Hadron Collider (LHC), allow us to create and study the properties of Quark Gluon Plasma (QGP) in the laboratory. The azimuthal anisotropy in particle production is, at these energies, an observable which provides experimental information on the Equation of State and the transport properties of QGP. This anisotropy is usually characterized by the Fourier flow-coefficients [1],

$$\begin{aligned} v_n &= \langle \cos[n(\varphi - \Psi_n)\rangle, \quad \text{or equivalently} \\ v_n &= \langle e^{in\varphi} e^{-in\Psi_n} \rangle, \end{aligned} \tag{1}$$

where φ is the azimuthal angle of the particles, Ψ_n is the $n^{\rm th}$-order flow plane (or named final state symmetry plane) angle and $\langle\ \rangle$ denotes an average over the selected particles and events. Experimental measurements together with comparisons to theoretical calculations show that anisotropic flow sheds new light on the initial conditions and is sensitive to the shear viscosity to entropy density ratio η/s of the QGP. Recent anisotropic flow measurements [2, 3, 4] in Pb–Pb collisions at $\sqrt{s_{\rm NN}} = 2.76$ TeV provide additional information to improve the theoretical descriptions of data. In these proceedings, we present the latest anisotropic flow measurements in Pb–Pb collisions at $\sqrt{s_{\rm NN}} = 5.02$ TeV in ALICE [5]. The comparisons to hydrodynamic and AMPT model calculations will be discussed below.

2. Analysis Details

The data used in this analysis were recorded with the ALICE detector in November 2015 in run 2 at the LHC with Pb-Pb collisions at $\sqrt{s_{\rm NN}} = 5.02$ TeV. About 140 k Pb–Pb events were recorded with a minimum-bias trigger, based on signals from two VZERO detectors ($-3.7<\eta<-1.7$ for VZERO-C and $2.8<\eta<5.1$ for VZERO-A) and on a Silicon Pixel Detector. The VZERO detectors were also used to determine the collision centrality in Pb–Pb collisions. Tracks of charged particles are reconstructed using a Inner Tracking System and a Time Projection Chamber with full azimuthal coverage for pseudo-rapidity range $|\eta|<0.8$.

3. Results and Discussions

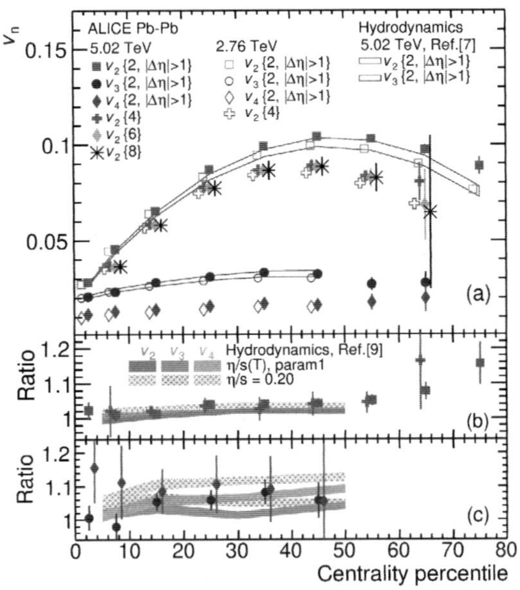

Figure 1. (color online) (a) Anisotropic flow v_n integrated over the p_T range $0.2 < p_T < 5.0$ GeV/c, as a function of event centrality, for the two–particle (with $|\Delta\eta| > 1$) and multi-particle cumulant methods. The ratios of v_n from Pb–Pb collisions at 5.02 TeV and 2.76 TeV, are presented in Fig. 1 (b) and (c). Various hydrodynamic calculations are also presented [7, 9].

The centrality dependence of v_2, v_3 and v_4 are presented in Fig. 1. Both two-particle correlations with pseudorapidity and multi-particle cumulants methods are used in these measurements. It is seen that $v_2\{2, |\Delta\eta| > 1\}$ increases from central to peripheral collisions, saturates in the 40–50% centrality class with a maximum value of 0.104 ± 0.001 (stat.) $\pm$ 0.002 (syst.). Similar centrality dependence is observed in the measurements of multi-particle cumulants $v_2\{4\}$, $v_2\{6\}$ and $v_2\{8\}$ and results of these multi-particle cumulant measurements are agree within 1%. Assuming that non-flow effects are suppressed by the pseudorapidity gap, the remaining differences between two- and multi-particle cumulants of v_2 can be attributed to the different sensitivities to elliptic flow fluctuations. Compared to v_2, higher harmonics i.e. v_3 and v_4 are smaller and the centrality dependence is much weaker. In addition, the hydrodynamic calculations [7] which combine the changes in initial spatial anisotropy and the hydrodynamic response are presented in Fig. 1. (a) for comparison. The calculations are found to be compatible with the measured anisotropic flow v_n coefficients. Furthermore, using a new set of Lund string fragmentation parameters and partonic cross section 3 mb, the AMPT model with string melting scenario can quantitatively reproduce the anisotropic flow measurements [8].

To better illustrate the energy evolution of anisotropic flow, Fig 1. (b) and (c) shows the ratios of v_n measured at 5.02 TeV to 2.76 TeV. It is observed that the increase of v_2 and v_3 from the two energies is rather moderate, while for v_4 it is more pronounced. Considering none of the ratios of flow harmonics from two energies exhibit a significant centrality dependence in the centrality range 0–50%, a fit with a constant value are performed for these ratios. An increase of (3.0±0.6)%, (4.3±1.4)% and (10.2±3.8)% over the centrality range 0–50% in Pb–Pb collisions is obtained for v_2, v_3 and v_4, respectively. This increase of anisotropic flow is in agreement with theoretical predictions in Ref [9]. The data seems to support a low value of η/s for the QGP created in Pb–Pb collisions at $\sqrt{s_{\rm NN}} = 5.02$ TeV and indicates that η/s does not change significantly with respect to Pb–Pb collisions at $\sqrt{s_{\rm NN}} = 2.76$ TeV.

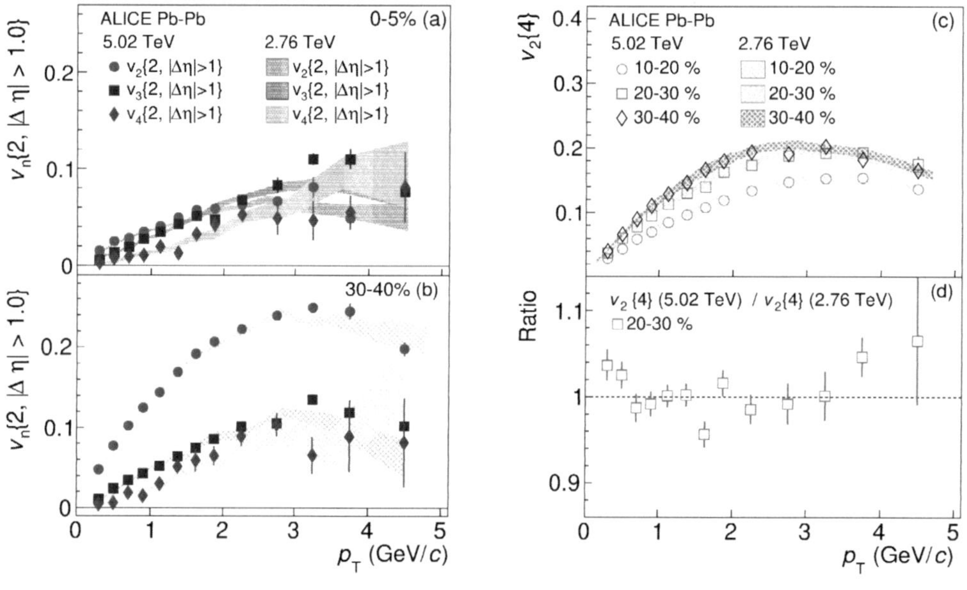

Figure 2. (color online) $v_n(p_{\rm T})$ for the centrality 10–20%, 20–30% and 30–40%. Measurements for Pb–Pb collisions at $\sqrt{s_{\rm NN}} = 2.76$ TeV are presented as shading. (d) The ratio of $v_2\{4\}$ in 20–30% from two collision energies is also shown here.

The transverse momentum ($p_{\rm T}$) dependent $v_2\{2, |\Delta\eta| > 1\}$, $v_3\{2, |\Delta\eta| > 1\}$ and $v_4\{2, |\Delta\eta| > 1\}$ are presented for the 0–5% and 30–40% centrality classes. In 0–5% centrality class, $v_3\{2\}$ is larger than $v_2\{2\}$ for $p_{\rm T} > 2$ GeV/c, and $v_4\{2\}$ is roughly compatible with $v_2\{2\}$. For the 30–40% centrality class, $v_2\{2\}$ is higher than $v_3\{2\}$ and $v_4\{2\}$ for the entire $p_{\rm T}$ range measured, with no crossing of the different order flow coefficients. The $p_{\rm T}$-differential $v_2\{4\}$ decreases from mid-central to central collisions over the $p_{\rm T}$ range measured. This measurement is consistent with the previous $p_{\rm T}$-differential $v_2\{4\}$ measurements from 2.76 TeV, as also illustrated by the ratio of $v_2\{4\}$ for the two energies in Fig. 2 (d). This suggests that the increase observed in the $p_{\rm T}$ integrated flow results seen in Fig. 1 could be due to stronger radial flow produced at 5.02 TeV. It leads to an increase of mean transverse momentum $\langle p_{\rm T}\rangle$ from 2.76 to 5.02 TeV. The measurements of $p_{\rm T}$-differential flow are expected to be more sensitive to initial conditions and η/s of the QGP. The results are found to be underestimated by AMPT simulations by 5-30%, meanwhile no hydrodynamic calculations are available yet. The future comparisons of p_T-differential flow between experimental measurements and hydrodynamic calculations will provide important information to constrain further details of the theoretical calculations, e.g. determination of radial flow and freeze-out conditions.

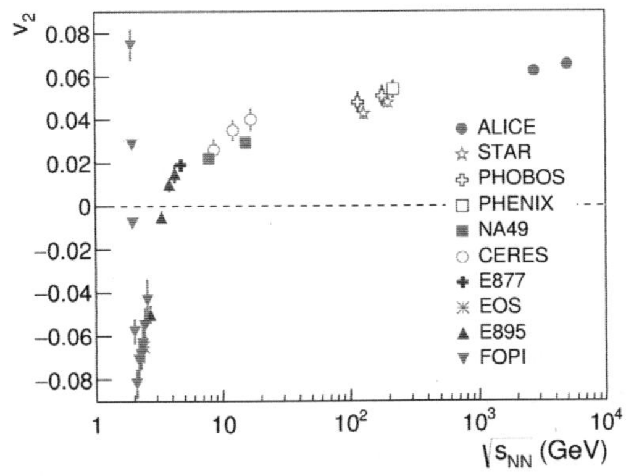

Figure 3. (color online) p_T integrated v_2 in centrality 20-30% as a function of collisions energy.

Figure 3 presents the collisions-energy dependence of the fully p_T-integrated v_2 measured in 20–30% centrality with different collision systems and energies. A continuous increase of anisotropic flow for this centrality has been observed from SPS/RHIC to LHC energies. For these fully p_T integrated coefficients, an increase of $4.9 \pm 1.9\%$ is observed going from $\sqrt{s_{NN}} = 2.76$ to 5.02 TeV, which is close to values of the previously-mentioned hydrodynamic calculations [7, 9].

4. Summary

In summary, the first anisotropic flow measurements of charged particles in Pb–Pb collisions at $\sqrt{s_{NN}} = 5.02$ TeV at the LHC are presented in this proceedings. An average increase of $(3.0\pm0.6)\%$, $(4.3\pm1.4)\%$ and $(10.2\pm3.8)\%$, is observed for the transverse momentum integrated elliptic, triangular and quadrangular flow, respectively, over the centrality range 0–50% going from 2.76 TeV to 5.02 TeV. The p_T-differential flow has also been investigated, it does not change appreciably between the two LHC energies. Therefore, the increase in integrated flow coefficients could be mainly due to an increase in average p_T. This agrees with predictions from hydrodynamic models [7, 9]. Further comparisons of p_T-differential flow measurements and theoretical calculations will provide extra constraints on the initial conditions and the transport properties of the QGP.

References

[1] S. Voloshin and Y. Zhang, Z. Phys. C **70**, 665 (1996)
[2] J. Adam *et al.* [ALICE Collaboration], arXiv:1604.07663 [nucl-ex].
[3] J. Adam *et al.* [ALICE Collaboration], arXiv:1605.02035 [nucl-ex].
[4] J. Adam *et al.* [ALICE Collaboration], arXiv:1606.06057 [nucl-ex].
[5] J. Adam *et al.* [ALICE Collaboration], Phys. Rev. Lett. **116**, no. 13, 132302 (2016)
[6] K. Aamodt *et al.* [ALICE Collaboration], JINST **3**, S08002 (2008).
[7] J. Noronha-Hostler, M. Luzum and J. Y. Ollitrault, Phys. Rev. C **93**, no. 3, 034912 (2016)
[8] Z. Feng, G. M. Huang and F. Liu, arXiv:1606.02416 [nucl-ex].
[9] H. Niemi, K. J. Eskola, R. Paatelainen and K. Tuominen, Phys. Rev. C **93**, no. 1, 014912 (2016)

Measurement of higher harmonic flow coefficients of identified hadrons in Pb-Pb collisions at $\sqrt{s_{\rm NN}} = 2.76$ TeV

Naghmeh Mohammadi for the ALICE collaboration
Nikhef, Science Park 105, 1098 XG, Amsterdam, The Netherlands
E-mail: naghmehmohammadi@cern.ch

Abstract. The latest ALICE results on the centrality and transverse momentum dependence of v_2, v_3, v_4 and v_5 for $\pi^{\pm}$, $\rm K^{\pm}$ and $\rm p(\bar{p})$ in Pb–Pb collisions at $\sqrt{s_{\rm NN}} = 2.76$ TeV are presented. The flow coefficients exhibit a clear mass ordering for $p_{\rm T} < 2.5$ GeV/c for all harmonics. For transverse momentum values larger than about 3 GeV/c, mesons exhibit distinctively lower flow values compared to baryons, suggesting that coalescence might be the relevant particle production mechanism in this region. The experimental data for $p_{\rm T} < 2.5$ GeV/c are described fairly well by the hydrodynamical model iEBE-VISHNU, which models the hydrodynamical expansion of the fireball using a value of $\eta/s = 0.08$, coupled to a hadronic cascade model (UrQMD). Finally, A Multi-Phase Transition Model (AMPT) highlights the significant contribution of the hadronic rescattering phase and of coalescence as the dominant particle production mechanism to the mass ordering at low $p_{\rm T}$ and the baryon-meson grouping at intermediate $p_{\rm T}$, respectively.
Keywords: QGP, ALICE, azimuthal anisotropy, shear viscosity, iEBE-VISHNU, UrQMD, AMPT

1. Introduction

The aim of the heavy ion collision experiments is to study the Quark Gluon Plasma (QGP) [1] and constrain its transport properties such as the ratio of shear viscosity to entropy density (η/s) and equation of state. One of the observables sensitive to the properties of the QGP is anisotropic flow which characterises the momentum anisotropy of the final state particles. This momentum anisotropy rises from the spatial anisotropy of the overlapping collision region through interactions between partons and at later stages between the produced particles. The momentum anistropy can be expressed in terms of a Fourier expansion of the particle distribution $dN/d(\varphi - \Psi_n)$ in azimuthal angle φ with respect to the $n^{\rm th}$-order symmetry plane angle Ψ_n [2], according to:

$$dN/d(\varphi - \Psi_n) \approx 1 + \sum_n 2v_n \cos[n(\varphi - \Psi_n)]. \quad (1)$$

In this equation, v_n is the $n^{\rm th}$-order flow coefficient and can be calculated according to $v_n = \langle \cos[n(\varphi - \Psi_n)] \rangle$, where the brackets denote an average over all events and particles.

In this article, we report the $p_{\rm T}$-differential elliptic (v_2), triangular (v_3), quadrangular (v_4) and pentagonal (v_5) flow for $\pi^{\pm}$, $\rm K^{\pm}$ and $\rm p(\bar{p})$ in two different centrality intervals of Pb–Pb collisions at $\sqrt{s_{\rm NN}} = 2.76$ TeV measured with the ALICE detector.

2. Analysis details

The analysis is performed on both Pb–Pb and pp collisions at $\sqrt{s_{\rm NN}}$ = 2.76 TeV recorded by ALICE during 2011 at the LHC [3]. For Pb–Pb collisions the minimum bias events are triggered by the coincidence between the signals from both sides of the V0 detectors. In addition, an online selection based on the V0 detectors are used in Pb–Pb collisions to increase the number of central (i.e. 0–10% centrality interval) and semi-central (i.e. 10–50%) collisions. For minimum bias pp collisions at least one hit is required in the SPD (the innermost layer of the Inner Tracking System) or V0 detectors.

The reconstructed primary vertex position along the z axis (z_{vtx}) for all analysed events are required to be within 10 cm from the nominal interaction point. In addition, events with multiple reconstructed primary vertices are rejected which leads to a negligible amount of pile-up events for all systems [4]. After applying all the selection critera, a filtered data sample with 25×10^6 events are analysed for Pb–Pb and 20×10^6 for pp collisions to produce the results presented in this article. The two centrality intervals presented in this article are 0–1% (i.e. smallest impact parameter) and 20–30% (i.e. larger impact parameter) centrality intervals in Pb–Pb collisions.

The track reconstruction is performed using the information from both the Time Projection Chamber (TPC) and the Inner Tracking System (ITS). The charged particle momenta is measured using both ITS and TPC with an azimuthal coverage of $|\eta| < 0.9$. The particle identification (PID) for pions ($\pi^{\pm}$), kaons ($K^{\pm}$) and protons ($p + \bar{p}$) used in this analysis is based on a procedure reported in [5] which combines the signals from the TPC and Time Of Flight (TOF) detectors. In this analysis, different particle species are identified by requiring a minimum identification probability of 90%. More details on the PID procedure followed can be found in [6].

The flow coefficients are measured with the Scalar Product method [7]. The identified particle of interest (POI), i.e. pions ($\pi^{\pm}$), kaons ($K^{\pm}$) and protons ($p + \bar{p}$), and the charged reference particles (RP) are selected from two non-overlapping subevents within the acceptance of the TPC detector ($-0.8 < \eta < 0$ and $0 < \eta < 0.8$). Correlations not related to the common symmetry plane, denoted as $\delta_n^{AA,pp}$, are estimated using minimum bias pp collisions and are subtracted from the measured flow coefficients in Pb-Pb collisions denoted as v_n^{AA} with a procedure described in [6]. In this article, the reported values, v_n^{sub}, are extracted according to $v_n^{sub} = v_n^{AA} - \delta_n^{AA,pp}$.

3. Results and discussions

Figure 1 presents the $p_{\rm T}$-differential v_2^{sub}, v_3^{sub}, v_4^{sub} and v_5^{sub} for charged pions, kaons and (anti-)protons for the 0–1% (upper row - ultra-central collisions) and 20–30% (bottom row - mid-central collisions) centrality intervals in Pb–Pb collisions at $\sqrt{s_{\rm NN}}$ = 2.76 TeV. In the ultra-central collisions, where the evolution of the system is mainly dominated by initial state fluctuations, one observes significant non-zero values for all harmonics and particle species. In addition, the magnitude of higher flow harmonics are comparable to that of elliptic flow. However, in mid-central collisions, the magnitude of elliptic flow is significantly larger than in ultra-central collisions and rises to become the dominant flow harmonic. The magnitude of higher harmonics also has increased with respect to the ultra-central collisions but this increase is not as significant as the one of elliptic flow. These observations confirm that elliptic flow is mainly driven by the anisotropy of the collision geometry. However, higher flow harmonics are less dependent on the geometry and more driven by the initial state fluctuations. This can be quantified by taking the ratio of v_n^{sub} in ultra- to mid-central collisions. For elliptic flow and for all particle species, this ratio is at most 25%, whereas for v_3^{sub} the ratio is >60%.

Furthermore, a clear mass ordering can be seen in all harmonics and centrality intervals in the low $p_{\rm T}$ region (i.e. $p_{\rm T} \leq 2.5$ GeV/c) which rises from the interplay between the anisotropic flow harmonics and radial flow. Radial flow creates a depletion in the low $p_{\rm T}$ region of the particle spectrum which is more pronounced in- than out-of-plane with increasing particle mass. This leads to a lower v_n^{sub} value for heavier particles. In addition, $v_{n(=2,3,4)}^{sub}$ show a clear baryon-meson grouping in the intermediate $p_{\rm T}$ region in which v_n^{sub} of baryons is larger than the values for mesons.

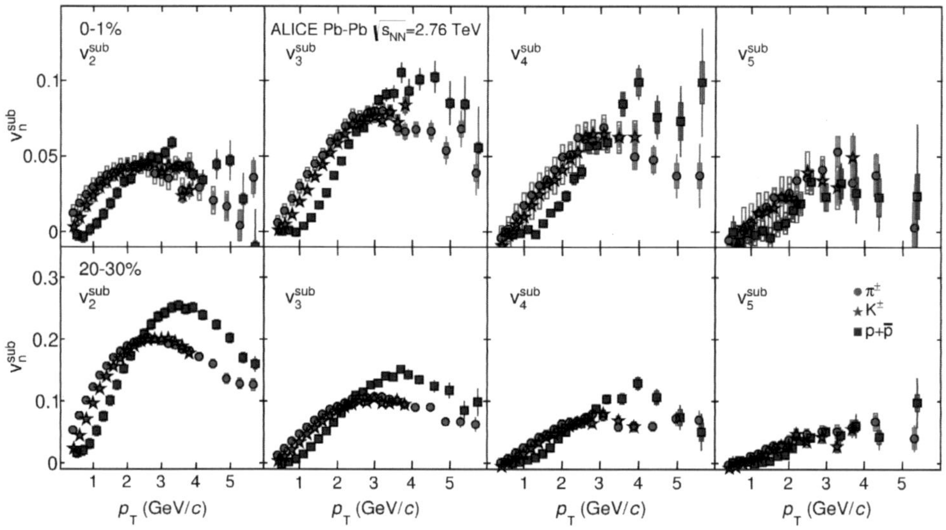

Figure 1. The $p_{\rm T}$-differential v_2^{sub} (first column), v_3^{sub} (second column), v_4^{sub} (third column), and v_5^{sub} (fourth column), for different particle species for the 0–1% (upper row) and the 20–30% (lower row) centrality intervals of Pb–Pb collisions at $\sqrt{s_{\rm NN}} = 2.76$ TeV [6].

A comparison is performed between v_n^{sub} $(n = 2 - 4)$ and recent v_n hydrodynamical calculations from [8]. These calculations are based on iEBE-VISHNU, an event-by-event version of the VISHNU hybrid model coupling 2+1 dimensional viscous hydrodynamics (VISH2+1) to a hadronic cascade model (UrQMD). In this model, the ratio of shear viscosity to entropy density is fixed to $\eta/s = 0.08$. The model describes the v_2^{sub} measurements for all particles and centrality intervals within 10-15%. The model predictions show a better agreement (within 5% deviations) with the measured v_3^{sub} and v_4^{sub}.

In this analysis, three different versions of A Multi-Phase Transport Model (AMPT) are investigated. AMPT can run over two configurations, namely the default and the string melting. In both configurations, the final state hadronic rescattering phase is implemented which also includes resonance decays. In the string melting configuration, final state hadrons are formed via a quark coalescence mechanism. The third version is based also on the string melting configuration with the hadronic rescattering phase switched off. More details on the AMPT configurations used in this analysis can be found in [6].

Figure 2 presents the $p_{\rm T}$-differential v_3 for charged pions, kaons and (anti-)protons for the 20-30% centrality interval measured with the Scalar Product method. Each panel presents the results of one of the three AMPT versions introduced above. The string melting version (left) predicts a clear mass ordering at low $p_{\rm T}$ as well as a clear baryon-meson grouping at the intermediate $p_{\rm T}$. The string melting version without hadronic rescattering (middle), however, can only reproduce the baryon-meson grouping and the default version (right) is only able to reproduce the mass ordering at low values of $p_{\rm T}$. These observations emphasize the effect of hadronic rescattering in the development of the mass ordering at the low $p_{\rm T}$ region and the

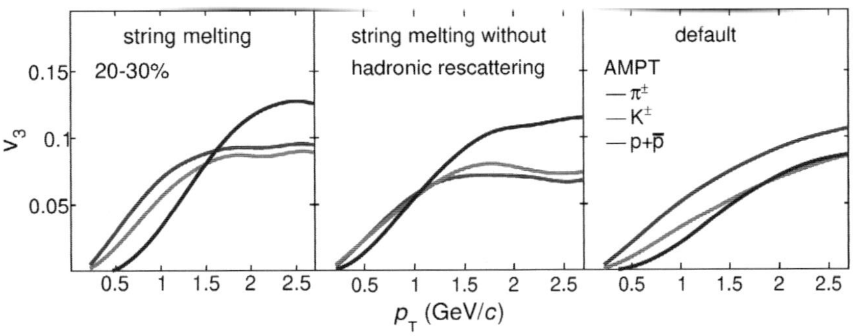

Figure 2. The $p_{\rm T}$-differential v_3 in 20-30% Pb–Pb collisions at $\sqrt{s_{\rm NN}} = 2.76$ TeV obtained using the string melting, with (left) and without (middle) hadronic rescattering, and the default (right) versions.

coalescence for the baryon-meson grouping at the intermediate $p_{\rm T}$ region.

4. Summary

In this article, the non-flow subtracted flow harmonics, v_2^{sub}, v_2^{sub}, v_2^{sub} and v_2^{sub} are reported as a function of transverse momentum for charged pions, kaons and (anti-)protons at 0-1% and 20-30% centrality intervals in Pb–Pb collisions at $\sqrt{s_{\rm NN}} = 2.76$ TeV. The v_n measurements are performed using the Scalar Product method where identified hadrons and reference charged particles are selected from two non-overlapping subevents. The non-flow contribution, i.e. correlations unrelated to the common symmetry plane, is estimated using pp collisions and then subtracted from the v_n measurements.

All flow harmonics exhibit an increase in magnitude in more peripheral collisions. This increase is more significant for elliptic flow compared to higher harmonics which indicates that v_2 is more dependent on the anisotropy of the collision geometry. On the other hand, higher harmonics are more driven by the initial state fluctuations. A distinct mass ordering is seen in low $p_{\rm T}$ region which is attributed to the interplay between radial flow and anisotropic flow harmonics. In addition, AMPT measurements highlight the importance of the final state hadronic rescattering phase in the development of the mass ordering of v_n in this $p_{\rm T}$ region. In the intermediate $p_{\rm T}$ values, a baryon-meson grouping is seen where baryons show a higher magnitude compared to mesons. This is observed also in the AMPT string melting version which demonstrates the effect of quark coalecence in this $p_{\rm T}$ region.

References

[1] S. A. Bass, M. Gyulassy, H. Stoecker and W. Greiner, J. Phys. G **25** (1999) R1 [hep-ph/9810281].
[2] S. Voloshin and Y. Zhang, Z. Phys. C **70** 665 (1996) [hep-ph/9407282].
[3] K. Aamodt *et al.* [ALICE Collaboration], JINST **3**, S08002 (2008).
[4] B. B. Abelev *et al.* [ALICE Collaboration], Int. J. Mod. Phys. A **29**, 1430044 (2014) [arXiv:1402.4476 [nucl-ex]].
[5] J. Adam *et al.* [ALICE Collaboration], Eur. Phys. J. Plus **131**, no. 5, 168 (2016) [arXiv:1602.01392 [physics.data-an]].
[6] J. Adam *et al.* [ALICE Collaboration], arXiv:1606.06057 [nucl-ex].
[7] S.A. Voloshin, A.M. Poskanzer, R. Snellings, in: Landolt-Boernstein, Relativistic Heavy Ion Physics, vol. 1/23, Springer-Verlag, pp. 554, 2010.
[8] H. j. Xu, Z. Li and H. Song, Phys. Rev. C **93**, no. 6, 064905 (2016) [arXiv:1602.02029 [nucl-th]].

IOP Conf. Series: Journal of Physics: Conf. Series **779** (2017) 012057 doi:10.1088/1742-6596/779/1/012057

Study of the long-range azimuthal correlations in pp and p+Pb collisions with the ATLAS detector at the LHC

Krzysztof W. Wozniak on behalf of the ATLAS Collaboration
Institute of Nuclear Physics, PAS, ul. Radzikowskiego 152, 31-342 Krakow, Poland
E-mail: krzysztof.wozniak@ifj.edu.pl

Abstract. ATLAS measurements of azimuthal correlations between particle pairs at large pseudorapidity separation in pp and p+Pb collisions are presented. The data were collected using a combination of the minimum-bias and high track-multiplicity triggers. The correlations in the "ridge" region are analysed using a new template fit method for extraction of flow harmonics. A comparison of this method with the previously used ZYAM method and a detailed study of the dependence of their results on the charged particle multiplicity and the transverse momentum of the particles forming a pair are shown.

1. Introduction

The measurements of Au+Au collisions and later of Pb+Pb collisions at very high energies revealed strong long-range azimuthal correlations between produced particles attributed to the creation of the Quark-Gluon Plasma (see [1] and references therein). The observation of similar in shape correlations (called "ridge"), though smaller in magnitude, in pp collisions at the LHC in events with very large multiplicities [2, 3] initiated more detailed studies [4] of pp and p+Pb collisions measured by the ATLAS detector [5], which are presented in this report.

The pp collisions measured at 2.76, 5.02 and 13 TeV and p+Pb collisions at 5.02 TeV are used in the analysis. As the most interesting phenomena are observed for events from the tail of the multilicity distribution, the minimum bias sample of events was suplemented with events obtained using several High Multiplicity Triggers, which accept events with the number of reconstructed tracks above some threshold.

2. Template fit method

The two-particle correlation function, $C(\Delta\eta, \Delta\phi)$, is parameterized by the difference of pseudorapidities and the difference of azimuthal angles of two reconstructed tracks:

$$C(\Delta\eta, \Delta\phi) = \frac{S(\Delta\eta, \Delta\phi)}{B(\Delta\eta, \Delta\phi)}, \quad (1)$$

where $S(\Delta\eta, \Delta\phi)$ and $B(\Delta\eta, \Delta\phi)$ denote the number of pairs in the measured events and that in mixed events, respectively. In the range $|\Delta\eta| < 2$ this correlation function is dominated by short-range correlations from jets, resonance decays and Bose-Einstein effects. In order to study the long-range correlations this function is integrated over the remaining particle-pair acceptance

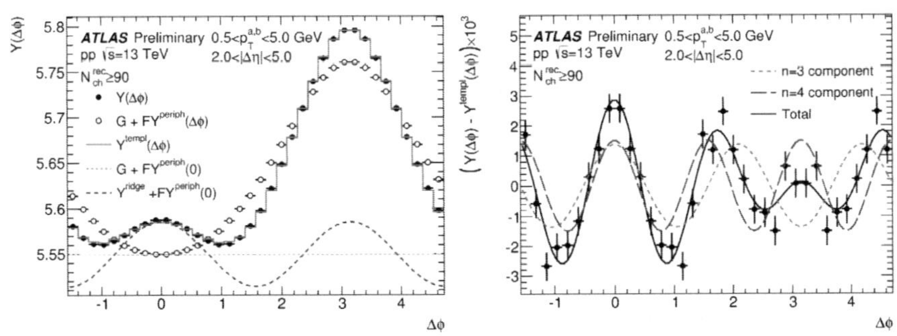

Figure 1. Left panel: Example of a template fit including only the 2nd order harmonic, $v_{2,2}$, to the per-trigger particle yields, $Y(\Delta\phi)$, in 13 TeV *pp* collisions [4]. Right panel: The difference between the $Y(\Delta\phi)$ and the template fit, showing the presence of $v_{3,3}$ and $v_{4,4}$ components (note the large difference in the verical scales).

of the ATLAS detector, i.e. $2 < |\Delta\eta| < 5$. The resulting one-dimensional function, $Y(\Delta\phi)$, is expected to contain contributions from hard processes (dijets), $Y^{hard}(\Delta\phi)$, and from the long-range correlation, $Y^{ridge}(\Delta\phi) = G[1 + 2v_{2,2}\cos(2\Delta\phi)]$. In previous analyses it was assumed that in the low-multiplicity events only effects of hard processes and a constant combinatoric contribution are present. In the ZYAM method (Zero Yield At Minimum) a constant pedestal was subtracted from the values of $Y(\Delta\phi)$ for peripheral, low-multiplicity events and this function was then subtracted from results obtained at larger multiplicities.

There are however indications that the modulated "ridge" component is present even in events with the lowest multiplicities, thus a new template-based method was proposed [3, 4]. The yields $Y(\Delta\phi)$ obtained for larger multiplicities are thus described by a template function:

$$Y^{templ}(\Delta\phi) = FY^{periph}(\Delta\phi) + G[1 + 2v_{2,2}\cos(2\Delta\phi)] \tag{2}$$

where F and $v_{2,2}$ are free parameters and G is fixed by normalization. The function with the second order harmonics very well approximates the data (Fig. 1 (left panel)), however in a closer examination of the difference between data points and the template, the contributions from higher harmonics are clearly visible (Fig. 1 (right panel)). These higher order terms were included in the template function and the values of v_n harmonics (n=2–4) are obtained assuming $v_n(p_{\mathrm{T}_1}) = v_{n,n}(p_{\mathrm{T}_1}, p_{\mathrm{T}_2})/\sqrt{v_{n,n}(p_{\mathrm{T}_2}, p_{\mathrm{T}_2})}$, where p_{T_1} and p_{T_2} are transverse momentum intervals of the trigger and associated particle, respecively.

In Fig. 2 the v_2 values obtained using the new template fit method and the ZYAM method are shown. For three different selections of the peripheral event sample, the template fit method gives consistent results, while the values of v_2 from the ZYAM method are varying and are always smaller than those from the template fit. Such differences are expected in the presence of the $\cos(2\Delta\phi)$ modulation in the peripheral events. In the ZYAM method such contribution is directly subtracted and thus the calculated $v_{2,2}^{ZYAM}$ is reduced, while in the template fit method presence of this contribution leads to a change of the G parameter without significant modification of $v_{2,2}^{templ}$.

3. Results and conclusions

In Figure 3 all important properties of the long-range "ridge" correlations for *pp* and *p*+Pb collisions are summarized. For the two energies of *pp* collisions considered, 5.02 and 13 TeV, the

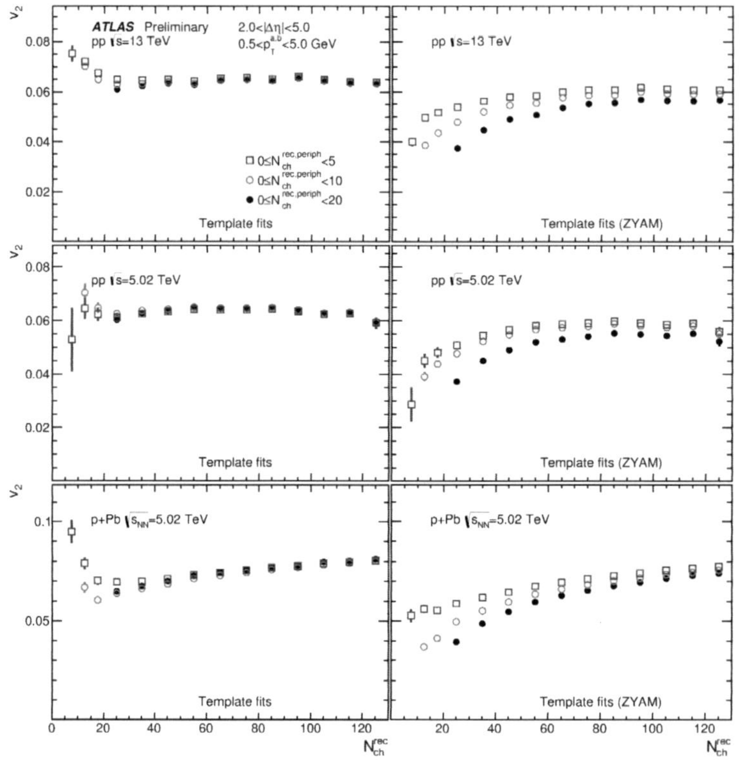

Figure 2. Dependence of v_2 on the choice of the peripheral bin for the template fitting method (left panels) and ZYAM-based method (right panels) s [4]. The top, middle and bottom panels correspond to 13 TeV *pp* collisions, 5.02 TeV *pp* collisions and 5.02 TeV *p*+Pb collisions, respectively.

v_2 values from the template method are very similar, both when presented as a function of event multiplicity and as a function of p_T of the trigger particle. In *pp* collisions there is no apparent dependence of v_2, v_3 or v_4 on event multiplicity (the last two shown for 13 TeV only). This is in contrast with the results obtained for *p*+Pb collisions at 5.02 TeV, where all harmonics are growing with increasing multiplicity. The values of v_2 and v_3 are similar for *pp* and *p*+Pb collisions only for the lowest multiplicities, while in the case of v_4 they seem to be similar in a wider multiplicity range.

The dependence of v_2 on p_T of the trigger particle for high multiplicity events both for *p*+Pb and *pp* collisions (Fig. 3 right panels) has a shape similar to that observed earlier in collisions of nuclei. However, the maximal value of $v_2(p_T)$ is about 1.5 times higher for *p*+Pb collisions than for *pp* collisions. The higher harmonics, v_3 and v_4, are increasing with p_T for both systems at least up to 4 GeV. Their values are higher for *p*+Pb collisions, but the difference is relatively smaller than for v_2. Other results of the template fit method are shown in [3, 4].

In summary, the new template fit method applied to the *pp* and *p*+Pb data allows to calculate the values of flow harmonics, v_n. They do not depend on the selection of peripheral events and are always slightly higher than those obtained from the ZYAM method. In *pp* collisions v_n values do not depend on event multiplicity, in *p*+Pb collisions they are growing with increasing event

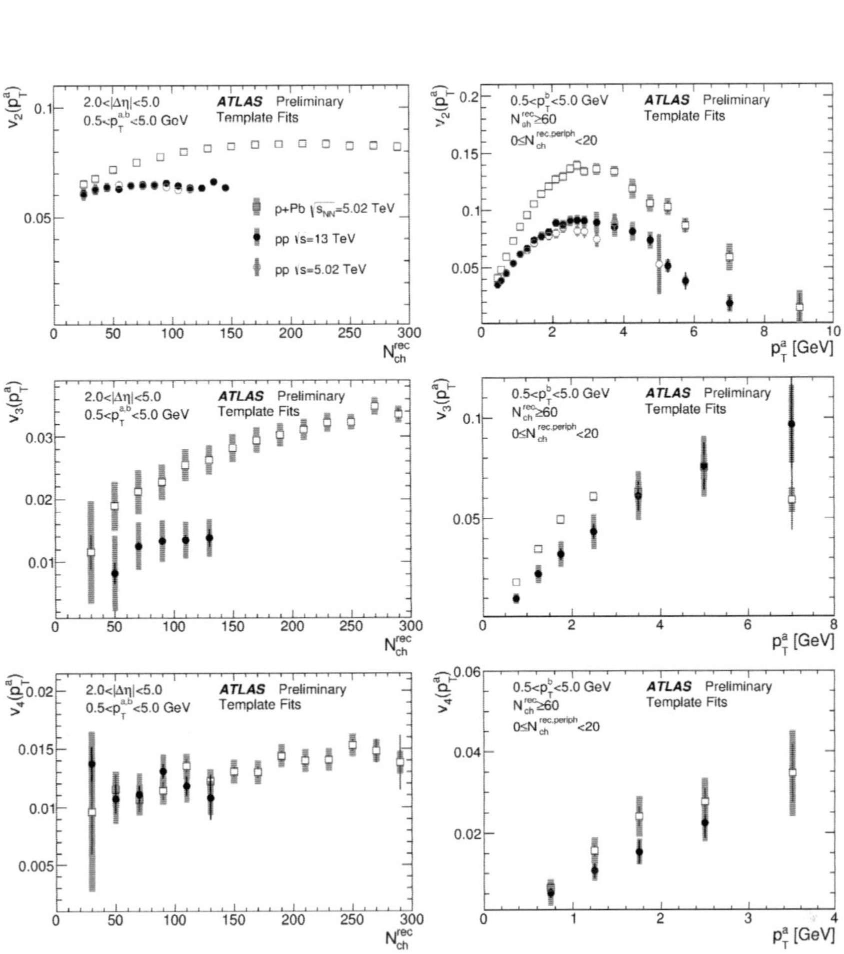

Figure 3. Comparison of the v_n obtained from the template fitting procedure in the 13 TeV *pp*, 5.02 TeV *pp*, and 5.02 TeV *p*+Pb data, as a function of multiplicity of reconstructed tracks, $N_{\rm ch}^{\rm rec}$, (left panels) and as a function of the transverse momentum, $p_{\rm T}$ (right panels) [4]. From top to bottom the rows correspond to $n = 2$, 3 and 4 respectively. The error bars and shaded bands indicate statistical and systematic uncertainties, respectively.

multiplicity and are always larger than in *pp* collisions.

Acknowledgments

This work was supported in part by the National Science Centre, Poland grant 2015/18/M/ST2/00087 and by PL-Grid Infrastructure.

References

[1] ATLAS Collaboration 2012 *Phys. Lett.* B **707** 330.
[2] CMS Collaboration 2010 *JHEP* **1009** 091.
[3] ATLAS Collaboration 2016 *Phys. Rev. Lett.* **116** 172301.
[4] ATLAS Collaboration 2016 ATLAS-CONF-2016-026 http://cds.cern.ch/record/2157690.
[5] ATLAS Collaboration 2008 *JINST* **3** S08003.

IOP Conf. Series: Journal of Physics: Conf. Series **779** (2017) 012058 doi:10.1088/1742-6596/779/1/012058

Collectivity of (non-)strange hadrons in high-multiplicity pp with CMS

George S. F. Stephans
For the CMS collaboration
Laboratory for Nuclear Science, Massachusetts Institute of Technology, Cambridge, MA USA
E-mail: gsfs@mit.edu

Abstract. Observation of an enhanced long-range, near-side, two-particle correlation (known as the "Ridge") in high-multiplicity pp and pPb collisions opened up new opportunities of exploring QCD dynamics in small collision systems. The CMS detector at the LHC has excellent capabilities of reconstructing weakly decaying strange hadrons such as K^0_S, Λ and Ξ. Studies of strange hadron production and correlations in small colliding systems provide additional insights into the physical origin of the observed collective phenomena. New results for p_T spectra and long-range two-particle correlations for charged particles and identified strange hadrons in high-multiplicity pp and pPb collisions are presented. The data at various collision energies for pp and pPb collisions are compared to those obtained for PbPb at similar multiplicities. Multi-particle cumulants for pp and pPb events are studied in order to more fully explore the collective nature of the long-range correlations.

1. Introduction

As all of the material presented in this talk has been submitted for publication, this proceeding will only briefly summarize the results and refer the reader to the final submitted papers. Full information about all heavy ion physics results from CMS, including figures, data tables, and links to the relevant submitted and/or published work, can be found at:
`https://twiki.cern.ch/twiki/bin/view/CMSPublic/PhysicsResultsHIN`.

2. Strange hadron production in pp, pPb, and PbPb: HIN-15-006

These results are presented in Ref. [1] and are summarized here.

Measurements of strange hadron (K^0_S, Λ, and Ξ) transverse momentum spectra in pp, pPb, and PbPb collisions are presented over a wide range of event charged-particle multiplicity and particle rapidity. The study is based on samples of pp collisions at $\sqrt{s} = 7$ TeV, pPb collisions at $\sqrt{s} = 5.02$ TeV, and PbPb collisions at $\sqrt{s_{NN}} = 2.76$ TeV, collected with the CMS detector at the LHC. In the context of hydrodynamic models, the measured particle spectra are fitted to a blast wave function, which describes an expanding fluid-like system. When comparing at a similar multiplicity, the extracted radial-flow velocity parameters are found to be larger in pp and pPb collisions than that in PbPb collisions. The average transverse kinetic energy $< \mathrm{KE_T} >$ of strange hadrons is observed to increase with multiplicity, with a stronger increase for heavier particles. At similar multiplicities, the difference in $< \mathrm{KE_T} >$ between the strange-particle species is larger in the smaller pp and pPb systems than in the PbPb system. For pPb collisions, $< KE_T >$ in the Pb-going direction for K^0_S (Λ) is 6% (12%) larger than in the p-going direction for events with the highest particle multiplicities.

3. Correlations of changed and strange hadrons in pp: HIN-16-010

These results are presented in Ref. [2] and are summarized here.

The CMS detector has been used to measure two- and multi-particle azimuthal correlations with K^0_S, Λ, and inclusive charged particles over a broad pseudorapidity and transverse momentum range in pp collisions at $\sqrt{s}$ =5, 7, and 13 TeV. With the implementation of high-multiplicity triggers during the LHC 2010 and 2015 pp runs, the correlation data are explored over a broad particle multiplicity range. The observed long-range ($|\Delta\eta| > 2$) correlations are quantified in terms of azimuthal anisotropy Fourier harmonics (v_n). The elliptic (v_2) and triangular (v_3) flow Fourier harmonics are extracted from long-range two-particle correlations. After subtracting contributions from back-to-back jet correlations estimated using low-multiplicity data, the v_2 and v_3 values are found to increase with multiplicity for $N^{offline}_{trk} \lesssim 100$, and reach a relatively constant value at higher values of $N^{offline}_{trk}$. The p_T dependence of the v_2 harmonics in high-multiplicity pp events is found to have no or very weak dependence on the collision energy. In low-multiplicity events, similar v_2 values as a function of p_T are observed for inclusive charged particles, K^0_S, and Λ, possibly reflecting a common back-to-back jet origin of the correlations for all particle species. Moving to the higher-multiplicity region, a particle species dependence of v_2 is observed with and without correcting for jet correlations. For $p_T \lesssim 2$ GeV/c, the v_2 of K^0_S is found to be larger than that of Λ. This behavior, which is consistent with predictions of hydrodynamic models, is similar to what was previously observed for identified particles produced in pPb and AA collisions at RHIC and the LHC. This mass ordering is reversed at higher p_T values. Finally, v_2 signals based on four- and six-particle correlations are observed for the first time in pp collisions. The v_2 values obtained with two-, four-, and six-particle correlations at $\sqrt{s}$ =13 TeV are found to be comparable within uncertainties. These observations provide strong evidence supporting the interpretation of a collective origin for the observed long-range correlations in high-multiplicity pp collisions.

References

[1] Khachatryan V, et al. (CMS Collaboration) 2016 Multiplicity and rapidity dependence of strange hadron production in pp, pPb, and PbPb collisions at the LHC *Preprint* nucl-ex/1605.06699
[2] Khachatryan V, et al. (CMS Collaboration) 2016 Evidence for collectivity in pp collisions at the LHC *Preprint* nucl-ex/1606.06198

IOP Conf. Series: Journal of Physics: Conf. Series **779** (2017) 012059 doi:10.1088/1742-6596/779/1/012059

Measurement of D^0 elliptic and triangular flow in Au+Au collisions at $\sqrt{s_{\rm NN}} = 200$ GeV at RHIC

Michael R. Lomnitz[1] **(STAR collaboration)**

[1] Nuclear Science Division, Lawrence Berkeley National Lab, MS70R0319, One Cyclotron Road, Berkeley, CA 94720, USA

E-mail: `mrlomnitz@lbl.gov`

Abstract. Due to their large masses, heavy quarks are predominantly produced through initial hard scatterings in heavy-ion collisions. As such, they experience the entire evolution of the hot and dense medium created in such collisions and are expected to thermalize much more slowly than light flavor quarks. For instance, the azimuthal anisotropy of charm quarks with respect to the reaction plane over a broad momentum range can provide insights into the degree of thermalization and the bulk properties of the system. Specifically at low transverse momenta we can examine the bulk properties in the strongly coupled regime.
In this talk we present the STAR measurement of elliptic (v_2) and triangular flow (v_3) of D^0 mesons in Au+Au collisions at $\sqrt{s_{\rm NN}} = 200$ GeV obtained from the first year of physics running with the new STAR Heavy Flavor Tracker. Comparison with the azimuthal anisotropy of other particle species and a series of model calculations will be shown, and the charm quark dynamics in the sQGP medium will be discussed.

1. Introduction

Heavy flavor quarks are suggested to be an excellent probe to study the strongly coupled quark-gluon plasma (sQGP) as they are predominantly produced early in heavy-ion collisions through hard scattering processes. They experience the full evolution of the system, while their large masses are mostly unaffected by the QCD medium. Furthermore, by studying how the heavy quarks are affected as they propagate through the sQGP we can access information about the transport properties of the medium, for example $2\pi T D_s$, which is given in terms of the temperature T and the charm spatial diffusion coefficient D_s [1].
Recent measurements at the Relativistic Heavy Ion Collider (RHIC) and the Large Hadron Collider (LHC) show that charmed hadron yields at high transverse momenta (p_T) in central collisions are considerably suppressed, suggesting strong charm quark interactions with the bulk, while the elliptic flow (v_2) of charmed hadrons measured at LHC is comparable to that of light hadrons [2, 3, 4]. The enhancement in the nuclear modification factor at intermediate p_T observed at RHIC [2] is suggestive of both charm flow and hadronization via coalescence. Studies of charm flow through semi-leptonic decay channels at RHIC suffer from large uncertainties and, as such, a precise measurement of charmed hadron v_2 is needed to fully understand the properties of the sQGP medium. Furthermore, some model calculations [5] including fluctuations in the initial conditions together with charm quark interactions with the medium have predicted a

non-zero value for charm quark triangular flow (v_3), providing another handle to study the early stages of the collisions.

2. Experimental set-up

The data used in these analyses were recorded in year 2014 by the STAR experiment at RHIC in Brookhaven National Laboratory, USA. The STAR experiment possesses full azimuthal coverage at mid-rapidity using the Time Projection Chamber (TPC) to reconstruct tracks inside a uniform 0.5 T magnetic field. The TPC can also provide excellent particle identification (PID) for hadrons over a broad range in p_T by combining information from the TPC and the Time Of Flight (TOF) detector.
The entire Heavy Flavor Tracker (HFT) was installed for the first time in 2014 and greatly improved STAR's tracking resolution. It provided a track pointing resolution of less than 50 μm for kaons with p_T = 750 MeV/c. About 780 million Au+Au collisions at $\sqrt{s_{\rm NN}} = 200$ GeV with a Minimum-Bias (MB) trigger were analyzed to reconstruct charmed hadrons. A cut on the reconstructed collision position, or primary vertex (PV), along the beam line ($|{\rm V_z}| < 6$ cm) is applied to ensure good acceptance in the HFT detector. D^0 mesons are fully reconstructed through the hadronic channel:

$$D^0(\overline{D^0}) \to K^{\pm}\pi^{\mp},\ c\tau \sim 120\ \mu m,\ \text{B.R. } 3.9\%$$

The D^0 daughter tracks are required to have: i) a minimum of 20 space points in the TPC, ii) at least one hit in each of the three inner layers of the HFT, iii) $p_T > 0.6$ GeV/c and iv) pseudorapidity $|\eta| < 1$. Particle identification is done using energy loss (dE/dx) from the TPC, i.e. selecting candidates within 2 to 3 standard deviations from the expected values. It is further enhanced using the TOF when available. The value of $1/\beta$ is estimated from the momentum and the timing measured in the TOF, and is required to be less than 0.03 from the expected values.
Once daughter candidates have been identified, the (displaced) decay vertex is reconstructed at the mid-point on their distance of closest approach (DCA). The combinatorial background is greatly suppressed by imposing criteria on the following topological variables: decay length (distance between primary and decay vertices), DCA between daughter tracks, DCA between reconstructed D^0 candidate and the PV, and DCA between daughter tracks and the PV.

3. Azimuthal anisotropy

Once the D^0 candidates have been selected, the second (v_2) and third (v_3) order azimuthal anisotropies are studied. In the case of v_2 the results obtained using two different methods were compared: the event plane method and the two particle correlation method, which will be discussed briefly in the following paragraphs. For the measurement of v_3 only the event plane method was used.
In the event plane method the second (third) order event plane, $\Psi_{2(3)}$, is reconstructed using TPC tracks and corrected for the non-uniform detector efficiency [6]. To reduce the contributions from non-flow effects originating from other two or multi-particle correlations, all tracks within a relative pseudorapidity gap $|\Delta\eta| \leq 0.15$ around the reconstructed D^0 candidates are removed from the event plane reconstruction. The azimuthal distribution of D^0 mesons with respect to the event plane $\phi - \Psi_n$ is obtained and weighted by $1/(\epsilon R)$, where ϵ is the D^0 reconstruction efficiency and R is the event plane resolution for each centrality. In each $\phi - \Psi_n$ bin, the mixed-event background is subtracted from the unlike-sign invariant mass spectrum. The D^0 yield is obtained by either the fitting or sideband method: at low p_T the invariant mass spectrum is fitted with a Gaussian function, representing the signal, and a first order polynomial function describing the correlated background; for the last p_T bin (5-10 GeV/c) fitting is limited by low background statistics, and the D^0 yield is obtained by subtracting scaled unlike-sign counts in

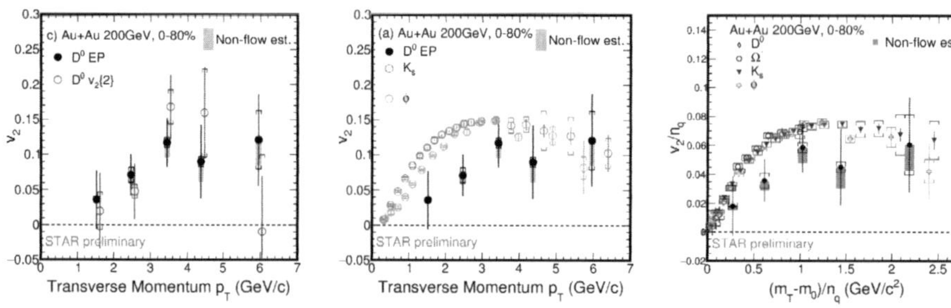

(a) (b) (c)

Figure 1: a) Comparison between v_2 from event plane and two particle correlation methods. b) Measured v_2 for D^0 compared to that of light hadrons. c) v_2/n_q vs $(m_T - m_0)/n_q$ for D^0 and other particle species.

two invariant mass regions around the signal region. The observed v_n^{obs} is then obtained by fitting the yield versus $\phi - \Psi_n$ with the functional form $A(1 + 2v_n \cos(n(\phi - \Psi_n)))$. Finally, the observed v_n^{obs} is corrected for the average event plane resolution $\langle 1/R \rangle$ to obtain the true value of the azimuthal anisotropy.

In the two particle correlation method, the average D^0-hadron correlation is calculated according to $V_2^{D-h} \equiv \langle \cos 2(\phi_D - \phi_h) \rangle$. Assuming that the D^0 and hadrons have no correlation other than through the event plane, it can be shown that $V_2^{D-h} = V_2^D V_2^h$. Finally, together with the hadron-hadron correlation, $V_2^{h_1-h_2} \equiv \langle \cos 2(\phi_{h1} - \phi_{h2}) \rangle$ the elliptic flow of D^0 can be obtained $v_2^D = V_2^{D-h}/(V_2^{h1-h2})^{1/2} = \langle \cos 2(\phi_D - \phi_h) \rangle / \langle \cos 2(\phi_{h1} - \phi_{h2}) \rangle^{1/2}$. The D^0 background is estimated by averaging both the like- and unlike-sign sidebands as well as the like-sign invariant mass in the D^0 signal range. As in the case of the event plane method, the contribution from non-flow effects is suppressed by introducing a gap $|\Delta\eta| > 0.2$ in the measurement of $D - h$ correlations, while only particles in opposite ends of the TPC are used in the hadron-hadron correlations (i.e. $h_1 \in \eta > 0$ and $h_2 \in \eta < 0$). The event plane and two particle correlation methods show good overall agreement within systematic uncertainties, as is shown in figure 1a. However, non-flow effects may still contribute to the measured v_2. This contribution is estimated by scaling the non-flow contribution estimated in p+p collisions to Au+Au collisions [7], in which case it can be written as $\langle \sum_i \cos(2(\phi_D - \phi_h)) \rangle / M v_2$ where M and v_2 are the average multiplicity and hadron v_2 measured in Au+Au collisions. Figure 1b shows the v_2 of D^0 mesons compared with that of other particle species [8]. The D^0 v_2 is non-zero for $p_T > 2$ GeV/c and is systematically below that of light hadrons in the range $1 < p_T < 4$ GeV/c within large uncertainties. To account for the different particle masses and number of constituent quarks (n_q), the comparison is done in figure 1c by plotting v_2/n_q vs $(m_T - m_0)/n_q$ where $m_T = \sqrt{p_T^2 + m_0^2}$. Once the scaling has been done the difference among particle species is reduced, although D^0 is still systematically below the light particle species in 0-80% centrality.

In figure 2a the results for D^0 are compared with four theoretical models. The TAMU model [9] uses a non-perturbative T-matrix approach assuming the two-body interactions can be described by a potential as a function of the transferred 4-momentum. Two curves from TAMU are shown: the scenario including charm diffusion (blue) agrees with the data while the predictions without charm diffusion (magenta) are systematically lower. The calculation by the SUBATECH group [5] employs pQCD with Hard Thermal Loop approximation for soft collisions and can also describe the measurement over the whole p_T range. In the model developed by the Duke group [10] the diffusion coefficient $2\pi T D_s$ is a free parameter which, in the case of the red curve shown,

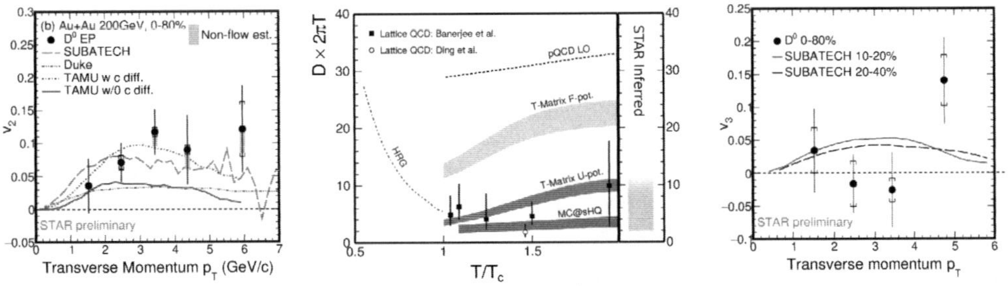

(a) (b) (c)

Figure 2: a) D^0 v_2 in 0-80% centrality compared to model calculations. b) Diffusion coefficient from model calculations and the inferred range from STAR measurements. c) D^0 v_3 measured in 0-80% centrality compared to model calculations in 10-20% and 20-40% central events.

has been constrained using the R_{AA} measured at LHC to be roughly 7. It is seen that this model underpredicts the v_2 observed in our data. Figure 2b shows the extracted diffusion coefficient from different model calculations compared to the yellow band on the far right showing the range of inferred values that are compatible with the measurement presented here. Figure 2c shows the D^0 triangular flow in 0-80% central events compared with model calculations in 10-20% and 20-40% central events from the SUBATECH group. The uncertainties in the current results are, however, too large to draw any firm conclusion.

4. Summary

STAR has carried out the first heavy flavor measurements in heavy-ion collisions using the newly installed, state-of-the-art vertexing detector, the HFT. The measured charmed meson v_2 in Au+Au collisions is found to be non-zero, but systematically below v_2 of light hadrons in 0-80% centrality. Comparison to a series of models shows that they are able to describe the data, favoring the scenario where charm quarks flow with the medium. The models infer a range of the charm diffusion coefficient $2\pi TD_s$ between 2 and ~12. An expected factor of 2-4 improvement in D^0 significance is expected from the reprocessed 2014 data as well as another factor of 2-3 from the 2016 data set. Together these should allow a more thorough centrality dependent study of the azimuthal anisotropy and may also provide an answer to whether the D^0 v_3 is non-zero.

Acknowledgments

This material is based upon work supported by the U.S. Department of Energy, Office of Science, Office of Workforce Development for Teachers and Scientists, Office of Science Graduate Student Research (SCGSR) program. The SCGSR program is administered by the Oak Ridge Institute for Science and Education for the DOE under contract number DE-AC05-06OR23100.

[1] Abreu L M, Cabrera D, Llanes-Estrada F J and Torres-Rincon J M 2011 *Annals of Physics* **326** 2737 – 2772
[2] Adamczyk L and et al ((STAR Collaboration)) 2014 *Phys. Rev. Lett.* **113**(14) 142301
[3] Abelev B and et al (ALICE Collaboration) 2013 *Phys. Rev. Lett.* **111**(10) 102301
[4] Adam J and et al (ALICE Collaboration) 2015 *arXiv:1509.06888*
[5] Nahrgang M, Aichelin J, Bass S, Gossiaux P B and Werner K 2015 *Phys. Rev. C* **91**(1) 014904
[6] Poskanzer A M and Voloshin S A 1998 *Phys. Rev. C* **58**(3) 1671–1678
[7] Adams J and et al (STAR Collaboration) 2004 *Phys. Rev. Lett.* **93**(25) 252301
[8] Abelev B I and et al (STAR Collaboration) 2008 *Phys. Rev. C* **77**(5) 054901
[9] He M, Fries R J and Rapp R 2012 *Phys. Rev. C* **86**(1) 014903
[10] Cao S, Qin G Y and Bass S A 2013 *Phys. Rev. C* **88**(4) 044907

IOP Conf. Series: Journal of Physics: Conf. Series **779** (2017) 012060 doi:10.1088/1742-6596/779/1/012060

Viscous Damping of Anisotropic Flow in $7.7 - 200$ GeV Au+Au Collisions

Niseem Magdy (For the STAR Collaboration)

Department of Chemistry, Stony Brook University, Stony Brook, NY, 11794-3400, USA

E-mail: niseemm@gmail.com

Abstract. Recent STAR measurements of the anisotropic flow coefficients v_n $(2 \leq n \leq 5)$ in Au+Au collisions at RHIC, are presented for the full span of energies ($7.7 - 200$ GeV) employed in beam energy scan I (BES-I). The measurements which can provide strong constraints for the baryon chemical potential (μ_B) and temperature (T) dependence of the specific shear viscosity η/s, indicate sizable dependencies on harmonic number n, p_T and centrality, with similar patterns [but different magnitudes] across the beam energies studied. An excitation function for the viscous coefficient, extracted via specific ratios of v_n for a fixed centrality, indicates a non-monotonic pattern which could be related to the onset of critical reaction dynamics in the BES-I energy range.

1. Introduction

Studies of the properties of the strongly interacting medium produced in heavy ion collisions is a major theme of current heavy ion research. Ongoing programs are underway at RHIC [1], SPS [2] and LHC [3], and new programs are slated for the future FAIR [4] and NICA [5] facilities. These programs are at the forefront of experimental efforts designed to map the QCD phase diagram and delineate the thermodynamic and transport properties of each QCD phase. The viscous coefficient, which is directly related to the specific shear viscosity of the QCD matter produced in heavy-ion collisions, can be probed via measurements of anisotropic flow [6, 7, 8, 9]. Initial works focused on the first and second flow harmonics, termed directed and elliptic flow respectively, as well as higher-order even harmonics [10]. Studies of the third harmonic [11, 12] received attention subsequent to the realization that event-by-event fluctuations [11, 12, 13, 14, 15] lead to non-zero magnitudes for the odd harmonics.

In this work, v_n [for $2 \leq n \leq 5$] measurements for Au+Au collisions, obtained in BES-I with the STAR detector, are presented and used to extract an excitation function for the viscous coefficient. Such an excitation function can give important insight on the temperature dependence of the specific viscosity $\eta/s(T)$, as well as an indication for a possible non-monotonic pattern which could signal critical reaction dynamics [16, 17].

2. Measurements

The two-particle azimuthal correlation function technique is employed for the present measurements. The correlation function $C(\Delta\phi, \Delta\eta)$ is given as,

$$C(\Delta\phi, \Delta\eta) = \frac{(dN/d\Delta\phi)_{Same}}{(dN/d\Delta\phi)_{Mix}}, \tag{1}$$

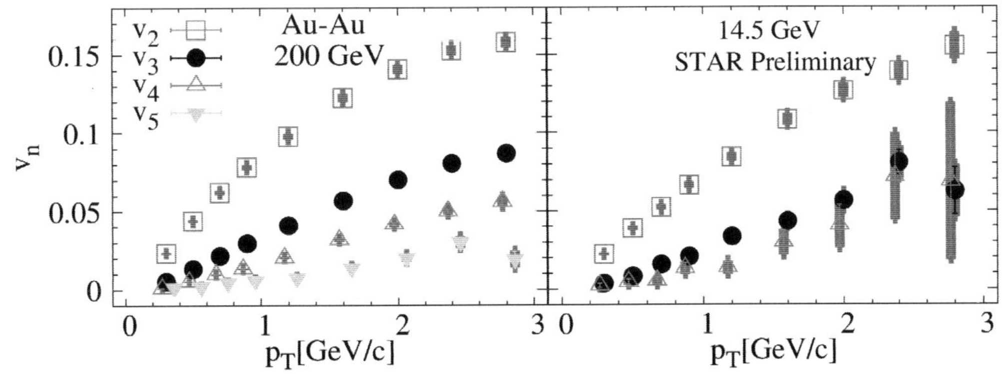

Figure 1. Illustrative examples of $v_n(p_T)$ for 0-40% central Au+Au collisions at two beam energies as indicated.

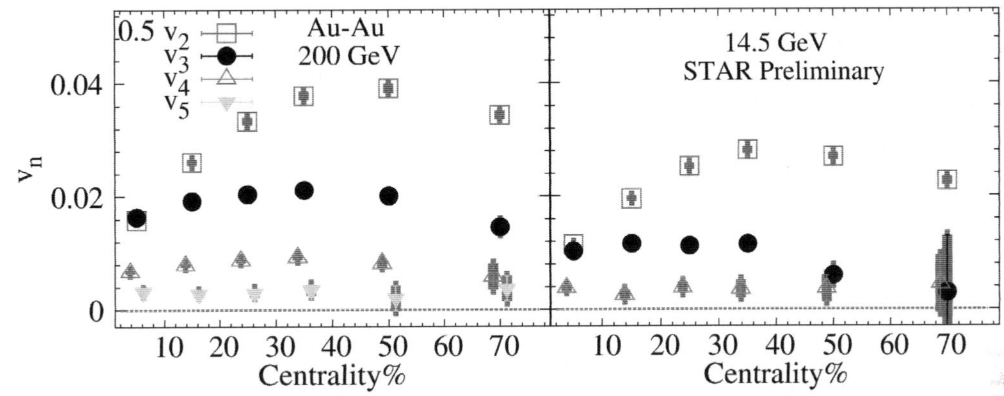

Figure 2. Centrality dependence of the p_T-integrated v_n values for Au+Au collisions for two beam energies as indicated.

where $(dN/d\Delta\phi)_{Same}$ and $(dN/d\Delta\phi)_{Mix}$ are the distributions for charged hadron pairs from the *same* and *mixed* events respectively; $\Delta\phi$ and $\Delta\eta$ are the azimuthal angle difference and the pseudorapidity difference between charged hadron pairs. The condition $\Delta\eta > 0.7$ was used to minimize the influence of the non-flow contributions. On average the relative contributions were found to be less than 3%.

The values for v_n are obtained via Fourier analysis the correlation function (cf. Eq.(1)) in conjunction with the factorization assumption;

$$v_n^2 = \left(\sum_{\Delta\phi} C(\Delta\phi, \Delta\eta)\cos(n\Delta\phi)\right) / \left(\sum_{\Delta\phi} C(\Delta\phi, \Delta\eta)\right). \quad (2)$$

$$v_n(p_T) \quad = \quad v_n^2(p_T, p_T^{ref})/\sqrt{v_n^2(p_T^{ref}, p_T^{ref})}, \quad (3)$$

where p_T^{ref} is the transverse momentum of the reference particle. The factorization assumption Eq.(3) was tested via several evaluations of $v_n(p_T)$ for different selections for p_T^{ref}.

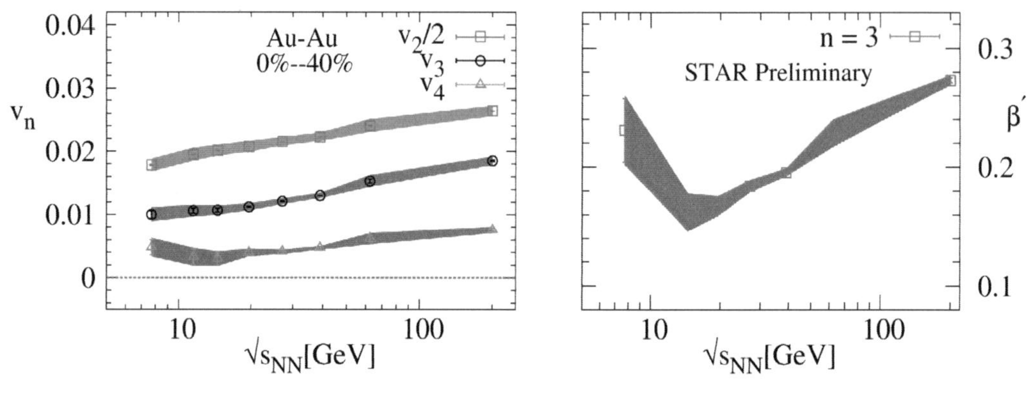

Figure 3. $\sqrt{s_{\rm NN}}$ dependence of the p_T-integrated v_n (left panel) and the viscous coefficient $\beta'' \propto \eta/s$ (right panel). Results are shown for 0-40% central Au+Au collisions; the shaded lines are the systematic uncertainty.

Figures 1 and 2 show representative examples of the p_T (0-40% central) and centrality dependence of v_n for $\sqrt{s_{\rm NN}} = 200$ and 14.5 GeV. These figures indicate a sizable dependence of the magnitude of v_n on p_T and the harmonic number n, with similar trends for the two beam energies shown. Similar patterns, albeit with different magnitudes, were observed for the other beam energies. Fig. 2 further suggests a weak centrality dependence for the higher harmonics, which all decrease with decreasing values of $\sqrt{s_{\rm NN}}$. These patterns may be related to the detailed dependence of the viscous effects in the created medium, which serve to attenuate the magnitude of v_n.

3. Extraction of the viscous coefficient

For a given beam energy, the viscous coefficient β'' which is related to η/s, can be estimated via the acoustic ansatz for anisotropic flow [19, 20, 21]. That is,

$$\frac{v_n(p_T, cent)}{\epsilon_n(cent)} \propto \exp(-n^2\beta'), \;\; \beta' \propto \frac{(\eta/s)}{RT}, \tag{4}$$

where RT and ϵ_n are the dimensionless size and the eccentricities of the collision zone respectively. For a given centrality, ϵ_n changes very slowly, if at all, with beam energy. Therefore, Eq. 4 can be further simplified as;

$$\frac{v_n^{1/n}}{v_{n'}^{1/n'}} \propto \exp(-(n-n')\beta'). \tag{5}$$

or

$$\ln(v_n^{1/n}/v_{n'}^{1/n'}) \;=\; -(n-n')\beta' + \ln(c), \tag{6}$$

where c is a constant and $n \neq n'$. Since the dimensionless size $RT \propto (dN/d\eta)^{1/3}$ [22], Eq. 6 reduces to;

$$\beta'' \;=\; (dN/d\eta)^{1/3}\ln(v_n^{1/n}/v_{n'}^{1/n'}) \;=\; -\,a\;(n-n')\;\eta/s \;+\; (dN/d\eta)^{1/3}\;\ln(c). \tag{7}$$

where a is a constant and $n' = 2$. Note that the viscous coefficient β'' is linearly related to η/s.

Estimates for β'' were made as a function of $\sqrt{s_{\rm NN}}$ to obtain constraints for its (T, μ_B) dependence, as well as to search for a possible non-monotonic pattern in its excitation function. Such a pattern could result from abrupt changes in the transport coefficients and relaxation rates due to anomalies in the expansion dynamics resulting from critical reaction dynamics [16, 17, 18].

The left panel of Fig. 3 shows the beam energy dependence of the p_T-integrated v_n for $0-40\%$ central Au+Au collisions. It indicates an essentially monotonic trend for v_2, v_3 and v_4 with $\sqrt{s_{\rm NN}}$ as might be expected for a temperature increase as $\sqrt{s_{\rm NN}}$ increases. In contrast, the excitation function for $\beta'' \propto \eta/s$, shown in the right panel of Fig. 3, indicates a non-monotonic trend over the same beam energy range. A similar trend has been indicated in hydrodynamical calculations [23].

4. Conclusion

In summary, we have presented a comprehensive set of STAR v_n measurements for Au+Au collisions spanning the BES-I beam energy scan. The measurements indicate a rich set of dependencies on harmonic number n, p_T and centrality across the beam energies studied, which can provide strong model constraints for detailed extraction of the μ_B and T dependence of the specific shear viscosity. An excitation function for the viscous coefficient, obtained from ratios of v_n for different values of n, indicates a non-monotonic pattern which could be related to the onset of critical reaction dynamics in the BES-I energy range.

References

[1] Experimental highlights of the RHIC program, P. Fachini, AIP Conf. Proc. **857** 62-75 (2006).
[2] The SPS heavy ion programme, Helmut Satz, Phys. Rept. **403-404** 33-50 (2004).
[3] Physics programme of ALICE experiment, K. Safarik, Nucl. Phys. A **749** 229-242 (2005).
[4] FAIR and its experimental program, W.F. Henning, J. Phys. G **34** S551-S557 (2007).
[5] Heavy-ion program at NICA/MPD at JINR, A. Sorin et al., Nucl. Phys. A **855** 510-513 (2011).
[6] D. Teaney, Phys. Rev. C **68**, 034913 (2003).
[7] Roy Lacey and A. Taranenko, PoS CFRNC **2006**, 021 (2006).
[8] T. Hirano et al., Phys. Lett. B **636**, 299 (2006).
[9] M. Luzum and P. Romatschke, Phys. Rev. Lett. **103**, 262302 (2009).
[10] P. F. Kolb, Phys. Rev. C **68**, 031902(R) (2003).
[11] A. P. Mishra, R. K. Mohapatra, P. S. Saumia and A. M. Srivastava, Phys. Rev. C **77**, 064902 (2008).
[12] B. Alver and G. Roland, Phys. Rev. C **81**, 054905 (2010).
[13] D. Teaney and L. Yan, Phys. Rev. C **83**, 064904 (2011).
[14] P. Sorensen, B. Bolliet, A. Mocsy, Y. Pandit and N. Pruthi, Phys. Lett. B **705**, 71 (2011).
[15] Roy A. Lacey, Rui Wei, N. N. Ajitanand, and A. Taranenko, Phys. Rev. C **83**, 044902 (2011).
[16] Laszlo P. Csernai, Joseph. I. Kapusta, Larry D. McLerran, Phys. Rev. Lett **97**, 152303 (2006).
[17] Roy A. Lacey et al., [arXiv:0708.3512].
[18] Roy A. Lacey, Phys. Rev. Lett. **114**, 142301 (2015).
[19] Roy A. Lacey et al., [arXiv:1301.0165].
[20] Edward Shuryak and Ismail Zahed, Phys. Rev. C **88**, 044915 (2013).
[21] Roy A. Lacey, A. Taranenko, N.N. Ajitanand and J.M. Alexander, [arXiv:1105.3782].
[22] Roy A. Lacey et al., [arXiv:1601.06001].
[23] Iu.A. Karpenko, P. Huovinen, H. Petersen and M. Bleicher, Phys. Rev. C **91**, 044915 (2015).

Insight from elliptic flow of open charm mesons using quark coalescence model at RHIC and LHC energies

Roli Esha*, Md. Nasim, Huan Zhong Huang
University of California, Los Angeles, CA 90095, USA
E-mail: *roliesha@physics.ucla.edu

Abstract. A study of elliptic flow of open charm mesons, D^0 and $D_S^{\pm}$ using quark coalescence as the mechanism of hadronization of heavy quarks will be presented. The coalescing partons are taken from a multi-phase transport model. The transverse momentum dependence of the elliptic flow parameter at mid-rapidity ($|y| < 1.0$) for minimum bias Au+Au collisions at $\sqrt{s_{NN}} = 200$ GeV (RHIC) and Pb+Pb collisions $\sqrt{s_{NN}} = 2.76$ TeV (LHC) for different values of partonic interaction cross-section and QCD coupling constant will be discussed. We have compared our calculations with the experimentally measured data at the LHC energy. We will also present the effect of specific viscosity on elliptic flow of open charm mesons within the transport model approach. Our study indicates that the elliptic flow of open charmed mesons is more sensitive to viscous properties of QGP medium compared to light hadrons.

1. Introduction

Heavy quarks are considered as an important probe to understand the properties of the Quark Gluon Plasma (QGP) created in relativistic heavy ion collisions as they are produced on a short time scale in hard partonic scatterings during the early stages of the nucleus-nucleus collision. As the probability of their thermal production is small, these are sensitive to medium dynamics. Heavy quarks decouple early in the evolution of QGP, thereby preserving the information from early stages of the system.

Elliptic flow, measured in heavy ion collisions, is the manifestation of the initial spatial anisotropy of collision geometry and the interaction among produced particles. The particle azimuthal distribution relative to reaction plane can be written in the form of Fourier series, the second coefficient of which is called the elliptic flow parameter, v_2 [1].

$$\frac{dN}{d\phi} \propto 1 + 2v_2 \cos(2(\phi - \Psi)); \qquad v_2 = \langle \cos(2(\phi - \Psi)) \rangle \tag{1}$$

where ϕ is the azimuthal angle of emitted particle and Ψ is the reaction plane. The magnitude and the conversion of geometrical eccentricity to elliptic flow depends on the dynamics and the equation of state of the medium.

This proceeding, which is about the study of elliptic of flow open charm mesons and its implications, is based on [2] and is organized in the following way. In Section 2, an introduction to the AMPT model with the prescription of the quark coalescence model used is discussed. A discussion on the elliptic flow of D-meson at RHIC (Au+Au at $\sqrt{s_{NN}} = 200$ GeV) and LHC (Pb+Pb at $\sqrt{s_{NN}} = 2.76$ TeV) energies and the effect of specific viscosity on elliptic flow of open charms is given in Section 3. We summarize in Section 4.

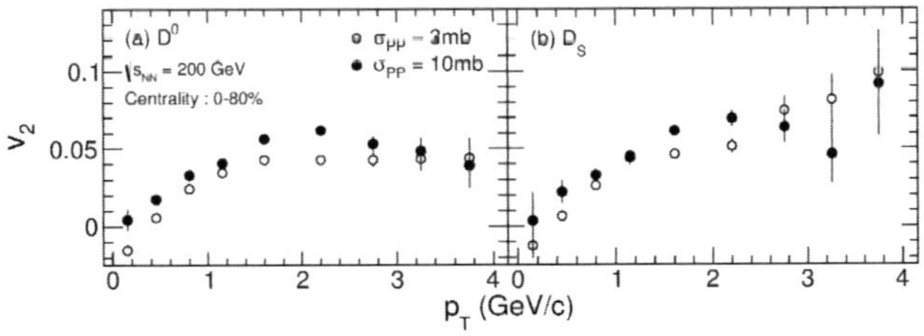

Figure 1. v_2 of D^0 and D_S as a function of p_T in 0-80% minimum bias Au+Au collisions at $\sqrt{s_{NN}} = 200$ GeV for $\sigma_{PP} = 3$ mb and 10 mb.

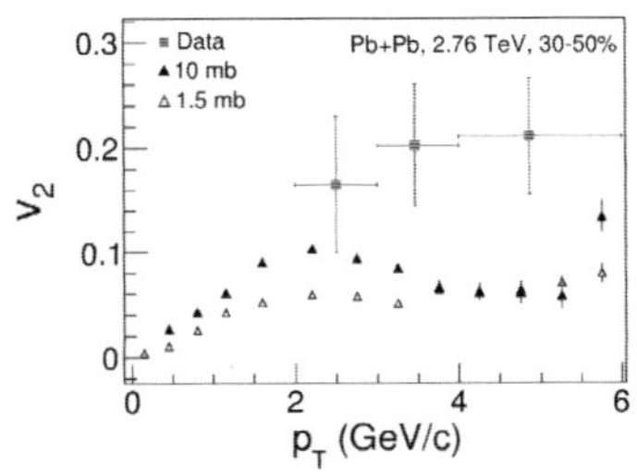

Figure 2. (Color online) Elliptic flow of D meson (average of D^0 and $D^{\pm}$) at mid-rapidity in Pb+Pb collision at $\sqrt{s_{NN}} = 2.76$ TeV for 30-50% centrality. Only statistical error is shown for ALICE data [5].

2. Quark Coalescence Mechanism

A Multi Phase Transport (AMPT) model consists of four main components: the initial conditions, partonic interactions, the conversion from the partonic to the hadronic matter, and hadronic interactions [3]. In the String melting version, a quark coalescence model is used instead to combine partons into hadrons. The dynamics of the subsequent hadronic matter is described by a hadronic cascade. However, hadronization of heavy quarks is not implemented in the string melting version of AMPT. Taking the partons from AMPT, we will implement the coalescence prescription described below to get the open charm mesons. Within the framework of the coalescence mechanism, the probability of producing a hadron from a soup of partons is determined by the overlap of the phase space distribution of partons at freeze-out with the parton Wigner phase space function inside the hadron under the assumption that the correlations between coalescing partons is weak and the binding energy of the formed hadron can be neglected. The Wigner phase space function for quarks inside a meson is obtained from its constituent quark wave function given by [4]

$$\rho^W(\mathbf{r}, \mathbf{k}) = \int \psi\left(\mathbf{r} + \frac{\mathbf{R}}{2}\right) \psi^\star\left(\mathbf{r} - \frac{\mathbf{R}}{2}\right) \exp(-i\mathbf{k} \cdot \mathbf{R}) d^3\mathbf{R} \tag{2}$$

$\mathbf{r}$ is the relative position and $\mathbf{k}$ is the relative momentum of the coalescing partons. The quark wave function is given by spherical harmonic oscillator.

3. Results and Discussions

The elliptic flow of D^0 and D_S mesons at mid-rapidity from the coalescence of heavy quarks in conjunction with AMPT model for different parton-parton interaction cross-section, σ_{PP} for

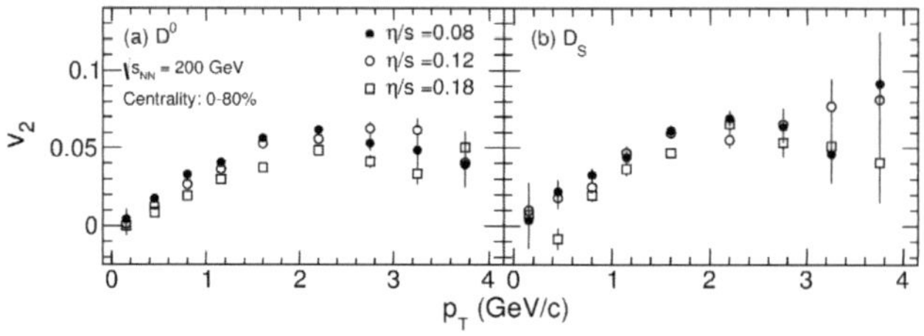

Figure 3. Elliptic flow of D^0 and D_S as a function of p_T for different specific viscous medium ($\eta_s/s = 0.08$, 0.12 and 0.18) in minimum bias Au+Au collisions at $\sqrt{s_{NN}} = 200$ GeV and σ_{PP} = 10 mb.

0-80% minimum bias Au+Au collisions at $\sqrt{s_{NN}} = 200$ GeV is shown in Fig. 1. The elliptic flow of open charm meson decreases with decrease in parton parton interaction cross-section, which is similar as observed for charged hadron. Fig. 2 shows a comparison between our model calculations for v_2 of D meson (average of D^0 and $D^{\pm}$) and measured D meson v_2 at 2.76 TeV for 30-50% central collisions by the ALICE experiment [5]. Our model calculations for both 1.5 mb (which describes charged hadrons well) and 10 mb under-predict the data for D-meson v_2. It has a different trend as well at high p_T. It would be very interesting to see the behavior of data at low p_T (below 2 GeV/c).

Transport coefficients play a major role in probing the properties of the soup of quarks and gluons created in high energy heavy ion collisions. In order to study this, we compare the behavior of v_2 of open charm mesons for different values of the ratio of shear viscosity, η_s, to entropy density, s, with p_T. In the AMPT model, this can be done by tuning the QCD coupling constant, α and screening mass, μ, keeping the parton scattering cross-section, σ_{PP} fixed, where $\sigma_{PP} \approx 9\pi\alpha^2/2\mu^2$. For a system of massless quarks and gluons at temperature T, the specific viscosity is given by [6]

$$\frac{\eta_s}{s} \approx \frac{3\pi}{40\alpha_s^2} \frac{1}{\left(9 + \frac{\mu^2}{T^2}\right) \ln\left(\frac{18+\mu^2/T^2}{\mu^2/T^2}\right) - 18} \tag{3}$$

Fig. 3 shows the change in v_2 of D^0 and D_S meson due to the variation of η_s/s in Au+Au collisions at $\sqrt{s_{NN}} = 200$ GeV. We show that v_2 decreases with increase in specific viscosity for low transverse momentum; the trend being more pronounced for D^0. This is consistent with the interpretation that increased shear viscosity reduces transverse expansion and hence reduces v_2. The production and propagation of open charm meson in QGP medium is expected to be different from light hadrons because of their large mass. Therefore, a study of the effect of specific viscosity on v_2 for both open charm mesons and charged hadrons will be useful to probe of the QGP medium. In addition, this can give us a constraint on the specific viscosity of the relatively early time scale of the medium created in relativistic heavy ion collisions.

The ratio of v_2 for $\eta_s/s = 0.08$ and for $\eta_s/s = 0.18$ is shown as function of p_T for 10–40% central collisions in Fig. 4. The solid red and open blue circles represents the results for charged hadrons and D^0 respectively. We can see that the change in v_2 for charged hadrons is relatively small as compared to D^0. Hence, we conclude that the elliptic flow of open charm meson is more sensitive to viscous properties of the QGP medium compared to the light hadrons. The ratio of v_2 of D^0 to the v_2 of charged hadrons is independent of analysis technique used to obtain elliptic flow. This is shown for Au+Au collisions at $\sqrt{s_{NN}} = 200$ GeV and Pb + Pb collisions at $\sqrt{s_{NN}}$ = 2.76 TeV in Fig. 5 for different η_s/s as a function of p_T. This ratio can be calculated in data and compared to model studies to constrain specific viscosity of the medium.

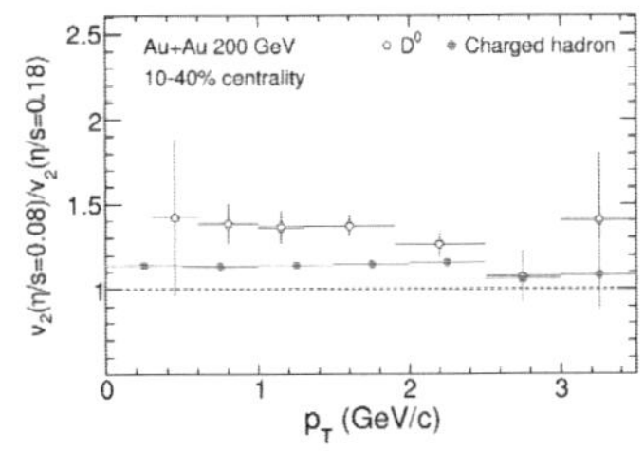

Figure 4. (Color online) Ratio between v_2 for $\eta_s/s = 0.08$ and for $\eta_s/s = 0.18$ as a function of p_T in Au+Au collisions at $\sqrt{s_{NN}} = 200$ GeV for charged hadrons and D^0 for 10–40% central collisions.

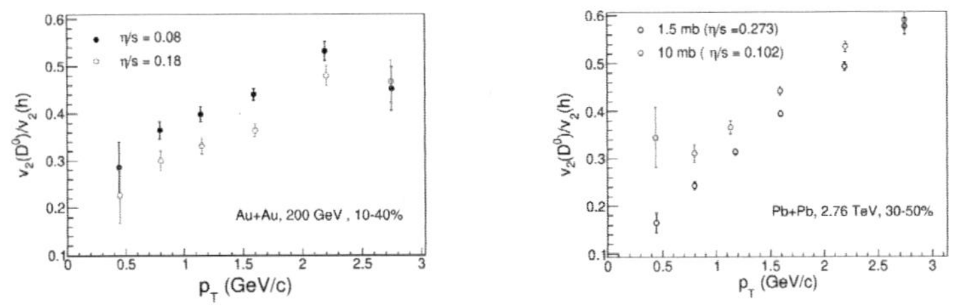

Figure 5. (Color online) Ratio between v_2 for D^0 to v_2 of charged hadrons for different η_s/s as a function of p_T in Au+Au collisions at $\sqrt{s_{NN}} = 200$ GeV (left) and Pb + Pb collisions at $\sqrt{s_{NN}} = 2.76$ TeV (right)

4. Conclusion

We have studied the elliptic flow of open charm mesons as a function of p_T for heavy ion collisions at $\sqrt{s_{NN}} = 200$ GeV and 2.76 TeV within the framework of A Multi Phase Transport model by using quark coalescence as a mechanism for hadron production from heavy quarks in this model study. We have given predictions for $v_2(p_T)$ of D^0 and $D_S^{\pm}$ meson for minimum-bias Au+Au collision at RHIC for different values of partonic interaction cross-sections. We find that v_2 increases with increase in the partonic interaction cross-section. We also notice that the parameters tuned to describe charged hadron multiplicity and v_2 at 2.76 TeV fail to reproduced D meson v_2 in Pb+Pb collisions at $\sqrt{s_{NN}} = 2.76$ TeV for $p_T > 2$ GeV/c.

In addition, we have presented a systematic study on the effect of specific viscosity on elliptic flow within the transport model approach. We have also shown that open charmed meson are more sensitive to viscous properties of QGP medium compared to light hadrons.

References

[1] Poskanzer A M and Voloshin S A 1998 *Phys. Rev.* **C 58** 1671.
[2] Esha R, Nasim M and Huang H Z 2016 *arXiv:1603.02700.*
[3] Lin Z W and Ko C M 2002 *Phys. Rev.* **C 65** 034904; Chen L W *et al.* 2005 *Phys. Lett.* **B 605** 95; Lin Z W *et al.* 2005 *Phys. Rev.* **C 72** 064901.
[4] Chen L W and Ko C M 2006 *Phys. Rev.* **C 73** 044903; Greco V, Ko C M and Rapp R 2004 *Phys. Lett.* **B 595** 202.
[5] Abelev B *et al.* (ALICE Collaboration) 2014 *Phys. Rev.***C 90**, 034904.
[6] Xu J and Ko C M 2011 *Phys. Rev.* **C 83**, 034904.

Investigating the correlations of flow harmonics in 2.76A TeV Pb–Pb collisions

Xiangrong Zhu[1,2,3], You Zhou[4], Haojie Xu[2,3], Huichao Song[2,3,5]

[1]School of Science, Huzhou University, Huzhou 313000, China
[2]Department of Physics and State Key Laboratory of Nuclear Physics and Technology, Peking University, Beijing 100871, China
[3]Collaborative Innovation Center of Quantum Matter, Beijing 100871, China
[4]Niels Bohr Institute, University of Copenhagen, Blegdamsvej 17, 2100 Copenhagen, Denmark
[5]Center for High Energy Physics, Peking University, Beijing 100871, China

E-mail: `Huichaosong@pku.edu.cn`

Abstract. This proceeding briefly summarizes our recent investigations on the correlations of flow harmonics in 2.76A TeV Pb–Pb collisions with viscous hydrodynamics `VISH2+1`. We calculated both the symmetric cumulants $SC^v(m,n)$ and the normalized symmetric cumulants $NSC^v(m,n)$, and found v_2 and v_4, v_2 and v_5, v_3 and v_5 are correlated, v_2 and v_3, v_3 and v_4 are anti-correlated. We also found $NSC^v(3,2)$ are insensitive to the QGP viscosity, which are mainly determined by the initial conditions.

1. Introduction

The ultra-relativistic heavy-ion collision programs at RHIC and LHC have been utilized to produce extreme conditions to create and study the strongly interacting Quark-Gluon Plasma (QGP), a deconfined state of quarks and gluons. One of the observables to probe the properties of the hot QCD matter is the azimuthal anisotropy in the momentum distribution of produced particles. The anisotropic flow coefficients V_n are generally defined through a Fourier decomposition of the emitted particle distribution as a function of the azimuthal angle φ, $P(\varphi) = \frac{1}{2\pi}\sum_{n=-\infty}^{+\infty} \overrightarrow{V_n}\, e^{-in\varphi}$ where $\overrightarrow{V_n} = v_n\, e^{in\Psi_n}$. The v_n is the n-th order anisotropic flow harmonics and Ψ_n is the symmetry plane angle. Recently, the correlations between different order $\overrightarrow{V_m}$ and $\overrightarrow{V_n}$ have been investigated both theoretically and experimentally, which not only focus on the correlations of the orientations of different flow-vector Ψ_n [1, 2, 3, 4, 5] but also on the correlations of the magnitudes of different flow-vector v_n [6, 7, 8, 9, 10, 11].

In this proceeding, we will briefly review our recent investigations on the correlations of flow harmonics in 2.76A TeV Pb–Pb collisions using the event-by-event viscous hydrodynamics `VISH2+1` with different initial conditions and the the value of shear viscosity [11].

2. Setup of the calculation

The `VISH2+1` is a (2+1)-d viscous hydrodynamic model to describe the fluid expansion of the QGP with longitudinal boost-invariance [12, 13]. In the following calculations, we use an equation of state (EoS) `s95p-PCE` [14], which matches the partially chemical equilibrium hadron resonance gas at low temperature and the lattice QCD data at high temperature. Three different initial conditions, `MC-Glauber`, `MC-KLN` [15, 16], and `AMPT` [17], are used in our calculations to study the influence of initial conditions on the correlations of flow harmonics. To explore the sensitivity of the QGP shear viscosity, we choose two values of η/s, the specific shear viscosity, for each initial condition. More specifically, η/s = 0.08 and 0.20, for the `MC-Glauber` and `MC-KLN` initial conditions, and η/s =0.08 and 0.16 for the `AMPT` initial conditions were utilised. The hydrodynamic output is converted to final hadron distributions along the freeze-out surface at the temperature T_{dec} = 120 MeV via the Cooper-Frye prescription [18, 19]. The initial time of hydrodynamic evolution τ_0 and the normalization factors of the initial entropy density profiles have been tuned to fit the 0-5% centrality data of $dN/d\eta$ and p_T spectra of π, K, and p [20]. The bulk viscosity, net baryon density, and the heat conductivity are set to zero to simplify the calculations.

3. Results and discussion

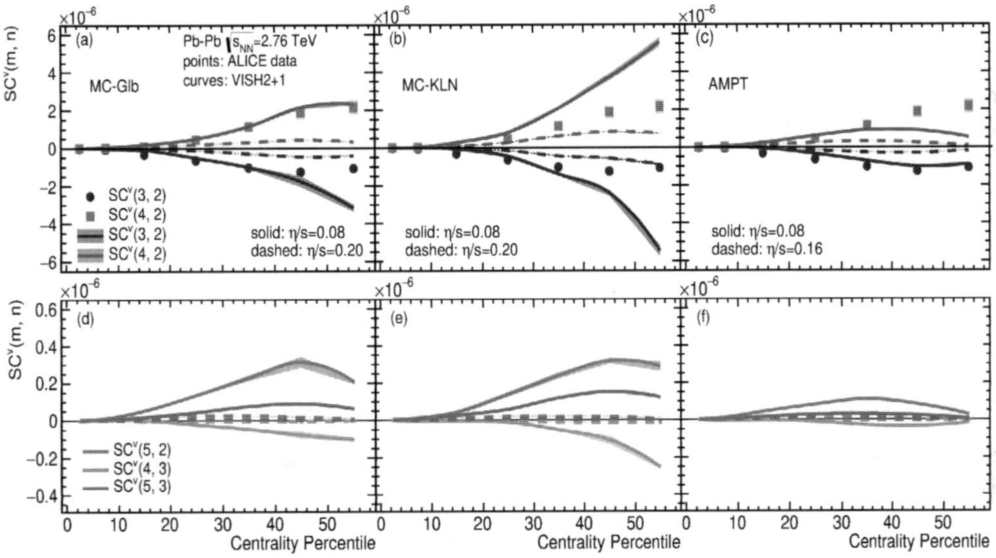

Figure 1. (Color online) Top: Symmetric cumulants $SC^v(3,2)$ and $SC^v(4,2)$ in 2.76A TeV Pb–Pb collisions. The ALICE measurements are taken from [6]. Bottom: Predicted symmetric cumulants $SC^v(5,2)$, $SC^v(5,3)$, and $SC^v(4,3)$ in 2.76A TeV Pb–Pb collisions.

Fig. 1 (top) shows the comparison between our calculations for the symmetric cumulants defined as $SC^v(m,n) = \langle v_m^2 v_n^2 \rangle - \langle v_m^2 \rangle \langle v_n^2 \rangle$ and the ALICE measurements [6]. We find that, for these initial conditions and different values of η/s, the `VISH2+1` calculations qualitatively capture the centrality dependence of the flow correlations. In particular, even though `VISH2+1` with `AMPT` initial conditions gives good descriptions for the integrated flow v_n ($n \leq 4$) [11], it fails to quantitatively reproduce the magnitude of the measured correlations of flow harmonics. This indicates that the correlations between different flow harmonics are more sensitive to the details of hydrodynamic calculations than the individual v_n coefficients alone.

Similar to the ALICE data [6], our model gives negative $SC^v(3,2)$ and positive $SC^v(4,2)$, which suggest that v_2 and v_3 are anti-correlated, while v_2 and v_4 are correlated. The results reveal that, for a given event, the case with an elliptic flow v_2 larger than the averaged $\langle v_2 \rangle$ enhances the probability of finding a triangular flow v_3 smaller than the averaged $\langle v_3 \rangle$ and the

IOP Conf. Series: Journal of Physics: Conf. Series **779** (2017) 012062 doi:10.1088/1742-6596/779/1/012062

probability of finding a quadrangular flow v_4 larger than the averaged $\langle v_4 \rangle$. The strengths of $SC^v(3,2)$ and $SC^v(4,2)$ are more suppressed with larger η/s for each initial condition, which suggests that both $SC^v(3,2)$ and $SC^v(4,2)$ are strongly influenced by the QGP viscosity. By comparing with the symmetric cumulants of the initial state, $SC^\varepsilon(m,n)$ in Ref. [11], we observe that the signs of $SC^v(3,2)$ and $SC^v(4,2)$ are determined by the signs of $SC^\varepsilon(3,2)$ and $SC^\varepsilon(4,3)$, respectively.

Figure 1 (bottom) presents our predictions for the centrality dependent $SC^v(m,n)$ with $(m,n) = (5,2)$, $(5,3)$, and $(4,3)$. We observe that, for each initial condition, the hydrodynamical calculations gives positive values for $SC^v(5,2)$ and $SC^v(5,3)$, and negative values for $SC^v(4,3)$. This reveals that v_2 and v_5, v_3 and v_5 are correlated, while v_3 and v_4 are anti-correlated. We also notice that their correlation strengths become weaker with increasing η/s. The signs of $SC^v(5,2)$ and $SC^v(5,3)$ are consistent with their initial state correlators $SC^\varepsilon(5,2)$ and $SC^\varepsilon(5,3)$ shown in Ref. [11]. However, $SC^v(4,3)$ and $SC^\varepsilon(4,3)$ show opposite signs for the `MC-Glauber` and `AMPT` initial conditions [11]. This can be well understood from the proposed relationship of $v_4 e^{i4\Phi} = a_0 \varepsilon_4 e^{i4\Psi_4} + a_1(\varepsilon_2 e^{i2\Psi_2})^2$ [21, 22], where the ε_2^2 term makes the dominant contributions in non-central collisions [23]. As a result, the signs of $SC^v(4,3)$ are affected by the correlation between ε_2 and ε_3 and the correlation between ε_3 and ε_4 rather than the correlation between ε_3 and ε_4 alone.

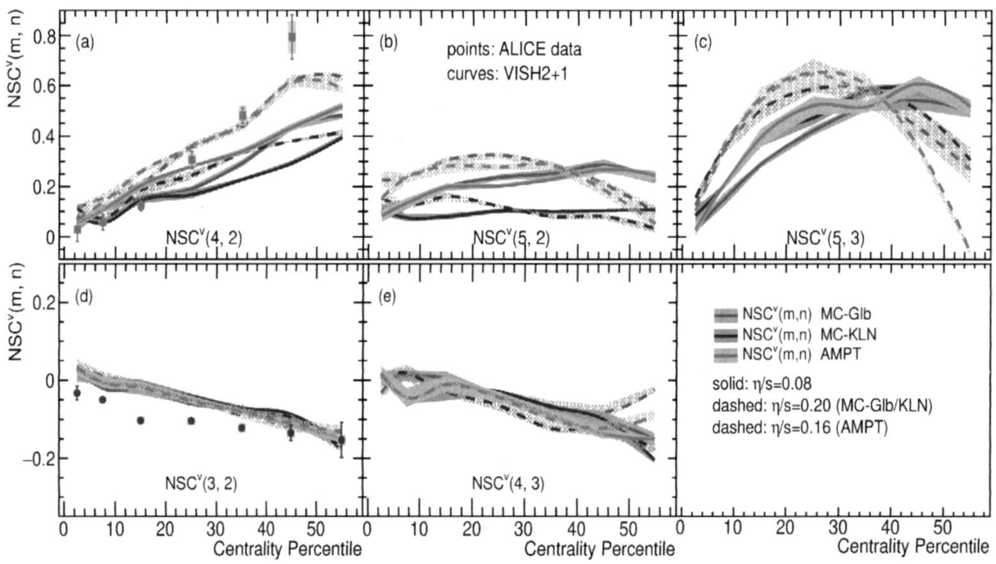

Figure 2. (Color online) Normalized symmetric cumulants $NSC^v(m,n)$ and normalized symmetric cumulants of the initial state $NSC^\varepsilon(m,n)$ in 2.76A TeV Pb–Pb collisions.

Figure 2 shows the normalized correlator of flow harmonics, which are defined as $NSC^v(m,n) = SC^v(m,n)/\langle v_m^2 \rangle \langle v_n^2 \rangle$. We find that $NSC^v(4,2)$, $NSC^v(5,2)$, and $NSC^v(5,3)$ are sensitive to both initial conditions and η/s. In Ref. [11], we found that their corresponding correlator for the initial state, $NSC^\varepsilon(m,n) = SC^\varepsilon(m,n)/\langle \varepsilon_m^2 \rangle \langle \varepsilon_n^2 \rangle$, are separated for different initial conditions. It can be seen that the NSC(4,2) captures the general centrality dependence of the measured points for `AMPT` initial conditions and $\eta/s = 0.16$ and for `MC-Glauber` initial conditions and $\eta/s = 0.2$. This indicates that the normalized symmetric cumulants can be used to constrain the QGP viscosity for different initial conditions. In addition, the $NSC^v(3,2)$

describe qualitatively the ALICE measurements fairly well regardless of the choice of initial conditions and values of η/s. Such η/s independent character of $NSC^v(3,2)$ can be naturally understood from the widely accepted relation $v_2 \approx k_1\varepsilon_2$ and $v_3 \approx k_2\varepsilon_3$, where k_1 and k_2 are the proportion coefficients. In Ref. [11], we also found that the $NSC^\varepsilon(3,2)$ from the three initial conditions used in our calculations also almost overlap from central to semi-central collisions. In contract, although the $NSC^\varepsilon(4,3)$ strongly depends on the initial conditions [11], the $NSC^v(4,3)$ almost overlap, making it insensitive to the initial conditions used in our calculation.

4. Summary

In summary, we investigated the correlations between flow harmonics in 2.76A TeV Pb–Pb collisions using the event-by-event viscous hydrodynamics `VISH2+1` with `MC-Glauber`, `MC-KLN`, and `AMPT` initial conditions. We found the symmetric cumulants $SC^v(m,n)$ are sensitive to both initial conditions and the QGP shear viscosity. The normalized symmetric cumulants $NSC^v(3,2)$ are mainly determined by the correlation in the initial state, which are insensitive to the QGP viscosity. In contrast, $NSC^v(4,2)$, $NSC^v(5,2)$, $NSC^v(5,3)$ are sensitive to both initial conditions and η/s. We found that the correlations of flow harmonics are more sensitive to the details of theoretical calculations than the individual flow harmonics. This could be used for further constraint the properties of the QGP.

Acknowledgments

This work is supported by the NSFC and the MOST under grant Nos.11435001 and 2015CB856900, and partially supported by China Postdoctoral Science Foundation under grant No. 2015M570878 and 2015M580908, by the Danish Council for Independent Research, Natural Sciences, and the Danish National Research Foundation (Danmarks Grundforskningsfond).

References

[1] G. Aad *et al.* [ATLAS Collaboration], Phys. Rev. C **90**, 024905 (2014).
[2] Z. Qiu and U. Heinz, Phys. Lett. B **717**, 261 (2012).
[3] D. Teaney and L. Yan, Nucl. Phys. A **904-905**, 365c (2013).
[4] J. Jia and D. Teaney, Eur. Phys. J. C **73**, 2558 (2013).
[5] H. Niemi, K. J. Eskola and R. Paatelainen, Phys. Rev. C **93**, 024907 (2016).
[6] J. Adam *et al.* [ALICE Collaboration], arXiv:1604.07663 [nucl-ex];
[7] G. Aad *et al.* [ATLAS Collaboration], Phys. Rev. C **92**, 034903 (2015).
[8] H. Niemi, G. S. Denicol, H. Holopainen and P. Huovinen, Phys. Rev. C **87**, 054901 (2013).
[9] G. Giacalone, L. Yan, J. Noronha-Hostler and J. Y. Ollitrault, arXiv:1605.08303 [nucl-th].
[10] J. Qian and U. Heinz, Phys. Rev. C **94**, 024910 (2016)
[11] X. Zhu, Y. Zhou, H. Xu and H. Song, arXiv:1608.05305 [nucl-th].
[12] H. Song and U. Heinz, Phys. Lett. **B658**, 279 (2008); Phys. Rev. C **77**, 064901 (2008); Phys. Rev. C **78**, 024902 (2008); H. Song, Ph.D Thesis, The Ohio State University, August 2009, arXiv:0908.3656 [nucl-th].
[13] C. Shen, Z. Qiu, H. Song, J. Bernhard, S. Bass and U. Heinz, Comput. Phys. Commun. **199**, 61 (2016).
[14] P. Huovinen and P. Petreczky, Nucl. Phys. **A837**, 26 (2010); C. Shen, U. Heinz, P. Huovinen and H. Song, Phys. Rev. C **82**, 054904 (2010).
[15] A. Adil, H. J. Drescher, A. Dumitru, A. Hayashigaki and Y. Nara, Phys. Rev. C **74** 044905 (2006); H. J. Drescher and Y. Nara, *ibid.* **76** 041903 (2007).
[16] T. Hirano and Y. Nara, Phys. Rev. C **79** 064904 (2009); and Nucl. Phys. **A830** 191c (2009).
[17] H. j. Xu, Z. Li and H. Song, Phys. Rev. C **93**, 064905 (2016).
[18] C. Shen, U. Heinz, P. Huovinen and H. Song, Phys. Rev. C **84**, 044903 (2011).
[19] H. Song, S. Bass and U. W. Heinz, Phys. Rev. C **89**, 034919 (2014); X. Zhu, F. Meng, H. Song and Y. X. Liu, Phys. Rev. C **91**, 034904 (2015).
[20] B. Abelev *et al.* [ALICE Collaboration], Phys. Rev. C **88**, 044910 (2013)
[21] F. G. Gardim, F. Grassi, M. Luzum and J. Y. Ollitrault, Phys. Rev. C **85**, 024908 (2012).
[22] D. Teaney and L. Yan, Phys. Rev. C **86**, 044908 (2012).
[23] L. Yan and J. Y. Ollitrault, Phys. Lett. B **744**, 82 (2015).

The physics mechanisms of light and heavy flavor v_2 and mass ordering in AMPT

Hanlin Li[1,2], **Zi-Wei Lin**[3,4], **Fuqiang Wang**[2]

[1]College of Science, Wuhan University of Science and Technology, Wuhan, Hubei 430065, China
[2]Department of Physics and Astronomy, Purdue University, West Lafayette, IN 47907, USA
[3]Key Laboratory of Quarks and Lepton Physics (MOE) and Institute of Particle Physics, Central China Normal University, Wuhan, 430079, China
[4]Department of Physics, East Carolina University, Greenville, NC 27858, USA

Abstract. A Multi-Phase Transport (AMPT) model has been shown to describe fairly well experimental data including the bulk properties of particle spectra and elliptic anisotropy (v_2) in heavy ion collisions. Recent studies have shown that AMPT describes the v_2 measurements in small system collisions as well. In these proceedings, we first investigate the origin of the mass ordering of identified hadrons v_2 in heavy ion as well as small system collisions. We then study the production mechanism of the charm v_2 in light of the escape mechanism for the light quark v_2.

1. Introduction

Relativistic heavy ion collisions aim to create the quark-gluon plasma (QGP) and study its properties at extreme conditions of high temperature and energy density [1, 2]. Collective flow is often used in experimental and theoretical investigations to study the QGP properties. Both hydrodynamics [3, 4] and transport theory [5] can describe the bulk data in heavy ion collisions. For example, the string melting version of A Multi-Phase Transport (AMPT) model [5, 6] reasonably reproduces particle yields, p_T spectra, and v_2 of low-p_T pions and kaons in central and mid-central Au+Au collisions at 200A GeV and Pb+Pb collisions at 2760A GeV [7]. Similar measurements in small systems can also be satisfactorily described by AMPT [8].

Recent studies have shown that light parton v_2 is mainly generated by the anisotropic parton escape from the collision zone and hydrodynamics may play only a minor role [9, 10]. It suggests that the mass ordering of v_2, commonly considered as a signature of hydrodynamic expansion of the system, may arise from other mechanisms. In these proceedings, we first investigate the physics mechanisms of this mass ordering in AMPT [11, 12]. We then study the production mechanism of the charm v_2 in light of the escape mechanism for the light quark v_2.

2. Model details and analysis method

We employ the string melting version of AMPT [5, 6] in our study. The model consists of four components: fluctuating initial conditions, parton elastic scatterings, quark coalescence for hadronization, and hadronic interactions. The parton interactions are modeled by Zhang's parton cascade (ZPC) [13]. We use Debye screened differential cross-section $d\sigma/dt \propto \alpha_s^2/(t-\mu_D^2)^2$ [5], with strong coupling constant $\alpha_s = 0.33$ and Debye screening mass $\mu_D = 2.265/\text{fm}$ (the total cross section is then $\sigma = 3$ mb) for all AMPT simulation in our work. Once partons stop interacting, a simple quark coalescence model is applied to combine two (three) nearest partons into a meson (baryon or antibaryon). The subsequent hadronic interactions are described by an extended ART model [5, 14]. We terminate the hadronic interactions at a cutoff time, when the observables of interest are stable with time; a cutoff time of 30 fm/c is used in this study.

We simulate three collision systems: Au+Au collisions with $b = 6.6$-8.1 fm at the nucleon-nucleon center-of-mass energy $\sqrt{s_{NN}} = 200$ GeV, d+Au collisions with $b = 0$ fm at $\sqrt{s_{NN}} = 200$ GeV, and p+Pb collisions with $b = 0$ fm at $\sqrt{s_{NN}} = 5$ TeV. We analyze the momentum-space azimuthal anisotropy of partons in the final state before hadronization, of hadrons right after hadronization but before hadronic rescattering takes place, and of freeze-out hadrons in the final state. The momentum anisotropy is characterized by the Fourier coefficients [15] according to $v_2 = \langle \cos 2(\phi - \psi_2^{(r)}) \rangle$, where ϕ is the azimuthal angle of a particle (parton or hadron). The $\psi_2^{(r)}$ is the harmonic plane of each event from its initial spatial configuration of all partons. All results shown are for particles within the pseudo-rapidity window $|\eta| < 1$.

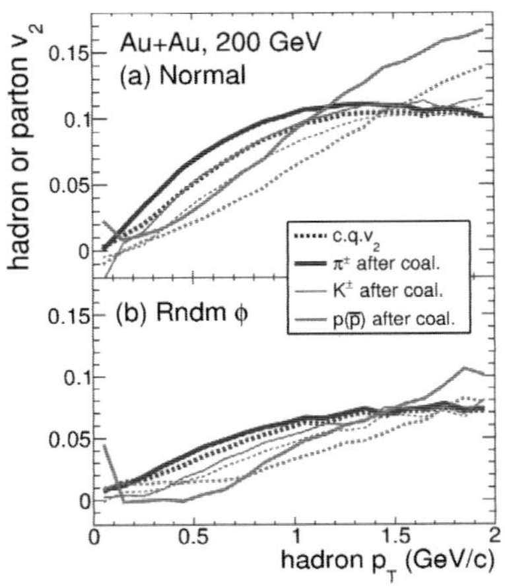

Figure 1. Mass splitting from coalescence. Constituent quark (c.q., dashed curves) and primordial hadron (solid curves) v_2 as a function of hadron p_T in Au+Au collisions by normal (a) and ϕ-randomized (b) AMPT.

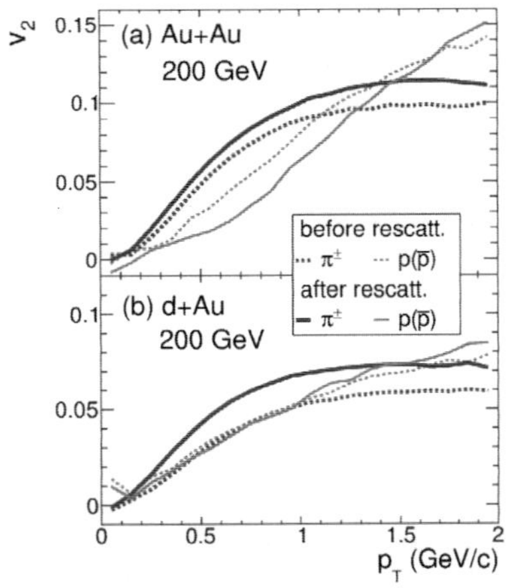

Figure 2. Effects of hadronic rescatterings. Pion and (anti) proton v_2 before (dashed) and after (solid) hadronic rescatterings in Au+Au (a) and d+Au (b) collisions by AMPT.

3. Mass ordering of v_2

We have found almost identical $v_2(p_T)$ values after parton cascade for up and down quarks and for quarks and antiquarks of the same flavor. Considering that pions and protons are made of only light constituent quarks, the difference between their v_2 should come from the hadronization process and/or hadronic rescatterings. Thus we first study the primordial hadrons (i.e. hadrons formed directly from quark coalescence before any decays or hadronic rescatterings take place). Figure 1(a) shows the v_2 of primordial charged pion, kaon, and (anti)proton as a function of p_T (solid curves). It demonstrates that the mass ordering of v_2 at low p_T in AMPT comes from the kinematics in coalescence such as the constituent quark number in a hadron and the finite opening angle between coalescence partners [11, 12]. The dynamical "selections" of constituent

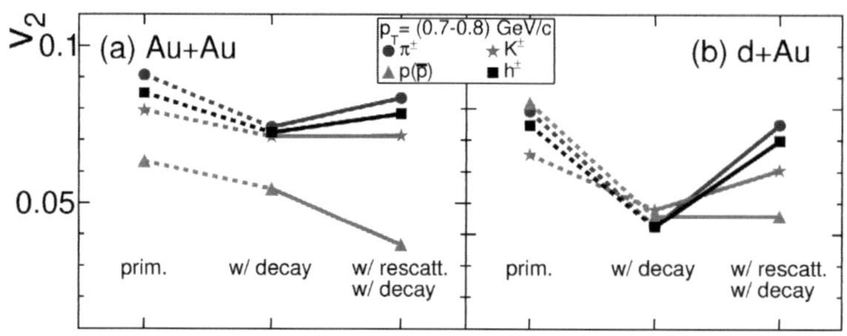

Figure 3. (Color online) Mid-rapidity ($|\eta| < 1$) v_2 of π, K, p($\bar{p}$) and charged hadrons ($h^{\pm}$) at $0.7 < p_T < 0.8$ GeV/c at different stages of system evolution in $b = 6.6$-8.1 fm Au+Au (left) and $b = 0$ fm d+Au (right) collisions at $\sqrt{s_{NN}}$=200 GeV by AMPT: initial v_2 of primordial hadrons (right after coalescence hadronization and before hadronic rescatterings), initial v_2 of all hadrons (including decays), and final v_2 of all hadrons (after hadronic rescatterings and decays).

quarks into pions, kaons, and protons are manifest in the constituent quark v_2 distributions shown by the dashed curves in Fig. 1(a), plotted at the respective *hadron* p_T.

Figure 1(b) shows the v_2 results by ϕ-randomized AMPT [9] for primordial hadrons right after the coalescence hadronization at the corresponding hadron p_T. No hydrodynamic anisotropic flow is present in the ϕ-randomized case [9], however, mass ordering is still present. This observation implies that the mass ordering is mainly due to the kinematics in the coalescence process. It is therefore not a unique signature of collective anisotropic flow or hydrodynamics.

After hadronization, particles undergo decays and rescatterings. We evaluate the v_2 of hadrons before hadronic rescatterings but including resonance decays. The results are represented in Fig. 2(a) by the dashed curves. The decay product v_2 is usually smaller than their parent v_2 [12]. By including decay products, the hadron v_2 is reduced from that of primordial hadrons (solid curves in Fig. 1(a)).

The v_2 values before hadronic rescatterings (but including resonance decay effects) can be considered as the initial v_2 for the hadronic evolution stage. The final-stage freezeout hadron v_2's (also including decay daughters) are shown in Fig. 2(a) by the solid curves. Pion v_2 increases while proton v_2 decreases after the hadronic rescattering phase. This may be due to interactions between pions and protons, after which they tend to flow together at the same velocity. Thus, the same-velocity pions and protons (i.e. small p_T pions and large p_T protons) will tend to have the same anisotropy. This will yield lower v_2 for protons and higher v_2 for pions at the same p_T value. This should happen regardless of the net change in the overall charged hadron v_2, which depends on the initial configuration geometry from which the hadronic evolution begins [12]. Similar results in d+Au collisions as shown in Fig. 2(b) [12].

To summarize the origin of v_2 mass splitting, we plot the v_2 of pions, kaons, and protons within $0.7 < p_T < 0.8$ GeV/c for different stages of the collision system evolution as shown in Fig. 3. The evolution stages contain: (i) right after coalescence hadronization including only primordial hadrons (labeled "prim."); (ii) right after coalescence hadronization but including decay products (labeled "w/ decay"); and (iii) at final freezeout (labeled "w/ rescatt. w/ decay). We can see the mass ordering actually comes from the interplay of several physics effects: coalescence, and more importantly, from the hadronic rescattering process.

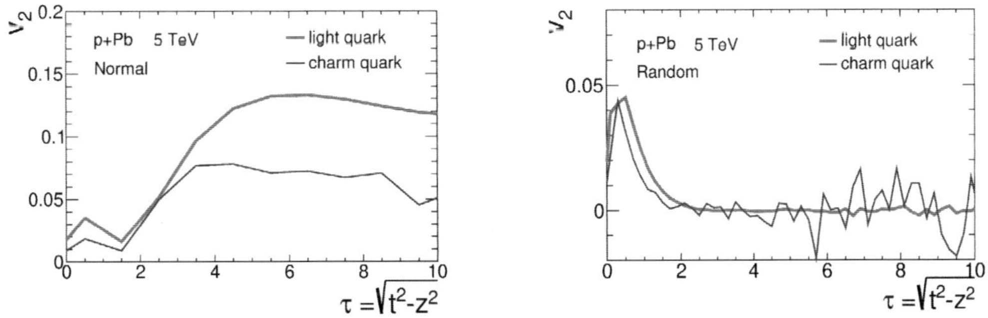

Figure 4. (Color online) The charm and light quark v_2 as a function of the freezeout proper time $\tau = \sqrt{t^2 - z^2}$ in p+Pb collisions with $b = 0$ fm. Both the normal (left) and ϕ-randomized (right) AMPT results are shown.

4. Charm v_2 mechanism

In this section we discuss the production mechanism of charm quark flow in pPb. We compare the light and charm quark v_2 at freezeout, integrated over all p_T, versus the freezeout proper time. This is shown in the left panel of Fig. 4. The charm v_2 is systematically lower than light quark v_2. This suggests that quark mass significantly affects the elliptic flow. The right panel of Fig. 4 shows the ϕ-randomized results. The charm and light quark v_2 are similar suggesting a common escape mechanism.

5. Summary

In these proceedings, we investigate the origin of mass splitting of identified hadron v_2. The mass splitting is due to coalescence and, more importantly, hadronic rescatterings. It is therefore not a unique signature of hydrodynamics. We also study the development of charm v_2 in pPb and found the escape mechanism to be the major contribution similar to light quarks.

Acknowledgments

This work is supported in part by US Department of Energy Grant No. DE-FG02-88ER40412 (FW) and National Natural Science Foundation of China grant No. 11628508 (ZL). HL acknowledges financial support from the China Scholarship Council.

References

[1] I. Arsene et al. BRAHMS Collaboration, *Nucl. Phys.* A **757**, 1 (2005).
[2] B. Back et al. PHOBOS Collaboration *Nucl. Phys.* A **757**, 28 (2005).
[3] U. Heinz and R. Snellings, Ann. *Rev. Nucl. Part. Sci.* **63**, 123 (2013).
[4] Zhu X, Meng F, Song H, and Liu Y X *Phys. Rev.* C **91**, 034904 (2015).
[5] Z.-W. Lin, C. M. Ko, B.-A. Li, B. Zhang and S. Pal , *Phys.Rev.* C **72**, 064901 (2005).
[6] Z.-W. Lin and C. M. Ko, *Phys.Rev.* C **65**, 034904 (2002).
[7] Z.-W. Lin, *Phys.Rev.* C **90**, 014904 (2014).
[8] A. Bzdak and G.-L. Ma, *Phys.Rev.* Lett. **113**, 252301 (2014).
[9] L. He, T. Edmonds, Z.-W. Lin, F. Liu, D. Molnar, and F. Wang, *Phys. Lett.* B **735**, 506 (2016).
[10] Z. -W. Lin, L. He, T. Edmonds, F. Liu, D. Molnar, and F. Wang, *Nucl. Phys.* A **956**, 316 (2016).
[11] H. Li, L. He, Z. W. Lin, D. Molnar, F. Wang and W. Xie, *Phys.Rev.* C **93**, 051901(R) (2016).
[12] H. Li, L. He, Z. W. Lin, D. Molnar, F. Wang and W. Xie, *arXiv:1604.07387* (2016).
[13] B. Zhang,*Comput.Phys.Commun.* **109**, 193 (1998).
[14] B. A. Li and C. M. Ko, *Phys. Rev.* C **52**, 2037 (1995).
[15] S. Voloshin and Y. Zhang, *Z.Phys.* **C70**,665 (1996).

The fluctuations of quadrangular flow

Giuliano Giacalone[1], Li Yan[1], Jacquelyn Noronha-Hostler[2] and Jean-Yves Ollitrault[1]

[1] Institut de physique théorique, Université Paris Saclay, CNRS, CEA, F-91191 Gif-sur-Yvette, France

[2] Department of Physics, University of Houston, Houston TX 77204, USA

E-mail: jean-yves.ollitrault@cea.fr

Abstract. The ATLAS Collaboration has measured for the first time the fourth cumulant of quadrangular flow, $v_4\{4\}^4$. Unlike the fourth cumulants of elliptic and triangular flows, it presents a change of sign above 30% centrality. We show that this change of sign is predicted by event-by-event hydrodynamics. We argue that it results from the combined effects of a nonlinear hydrodynamic response, which couples quadrangular flow to elliptic flow, and elliptic flow fluctuations.

1. Introduction

The bulk of particle production in ultrarelativistic nucleus-nucleus collisions is described by the flow paradigm [1], which states that particles are emitted independently from an underlying probability distribution. In particular, the flow paradigm naturally explains the long-range azimuthal correlations, which are a salient feature of heavy-ion collisions, as resulting from the fluctuations of the underlying azimuthal probability distribution $P(\varphi)$ [2]. This azimuthal distribution is traditionally written as a Fourier series:

$$P(\varphi) = \frac{1}{2\pi} \sum_{n=-\infty}^{+\infty} V_n e^{-in\varphi}, \tag{1}$$

where $V_n = v_n \exp(in\Psi_n)$ is the (complex) anisotropic flow coefficient in the nth harmonic and $V_{-n} = V_n^*$. Both the magnitude and phase of V_n fluctuate event to event [3]. Experimental observables involving anisotropic flow can be recast as statistical properties of the distribution of V_n. For instance, the cumulants $v_n\{2\}^2$ and $v_n\{4\}^4$ are defined by [4]:

$$\begin{aligned} v_n\{2\}^2 &\equiv \langle v_n^2 \rangle, \\ v_n\{4\}^4 &\equiv 2\langle v_n^2 \rangle^2 - \langle v_n^4 \rangle, \end{aligned} \tag{2}$$

where angular brackets denote an average over events in a centrality class. The lowest order cumulant, $v_n\{2\}^2$, is simply the mean square value of v_n, while the fourth cumulant, $v_n\{4\}^4$, is a nontrivial combination of moments. In particular, despite the unfortunate notation, $v_n\{4\}^4$ can be either positive or negative, depending on the probability distribution of v_n. Both $v_2\{4\}^4$ [5] and $v_3\{4\}^4$ [6] are observed to be positive across all centralities in Pb+Pb collisions at the LHC. However, a striking observation, which seems to have received little attention from the theory community, is that $v_4\{4\}^4$ changes sign [7]: it is positive only up to $\sim$ 30% centrality.

In this article, we show that the change of sign of $v_4\{4\}^4$ is predicted by hydrodynamics. In hydrodynamics, the fluctuations of the anisotropic flow coefficients, V_n, result from the fluctuations of the energy density profile released after the collisions [8]. v_2 and v_3 are determined to a good approximation [9, 10] by linear response to the initial anisotropies in the corresponding harmonics. This, in turn, explains why both $v_2\{4\}^4$ and $v_3\{4\}^4$ are positive, even though they originate from different mechanisms: $v_2\{4\}$ is driven by the reaction-plane eccentricity [11] while $v_3\{4\}$ is driven by non-Gaussian fluctuations of the initial triangularity [12–14]. By contrast, simple linear response to the initial anisotropy in the fourth harmonic is unable to explain the observed fluctuations of v_4 [15, 16]. In Sec. 2, we recall why linear response does not apply to v_4 and how a significant nonlinear response can be taken into account [17]. We infer the nonlinear response from experimental data and we show that its magnitude is correctly predicted by hydrodynamics. In Sec. 3, we calculate $v_4\{4\}^4$ in hydrodynamics.

2. Linear and nonlinear hydrodynamic response

V_4 and $(V_2)^2$ transform identically under azimuthal rotations. Therefore, rotational symmetry allows for a coupling between these two quantities, which is indeed predicted by hydrodynamics [18]. We take this coupling into account by writing V_4 as the sum of a term proportional to $(V_2)^2$ (the nonlinear response) and a term uncorrelated with $(V_2)^2$, which we dub the linear part, V_{4L} [19, 20]:

$$V_4 = V_{4L} + \chi_4 (V_2)^2. \tag{3}$$

The condition that linear and nonlinear parts are uncorrelated, $\langle V_{4L}(V_2)^{*2}\rangle = 0$, uniquely defines the proportionality coefficient, χ_4, i.e.,

$$\chi_4 = \frac{\langle V_4 (V_2^*)^2\rangle}{\langle |V_2|^4\rangle}. \tag{4}$$

Note that the decomposition defined in Eqs. (3) and (4) is purely mathematical and holds irrespective of the hydrodynamical setup.[1] The nonlinear part thus defined can be isolated by analyzing V_4 with respect to the direction of elliptic flow [22, 23]. The resulting observable is dubbed $v_4\{\Psi_2\}$ [24] and is defined by [25]:

$$v_4\{\Psi_2\} \equiv \frac{\langle V_4 (V_2^*)^2\rangle}{\sqrt{\langle |V_2|^4\rangle}}. \tag{5}$$

From Eqs. (4) and (5), one obtains $v_4\{\Psi_2\} = \chi_4\sqrt{\langle |V_2|^4\rangle}$, that is, $v_4\{\Psi_2\}$ is the rms value of the nonlinear contribution to V_4 in Eq. (3). The triangular inequality guarantees that $v_4\{\Psi_2\} \leq v_4\{2\}$ [20], where $v_4\{2\}$ is defined in Eq. (2). The mean square value of V_{4L} can be obtained by subtracting $v_4\{\Psi_2\}^2$ from $v_4\{2\}^2$ [26, 27]. Figure 1 displays CMS data [28] for $v_4\{\Psi_2\}$ and $v_4\{2\}$. They satisfy $v_4\{\Psi_2\} < v_4\{2\}$ for all centralities, which is a nontrivial test of the flow paradigm. Further, because of the increase of elliptic flow in the reaction plane, the relative weight of the nonlinear contribution, $v_4\{\Psi_2\}$, increases with centrality. In order to illustrate that these trends are captured by hydrodynamics [29–31], we carry out event-by-event viscous relativistic hydrodynamic calculations within the code v-USPhydro [32, 33]. Initial conditions are given by the Monte Carlo Glauber model [34]. The setup is the same as in Ref. [35]: In particular, the shear viscosity over entropy ratio is $\eta/s = 0.08$ [36]. Anisotropic flow, V_n, is calculated at freeze-out [37] for pions in every event. We calculate both $v_4\{2\}$ and $v_4\{\Psi_2\}$ by averaging over events. Results are shown in Fig. 1. Our calculation matches

[1] As an example, hydrodynamics predicts a similar decomposition [21], where the linear part results from initial fluctuations in the fourth harmonic. In this case, there is no requirement concerning how the linear and the nonlinear part are correlated.

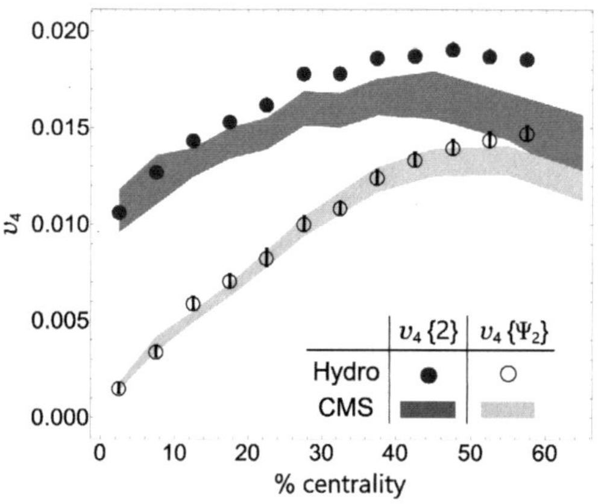

Figure 1. $v_4\{2\}$ (full symbols and dark shaded band) and $v_4\{\Psi_2\}$ (open symbols and light shaded band) as a function of centrality percentile in Pb+Pb collisions at 2.76 TeV. Bands: CMS data for charged particles in the p_t range $0.3 < p_t < 3$ GeV [28]. Symbols: hydrodynamic calculations (pions, same p_t range).

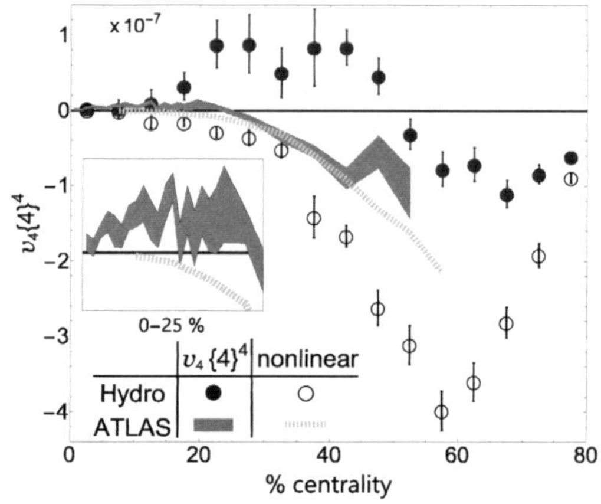

Figure 2. Dark shaded band: ATLAS data [7] for $v_4\{4\}^4$ as a function of centrality percentile for charged particles in the p_t range $0.5 < p_t < 5$ GeV. Full symbols: hydrodynamic calculation (pions, $0.5 < p_t < 3$ GeV). Dashed line and open symbols: last term of Eq. (6), from ATLAS data and hydrodynamic calculations.

experimental data on $v_4\{\Psi_2\}$ but slightly overestimates $v_4\{2\}$, meaning that our hydrodynamical setup overestimates the linear part, V_{4L}. We stress that, in our calculation, we implement a very low value of η/s and that the linear part, V_{4L}, is more strongly damped by viscosity than the nonlinear part [21]. Agreement with data is likely to be improved with a larger value of η/s.

3. Explaining $v_4\{4\}^4$

Figure 2 presents $v_4\{4\}^4$, as defined in Eq. (2), from ATLAS data [7] (the plotted quantity is $c_4\{4\} \equiv -v_4\{4\}^4$). It is positive up to 25% centrality (see inset in Fig. 2) and then negative. This change of sign is also observed in hydrodynamic calculations (full symbols), although it occurs around 50% centrality. We now argue that the change of sign is driven by the nonlinear response. Neglecting the linear part, V_{4L}, in Eq. (3), one obtains

$$v_4\{4\}^4 = \chi_4^4 \left(2\langle v_2^4\rangle^2 - \langle v_2^8\rangle\right) = v_4\{\Psi_2\}^4 \left(2 - \frac{\langle v_2^8\rangle}{\langle v_2^4\rangle^2}\right), \quad (6)$$

where, in the last equality, we have used Eqs. (4) and (5). We compute the last term of Eq. (6) both in hydrodynamics and using experimental data. The estimate of hydrodynamics is shown as open symbols in Fig. 2. As for experimental data, we employ the relation $v_4\{\Psi_2\} \equiv v_4\{2\}\langle\cos(4(\Phi_4 - \Phi_2))\rangle_w$, where $\langle\cos(4(\Phi_4 - \Phi_2))\rangle_w$ is the event-plane correlation [19, 38] measured by the ATLAS Collaboration. Higher-order moments of v_2 are instead obtained from the measured higher-order cumulants of elliptic flow [7, 19]. The resulting estimate is plotted as a dashed line in Fig. 2. Both estimates show that large fluctuations of v_2 lift the value of $\langle v_2^8\rangle/\langle v_2^4\rangle^2$, causing the contribution of the nonlinear term to be negative for all centralities. The nonlinear term increases in magnitude as a function of centrality percentile and drives the change of sign of $v_4\{4\}^4$. We find that the hydrodynamic calculation overestimates both $v_4\{4\}^4$

and the difference between $v_4\{4\}^4$ and the nonlinear contribution. Moreover, the change of sign of $v_4\{4\}^4$ occurs at a centrality percentile which is too large. These issues are consistent with the conclusion drawn in Fig. 1: Our hydrodynamical setup overestimates the linear part, V_{4L}.

We have shown that the peculiar centrality dependence of $v_4\{4\}^4$ observed in Pb+Pb collisions at the LHC is understood in hydrodynamics as resulting from the combined effects of a nonlinear hydrodynamic response coupling v_4 to v_2, and large v_2 fluctuations. This provides further evidence for a fluidlike behavior of the matter created in ultrarelativistic Pb+Pb collisions.

Acknowledgments

This work is supported by the European Research Council under the Advanced Investigator Grant ERC-AD-267258. JNH was supported by the National Science Foundation under grant no. PHY-1513864 and underneath FAPESP grant: 2016/03274-2. The authors wish to thank the ATLAS Collaboration for providing experimental data on $c_4\{4\}$.

References

[1] Luzum M 2011 *J. Phys.* **G38** 124026 (*Preprint* 1107.0592)
[2] Alver B and Roland G 2010 *Phys. Rev.* **C81** 054905 [Erratum: Phys. Rev.C82,039903(2010)] (*Preprint* 1003.0194)
[3] Alver B *et al.* (PHOBOS) 2007 *Phys. Rev. Lett.* **98** 242302 (*Preprint* nucl-ex/0610037)
[4] Borghini N, Dinh P M and Ollitrault J Y 2001 *Phys. Rev.* **C64** 054901 (*Preprint* nucl-th/0105040)
[5] Aamodt K *et al.* (ALICE) 2010 *Phys. Rev. Lett.* **105** 252302 (*Preprint* 1011.3914)
[6] Aamodt K *et al.* (ALICE) 2011 *Phys. Rev. Lett.* **107** 032301 (*Preprint* 1105.3865)
[7] Aad G *et al.* (ATLAS) 2014 *Eur. Phys. J.* **C74** 3157 (*Preprint* 1408.4342)
[8] Teaney D and Yan L 2011 *Phys. Rev.* **C83** 064904 (*Preprint* 1010.1876)
[9] Niemi H, Denicol G S, Holopainen H and Huovinen P 2013 *Phys. Rev.* **C87** 054901 (*Preprint* 1212.1008)
[10] Gardim F G, Noronha-Hostler J, Luzum M and Grassi F 2015 *Phys. Rev.* **C91** 034902 (*Preprint* 1411.2574)
[11] Voloshin S A, Poskanzer A M, Tang A and Wang G 2008 *Phys. Lett.* **B659** 537–541 (*Preprint* 0708.0800)
[12] Bhalerao R S, Luzum M and Ollitrault J Y 2011 *Phys. Rev.* **C84** 054901 (*Preprint* 1107.5485)
[13] Yan L and Ollitrault J Y 2014 *Phys. Rev. Lett.* **112** 082301 (*Preprint* 1312.6555)
[14] Grönqvist H, Blaizot J P and Ollitrault J Y 2016 (*Preprint* 1604.07230)
[15] Rybczynski M and Broniowski W 2016 *Acta Phys. Polon.* **B47** 1033 (*Preprint* 1510.08242)
[16] Ghosh S, Singh S K, Chatterjee S, Alam J and Sarkar S 2016 *Phys. Rev.* **C93** 054904 (*Preprint* 1601.03971)
[17] Gardim F G, Grassi F, Luzum M and Ollitrault J Y 2012 *Phys. Rev.* **C85** 024908 (*Preprint* 1111.6538)
[18] Borghini N and Ollitrault J Y 2006 *Phys. Lett.* **B642** 227–231 (*Preprint* nucl-th/0506045)
[19] Yan L and Ollitrault J Y 2015 *Phys. Lett.* **B744** 82–87 (*Preprint* 1502.02502)
[20] Giacalone G, Yan L, Noronha-Hostler J and Ollitrault J Y 2016 *Phys. Rev.* **C94** 014906 (*Preprint* 1605.08303)
[21] Teaney D and Yan L 2012 *Phys. Rev.* **C86** 044908 (*Preprint* 1206.1905)
[22] Adams J *et al.* (STAR) 2004 *Phys. Rev. Lett.* **92** 062301 (*Preprint* nucl-ex/0310029)
[23] Adare A *et al.* (PHENIX) 2010 *Phys. Rev. Lett.* **105** 062301 (*Preprint* 1003.5586)
[24] Gardim F G, Grassi F, Luzum M and Ollitrault J Y 2012 *Phys. Rev. Lett.* **109** 202302 (*Preprint* 1203.2882)
[25] Luzum M and Ollitrault J Y 2013 *Phys. Rev.* **C87** 044907 (*Preprint* 1209.2323)
[26] Jia J 2014 *J. Phys.* **G41** 124003 (*Preprint* 1407.6057)
[27] Qian J and Heinz U 2016 (*Preprint* 1607.01732)
[28] Chatrchyan S *et al.* (CMS) 2014 *Phys. Rev.* **C89** 044906 (*Preprint* 1310.8651)
[29] Qiu Z and Heinz U 2012 *Phys. Lett.* **B717** 261–265 (*Preprint* 1208.1200)
[30] Teaney D and Yan L 2014 *Phys. Rev.* **C90** 024902 (*Preprint* 1312.3689)
[31] Niemi H, Eskola K J and Paatelainen R 2016 *Phys. Rev.* **C93** 024907 (*Preprint* 1505.02677)
[32] Noronha-Hostler J, Denicol G S, Noronha J, Andrade R P G and Grassi F 2013 *Phys. Rev.* **C88** 044916 (*Preprint* 1305.1981)
[33] Noronha-Hostler J, Noronha J and Grassi F 2014 *Phys. Rev.* **C90** 034907 (*Preprint* 1406.3333)
[34] Alver B, Baker M, Loizides C and Steinberg P 2008 (*Preprint* 0805.4411)
[35] Noronha-Hostler J, Yan L, Gardim F G and Ollitrault J Y 2016 *Phys. Rev.* **C93** 014909 (*Preprint* 1511.03896)
[36] Policastro G, Son D T and Starinets A O 2001 *Phys. Rev. Lett.* **87** 081601 (*Preprint* hep-th/0104066)
[37] Teaney D 2003 *Phys. Rev.* **C68** 034913 (*Preprint* nucl-th/0301099)
[38] Aad G *et al.* (ATLAS) 2014 *Phys. Rev.* **C90** 024905 (*Preprint* 1403.0489)

A viscous blast-wave model for heavy-ion collisions

Amaresh Jaiswal[1] **and Volker Koch**[2]

[1]GSI, Helmholtzzentrum für Schwerionenforschung, Planckstrasse 1, D-64291 Darmstadt, Germany

[2]Lawrence Berkeley National Laboratory, Nuclear Science Division, MS 70R0319, Berkeley, California 94720, USA

E-mail: `jaiswal.amaresh@gmail.com, vkoch@lbl.gov`

Abstract. We present a generalization of the blast-wave model by incorporating viscous effects in the fluid velocity profile as well as in the Cooper-Frye freeze-out. We apply this model to study the identified particles spectra and anisotropic flow at the Large Hadron Collider (LHC). We show that this improved viscous blast-wave model leads to good description of the transverse momentum distribution of particle multiplicities and elliptic as well as triangular flow. Within this model, we estimate the shear viscosity to entropy density ratio $\eta/s \simeq 0.24$ at the LHC.

1. Introduction

The quark-gluon plasma (QGP) formed in relativistic heavy-ion collisions exhibit strong collective behaviour and hence can be studied within the framework of relativistic hydrodynamics. The hydrodynamical modelling of heavy-ion collisions suggests that the QGP behaves like a nearly perfect fluid having an extremely small shear viscosity to entropy density ratio η/s [1]. Apart from hydrodynamics, the collective behaviour of QGP can also be studied within the so-called blast-wave model [2, 3, 4, 5, 6]. Here we generalize the blast-wave model to include viscous effects by using a viscosity-based survival scale for initial geometrical anisotropies, formed in relativistic heavy-ion collisions, to parametrize the radial flow velocity. We employ this viscous blast-wave model to obtain the transverse momentum dependence of particle yields and flow harmonics for the Large Hadron Collider (LHC). We fix the model parameters by fitting the transverse momentum distribution of identified particle spectra. Subsequently, we demonstrate that this leads to reasonably good agreement with transverse momentum dependence of elliptic and triangular flow for various centralities. Within the present model, we estimate the shear viscosity to entropy density ratio $\eta/s \simeq 0.24$ at the LHC.

2. The model

We work in the Milne co-ordinate system where, $\tau = \sqrt{t^2 - z^2}$, $\eta_s = \tanh^{-1}(z/t)$, $r = \sqrt{x^2 + y^2}$, $\varphi = \mathrm{atan2}(y, x)$. The most important feature of the blast-wave model is the Hubble like parametrization of transverse velocity, $u^r \sim r$, which is found to be in agreement with hydro results [7]. In addition, we include angular anisotropy in the radial fluid velocity profile in the transverse plane. The hydrodynamic fields are parametrized as [8]

$$T = T_f, \quad u^r = u_0 \frac{r}{R}\left[1 + 2\sum_{n=1}^{\infty} u_n \cos[n(\varphi - \psi_n)]\right], \quad u^\varphi = u^{\eta_s} = 0, \quad u^\tau = \sqrt{1 + (u^r)^2}, \quad (1)$$

where R is the transverse radius of the fireball at freeze-out. The condition $u^{\mu}u_{\mu} = 1$ leads to the expression for u^{t}. In order to determine u_n, we use the fact that the initial geometrical anisotropies eventually converts to anisotropies in the radial fluid velocity. The participant anisotropies, ε_n, is defined in terms of the Fourier expansion for a single-particle distribution as

$$f(\varphi) = \frac{1}{2\pi}\left[1 + 2\sum_{n=1}^{\infty} \varepsilon_n \cos[n(\varphi - \psi_n)]\right], \tag{2}$$

where ψ_n is the angle between the x axis and the major axis of the participant distribution. Next, we determine the conversion efficiency of the initial geometrical eccentricity to final anisotropy in the radial fluid velocity, i.e., u_n/ε_n.

We start from the dispersion relation for sound in a viscous medium, $\omega = c_s k + ik^2\frac{1}{T}\left(\frac{2}{3}\frac{\eta}{s}\right)$, where c_s is the speed of sound in the medium, η is the coefficient of shear viscosity and $s = (\epsilon + P)/T$ is the entropy density. Using a Fourier ansatz, one finds that the amplitudes of the stress tensor harmonics, with momentum k, are attenuated by a factor [9]

$$\delta T^{\mu\nu}(t,k) = \exp\left[-\left(\frac{2}{3}\frac{\eta}{s}\right)\frac{k^2 t}{T}\right]\delta T^{\mu\nu}(0,k), \tag{3}$$

where the oscillatory pre-factor has been ignored. Each harmonics is a damped oscillator and form standing waves on the fireball circumference throughout the evolution, i.e., $k = n/R$. Note that the presence of momentum squared in the exponent results in enhanced effect of viscosity for the higher harmonics. Therefore, at freeze-out time t_f, the wave amplitude reaction is [10]

$$\frac{\delta T^{\mu\nu}|_{t=t_f}}{\delta T^{\mu\nu}|_{t=0}} = \exp\left[-n^2\left(\frac{2}{3}\frac{\eta}{s}\right)\frac{t_f}{R^2 T_f}\right], \tag{4}$$

where T_f is the freeze-out temperature. The above equation indicates a viscosity-based survival scale for anisotropic structures formed by point like perturbations. The final radial fluid velocity is due to initial geometrical perturbations and, in the absence of viscosity, the conversion efficiency remains the same for all harmonics. Therefore, for a viscous medium, the conversion efficiency must be proportional to the wave amplitude reaction,

$$\frac{u_n}{\varepsilon_n} = \alpha_0 \exp\left[-n^2\left(\frac{2}{3}\frac{\eta}{s}\right)\frac{t_f}{R^2 T_f}\right], \tag{5}$$

where α_0 is the constant of proportionality. The above equation, along with Eq. (1), leads to the fluid velocity profile which is used in the freeze-out prescription to obtain particle spectra and anisotropic flow coefficients.

3. Particle spectra and anisotropic flow

Hadron spectra can be obtained using the Cooper-Frye prescription for particle production [11]

$$\frac{dN}{d^2p_T dy} = \frac{1}{(2\pi)^3}\int p_\mu d\Sigma^\mu f(x,p), \tag{6}$$

where $d\Sigma_\mu$ is the freeze-out hyper-surface and $f(x,p)$ is the distribution function of the particles. For a system close to equilibrium, $f = f_0 + \delta f$ where $\delta f \ll f_0$. For the equilibrium distribution function, f_0, we use Fermi-Dirac statistics ($a = +1$) for baryons and Bose-Einstein statistics ($a = -1$) for mesons. For the non-equilibrium part, we employ the Grad's 14-moment approximation with viscous corrections up to first-order [12]

$$\delta f_1 = \frac{f_0\tilde{f}_0}{T^3}\left(\frac{\eta}{s}\right)p^\alpha p^\beta \nabla_{\langle\alpha}u_{\beta\rangle}. \tag{7}$$

where $\tilde{f}_0 = 1 - af_0$ and the angular brackets denote traceless symmetric projection orthogonal to u^μ. In the case of blast-wave model, the form of $p^\alpha p^\beta \nabla_{\langle\alpha}u_{\beta\rangle}$ is explicitly calculated in Refs. [5, 8].

The anisotropic flow is defined as

$$v_n(p_T) \equiv \frac{\displaystyle\int_{-\pi}^{\pi} d\phi \, \cos[n(\phi - \Psi_n)] \, \frac{dN}{dy\, p_t\, dp_T\, d\phi}}{\displaystyle\int_{-\pi}^{\pi} d\phi \, \frac{dN}{dy\, p_t\, dp_T\, d\phi}}, \tag{8}$$

where Ψ_n is the event-plane angle for the n-th harmonic. Up to first order in viscosity [5],

$$v_n(p_T) = v_n^{(0)}(p_T)\left(1 - \frac{\displaystyle\int d\phi \, \frac{dN^{(1)}}{dy\, p_t\, dp_T\, d\phi}}{\displaystyle\int d\phi \, \frac{dN^{(0)}}{dy\, p_t\, dp_T\, d\phi}}\right) + \frac{\displaystyle\int d\phi \, \cos[n(\phi - \Psi_n)] \, \frac{dN^{(1)}}{dy\, p_t\, dp_T\, d\phi}}{\displaystyle\int d\phi \, \frac{dN^{(0)}}{dy\, p_t\, dp_T\, d\phi}}, \tag{9}$$

where the superscript '(0)' and '(1)' denote quantities calculated using the ideal distribution function and first-order viscous correction, Eq. (7), respectively.

4. Initial conditions and numerical results

We consider two identical relativistic nuclei with mass number A colliding with impact parameter b. The transverse expansion of the fireball is obtained by using the unperturbed radial velocity,

$$u^r \equiv \frac{dr}{d\tau} = u_0 \frac{r}{R} \quad \Rightarrow \quad \int_{r_0}^{R} \frac{dr}{r} = \int_0^{\tau_f} \frac{u_0}{R}\, d\tau \quad \Rightarrow \quad R = r_0 \exp\left(\frac{u_0\, \tau_f}{R}\right), \tag{10}$$

where $r_0 = \frac{1}{2}\left(b^2 - 2\, bR_0\sqrt{2 + b/R_0} + 4R_0^2\right)^{1/2}$ and $R_0 = 1.25 A^{1/3}$ fm is the radius of each colliding nuclei [8]. In order to determine the freeze-out times for non-central collisions, we employ the analytical result of Bjorken expansion $\epsilon \propto \tau^{-4/3}$. Assuming the initial thermalization time and the freeze-out energy density to be same for all collisions, we get

$$\tau_f = \tau_{f0}\left(\frac{\epsilon_i}{\epsilon_{i0}}\right)^{3/4}, \tag{11}$$

where τ_{f0} is the freeze-out time for most central collisions which is fitted to match the transverse momentum distribution of particle multiplicities. The ratio ϵ_i/ϵ_{i0} is the initial central energy density for all centrality, scaled by its corresponding value in most central collisions. While the Bjorken estimate for the absolute values of the freeze-out times is rather crude, we have used the above equation to fix the freeze-out time for non-central collisions relative to that of the most central ones. In the present case, it seems to be a reliable approximation as is evident from Fig. 1. Therefore, in order to fit the transverse momentum spectra, the parameters that needs to be fixed within the viscous-blast wave model are the freeze-out temperature T_f, the freeze-out time for central collision τ_{f0} and the maximum radial flow velocity u_0. An interplay between α_0 in Eq. (5) and η/s is important to reproduce the transverse momentum dependence of flow harmonics.

Figure 1 shows the transverse momentum distribution of pions, kaons, and protons spectra for Pb+Pb collisions at $\sqrt{s_{NN}} = 2.76$ TeV in (a): $0-5\%$ and (b): $20-30\%$ centrality. We observe that, for a freeze-out temperature of 120 MeV, the spectra for π^+ and K^+ from the viscous blast-wave model are in good overall agreement with the ALICE data [13]. On the other hand, the freeze-out temperature for protons is considered to be 135 MeV in order to

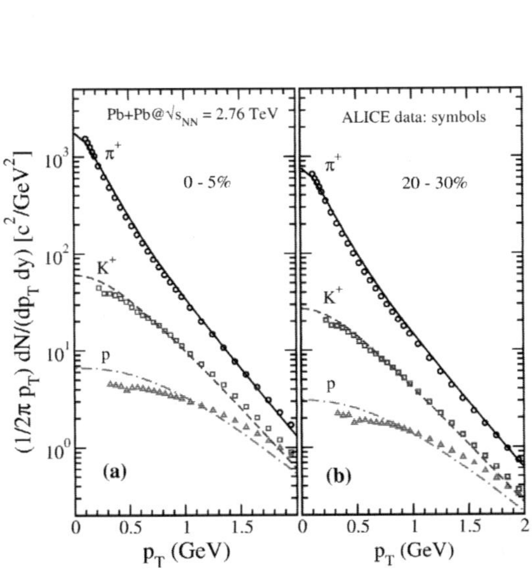

Figure 1. Transverse momentum distribution of π^+, K^+ and p multiplicities in Pb+Pb collisions at $\sqrt{s_{NN}} = 2.76$ TeV in two centrality ranges, (a): $0-5\%$ and (b): $20-30\%$. The symbols represent ALICE data [13] at mid-rapidity and the lines correspond to viscous blast-wave results.

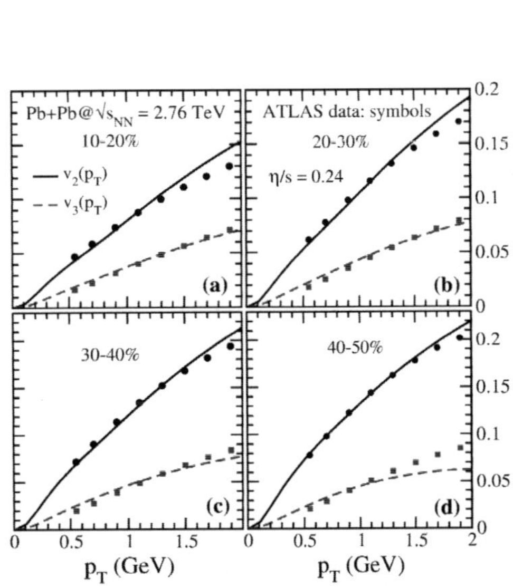

Figure 2. Transverse momentum dependence of the anisotropic flow coefficients $v_n(p_T)$ of charged hadrons, for $n = 2$ and 3, calculated at various centralities in Pb+Pb collisions at $\sqrt{s_{NN}} = 2.76$ TeV in the viscous blast-wave model (lines) with $\eta/s = 0.24$ as compared to the ATLAS data [14] (symbols).

fit the experimental data. Figure 2 shows our results for the $v_n(p_T)$, for various centralities, in comparison with the ATLAS data [14]. For $\alpha_0 = 0.4$ and $\eta/s = 0.24$, we find fairly good agreement with the experimental data for elliptic flow, $v_2(p_T)$, and triangular flow, $v_3(p_T)$, at the LHC. We have used the root-mean square values of initial eccentricities, ε_n, obtained from the Monte-Carlo Glauber model [15].

Acknowledgements:- A.J. was supported by the Frankfurt Institute for Advanced Studies (FIAS), Germany. The work of V.K. was supported by the ExtreMe Matter Institute (EMMI) and by the Office of Nuclear Physics in the US Department of Energy's Office of Science under Contract No. DE-AC02-05CH11231.

References

[1] U. Heinz and R. Snellings, Ann. Rev. Nucl. Part. Sci. **63**, 123 (2013).
[2] E. Schnedermann, J. Sollfrank and U. W. Heinz, Phys. Rev. C **48**, 2462 (1993).
[3] P. Huovinen, P. F. Kolb, U. W. Heinz, P. V. Ruuskanen and S. A. Voloshin, Phys. Lett. B **503**, 58 (2001).
[4] C. Adler *et al.* [STAR Collaboration], Phys. Rev. Lett. **87**, 182301 (2001).
[5] D. Teaney, Phys. Rev. C **68**, 034913 (2003).
[6] X. Sun, H. Masui, A. M. Poskanzer and A. Schmah, Phys. Rev. C **91**, 024903 (2015).
[7] D. Teaney, J. Lauret and E. V. Shuryak, nucl-th/0110037.
[8] A. Jaiswal and V. Koch, arXiv:1508.05878 [nucl-th]; EPJ Web Conf. **120**, 06001 (2016).
[9] P. Staig and E. Shuryak, Phys. Rev. C **84**, 034908 (2011).
[10] E. Shuryak and I. Zahed, Phys. Rev. C **88**, 044915 (2013).
[11] F. Cooper and G. Frye, Phys. Rev. D **10**, 186 (1974).
[12] H. Grad, Comm. Pure Appl. Math. **2**, 331 (1949).
[13] B. Abelev *et al.* [ALICE Collaboration], Phys. Rev. C **88**, 044910 (2013).
[14] G. Aad *et al.* [ATLAS Collaboration], Phys. Rev. C **86**, 014907 (2012).
[15] E. Retinskaya, M. Luzum and J. Y. Ollitrault, Phys. Rev. C **89**, 014902 (2014).

Directed flow of Λ, $\bar{\Lambda}$, $K^{\pm}$, K^0_S and ϕ mesons from Beam Energy Scan Au+Au collisions using the STAR experiment

Subhash Singha (for the STAR collaboration)

Department of Physics, Kent State University, Ohio 44242, USA

E-mail: `subhash@rcf.rhic.bnl.gov`

Abstract. We report the results of v_1 and dv_1/dy near mid-rapidity for Λ, $\bar{\Lambda}$, $K^{\pm}$, K^0_s and ϕ in Au+Au collisions at $\sqrt{s_{\rm NN}}$ = 7.7, 11.5, 14.5, 19.6, 27 and 39 GeV using the STAR detector at RHIC. The dv_1/dy of Λ is found to be consistent with that of the proton and shows a change in sign near $\sqrt{s_{\rm NN}}$ = 11.5 GeV. The v_1 slope for $\bar{\Lambda}$, $\bar{p}$ and ϕ shows a similar trend for $\sqrt{s_{\rm NN}}$ > 14.5 GeV, while below 14.5 GeV, ϕ v_1 is consistent with zero but with a large uncertainty. The dv_1/dy for net protons and net kaons is similar for $\sqrt{s_{\rm NN}}$ > 14.5 GeV, but they deviate at lower beam energies.

1. Introduction

The measurement of collective flow in relativistic heavy-ion collisions can offer an insight into the Equation of State (EoS) of the QCD matter produced in early stages of the collision [1]. The directed flow (v_1) is characterized by the first harmonic coefficient in the Fourier expansion of the azimuthal distribution of the produced particles with respect to the reaction plane (Ψ_R), i.e, $v_1 \equiv \langle \cos(\varphi - \Psi_R) \rangle$, where φ denotes the azimuthal angle of the produced particles [2]. Both hydrodynamic [3, 4] and transport model [5] calculations imply that v_1 is sensitive to the EoS and is a promising observable to explore the QCD phase diagram. One of the main goals of the Beam Energy Scan (BES) program at RHIC is to search for signatures of a possible QCD critical point and first-order phase transition [6, 7, 8]. Recently, STAR published the beam energy dependence of the slope of directed flow (dv_1/dy) for protons and net protons near mid-rapidity [9]. The observation of a minimum in the slope of directed flow for protons and net protons around $\sqrt{s_{\rm NN}}$ = 10-20 GeV, and a double sign change in this observable for net protons, point to a possible softening of the QCD equation of state [10]. However, other recent model calculations, with and without crossover and first-order phase transitions, show large discrepancies with the STAR measurements [10, 11, 12, 13].

A new set of measurements with different hadron species (Λ, $\bar{\Lambda}$, $K^{\pm}$, K^0_S and ϕ), and hence different constituent quarks, will not only help to understand the QCD phase transition, but in addition will allow to disentangle the role of produced and transported quarks in heavy-ion collisions. In particular, the ϕ meson offers a unique advantage because its mass is similar to the proton mass, yet it is a vector meson. Moreover ϕ meson is minimally affected by late-stage hadronic interactions.

IOP Conf. Series: Journal of Physics: Conf. Series **779** (2017) 012066 doi:10.1088/1742-6596/779/1/012066

2. Analysis Details

The v_1 analysis is applied to data taken by the STAR detector during 2010, 2011 and 2014. The STAR detector offers uniform acceptance, full azimuthal coverage, and excellent particle identification. The Time Projection Chamber (TPC) [14] performed charged particle tracking near mid-rapidity. The collision centrality is estimated from the charged particle multiplicity within the pseudorapidity region $|\eta| < 0.5$. Particles are identified using both the Time Projection Chamber (TPC) and Time of Flight (TOF) [15] detectors. Two Beam Beam Counters (BBC) [16], covering $3.3 < |\eta| < 5.2$, are used to reconstruct the event plane. The BBC event plane is based on the first harmonic, since the v_1 signal is strong in the BBC acceptance region at BES energies. Moreover, a large η-gap relative to the TPC reduces non-flow effects [2] in the v_1 measurements. The uncharged particle species are measured through the reconstruction of the topology of their weak decays into charged particles: $K_s^0 \to \pi^+\pi^-$, $\Lambda \to p\pi$ (and charge conjugates for anti-particles). Topological selection cuts are applied in order to reduce backgrounds without much loss of the signal. The ϕ meson is measured using its hadronic decay channel: $\phi \to K^+K^-$. The v_1 for these particles is extracted by an invariant mass method [17].

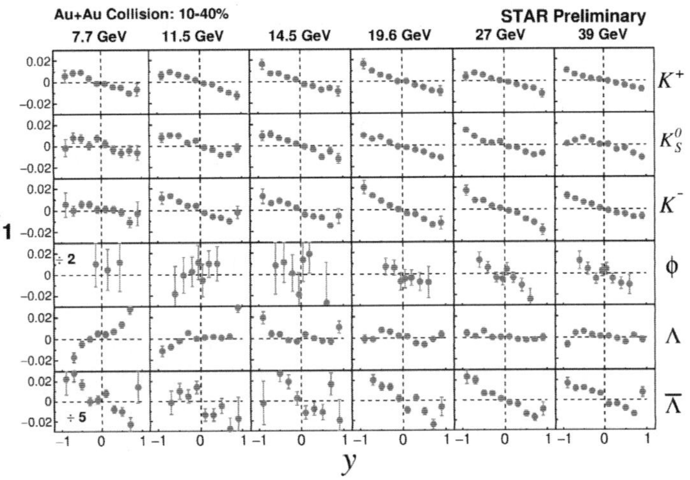

Figure 1. Rapidity dependence of v_1 for $K^{\pm}$, K_s^0, ϕ, Λ and $\bar{\Lambda}$ in 10-40% Au+Au collisions at $\sqrt{s_{NN}}$ = 7.7, 11.5, 14.5, 19.6, 27 and 39 GeV.

3. Results and Discussion

Figure 1 presents v_1 as a function of rapidity for $K^{\pm}$, K_s^0, ϕ, Λ and $\bar{\Lambda}$ in 10-40% Au+Au collisions at $\sqrt{s_{NN}}$ = 7.7, 11.5, 14.5, 19.6, 27 and 39 GeV. The data points for $\bar{\Lambda}$ and ϕ at 7.7 GeV are divided by factors of 5 and 2, respectively, in order to use the same vertical scale in all panels. In a previous analysis, the $v_1(y)$ slope parameter near mid-rapidity was extracted by fitting it with a cubic function [9]. Although the cubic fit reduces sensitivity to the rapidity range where the fit is performed, it is unstable for particles with poor statistics (e.g. ϕ and $\bar{\Lambda}$). Therefore in Ref. [18] and the present analysis, the v_1 slope is based on a linear fit within the range $|y| < 0.8$ for all particle species and all energies. All kaons and $\bar{\Lambda}$ show a negative slope at all beam energies. The ϕ meson shows a negative slope at $\sqrt{s_{NN}} > 14.5$ GeV. The dv_1/dy for all the measured strange hadrons is shown in Fig. 2. The dv_1/dy of Λ is consistent with

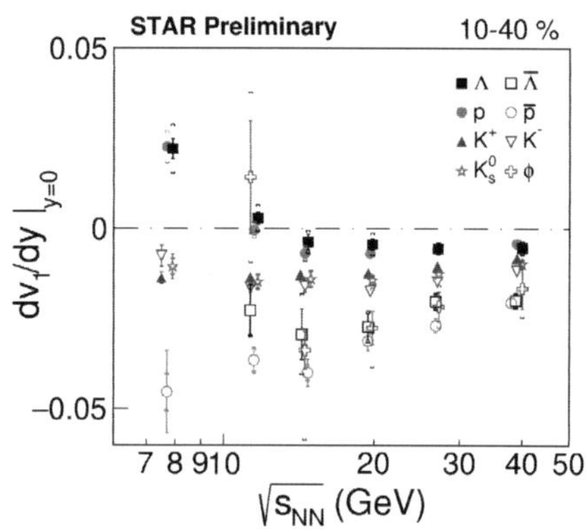

Figure 2. (Color online) Beam energy dependence of dv_1/dy for p, $\bar{p}$, $K^{\pm}$, K^0_s, ϕ, Λ and $\bar{\Lambda}$ in 10-40% Au+Au collisions.

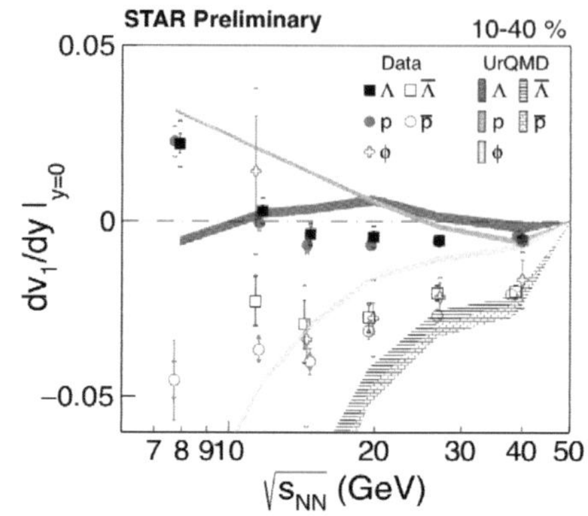

Figure 3. (Color online) Beam energy dependence of dv_1/dy for p, $\bar{p}$, ϕ, Λ and $\bar{\Lambda}$ in 10-40% Au+Au collisions. STAR data are compared with UrQMD [5] model calculations.

that of protons at all beam energies, while the $\bar{\Lambda}$ follows the trend of antiprotons. The dv_1/dy for K^+ is closer to zero than that for K^- for $\sqrt{s_{\rm NN}} > 11.5$ GeV, while the trend is reversed at 7.7 GeV. The K^0_s dv_1/dy lies between K^+ and K^- at all beam energies. Theorists [19] have argued that kaon-nucleon potentials influence directed flow at these energies, where the baryon density is high. The dv_1/dy for ϕ shows similar trends as $\bar{\Lambda}$ and $\bar{p}$ for $\sqrt{s_{\rm NN}} > 14.5$ GeV, while at 11.5 GeV, dv_1/dy for ϕ is consistent with zero with large uncertainty. In the higher energy region where antibaryon production is significant, $\bar{p}$, ϕ and $\bar{\Lambda}$ show similar energy dependence. Figure 3 presents a comparison between the measurements and UrQMD [5] for p, Λ and ϕ. UrQMD seems to follow the trend of the data for higher beam energies, but deviates at lower energies. Figure 4 shows the dv_1/dy of p, K^+, π^- and K^0_S on the left panel, and corresponding anti-particles in the right panel, as a function of beam energy. Here the particles in the left panel are expected to have more quarks from stopped initial-state nucleons than the antiparticles on the right. The charged kaons and K^0_S are compared with UrQMD [5] and HSD [13] model calculations. Both models qualitatively describe the data at higher energies, but fail to follow the trend of the data at lower energies. Figure 5 shows a comparison of the beam energy dependence of dv_1/dy for net protons and net kaons. Study of net particle v_1 is motivated by the goal of separating the contributions from produced quarks versus quarks that are transported from the initial state. We define

$$F_p = r_1(y)F_{\bar{p}} + (1 + r_1(y))F_{\text{net-}p} \tag{1}$$

$$F_{K+} = r_2(y)F_{K-} + (1 + r_2(y))F_{\text{net-}K} \tag{2}$$

where F denotes dv_1/dy for the indicated particle species; $r_1(y)$ and $r_2(y)$ are the rapidity dependence of the ratios of the corresponding antiparticles to particles. It is observed that net-p and net-K dv_1/dy follow a similar trend for $\sqrt{s_{\rm NN}} = 14.5$ - 200 GeV, while they deviate from each other strongly at 7.7 GeV.

4. Summary

In these proceedings, we present v_1 and dv_1/dy measurements of strange hadrons from the STAR experiment at RHIC. The Λ dv_1/dy is consistent with the proton dv_1/dy and both show a sign change near $\sqrt{s_{\rm NN}} = 11.5$ GeV. Charged kaons and K^0_s show negative v_1 for all studied beam

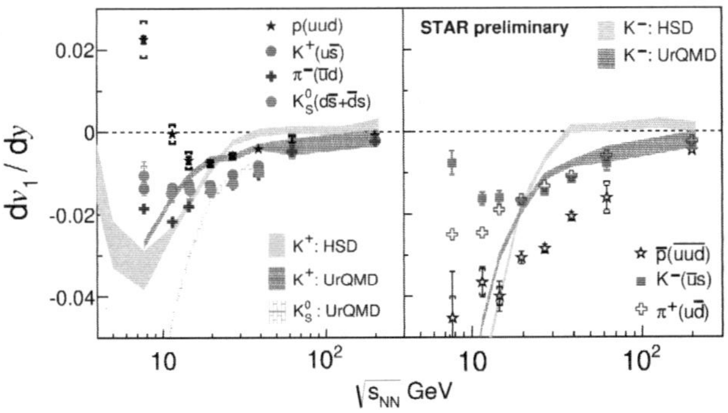

Figure 4. (Color online) Left Panel: Beam energy dependence of dv_1/dy for p, K^+, π^- and K^0_s in 10-40% Au+Au collisions. Right panel: Beam energy dependence of dv_1/dy for $\bar{p}$, K^-, π^+ in 10-40% Au+Au collisions. HSD [13] and UrQMD [5] model calculations are shown for $K^\pm$ and K^0_s.

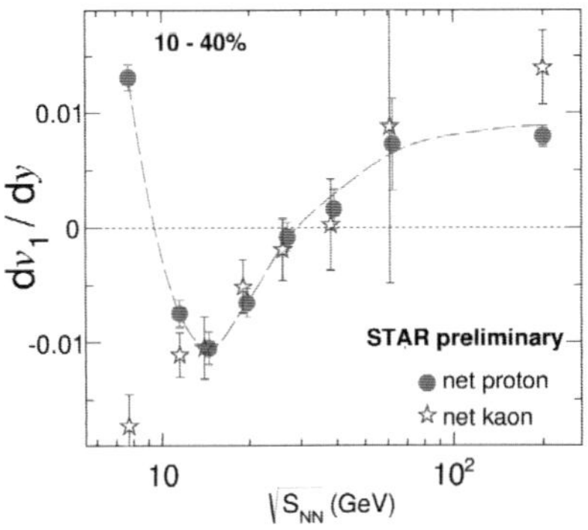

Figure 5. (Color online) Beam Energy dependence of net-particle dv_1/dy in 10-40% Au+Au collisions.

energies, while the K^0_s lies between the charged kaons. $\bar{\Lambda}$, $\bar{p}$ and ϕ show similar v_1 slope for $\sqrt{s_{\rm NN}} > 11.5$ GeV. The UrQMD and HSD results qualitatively explain the data at higher beam energies, but fail to describe the trend observed at lower energies.

References

[1] W. Reisdorf and H. G. Ritter, Annu. Rev. Nucl. Part. Sci. **47**, 663 (1997); H. Sorge, Phys. Rev. Lett. **78**, 2309 (1997).
[2] A. M. Poskanzer and S. A. Voloshin, Phys. Rev. C **58**, 1671 (1998); S. Voloshin and Y. Zhang, Z. Phys. C **70**, 665 (1996).
[3] U. W. Heinz, in *Relativistic Heavy Ion Physics*, Landolt-Boernstein New Series, Vol. I/23, edited by R. Stock (Springer Verlag, New York, 2010).
[4] H. Stöcker, Nucl. Phys. **A750**, 121 (2005).
[5] S. A. Bass et al., Prog. Part. Nucl. Phys. **41**, 255 (1998); M. Bleicher, E. Zabrodin, C. Spieles, S. A. Bass, C. Ernst, S. Soff, L. Bravina, M. Belkacem, H. Weber, H. Stöcker and W. Greiner, J. Phys. G **25**, 1859 (1999).
[6] B. I. Abelev *et al.* (STAR collaboration), STAR Note SN0493 (2009).
[7] M. M. Aggarwal *et al.* (STAR collaboration), arXiv:1007.2613.
[8] STAR collaboration, BES Phase-II Whitepaper, STAR Note SN0598 (2014).
[9] L. Adamczyk *et al.* (STAR Collaboration), Phys. Rev. Lett. **112**, 162301 (2014).
[10] Y. Nara, H. Niemi, A. Ohnishi, and H. Stöcker, arXiv:1601.07692 [hep-ph].
[11] J. Steinheimer, J. Auvinen, H. Petersen, M. Bleicher and H. Stöcker, Phys. Rev. C **89**, 054913 (2014).
[12] Yu. B. Ivanov, V. N. Russkikh, and V. D. Toneev, Phys. Rev. C **73**, 044904 (2006).
[13] V. P. Konchakovski, W. Cassing, Yu. B. Ivanov and V. D. Toneev, Phys. Rev. C **90**, 014903 (2014).
[14] M. Anderson *et al.*, Nucl. Instr. Meth. A **499**, (2003) 659 .
[15] W.J. Llope (STAR Collaboration), Nucl. Inst. and Methods A **661**, (2012) S110
[16] C. A. Whitten (STAR Collaboration), AIP Conf. Proc. **980**, (2008) 390.
[17] N. Borghini and J. Y. Ollitrault, Phys. Rev. C **70**, 064905 (2004).
[18] P. Shanmuganathan for the STAR Collaboration, Proc. of Quark Matter 2015, Kobe, Japan, Nucl. Phys. **A** (in press); arXiv:1512.09009v1.
[19] W. Cassing, V. P. Konchakovski, A. Palmese, V. D. Toneev, E. L. Bratkovskaya, EPJ Web Conf. **95**, 01004 (2015), arXiv:1408.4313, and references therein.

IOP Conf. Series: Journal of Physics: Conf. Series **779** (2017) 012067 doi:10.1088/1742-6596/779/1/012067

Systematic Searches for the Chiral Magnetic Effect and the Chiral Vortical Effect Using Identified Particles at RHIC/STAR

Liwen Wen(for the STAR Collaboration)
University of California, Los Angeles, California, USA
E-mail: lwen@physics.ucla.edu

Abstract. STAR measurements of identified-particle correlations in Au+Au collisions at $\sqrt{s_{\rm NN}} = 200$ GeV are used to explore possible hierarchical structures in the particle-dependent correlations due to the chiral magnetic effect (CME) and the chiral vortical effect (CVE). Correlations of p-π (p-Λ), sensitive to the possible CME (CVE), are reported and compared with hadron-hadron (h-h) correlation, yielding a charge separation ordering (p-$\Lambda > p$-$\pi \sim h$-h). The estimated background strength from transverse momentum conservation (TMC) hardly shows any centrality dependence. For Au+Au collisions at 19.6, 39 and 200 GeV, clear signals above the background are observed for mid-central and peripheral collisions while results from central collisions ($< 15\%$) can not separate from the background. For the results from collisions at 7.7 GeV, all are consistent with the TMC background although the errors are large.

1. Introduction

Chirality imbalance in a quark-gluon plasma could occur on an event-by-event basis. In the presence of a strong magnetic field (vorticity) produced in noncentral heavy-ion collisions, the chirality imbalance may lead to novel non-dissipative transport phenomena, among which are the CME and the CVE [1]. As the experimental manifestation of the CME, the signals of electric charge separation across the reaction plane have been extensively searched for using charged-hadron correlations at the Relativistic Heavy Ion Collider (RHIC) [2] and the Large Hadron Collider (LHC) [3]. Similarly, baryonic charge separation is considered to be the evidence for the CVE. However, theoretical predictions on the magnitude of the CME (CVE) and the particle-type dependence are still limited. In addition, previous charged-hadron correlations may contain the CVE (because protons are baryons but not eliminated), so experimental efforts to study these two effects respectively become of great importance. Identified-particle correlations can probe a specific underlying mechanism for the electric (baryonic) charge separation. In this article, we report γ correlations of p-Λ and p-π for Au+Au collisions at $\sqrt{s_{\rm NN}} = 200$ GeV at STAR/RHIC. The results are compared with h-h correlations. Although the γ correlator has scaled down flow-related background by approximately a factor of elliptic flow (v_2) [1], the background contribution cannot be completely ignored. In Sec. 4, we present our recent estimation of one flow-related background source, transverse momentum conservation (TMC), and rearrange our published BES results [4] to study the background effects.

2. Data analysis

Around 200 million minimum-bias events of Au+Au collisions at $\sqrt{s_{\rm NN}}$ =200 GeV from RHIC Run11 have been used in this analysis. With the Time Projection Chamber (TPC), the deviation of specific energy loss rate (dE/dx) from the expected value is used to identify electrons, pions, kaons and protons. The Time-of-Flight (TOF) detector provides the mass information to further select particles of interest. A pseudorapidity cut of $|\eta| < 1$ is applied to all the measured particles. Standard topological cuts are used to reconstruct Λ via $\Lambda \to \pi^- + p$ ($B.R. = 63.9 \pm 0.5\%$ [5]). Protons with $0.4 < p_T < 2.0$ GeV/c, Λ's with $0.5 < p_T < 5.0$ GeV/c and other hadrons with $0.15 < p_T < 2.0$ GeV/c are used in the corresponding correlation measurements. The second-order event plane, constructed from TPC tracks, is used to approximate the real reaction plane. The final event plane orientation is corrected for the event plane resolution, and the shifting method is used to eliminate detector effects.

The three-point correlator γ, sensitive to charge separation fluctuations, is used to search for the chiral effects [6]:

$$\gamma \equiv \langle \phi_\alpha + \phi_\beta - 2\Psi_{\rm RP} \rangle, \qquad (1)$$

where $\phi_{\alpha,\beta}$ refers to the azimuthal angle of the correlated particles, and $\Psi_{\rm RP}$ denotes the reaction plane orientation.

3. Gamma correlation results

Assuming s quarks take part in chiral dynamics the same way as u, d quarks do, p-Λ correlations are proposed to search for the CVE with limited contamination from the electric charge separation signal, because of the electric neutrality of Λ. Figures 1 and 2 show opposite- and same-sign correlations of p-Λ (left plot), as well as their difference (right plot) in Au+Au collisions at 200 GeV. Significant separation signals are observed in mid-central to peripheral collisions, which qualitatively meets the theoretical expectation for the CVE [1]. As a systematic check of h-h correlations, p-π correlations (Figs. 3 and 4) in 200 GeV Au+Au show that the electric charge separation signals have the magnitudes close to h-h correlations, which suggests a similar underlying mechanism (CME) as expected.

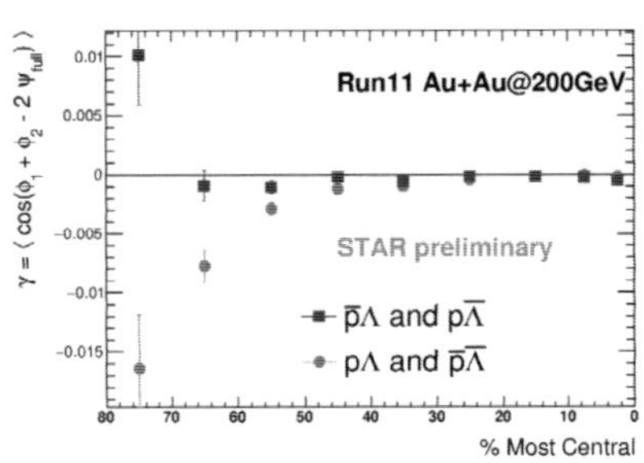

Figure 1. Opposite- and same-sign γ correlations of p-Λ pairs for Au+Au collisions at 200 GeV.

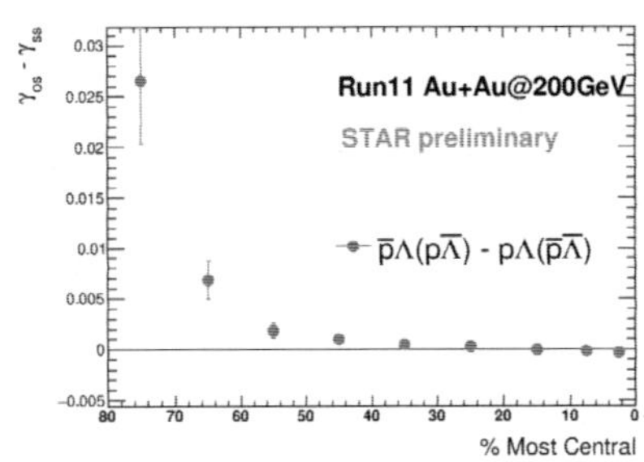

Figure 2. Difference between opposite- and same-sign γ correlations of p-Λ for Au+Au collisions at 200 GeV.

Figure 5 shows several identified-particle correlations under study, and exhibits a hierarchical structure of possible chiral effects. The large difference between baryonic and electric charge separation may suggest two differently driven mechanisms (magnetic field and vorticity), but more investigations are needed to reveal a true relationship between these two types of correlations. In Fig. 6, N_{part}, number of participants, scaled results are plotted to effectively eliminate the dilution effect, and show that in most central collisions, the signals are consistent with zero.

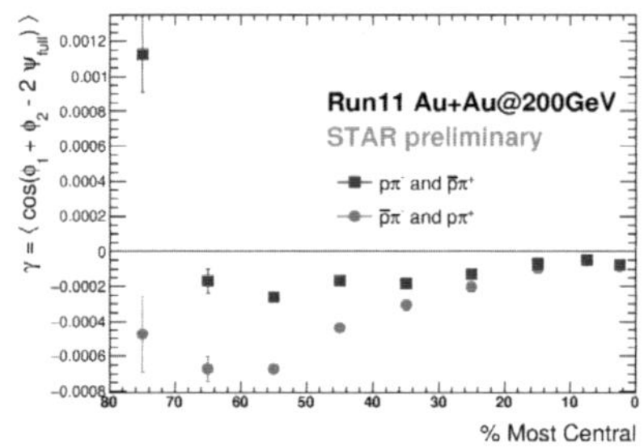

Figure 3. Opposite- and same-sign γ correlations of p-π pairs for Au+Au collisions at 200 GeV.

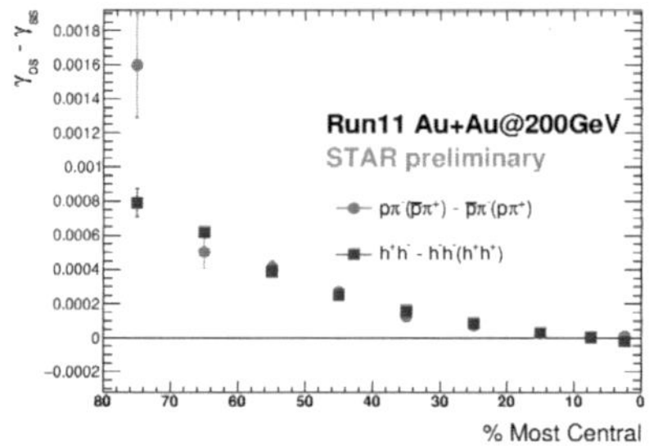

Figure 4. Difference between opposite- and same-sign γ correlations of p-π for Au+Au collisions at 200 GeV, compared to the h-h result.

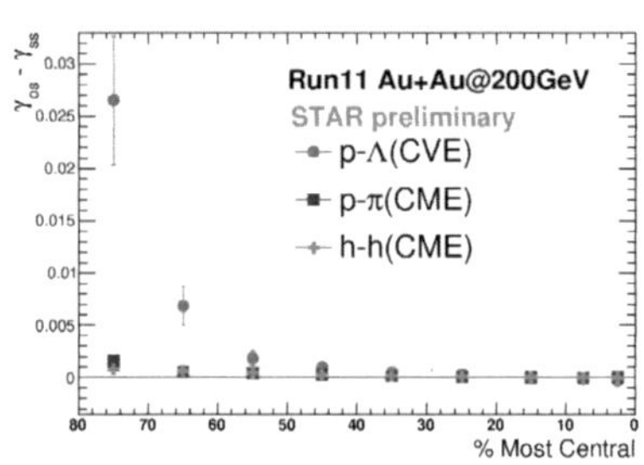

Figure 5. Comparison of $\gamma_{os} - \gamma_{ss}$ between p-Λ, p-π, h-h pairs for Au+Au collisions at 200 GeV.

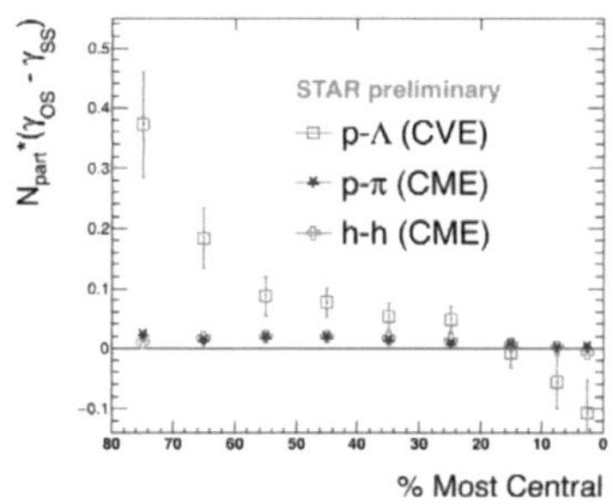

Figure 6. $N_{\rm part}$ scaled $\gamma_{os} - \gamma_{ss}$ of p-Λ, p-π, h-h for Au+Au collisions at 200 GeV.

4. Flow background estimation

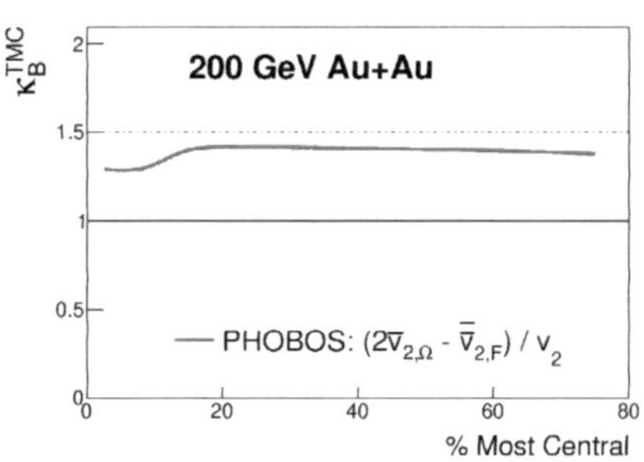

Figure 7. Estimated $\kappa_{\rm B}^{\rm TMC}$ as a function of centrality for Au+Au collisions at 200 GeV.

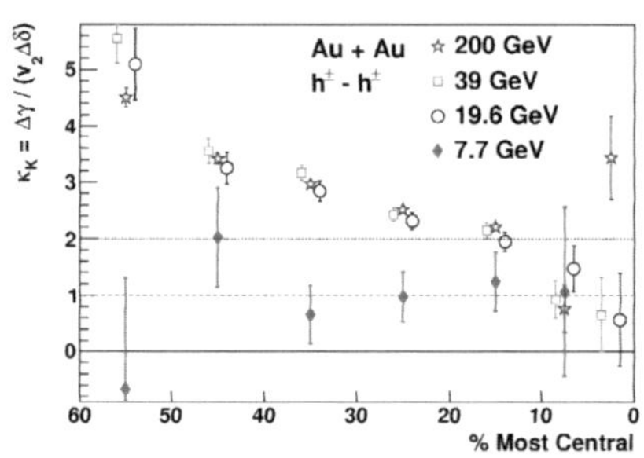

Figure 8. Centrality dependence of $\kappa_{\rm K}$ for Au+Au collisions at different energies (7.7, 19.6, 39 and 200 GeV).

The flow-related background is significantly suppressed in the construction of the γ correlator [1], but its effect on existing measurements is still non-negligible. Disentanglement of the background and the CME/CVE signal has been the recent focal point of theoretical and experimental efforts. In Ref. [7], a modified correlator, $H = (\kappa_{\rm B} v_2 \delta - \gamma)/(1 + \kappa_{\rm B} v_2)$, is proposed to remove the flow-induced background in an empirical approach. Here $\delta \equiv \langle \cos(\phi_\alpha - \phi_\beta) \rangle$, represents the two-particle correlation that includes contributions from TMC, local charge conservation (LCC), flowing resonances and so on. The coefficient, $\kappa_{\rm B}$, quantifies the background strength, and requires a practical approach to gauge its range in our correlation measurements.

Theoretically, the TMC effect on the γ correlator can be formulated and the corresponding $\kappa_{\rm B}^{\rm TMC}$ is approximated by $\kappa_{\rm B}^{\rm TMC} \approx 2 - \bar{v}_{2,F}/\bar{\bar{v}}_{2,\Omega}$ [7]. The $\bar{v}_{2,F}$ and $\bar{\bar{v}}_{2,\Omega}$ denote the p_T-weighted and p_T^2-weighted v_2 of particles of interest in the full space and the detector acceptance, respectively. The PHOBOS Collaboration at RHIC provides a v_2 measurement for charged particles over a broad phase space coverage in pseudo-rapidity [8]. Combined with transverse momentum spectrum information, $\kappa_{\rm B}^{\rm TMC}$ as a function of centrality is estimated and shown in Fig. 7. The magnitude is almost constant across different centralities and slightly smaller than 1.4. We can use this result as a reference to check our previous h-h correlations.

We define a "normalized" signal+background, $\kappa_{\rm K}$:

$$\kappa_{\rm K} = \Delta\gamma/(v_2 \Delta\delta), \tag{2}$$

which denotes the minimum $\kappa_{\rm B}$ value needed to saturate the gamma correlator strength with the background. Any result that is larger than real κ_B might represent the genuine physics. In Fig. 8, the h-h correlations in BES I [4] show that except for the 7.7 GeV data, $\kappa_{\rm K}$ for higher energies in mid-central collisions is always above the empirically allowed range of $\kappa_{\rm B}$ (between 1 and 2), where the $\kappa_{\rm B}$ is only constructed from TMC as a test case. This means that our measured H correlation strength cannot be entirely caused by the flow-induced background: the signal+background is larger than the background in the "κ" unit. There must be additional contributions which may be related to the CME. On the other hand, $\kappa_{\rm K}$ in Au+Au collisions at 7.7 GeV is consistent with unity, indicating a different behavior than that in collisions at higher energies.

5. Summary and outlook

γ correlations of p-Λ and p-π have been presented and compared with hadron-hadron(h-h) correlations in Au+Au collisions at 200 GeV. The p-π correlations show similar signal magnitudes as h-h, while the p-Λ correlations have much larger signals in mid-central to peripheral collisions. The $\kappa_{\rm B}^{\rm TMC}$ value has been estimated. Except for the 7.7 GeV data, the CME observables for higher energies are identified to be above the empirical background. In the future, isobaric collisions may provide a good control on the flow dependent background. Since the flow background will be the same in $^{96}_{40}$Zr+$^{96}_{40}$Zr and $^{96}_{44}$Ru+$^{96}_{44}$Ru collisions, a potential difference of particle correlation between these two colliding systems will allow us to relate the correlation measurements with the strength of the magnetic field, which is an essential step toward the discovery of the CME [9].

References

[1] Kharzeev D E *et al.* 2016 *Prog. Part. Nucl. Phys.* **88**, 1
[2] Abelev B I *et al.* (STAR Collaboration) 2009 *Phys. Rev. Lett.* **103** 251601; 2010 *Phys. Rev. C* **81** 054908
[3] Adam J *et al.* (ALICE Collaboration) 2014 *Phys. Rev. C* **93** 044903
[4] Adamczyk L *et al.* (STAR Collaboration) 2014 *Phys. Rev. Lett.* **113** 052302
[5] Olive K A *et al.* (Particle Data Group) 2014 *Chin. Phys. C* **38** 090001 and 2015 update.
[6] Voloshin S A 2004 *Phys. Rev. C* **70** 057901
[7] Bzdak A, Koch V and Liao J 2011 *Phys. Rev. C* **83** 014905
[8] Alver B *et al.* 2005 *Phys. Rev. C* **83** 024913; *Phys. Rev. C* **72** 014904
[9] Deng W-T, Huang X-G, Ma G-L and Wang G 2016 arXiv:1607.04697

IOP Conf. Series: Journal of Physics: Conf. Series **779** (2017) 012068 doi:10.1088/1742-6596/779/1/012068

Vorticity in the QGP liquid and hyperon polarization at the RHIC BES energies

Iurii Karpenko[1,2] **and Francesco Becattini**[1,3]

[1] INFN - Sezione di Firenze, Via G. Sansone 1, I-50019 Sesto Fiorentino (Firenze), Italy
[2] Bogolyubov Institute for Theoretical Physics, 14-b, Metrolohichna st., 03680 Kiev, Ukraine
[3] Universitá di Firenze, Via G. Sansone 1, I-50019 Sesto Fiorentino (Firenze), Italy

E-mail: yu.karpenko@gmail.com

Abstract. We calculate the polarization of Λ hyperons in Au-Au collisions at RHIC Beam Energy Scan range $\sqrt{s_{\rm NN}} = 7.7, \dots, 200$ GeV in a state-of-the-art 3+1 dimensional cascade + viscous hydro model vHLLE+UrQMD. We find that the polarization of Λ in the out-of-plane direction decreases substantially with collision energy. We explore the connection between the polarization signal and thermal vorticity and discuss the feed-down and hadronic rescattering effects on the mean polarization of all produced Λ hyperons.

1. Introduction

Signatures of the Quark-Gluon Plasma (QGP) formation have been extensively studied via momentum space distributions of produced hadrons, but it is also natural to assume that finite angular momentum of the QGP droplets should be translated to alignment of final hadron's spins. Hydrodynamical approach is a well-established tool for the QGP studies, however, polarization of produced hadrons had not been systematically assessed in it. In particular, the widely used Cooper-Frye formula assumes no distinction between different spin projections of the produced hadrons. Spin only enters in the distribution function part of the Cooper-Frye formula as a $(2J+1)$ multiplier. However, it was shown in [1] that a more subtle effect is present and can play a role: spin-vorticity coupling can induce non-vanishing polarization of hadrons with nonzero spin produced in local thermal equilibrium. This is a purely thermo-mechanical effect, which polarizes both particles and anti-particles in the same direction defined by local thermal vorticity of the medium.

Up to now, few results for the polarization of Λ baryons based on this effect have been reported. In [2], simulation of noncentral $\sqrt{s_{\rm NN}} = 200$ GeV Au-Au collisions at RHIC with initial state from Yang-Mills dynamics followed by a 3 dimensional ideal hydro expansion resulted in polarization of low-p_T Λ of few percent magnitude, parallel to the direction of total angular momentum of the fireball, whereas for few GeV p_T it reaches 9%. In another work [3] aiming at the same collision energy, a 3 dimensional hydrodynamic expansion with initial state from optical Glauber model with parametrized rapidity dependence chosen to reproduce directed flow results [4] resulted in much lower values of polarization: about 0.2% for low-p_T Λ and up to 1.5% for z component of polarization of high-p_T Λ. In a more recent work [5] an event-by-event 3 dimensional viscous hydrodynamics with initial state from AMPT model results in similar few per mille average polarization for A+A collisions at $\sqrt{s_{\rm NN}} = 62.4, 200$ and 2760 GeV.

The abovementioned results confront the experimental analysis of Au-Au collisions at $\sqrt{s_{\rm NN}} = 200$ GeV by STAR [6] which only puts an upper limit of 0.02 on the polarization of Λ baryons. At lower collision energies, because of emerging baryon stopping effect one may expect larger local vorticity and, as a result, larger polarization signal. Indeed, preliminary STAR results [7] show that the mean polarization of hyperons can reach as much as several percents in Au-Au collisions at lower RHIC Beam Energy Scan (BES) energies. In this report we give an assessment of the polarization signal in a state-of-the-art hybrid model, which is tuned to reproduce basic hadron observables for heavy ion collisions in the BES program: (pseudo)rapidity, transverse momentum distributions and elliptic flow coefficients.

2. Polarization observable in the cascade+viscous hydrodynamic model

We simulate heavy ion collisions at RHIC Beam Energy Scan energies in `vHLLE+UrQMD` hybrid [8]. Herein we summarize its main features. The initial state is modeled with `UrQMD` cascade [9, 10]. At a hypersurface $\tau = \sqrt{t^2 - z^2} = \tau_0$ the fluidization is imposed. At lower BES energies τ_0 corresponds to a moment when two nuclei have completely passed through each other, $\tau_0 = 2R/(\gamma v_z)$. Therefore the duration of pre-hydro stage changes considerably, from 3.2 fm/c at $\sqrt{s_{\rm NN}} = 7.7$ GeV to 0.4 fm/c at $\sqrt{s_{\rm NN}} = 200$ GeV. Following 3 dimensional hydrodynamic expansion is numerically solved with `vHLLE` code [11]. In the hydrodynamic phase, a finite effective value of shear viscosity over entropy density η/s is taken into account, and the equation of state of the medium with finite $\mu_{\rm B}$ is based on Chiral model [12].

It was shown in [1] that the mean relativistic spin vector of spin ½ particles of sort i with four-momentum p, produced around point x on particlization hypersurface is:

$$\Pi_i^\mu(x,p) = \frac{1}{8m_i}(1 - f_i(x,p))\epsilon^{\mu\nu\rho\sigma}p_\sigma\varpi_{\sigma\nu} \tag{1}$$

where $f_i(x,p)$ is (Fermi-Dirac) distribution function of the particles and $\varpi_{\mu\nu} = -1/2(\partial_\mu\beta_\nu - \partial_\nu\beta_\mu)$ is thermal vorticity, equal to (minus) the antisymmetric part of the gradient of the four-temperature field $\beta^\mu = (1/T)u^\mu$, where T is the proper temperature and u^μ is the hydrodynamic 4-velocity. The mean spin vector of the hadron species i, produced at particlization surface (which we will call *direct* hadrons), can be then calculated using an extension of the Cooper-Frye formula:

$$\langle\Pi_i^{*\mu}\rangle = \frac{1}{N_i}\int\frac{d^3p}{p^0}\int d\Sigma_\lambda p^\lambda f_i(x,p)\Pi_i^{*\mu}(x,p) \tag{2}$$

where $N_i = \int\frac{d^3p}{p^0}\int d\Sigma_\lambda p^\lambda f_i(x,p)$ is the average number of the hadrons produced, and Lorentz transformation is made to the rest frames of the particles, denoted with the asterisk. Finally we note that the experimentally observable quantity is mean polarization vector $P^\mu = \Pi^\mu/S$, where S is the spin of particle.

3. Results and conclusions

The resulting components of the mean polarization vector of direct Λ baryons at midrapidity, calculated in the model for 20-50% central Au-Au collisions at RHIC BES energies is presented on the left panel of Fig. 1 as a solid line. The P_J is a component perpendicular to the reaction plane, and P_b is the one parallel to the impact parameter. We observe a rapid decrease of the mean Λ polarization with collision energy from a maximal value of about 1.8% at $\sqrt{s_{\rm NN}} = 7.7$ GeV. While qualitatively one expects that the polarization is related to the angular momentum of the fireball, we argue that quantitatively it is not the case. In fact the total angular momentum of the fireball increases with increasing collision energy as one can see on the right panel of Fig. 2, and the ratio of total angular momentum to total energy J/E exhibits only a mild decrease.

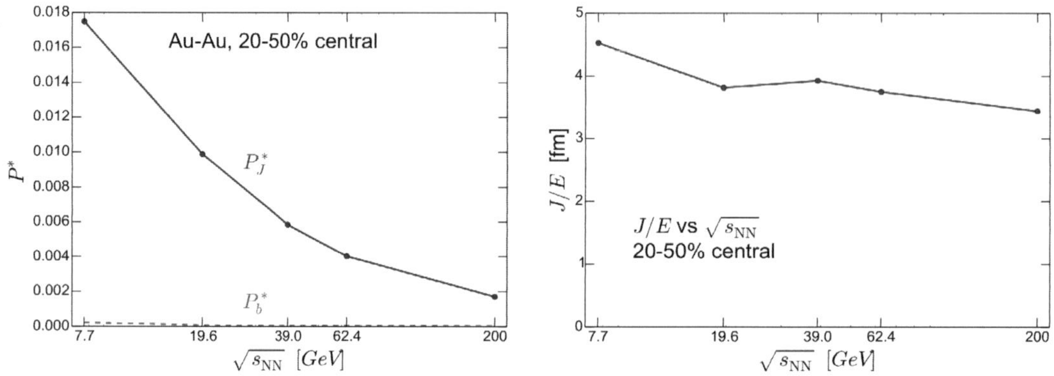

Figure 1. Components of mean polarization vector of midrapidity Λ produced at the particlization surface (left) and ratio of total angular momentum to total energy of the fireball (right), calculated in the model for 20-50% central Au-Au collisions at $\sqrt{s_{NN}} = 7.7, ..., 200$ GeV.

As one can see from Eq. 1, the actual quantity generating local polarization of hadrons is the local thermal vorticity at the points of their production. Having an access to the space-time flow structure in the model, we can explore how the vorticity changes with collision energy. On the right panel of fig. 2 we show a time evolution of the xz component of thermal vorticity, which generates the dominant contribution to the out-of-plane component of polarization P_J, at two different collision energies. We find that (1) for lower collision energy, $\sqrt{s_{NN}} = 7.7$ GeV the hydrodynamic expansion starts with higher value of ϖ_{xz}, which is a result of shear flow structure (plotted with white vectors on the left panel plot) at midrapidity formed by baryon stopping. Also, (2) shorter lifetime of hydro phase (as compared to the simulation at $\sqrt{s_{NN}} = 62.4$ GeV) dilutes the initial vorticity less, resulting in larger vorticity at the end of hydrodynamic phase. As the low-p_T Λ are more likely to be emitted at the end of the hydrodynamic phase, the abovementioned factors explain the steep decrease of the mean Λ polarization with collision energy. Similar trends - increase of angular momentum of the fireball and decrease of the mean vorticity with increasing collision energy - were observed in AMPT microscopic model [13].

Large fraction of observed Λ come from decays of higher mass hyperon resonances, which possess polarization as well. The most abundant ones are strong decay of $\Sigma(1385)$ and electromagnetic decay of Σ^0. Preliminary analysis of the feed-down contributions and polarization transfer in the hyperon decays shows us that feed-down contributions from Σ^0 and $\Sigma(1385)$ decrease the resulting mean polarization by only few percents. However, extending the list of decays up to $\Sigma(1670)$ results in about 15% suppression of the overall Λ polarization, as Λ from decays of heavier resonances acquire alternating signs of polarization [14].

In addition, in UrQMD cascade which is used to treat interactions after particlization, cross sections of Λ and Σ^0 with mostly abundant mesons and baryons, which are calculated according to Additive Quark Model, are comparable to those of nucleons [10]. This implies that Λ actively rescatter in the hadronic phase. Scatterings will presumably randomize the spin direction of primary as well as secondary particles, thus decreasing the estimated mean global polarization. Therefore the values of polarization on Fig. 1 should serve only as an upper bound for the polarization of all Λ in a complete hydro+cascade model with given initial conditions.

4. Acknowledgements
This work was partly supported by the University of Florence grant *Fisica dei plasmi relativistici: teoria e applicazioni moderne*.

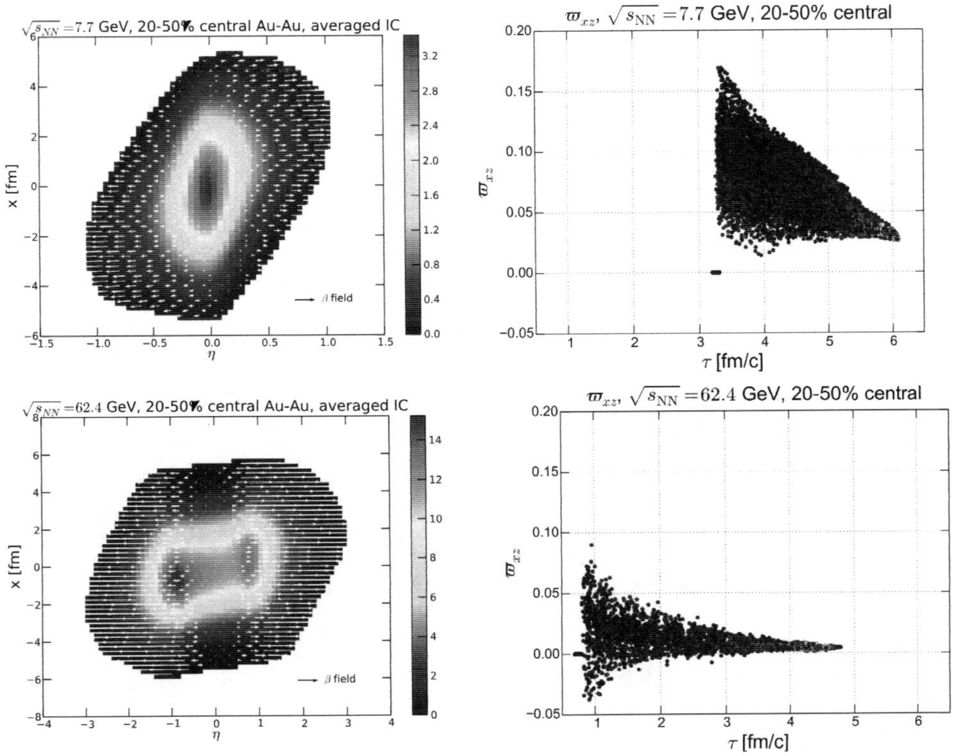

Figure 2. Initial energy density profiles for hydrodynamic stage with arrows depicting initial beta field superimposed (left column) and ϖ_{xz} over midrapidity ($|y| < 0.3$) slice of particlization surface, projected onto time axis (right column). The hydrodynamic evolutions start from averaged initial states corresponding to 20-50% central Au-Au collisions at $\sqrt{s_{\rm NN}} = 7.7$ (top row) and 62.4 GeV (bottom row).

References

[1] F. Becattini, V. Chandra, L. Del Zanna and E. Grossi, Annals Phys. **338** (2013) 32 [arXiv:1303.3431 [nucl-th]].
[2] F. Becattini, L. Csernai and D. J. Wang, Phys. Rev. C **88**, no. 3, 034905 (2013) Erratum: [Phys. Rev. C **93**, no. 6, 069901 (2016)] [arXiv:1304.4427 [nucl-th]].
[3] F. Becattini *et al.*, Eur. Phys. J. C **75**, no. 9, 406 (2015) [arXiv:1501.04468 [nucl-th]].
[4] P. Bozek and I. Wyskiel, Phys. Rev. C **81**, 054902 (2010) [arXiv:1002.4999 [nucl-th]].
[5] L. G. Pang, H. Petersen, Q. Wang and X. N. Wang, arXiv:1605.04024 [hep-ph].
[6] B. I. Abelev *et al.* [STAR Collaboration], Phys. Rev. C **76**, 024915 (2007) [arXiv:0705.1691 [nucl-ex]].
[7] Talk of M. Lisa at QCD Chirality Workshop 2016, `http://starmeetings.physics.ucla.edu/sites/default/files/mike_lisa.pdf`
[8] I. A. Karpenko, P. Huovinen, H. Petersen and M. Bleicher, Phys. Rev. C **91** (2015) 6, 064901 [arXiv:1502.01978 [nucl-th]].
[9] S. A. Bass *et al.*, Prog. Part. Nucl. Phys. **41** (1998) 255 [nucl-th/9803035].
[10] M. Bleicher *et al.*, J. Phys. G **25** (1999) 1859 [hep-ph/9909407].
[11] I. Karpenko, P. Huovinen and M. Bleicher, Comput. Phys. Commun. **185** (2014) 3016
[12] J. Steinheimer, S. Schramm and H. Stocker, J. Phys. G **38** (2011) 035001 [arXiv:1009.5239 [hep-ph]].
[13] Y. Jiang, Z. W. Lin and J. Liao, arXiv:1602.06580 [hep-ph].
[14] F. Becattini, I. Karpenko, M. Lisa, I. Upsal and S. Voloshin, arXiv:1610.02506 [nucl-th].

Vorticity and Λ polarization in event-by-event (3+1)D viscous hydrodynamics

Long-Gang Pang[1], Ren-Hong Fang[4], Hannah Petersen[1,2,3], Qun Wang[4], Xin-Nian Wang[5,6]

[1]Frankfurt Institute for Advanced Studies, Frankfurt, Germany
[2]Institute for Theoretical Physics, Goethe University, Frankfurt, Germany
[3]GSI Helmholtzzentrum for Schwerionenforschung, Darmstadt, Germany
[4] Interdisciplinary Center for Theoretical Study and Department of Modern Physics, University of Science and Technology of China, Hefei, China
[5]Key Laboratory of Quark & Lepton Physics (MOE) and Institute of Particle Physics, Central China Normal University, Wuhan, China
[6]Nuclear Science Division MS70R0319, Lawrence Berkeley National Laboratory, Berkeley, USA

E-mail: pang@fias.uni-frankfurt.de

Abstract. We visualized the vortical fluid in fluctuating QGP using (3+1)D viscous hydrodynamics, computed the spin distribution and correlation of hyperons and estimated the polarization splitting between Λ and $\bar{\Lambda}$.

1. Introduction

In high energy heavy ion collisions, the large fluctuations and fluid velocity shear present in the colliding system lead to non-vanishing local vorticity in the hot and dense sQGP. The local vorticity couples to the large orbital angular momentum and strong magnetic field in non-central collisions [1, 2, 3, 4, 5, 6, 7, 8, 9]. As a result of spin-vorticity and spin-magnetic coupling, quarks and anti-quarks become polarized along the normal direction of the reaction plane [1, 2, 5]. After hadronization the hyperons and vector mesons also become locally and globally polarized in the final state [6, 7, 8, 9, 10, 11, 12]. Recently STAR collaboration measured the Λ polarization in Au+Au collisions at various beam energies [13], and showed that (1) the Λ polarization is stronger at low beam energies than high beam energies. (2) the polarization for $\bar{\Lambda}$ is always stronger than that for Λ. In this paper, we visualize the rotational fluid velocity vectors in transverse and reaction plane, and propose to measure the azimuthal angle distribution and correlation of Λ spins (the bridge between collective flow and quantum properties of individual particles). We also estimate quantitatively the splitting of Λ and $\bar{\Lambda}$ polarization by spin-vorticity and spin-magnetic coupling.

2. Vortical fluid in event-by-event (3+1)D viscous hydrodynamics

The rotational fluid velocity vectors are obtained from Helmholtz-Hodge decomposition [14] of the fluid four-velocity vector u^{μ}, which is given in CLVisc [15, 11], through numerically solving second-order viscous hydrodynamic equations. We employ AMPT [16] to generate fluctuating

IOP Conf. Series: Journal of Physics: Conf. Series **779** (2017) 012069 doi:10.1088/1742-6596/779/1/012069

initial conditions, where angular momentum is given by the asymmetric distribution of forward-backward going participants and their associated strings whose lengths fluctuate strongly [17].

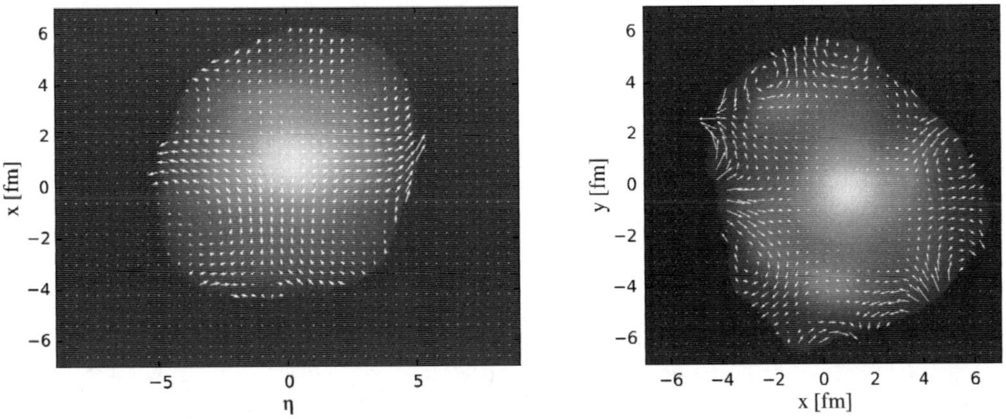

Figure 1: (color online) The rotational fluid velocity vectors in reaction plane (left) and transverse plane (right), at proper time $\tau = 3.4$ fm, for (3+1)D viscous hydrodynamic simulations with fluctuating initial condition given by AMPT.

As shown in Fig. 1, the transverse vorticity of the sQGP (left) has a circular structure around the beam direction due to the convective longitudinal flow in addition to the global alignment along the direction of the orbital angular momentum of non-central collisions. The longitudinal vorticity (right), however, has a vortex-pairing structure in a given transverse plane due to the convective radial flow of hot spots. Fermions are locally polarized on the freeze-out hypersurface given by [18, 10],

$$P^{\mu} \equiv \frac{d\Pi^{\mu}(p)/d^3p}{dN/d^3p} = \frac{^{-}h}{4m}\frac{\int d\Sigma_{\alpha}p^{\alpha}\Omega^{\mu\nu}p_{\nu}f(1-f)}{\int d\Sigma_{\alpha}p^{\alpha}f}, \tag{1}$$

where f is the Fermion distribution function, $\Omega^{\mu\nu} = \frac{1}{2}\epsilon^{\mu\nu\rho\sigma}\partial_{\rho}(\beta u_{\sigma})$ the thermal vorticity consisting of not only convective flow (spatial gradients of fluid velocity), but also acceleration (temporal gradients of fluid velocity) and conduction (space-time gradients of the temperature), with the temperature $T = 1/\beta$. Because of strong collectivity, four momentum of hadrons particalized on the freeze-out hypersurface have strong correlation with their space-time coordinates and their local polarization. This correlation brings azimuthal angle distribution of Λ spins. It is thus feasible to measure not only the global polarization, but also the spin correlation of two hyperons, to study these vortical structures of sQGP. The detailed studies on the hyperon spin correlations and their dependence on collision energy, rapidity, centrality and the shear viscosity are given in Ref. [11] .

3. Λ polarization from spin-vorticity and spin-magnetic coupling

Owing to spin-vorticity and spin-magnetic coupling, the energy levels of the fermions are shifted according to their spin states [19], as a result, there are more fermions or anti-fermions staying in lower energy states, which brings spin polarization that is defined as the number difference between different spin states,

$$\Pi = \frac{1}{2}\int\frac{d^3p}{(2\pi)^3}\left[f\left(E_p-\delta E_p\right)-f\left(E_p+\delta E_p\right)\right] \approx -\int\frac{d^3p}{(2\pi)^3}\delta E_p\frac{\partial f(E_p)}{\partial E_p}, \tag{2}$$

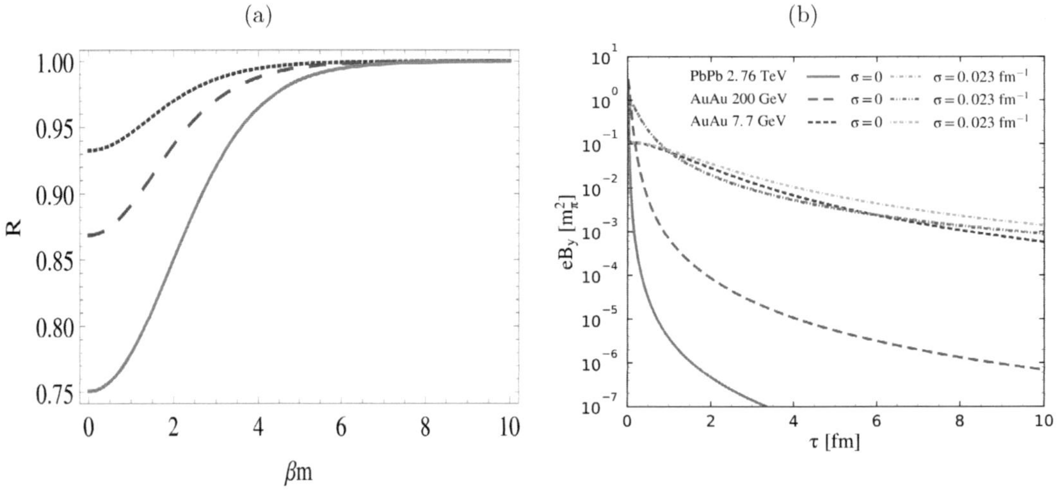

Figure 2: (color online) (a) The ratio R as functions of βm at three values $\beta\mu = 0.5, 1, 2$ corresponding to short-dashed, long-dashed and solid lines respectively. (b) The time evolution of magnetic field for Pb+Pb 2.76 TeV (solid line for $\sigma = 0$ and dash-dot line for $\sigma = 0.023$ fm^{-1}), Au+Au 200 GeV (long dashed line for $\sigma = 0$ and dash-dot-dot line for $\sigma = 0.023$ fm^{-1}), and Au+Au 7.7 GeV collisions (short dashed line for $\sigma = 0$ and dot-dash-dash line for $\sigma = 0.023$ fm^{-1}).

where $\delta E_p = \frac{\hbar}{2}\mathrm{n}\cdot\omega$ for spin-vorticity coupling and $\delta E_p = e_q\hbar\frac{\mathrm{n}\cdot\mathrm{B}}{E_p}$ for spin-magnetic coupling. The $\frac{\partial f(E_p)}{\partial E_p} = -\beta f(1-f)$ term is called susceptibility which describes how difficult it is for the fermions/anti-fermions to be polarized. The polarization per fermion for Λ and $\bar{\Lambda}$ are,

$$P_\Lambda = \beta \int \frac{d^3p}{(2\pi)^3}(\omega/2 + \mu_\Lambda B) f(1-f) / \int \frac{d^3p}{(2\pi)^3} f, \tag{3}$$

$$P_{\bar{\Lambda}} = \beta \int \frac{d^3p}{(2\pi)^3}(\omega/2 - \mu_\Lambda B) \bar{f}(1-\bar{f}) / \int \frac{d^3p}{(2\pi)^3} \bar{f}, \tag{4}$$

where f and $\bar{f}$ are distribution functions for Λ and $\bar{\Lambda}$ respectively.

The splitting between Λ and $\bar{\Lambda}$ polarization is caused by two reasons: one is the non-zero baryon chemical potential and the other is the magnetic field on the freeze-out hypersurface. If we only consider spin-vorticity coupling, the anti-fermions are much easier to be polarized than fermions due to Pauli blocking, considering that there are more fermions than anti-fermions with positive baryon chemical potential in the distribution function. The ratio of integrated polarization per particle $R = P_\Lambda / P_{\bar{\Lambda}}$ is computed as functions of βm and $\beta\mu$ where μ is the baryon chemical potential [10]. As shown in Fig. 2a, the ratio is very sensitive to baryon chemical potential and fermion mass. However, using the mass of Λ baryon brings a big βm and a very small splitting. On the other hand, spin-magnetic coupling also brings splitting between Λ and $\bar{\Lambda}$ polarization,

$$\Delta P = P_{\bar{\Lambda}} - P_\Lambda = -2\mu_\Lambda B\beta \int \frac{d^3p}{(2\pi)^3} f(1-f) / \int \frac{d^3p}{(2\pi)^3} f \approx -2\mu_\Lambda B\beta, \tag{5}$$

where $\mu_\Lambda = -0.613\mu_N$ is the Λ magnetic moment, B the strength of the magnetic field and f the distribution function for both Λ and $\bar{\Lambda}$ neglecting baryon chemical potential. The average splitting between Λ and $\bar{\Lambda}$ polarization from STAR experiment is about 0.03 at beam energy scan region ($\sqrt{s_{NN}} = 7.7$ GeV). From this the magnetic field along y direction estimated from $B = \frac{\Delta P}{-2\mu_\Lambda \beta}$ is about $1.05 m_\pi^2$ [20]. It worth noting that the current splitting measured by STAR bears an error bar of 0.02, which may increase or decrease the strength of estimated magnetic field by 60%. As a comparison, we computed the magnetic field in vacuum and a conducting medium with electric conductivity $\sigma = 0.023$ fm^{-1} following [21]. As shown in Fig. 2b, the magnetic field for Au+Au $\sqrt{s_{NN}} = 7.7$ GeV collisions is stronger than that for Au+Au 200 GeV and Pb+Pb 2.76 TeV collisions, after $\tau = 1$ fm, which qualitatively agrees with the experimental observation that the splitting at lower beam energies is bigger than top RHIC and LHC energies [20]. Quantitatively, the maximum magnetic field strength for Au+Au $\sqrt{s_{NN}} = 7.7$ GeV collisions is smaller than 0.1 m_π^2, from which the estimated splitting is 10 times smaller than what one gets from STAR experiment. Considering that hadronization usually happens at large radius and later time where magnetic field is very small, the splitting from spin-magnetic coupling would be even smaller. Notice that the electric conductivity at low beam energies does not affect the magnetic field so strongly, because the spectators fly away from each other much slower than top RHIC and LHC energies. For a fully consistent study, event-by-event (3+1)D relativistic magnetohydrodynamics is required to provide decent knowledge of the magnetic field.

Acknowledgments We thank S. Pu, M. Lisa, F. Becattini, X.G. Huang and I. Karpenko for helpful discussions. This work is supported in part by NSFC under the Grant Nos. 11221504 and 11535012, by MOST of China under Grant No. 2014DFG02050, by the Major State Basic Research De- velopment Program (MSBRD) in China under the Grant No. 2015CB856902, 2014CB845404, and 2014CB845406, by U.S. DOE under Contract No. DE-AC02-05CH11231, by Helmholtz Young Investigator Group VH-NG-822 from the Helmholtz Association and GSI and by HIC for FAIR within the framework of the Landes-Offensive zur Entwicklung Wissenschaftlich-Oekonomischer Exzel- lenz (LOEWE) program launched by the State of Hesse. Computations are performed at the Green Cube at GSI and GPU workstations at CCNU.

References

[1] Zuo-Tang Liang and Xin-Nian Wang. *Phys. Rev. Lett.*, 94:102301, 2005. [Erratum: Phys. Rev. Lett.96,039901(2006)].
[2] Zuo-Tang Liang and Xin-Nian Wang. *Phys. Lett.*, B629:20–26, 2005.
[3] F. Becattini, F. Piccinini, and J. Rizzo. *Phys. Rev.*, C77:024906, 2008.
[4] Barbara Betz, Miklos Gyulassy, and Giorgio Torrieri. *Phys. Rev.*, C76:044901, 2007.
[5] Jian-Hua Gao, Shou-Wan Chen, Wei-tian Deng, Zuo-Tang Liang, Qun Wang, and Xin-Nian Wang. *Phys. Rev.*, C77:044902, 2008.
[6] L. P. Csernai, V. K. Magas, and D. J. Wang. *Phys. Rev.*, C87(3):034906, 2013.
[7] F. Becattini, G. Inghirami, V. Rolando, A. Beraudo, L. Del Zanna, A. De Pace, M. Nardi, G. Pagliara, and V. Chandra. *Eur. Phys. J.*, C75(9):406, 2015.
[8] Yin Jiang, Zi-Wei Lin, and Jinfeng Liao. 2016.
[9] Wei-Tian Deng and Xu-Guang Huang. *Phys. Rev.*, C93(6):064907, 2016.
[10] Ren-hong Fang, Long-gang Pang, Qun Wang, and Xin-nian Wang. *Phys. Rev.*, C94(2):024904, 2016.
[11] Long-Gang Pang, Hannah Petersen, Qun Wang, and Xin-Nian Wang. 2016.
[12] I. Karpenko. Proceeding for SQM2016, UC Berkeley, June 27th-July 1st, 2016.
[13] Isaac Upsal. *Journal of Physics: Conference Series*, 736(1):012016, 2016.
[14] H. Helmholtz. *Journal fur die reine und angewandte Mathematik*, 55:25–55, 1858.
[15] Long-Gang Pang, Yoshitaka Hatta, Xin-Nian Wang, and Bo-Wen Xiao. *Phys. Rev.*, D91(7):074027, 2015.
[16] Zi-Wei Lin, Che Ming Ko, Bao-An Li, Bin Zhang, and Subrata Pal. *Phys. Rev.*, C72:064901, 2005.
[17] Long-Gang Pang, Hannah Petersen, Guang-You Qin, Victor Roy, and Xin-Nian Wang. *Eur. Phys. J.*, A52(4):97, 2016.
[18] F. Becattini, V. Chandra, L. Del Zanna, and E. Grossi. *Annals Phys.*, 338:32–49, 2013.
[19] Jian-hua Gao and Qun Wang. *Phys. Lett.*, B749:542–546, 2015.
[20] M. Lisa. Preliminary results by STAR, on SQM2016, UC Berkeley, June 27th-July 1st, 2016.
[21] Umut Gursoy, Dmitri Kharzeev, and Krishna Rajagopal. *Phys. Rev.*, C89(5):054905, 2014.

Event-by-event generation of vorticity in heavy-ion collisions

Wei-Tian Deng

School of physics, Huazhong University of Science and Technology, Wuhan 430074, China

Xu-Guang Huang

Physics Department and Center for Particle Physics and Field Theory, Fudan University, Shanghai 200433, China

Department of Physics, Brookhaven National Laboratory, Upton, New York 11973-5000, USA

E-mail: huangxuguang@fudan.edu.cn

Abstract. In a noncentral heavy-ion collision, the two colliding nuclei have a finite angular momentum in the direction perpendicular to the reaction plane. After the collision, a fraction of the total angular momentum is retained in the produced hot quark-gluon matter and is manifested in the form of fluid shear. Such fluid shear creates finite flow vorticity. We study some features of such generated vorticity, including its strength, beam energy dependence, centrality dependence, and spatial distribution.

1. Introduction

Relativistic heavy-ion collisions provide us with the environments in which we can study the strongly interacting matter under unusual conditions, like extremely high temperature [1] and extremely strong magnetic field [2, 3]. Recently, it was realized that noncentral heavy-ion collisions can also generate finite flow voriticty and thus provide us with a chance to study quark-gluon matter under local rotation [4, 5, 6, 7]. The mechanism of the generation of finite vorticity is simple. In a noncentral heavy-ion collision, the two colliding nuclei have a finite angular momentum in the direction perpendicular to the reaction plane, $J_0 \sim Ab\sqrt{s_{\rm NN}}/2$, with A the number of nucleons in one nucleus, b the impact parameter, and $\sqrt{s_{\rm NN}}$ the center-of-mass energy of a pair of colliding nucleons. After the collision, a fraction of the total angular momentum is retained in the produced partonic matter which we will call the quark-gluon plasma (QGP). This fraction of angular momentum manifests itself as a shear of the longitudinal momentum density and thus nonzero local vorticity arises. In hydrodynamics, the vorticity represents the local angular velocity, and its existence in heavy-ion collisions may induce a number of intriguing phenomena, for example, the chiral vortical effect (CVE) [8], the chiral vortical wave (CVW) [9], and the global polarization of quarks and Λ baryons [10, 11, 12, 13, 14, 15].

We present detailed numerical study of the event-by-event generation of vorticity in heavy-ion collisions by using the HIJING event generator [7].

2. Numerical setups

The coordinate system is setup as the following. We choose the z axis to be along the beam direction of the projectile, x axis to be along the impact parameter $\boldsymbol{b}$ which points from the target to the projectile, and y axis to be perpendicular to the reaction plane. The origin of the time axis, $t = 0$, is set to the time when the two colliding nuclei overlap maximally.

To calculate the vorticity field, we need to know the velocity field $\boldsymbol{v}(x)$. In a relativistic system, the definition of the velocity is not unique. We define two types of velocity fields, $\boldsymbol{v}_1$ and $\boldsymbol{v}_2$, which represent the velocity of the particle flow and the velocity of the energy flow, respectively,

$$v_1^a(x) = \frac{1}{\sum_i \Phi(x, x_i)} \sum_i \frac{p_i^a}{p_i^0} \Phi(x, x_i), \tag{1}$$

$$v_2^a(x) = \frac{\sum_i p_i^a \Phi(x, x_i)}{\sum_i [p_i^0 + (p_i^a)^2/p_i^0] \Phi(x, x_i)}, \tag{2}$$

where $a = 1, 2$ or 3 is the spatial index, $\boldsymbol{p}_i$ and p_i^0 are the momentum and energy of the ith particle, and the summation is over all the particles. In our simulation, p_i and x_i in each event are generated by HIJING. The function $\Phi(x, x_i)$ is to smear a physical quantity, e.g., energy or momentum, carried by the ith particle at x_i to another coordinate point x. Therefore, $\Phi(x, x_i)$ somehow represents the quantum wave packet of the ith particle whose functional form at $\tau = \tau_0$ is chosen to be a Gaussian [16],

$$\Phi_{\rm G}(x, x_i) = \frac{K}{\tau_0 \sqrt{2\pi\sigma_\eta^2 2\pi\sigma_r^2}} \exp\left[-\frac{(x - x_i)^2 + (y - y_i)^2}{2\sigma_r^2} - \frac{(\eta - \eta_i)^2}{2\sigma_\eta^2} \right], \tag{3}$$

where σ_r and σ_η are two width parameters and K is a scale factor. The parameters are given in Refs. [7].

Once the specific definition of the velocity field is given in numerical setup, the vorticity is calculated according to

$$\boldsymbol{\omega}_1 = \boldsymbol{\nabla} \times \boldsymbol{v}, \tag{4}$$

$$\boldsymbol{\omega}_2 = \gamma^2 \boldsymbol{\nabla} \times \boldsymbol{v}, \tag{5}$$

where $\boldsymbol{\omega}_1$ is the usual nonrelativistic definition and $\boldsymbol{\omega}_2$ is the spatial components of the relativistic definition, $\omega^\mu = \epsilon^{\mu\nu\rho\sigma} u_\nu \partial_\rho u_\sigma$ with $O(v^2)$ term omitted except for γ.

3. Numerical results

In Fig. 1, we show the space-averaged y-component of the vorticities at $\tau = \tau_0$ and $\eta = 0$ averaged over 10^5 events. The spatial average is performed by using number density as weight if the vorticity is computed with the particle flow velocity $\boldsymbol{v}_1$, or by using ε as weight if the vorticity is computed with the energy flow velocity $\boldsymbol{v}_2$. For $b \lesssim 2R_A$ (R_A is the radius of the nucleus), the vorticity increases with the impact parameter. The magnitude of the vorticity at $\sqrt{s_{\rm NN}} = 200$ GeV is big in the sense that it may be comparable to the magnetic field, $T^2\omega \sim \mu e B$ (for, say, temperature $T \sim 350$ MeV and baryon chemical potential $\mu = 10$ MeV).

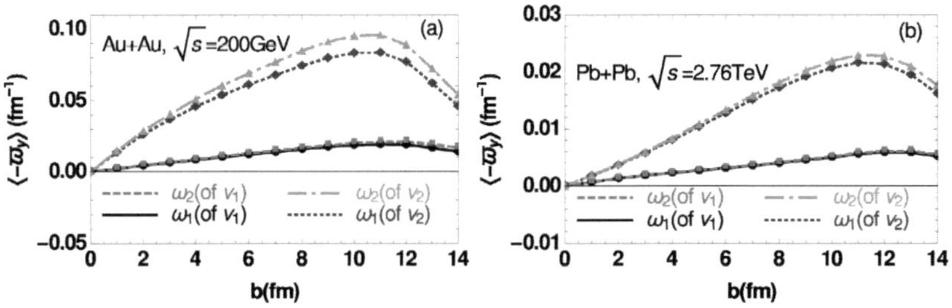

Figure 1. The space-averaged vorticity at $\tau = \tau_0$ and $\eta = 0$ averaged over 10^5 events for RHIC Au + Au collisions at $\sqrt{s_{\rm NN}} = 200$ GeV (Left) and LHC Pb + Pb collisions at $\sqrt{s_{\rm NN}} = 2.76$ TeV (Right).

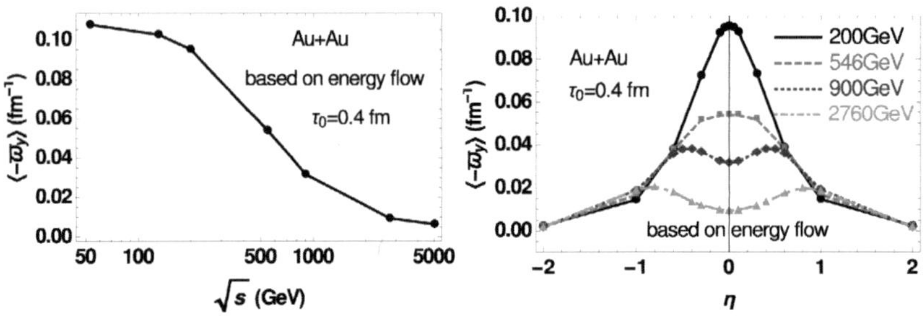

Figure 2. The collision energy dependence of the vorticity at $\eta = 0$ (Left) and the spacetime rapidity dependence (Right) at various collision energies. Proper time is fixed $\tau = 0.4$ fm and impact parameter is $b = 10$ fm.

In Fig. 2 (Left) we show $\langle \bar{\omega}_y \rangle$ at mid-rapidity as a function of collision energy $\sqrt{s_{\rm NN}}$. Clearly, the magnitude of $\langle \bar{\omega}_y \rangle$ decreases when $\sqrt{s_{\rm NN}}$ increases. This, at first sight, may seem counter-intuitive as the angular momentum increases with $\sqrt{s_{\rm NN}}$. However, with increasing $\sqrt{s_{\rm NN}}$, the moment of inertia grows more rapidly than the increasing of the total angular momentum of QGP, and can make the vorticity decrease. More importantly, with increasing collision energy, more angular momentum is carried by particles at finite rapidity and thus the vorticity at $\eta = 0$ is relatively weakened (see Fig. 2 (Right)). This reflects the fact that at higher collision energy, the system at the mid-rapidity region behaves closer to the Bjorken boost invariant picture and thus allows smaller vorticity.

The spatial distribution of the vorticity (we present only $\langle \omega_{2y} \rangle$ of $\boldsymbol{v}_2$ as an example) in the transverse plane is shown in Fig. 3 (Left). Notice that $\langle \omega_{2y} \rangle$ varies more steeply along the x direction than along the y direction in consistence with the elliptic shape of the overlapping region. The spatial distribution of the T^2-weighted flow helicity in the transverse plane is shown in Fig. 3 (Right). Clearly, the reaction plane separates the region with positive helicity from the region with negative helicity. The flow helicity separation may have interesting experimental implication, for example, it may be related to the chiral charge separation via the CVE [17, 18].

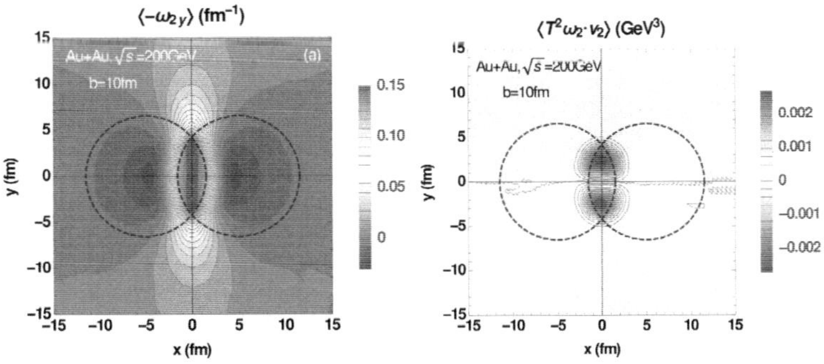

Figure 3. The spatial distributions of $\langle\omega_{2y}\rangle$ (Left) and the event-averaged helicity, $\langle T^2\boldsymbol{\omega}_2\cdot\boldsymbol{v}_2\rangle$ (Right), in the transverse plane for RHIC Au + Au collisions at $\sqrt{s_{\rm NN}} = 200$ GeV.

4. Summary

In summary, we have performed an event-by-event study of a number of interesting features of the vorticity generated in heavy-ion collisions, including its strength, beam energy dependence, centrality dependence, and spatial distribution. More results can be found in Ref. [7]. Our study may be useful to the numerical simulation of the CVE, the CVW, and the Λ baryon spin polarization.

Acknowledgments: W.-T.D is supported by the Independent Innovation Research Foundation of Huazhong University of Science and Technology (Grant No. 2014QN190) and the NSFC with Grant No. 11405066. X.-G.H. is supported by NSFC with Grant No. 11535012 and the One Thousand Young Talents Program of China.

References

[1] X. N. Wang, *Quark-Gluon Plasma 5*, World Scientific, 2016.
[2] X. G. Huang, Rept. Prog. Phys. **79** (2016), 076302 [arXiv:1509.04073 [nucl-th]].
[3] K. Hattori and X. G. Huang, arXiv:1609.00747 [nucl-th].
[4] L. P. Csernai, V. K. Magas and D. J. Wang, Phys. Rev. C **87** (2013), 034906 [arXiv:1302.5310 [nucl-th]].
[5] F. Becattini *et al.*, Eur. Phys. J. C **75** (2015), 406 [arXiv:1501.04468 [nucl-th]].
[6] Y. Jiang, Z. W. Lin and J. Liao, arXiv:1602.06580 [hep-ph].
[7] W. T. Deng and X. G. Huang, Phys. Rev. C **93** (2016), 064907 [arXiv:1603.06117 [nucl-th]].
[8] D. E. Kharzeev, J. Liao, S. A. Voloshin and G. Wang, Prog. Part. Nucl. Phys. **88** (2016), 1 [arXiv:1511.04050 [hep-ph]].
[9] Y. Jiang, X. G. Huang and J. Liao, Phys. Rev. D **92** (2015), 071501 [arXiv:1504.03201 [hep-ph]].
[10] Z. T. Liang and X. N. Wang, Phys. Rev. Lett. **94** (2005), 102301 Erratum: [Phys. Rev. Lett. **96** (2006), 039901] [nucl-th/0410079].
[11] X. G. Huang, P. Huovinen and X. N. Wang, Phys. Rev. C **84** (2011), 054910 [arXiv:1108.5649 [nucl-th]].
[12] F. Becattini, V. Chandra, L. Del Zanna and E. Grossi, Annals Phys. **338** (2013), 32 [arXiv:1303.3431 [nucl-th]].
[13] L. G. Pang, H. Petersen, Q. Wang and X. N. Wang, arXiv:1605.04024 [hep-ph].
[14] R. H. Fang, L. G. Pang, Q. Wang and X. N. Wang, Phys. Rev. C **94** (2016), 024904 [arXiv:1604.04036 [nucl-th]].
[15] M. Lisa, talk given at Strangeness in Quark Matter 2016 Conference, Berkeley, 2016.
[16] L. Pang, Q. Wang and X. N. Wang, Phys. Rev. C **86** (2012), 024911 [arXiv:1205.5019 [nucl-th]].
[17] A. Avdoshkin, V. P. Kirilin, A. V. Sadofyev and V. I. Zakharov, Phys. Lett. B **755** (2016), 1 [arXiv:1402.3587 [hep-th]].
[18] N. Yamamoto, Phys. Rev. D **93** (2016), 065017 [arXiv:1511.00933 [astro-ph.HE]].

Identified particle production in pp collisions at $\sqrt{s} = 7$ and 13 TeV measured with ALICE

R. Derradi de Souza (for the ALICE Collaboration)
Instituto de Física Gleb Wataghin, UNICAMP, 13083-859 Campinas SP, Brazil
E-mail: rderradi@ifi.unicamp.br

Abstract. Proton-proton (pp) collisions have been used extensively as a reference for the study of interactions of larger colliding systems at the LHC. Recent measurements performed in high-multiplicity pp and proton-lead (p-Pb) collisions have shown features that are reminiscent of those observed in lead-lead (Pb-Pb) collisions. In this context, the study of identified particle spectra and yields as a function of multiplicity is a key tool for the understanding of similarities and differences between small and large systems. We report on the production of pions, kaons, protons, K^0_S, Λ, Ξ, Ω and K^{*0} as a function of multiplicity in pp collisions at $\sqrt{s} = 7$ TeV measured with the ALICE experiment. The work presented here represents the most comprehensive set of results on identified particle production in pp collisions at the LHC. Spectral shapes, studied both for individual particles and via particle ratios as a function of p_T, exhibit an evolution with charged particle multiplicity that is similar to the one observed in larger systems. In addition, results on the production of light flavour hadrons in pp collisions at $\sqrt{s} = 13$ TeV, the highest centre-of-mass energy ever reached in the laboratory, are also presented and compared with previous, lower energy results.

1. Introduction

Recent results of proton-proton (pp) and proton-lead (p-Pb) collisions at the LHC energies have shown features that are similar to those observed in collisions of larger systems such as lead-lead (Pb-Pb). Notably, these findings include the observation of the double ridge structure in two-particle correlation studies [1, 2, 3] and the hardening of the transverse momentum spectra of identified particles [4] (see also Ref. [5] for a short summary of experimental results from small system measurements at the LHC). Such results have motivated several theoretical investigations [6, 7, 8] as well as an effort on the experimental side in order to provide further and more comprehensive data from small systems collisions. In this context, studying identified particle production systematically as a function of charged particle multiplicity in pp may provide further insights into the dynamics of small systems.

2. Analysis Description

Light flavor hadrons (π, K, p) are identified in the ALICE experiment [9] using combined information from ITS, TPC and TOF detectors [10]. An extended p_T range (up to ~20 GeV/c) is achieved with statistical particle identification via the relativistic rise of the dE/dx in the TPC [11]. The measurements of strange and multi-strange hadrons (K^0_S, Λ, Ξ and Ω) are performed via the invariant mass distribution obtained from their respective decay products in the following channels (branching ratios): $K^0_S \rightarrow \pi\pi$ (69.2%), $\Lambda \rightarrow p\pi$ (63.9%), $\Xi \rightarrow \Lambda\pi$ (99.9%) and $\Omega \rightarrow \Lambda K$

(67.8%). A set of topological cuts is applied to the candidates so that those which do not fit into the expected decay are eliminated and the extraction of the raw yield is performed via bin counting technique as done in previous work [12]. Finally, short lived hadronic resonances (K^{*0}, $\bar{K}^{*0}$) are also measured via the invariant mass reconstruction of their decay products ($K\pi$). In this case, combinatorial background is estimated using a mixed-event technique and subtracted from the original distribution. The raw yield is then extracted from a fit of the signal peak [13]. In all analyses, the transverse momentum spectra are corrected for acceptance and efficiency, which are determined with Monte Carlo simulations of the ALICE detector response.

Multiplicity classes are defined based on the total charge deposited in the V0A and V0C detectors located at forward ($2.8 < \eta < 5.1$) and backward ($-3.7 < \eta < -1.7$) pseudorapidity regions, respectively. The average charged particle density, $\langle dN_{ch}/d\eta \rangle$, is estimated within each multiplicity class by the average number of charged tracks in the region $|\eta| < 0.5$.

3. Results

Figure 1 shows the final transverse momentum spectra of the identified hadrons measured in pp collisions at 7 TeV for several event multiplicity classes (indicated by Roman numerals). The average charged particle densities in the highest (class I) and lowest (class X) multiplicity classes correspond to approximately 3.5 and 0.4 times the value in the integrated sample ($\langle dN_{ch}/d\eta \rangle \approx 6.0$), respectively. A clear hardening of the p_T spectra is observed for all species

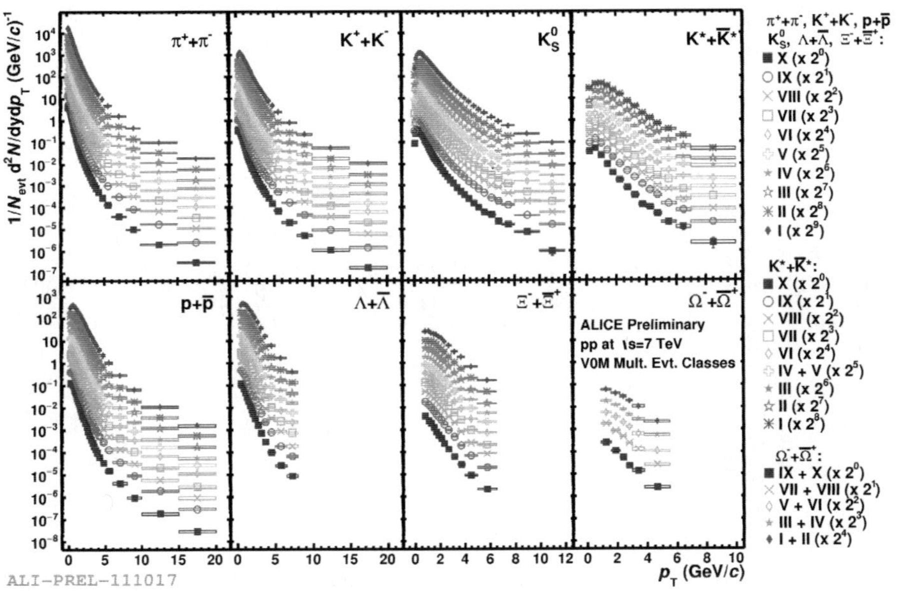

Figure 1. p_T-spectra of identified particles measured in pp collisions at $\sqrt{s} = 7$ TeV for several multiplicity classes, labeled by Roman numerals from high (class I) to low (class X) multiplicities.

from low to high multiplicity events. Moreover, this effect seems to be slightly more pronounced for baryons than for mesons. In fact, the baryon-to-meson ratio in pp collisions has been recently shown to follow a similar trend between lowest and highest multiplicity classes as observed in p-Pb and Pb-Pb collisions [14]. In order to further explore the evolution of the different physics mechanisms related to baryon and meson production at different regions of the p_T-spectra, the ratio of proton-to-pion was studied at low-, mid- and high-p_T regions as a function of $\langle dN_{ch}/d\eta \rangle$, as shown in Fig. 2. The continuity observed across the various collision systems is remarkable

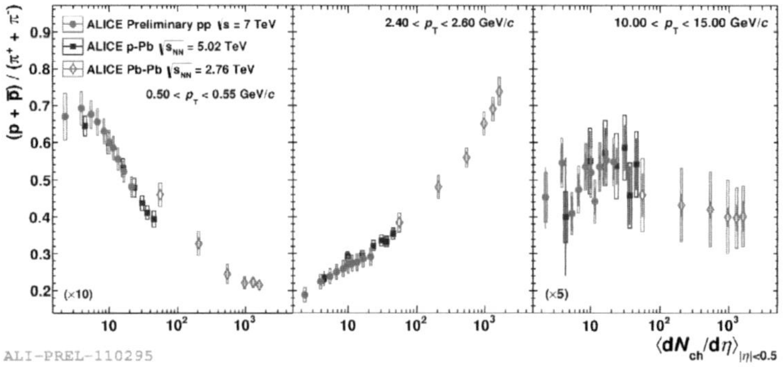

Figure 2. Proton-to-pion ratio at low (left panel), intermediate (middle panel) and high (right panel) transverse momentum as a function of the average charged particle density at midrapidity for pp, p-Pb and Pb-Pb.

for all three p_{T} regions. The smooth evolution of particle ratios suggests the similarity of the underlying mechanisms despite the differences in the initial colliding systems. Comparisons with predictions from QCD-inspired models such as PYTHIA8 [15], DIPSY [16, 17, 18] and HERWIG7 [19, 20] have also been performed. The color reconnection mechanism in PYTHIA8

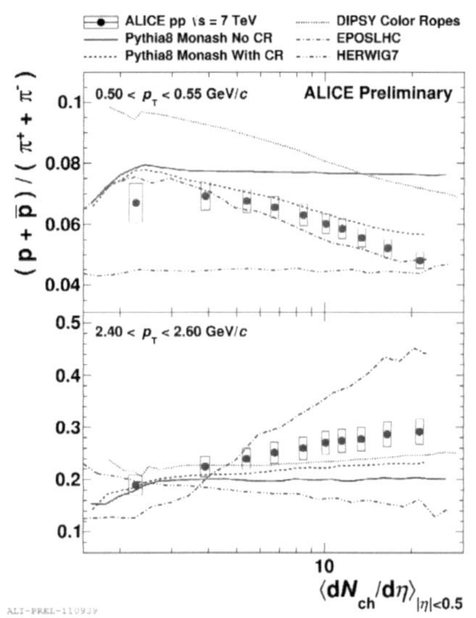

Figure 3. Proton-to-pion ratio at low (top panel) and intermediate (bottom panel) transverse momentum as a function of the average charged particle density at midrapidity for pp collisions at $\sqrt{s}$ = 7 TeV (symbols) compared to model predictions (lines).

as well as the color ropes implementation in DIPSY seem to qualitatively reproduce the general trend observed at low and intermediate p_{T} (see Fig. 3). An alternative description is provided by EPOS LHC [21], which implements collective radial flow using hydrodynamics, and also follows the data qualitatively. However, none of the models tested were able to quantitatively describe the data for the two p_{T} regions simultaneously.

The evolution of particle ratios with the collision energy is another important aspect which may reveal essential details concerning the underlying particle production mechanisms. Figures 4 and 5 show the ratios with respect to pions for p_{T}-integrated yields of protons and hyperons, respectively, as a function of the collision energy. The yields of individual species are obtained as done in previous analysis [10, 22]. The values observed for proton-to-pion ratio seem to saturate at the LHC energies, with the latest point added at 13 TeV following the trend of previous measurements. On the other hand, the hyperon-to-pion ratios present a slight increase at 13 TeV with respect to lower collision energies.

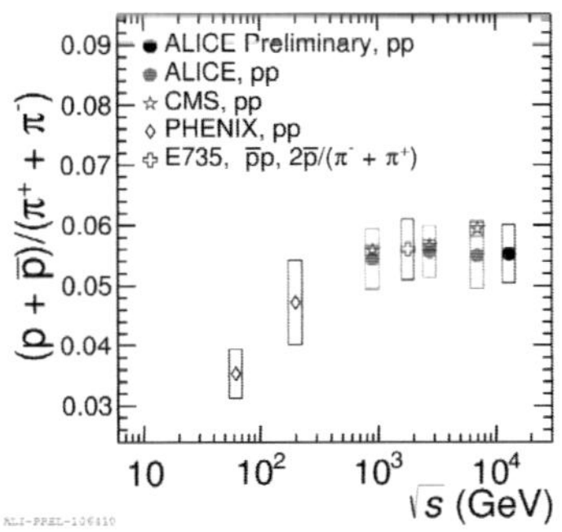

Figure 4. Proton-to-pion ratios as a function of $\sqrt{s}$.

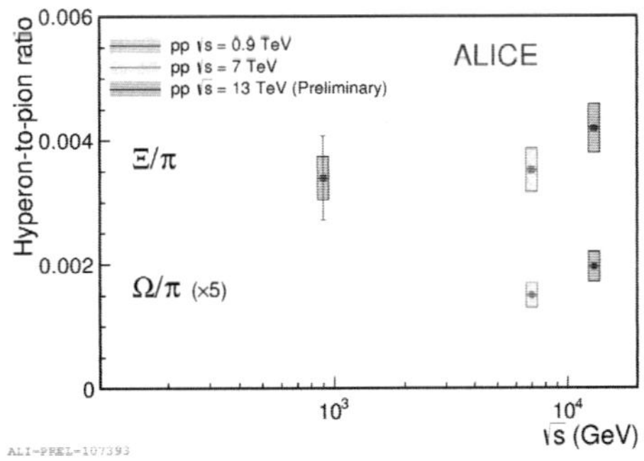

Figure 5. Hyperon-to-pion ratios as a function of $\sqrt{s}$ at LHC energies.

4. Summary and Outlook

A comprehensive set of measurements of identified hadron production in pp collisions is presented and a detailed investigation of the underlying particle production mechanisms is currently ongoing. Furthermore, the increase as a function of collision energy of the hyperon-to-pion ratios at 13 TeV may also be related to the enhanced production of strange and multi-strange hadrons as a function of event multiplicity observed in pp collisions at 7 TeV [23], as the mean charged particle multiplicity density also increases with collision energy [24]. The study of the multiplicity dependence of particle ratios at 13 TeV in combination with the reported results at 7 TeV will provide crucial input to disentangle the dependence of particle production with collision energy and event multiplicity.

References

[1] Khachatryan V *et al.* (CMS) 2010 *JHEP* **09** 091 (*Preprint* 1009.4122)
[2] Abelev B *et al.* (ALICE) 2013 *Phys. Lett.* **B719** 29–41 (*Preprint* 1212.2001)
[3] Aad G *et al.* (ATLAS) 2016 *Phys. Rev. Lett.* **116** 172301 (*Preprint* 1509.04776)
[4] Abelev B B *et al.* (ALICE) 2014 *Phys. Lett.* **B728** 25–38 (*Preprint* 1307.6796)
[5] Loizides C 2016 *25th International Conference on Ultra-Relativistic Nucleus-Nucleus Collisions (Quark Matter 2015) Kobe, Japan, September 27-October 3, 2015* arXiv:1602.09138
[6] Dusling K and Venugopalan R 2013 *Phys. Rev.* **D87** 094034 (*Preprint* 1302.7018)
[7] Ortiz Velasquez A *et al.* 2013 *Phys. Rev. Lett.* **111** 042001 (*Preprint* 1303.6326)
[8] Bozek P and Broniowski W 2013 *Phys. Rev.* **C88** 014903 (*Preprint* 1304.3044)
[9] Aamodt K *et al.* (ALICE) 2008 *JINST* **3** S08002
[10] Adam J *et al.* (ALICE) 2015 *Eur. Phys. J.* **C75** 226 (*Preprint* 1504.00024)
[11] Abelev B B *et al.* (ALICE) 2014 *Int. J. Mod. Phys.* **A29** 1430044 (*Preprint* 1402.4476)
[12] Aamodt K *et al.* (ALICE) 2011 *Eur. Phys. J.* **C71** 1594 (*Preprint* 1012.3257)
[13] Abelev B *et al.* (ALICE) 2012 *Eur. Phys. J.* **C72** 2183 (*Preprint* 1208.5717)
[14] Bianchi L (ALICE) 2016 ArXiv:1604.06736
[15] Sjostrand T, Mrenna S and Skands P Z 2008 *Comput. Phys. Commun.* **178** 852–867 (*Preprint* 0710.3820)
[16] Flensburg C, Gustafson G and Lonnblad L 2011 *JHEP* **08** 103 (*Preprint* 1103.4321)
[17] Bierlich C, Gustafson G, Lnnblad L and Tarasov A 2015 *JHEP* **03** 148 (*Preprint* 1412.6259)
[18] Bierlich C and Christiansen J R 2015 *Phys. Rev.* **D92** 094010 (*Preprint* 1507.02091)
[19] Bahr M *et al.* 2008 *Eur. Phys. J.* **C58** 639–707 (*Preprint* 0803.0883)
[20] Bellm J *et al.* 2016 *Eur. Phys. J.* **C76** 196 (*Preprint* 1512.01178)
[21] Pierog T *et al.* 2015 *Phys. Rev.* **C92** 034906 (*Preprint* 1306.0121)
[22] Abelev B *et al.* (ALICE) 2012 *Phys. Lett.* **B712** 309–318 (*Preprint* 1204.0282)
[23] Adam J *et al.* (ALICE) 2016 ArXiv:1606.07424
[24] Adam J *et al.* (ALICE) 2016 *Phys. Lett.* **B753** 319–329 (*Preprint* 1509.08734)

Recent hadronic resonance measurements at ALICE

A G Knospe (for the ALICE Collaboration)

Department of Physics, The University of Houston, Houston, TX, USA

E-mail: anders.knospe@cern.ch

Abstract. In heavy-ion physics, measurements of short-lived hadronic resonances allow the properties of the hadronic phase of the collision to be studied. In addition, resonances can be used along with stable hadrons to study parton energy loss in the quark-gluon plasma and the mechanisms that shape hadron p_T spectra at intermediate transverse momenta. Resonance measurements in small systems serve as a reference for heavy-ion collisions and contribute to searches for collective effects. An overview of recent results on hadronic resonance production measured in ALICE is presented. These results include the p_T spectra and yields of the $\rho(770)^0$, $K^*(892)^0$, and $\phi(1020)$ mesons in pp, p–Pb, and Pb–Pb collisions at different energies as well as the $\Sigma(1385)^{\pm}$ and $\Xi(1530)^0$ baryons in pp and p–Pb collisions.

Hadronic resonances, along with stable hadrons, allow the study of properties of heavy-ion collisions, both in the early (quark-gluon plasma) and late (hadronic) stages of their evolution. Energy loss of partons in nuclear matter can be studied by measuring the nuclear modification factors R_{AA} and R_{pPb}. The yields of resonances may be modified even after chemical freeze-out by re-scattering and regeneration, which are (pseudo-)elastic scattering processes expected to have their greatest strength at low p_T ($\lesssim$ 2 GeV/c) [1, 2, 3]. Measurements of resonance and stable-hadron yields can be used along with theoretical models [2, 4, 5] to estimate the properties (temperature and lifetime) of the hadronic phase of heavy-ion collisions. The various mechanisms that may determine the shapes of particle p_T spectra, including vacuum fragmentation, quark recombination, hydrodynamic flow, re-scattering, and regeneration, can be studied through comparison of different measurements of multiple particle species (including hadronic resonances) with differing masses and quark content. These proceedings present measurements from the ALICE experiment that are related to these topics.

The $\rho(770)^0$, $K^*(892)^0$, and $\phi(1020)$ mesons (hereafter ρ^0, K^{*0}, and ϕ) and the $\Sigma(1385)^{\pm}$ and $\Xi(1530)^0$ baryons are measured using invariant-mass analyses to reconstruct their hadronic decays ($\rho^0 \rightarrow \pi^-\pi^+$, $K^{*0} \rightarrow \pi^{\pm}K^{\mp}$, $\phi \rightarrow K^-K^+$, $\Sigma(1385)^{\pm} \rightarrow \Lambda\pi^{\pm}$, and $\Xi(1530)^0 \rightarrow \Xi^-\pi^+$). For the baryonic resonances, the intermediate decay daughters Λ and Ξ^- are identified through selections based on their masses and decay topologies. Combinatorial backgrounds, constructed using either like-charge pairs or event mixing, are subtracted from the distributions of unlike-charge pairs for multiple p_T and centrality/multiplicity intervals. The resulting invariant-mass distributions are then fitted. For the ρ^0, a cocktail is used, including a smooth function to describe a continuum residual background, plus peaks accounting for the contributions of the K^0_S, $\omega(782)$, K^{*0}, $f_0(980)$, and $f_2(1270)$. The shape of the ρ^0 peak is described by the product of a relativistic p-wave Breit-Wigner function, a phase-space factor, a mass-dependent reconstruction efficiency, and a Söding interference term [12]. For the other resonances, the background-subtracted invariant-mass distributions are fitted using Breit-Wigner or Voigtian peaks added

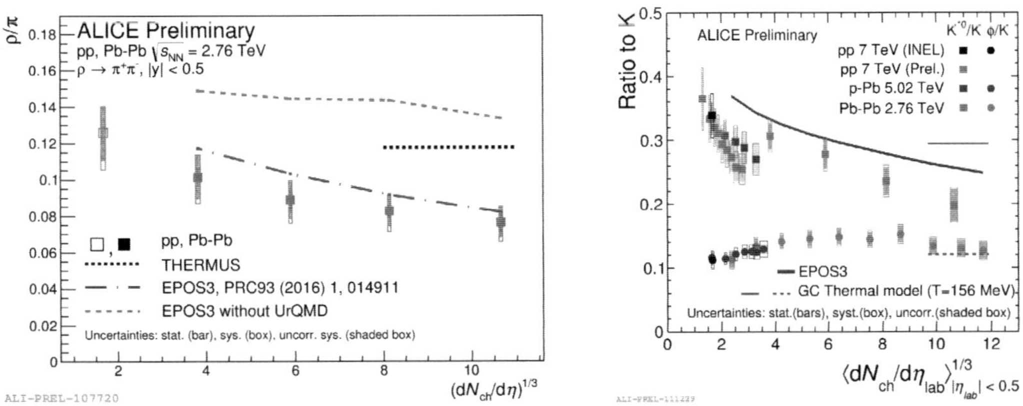

Figure 1. Ratios $\rho^0/\langle\pi^\pm\rangle$, K^{*0}/K, and ϕ/K [6, 7, 8] as functions of $\langle dN_{ch}/d\eta\rangle^{1/3}$ for various collision systems, with grand-canonical thermal-model [9, 10] and EPOS [11] calculations.

to first- or second-order polynomials to describe the residual backgrounds; see [6, 13] for further details. The p_T spectra, which are corrected for efficiency, acceptance, and branching ratios, are fitted using Lévy-Tsallis functions (for pp and p–Pb collisions) or Boltzmann-Gibbs blast-wave functions (for Pb–Pb collisions) so that the total integrated yields and mean transverse momenta $\langle p_T\rangle$ can be extracted. The p_T spectra and yields of the ρ^0 have been measured in pp and Pb–Pb collisions at $\sqrt{s_{NN}} = 2.76$ TeV. The K^{*0} and ϕ have been measured in pp, p–Pb, and Pb–Pb collisions at various energies in different multiplicity or centrality intervals. The $\Sigma(1385)^\pm$ and $\Xi(1530)^0$ have been measured in pp collisions at $\sqrt{s} = 7$ TeV and in p–Pb collisions at $\sqrt{s_{NN}} = 5.02$ TeV.

Figure 1 shows the ratios of integrated yields $\rho^0/\langle\pi^\pm\rangle$, K^{*0}/K, and ϕ/K for the various collision systems [6, 7, 8]. A centrality-dependent suppression of the $\rho^0/\langle\pi^\pm\rangle$ and K^{*0}/K ratios has been observed in Pb–Pb collisions. These ratios in central Pb–Pb collisions are also suppressed with respect to grand-canonical thermal-model calculations [9, 10] with a chemical freeze-out temperature of 156 MeV. This suppression may be the result of re-scattering of the ρ^0 and K^{*0} decay products in the hadronic phase of the medium (re-scattering being dominant over regeneration of these resonances) and is at least qualitatively reproduced by calculations using the EPOS model [11]. In contrast, the ϕ lives 10 times longer than the K^{*0} and 35 times longer than the ρ^0; it decays predominantly after the end of the hadronic phase and is not significantly affected by re-scattering or regeneration. Furthermore, an apparent multiplicity-dependent suppression of the K^{*0}/K ratio has been observed in pp and p–Pb collisions, which might be an indication of a hadron-gas phase with non-zero lifetime in high-multiplicity pp and p–Pb collisions. The K^{*0} and ϕ mesons have also been measured in inelastic pp collisions at $\sqrt{s} = 13$ TeV. No energy evolution in the p_T-integrated K^{*0}/K and ϕ/K ratios is observed from RHIC to the top LHC energy. Furthermore, the $\Sigma(1385)^\pm/\Lambda$ and $\Xi(1530)^0/\Xi^-$ ratios in p–Pb collisions are consistent with the value previously measured by ALICE in minimum-bias pp collisions at $\sqrt{s} = 7$ TeV [13]. No multiplicity dependence is observed for these ratios in p–Pb collisions. This implies that the $\Sigma(1385)^\pm/\pi$ ratio has the same evolution with multiplicity as the Λ/π ratio, and similarly that $\Xi(1530)^0/\pi$ evolves like the Ξ^-/π ratio. The Λ/π and Ξ^-/π ratios are enhanced as functions of multiplicity in p–Pb collisions [14]; the evolution of the $\Sigma(1385)^\pm/\pi$ and $\Xi(1530)^0/\pi$ suggests that the enhancement depends on their strangeness content and not on their masses.

IOP Conf. Series: Journal of Physics: Conf. Series **779** (2017) 012072 doi:10.1088/1742-6596/779/1/012072

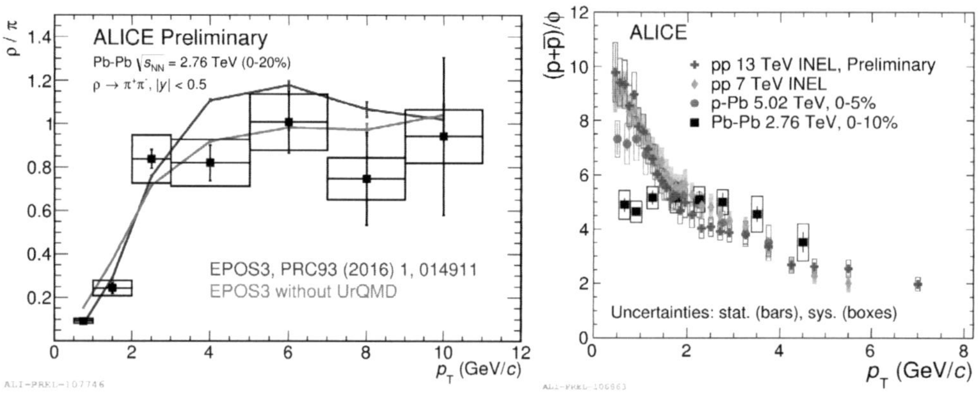

Figure 2. Ratios ρ^0/π and p/ϕ as functions of p_T for various collision systems [6, 7, 8]. EPOS calculations [11] of the ρ^0/π ratio are also shown.

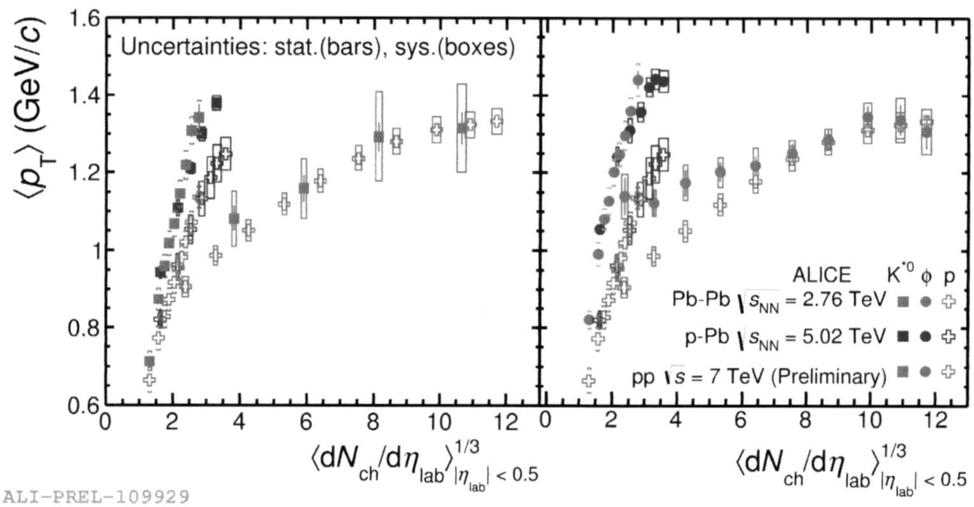

Figure 3. Mean transverse momentum $\langle p_T \rangle$ values of K^{*0}, p, and ϕ in various collision systems [6, 7, 15, 16].

The p_T dependent ρ^0/π ratio has also been calculated in Pb–Pb collisions, see Fig. 2(a). In central collisions, this ratio is fairly well described by EPOS [11] with its UrQMD module turned on, but is overestimated by at least 30% for $p_T < 2$ GeV/c if UrQMD is turned off. Both EPOS calculations describe the ratio in peripheral Pb–Pb collisions well. This suggests that the hadronic phase, which includes re-scattering effects in UrQMD, may indeed be responsible for the observed suppression of the ρ^0 yield in central Pb–Pb collisions.

Baryon-to-meson ratios (*e.g.*, p/π and Λ/K^0_S) have been observed to be enhanced in Pb–Pb collisions with respect to pp collisions at intermediate p_T ($1.5 \lesssim p_T \lesssim 6$ GeV/c) [15, 17]. This enhancement may be attributed to hadronization by recombination [18, 19] or to hydrodynamic flow that modifies the particle p_T spectra [20, 21, 22]. The p/ϕ ratio, in which the baryon numerator and meson denominator particles have very similar masses, can be used to further study these effects. This ratio is constant in central Pb–Pb collisions for $p_T < 4$ GeV/c [6], but becomes sloped for peripheral Pb–Pb, p–Pb, and pp collisions, as shown in Fig. 2(b).

The constant behavior in central Pb–Pb collisions is consistent with the basic hydrodynamic assumption, although the behavior is also reproduced by some recombination models [23].

The dependence of the shapes of p_{T} spectra on mass and quark content can also be studied using the mean transverse momentum $\langle p_{\mathrm{T}} \rangle$. The $\langle p_{\mathrm{T}} \rangle$ values of the K^{*0}, p, and ϕ (which all have similar masses) are shown in Fig. 3 for different collision systems [6]. For central Pb–Pb collisions, the $\langle p_{\mathrm{T}} \rangle$ values for these three particles are consistent. This suggests that the shapes of the p_{T} spectra are determined primarily by the particle masses, which would be expected for hydrodynamic behavior. However, this behavior is not observed for smaller collision systems, where the proton is observed to have lower $\langle p_{\mathrm{T}} \rangle$ values than the two mesonic resonances. The $\langle p_{\mathrm{T}} \rangle$ values in pp and p–Pb collisions also follow different trends and rise faster with multiplicity than in Pb–Pb collisions. In pp and p–Pb collisions, the $\langle p_{\mathrm{T}} \rangle$ values approach or even exceed the values measured in central Pb–Pb collisions.

In summary, the short lifetimes of hadronic resonances make them useful probes for the study of the hadronic phase of heavy-ion collisions. Resonances can also be used along with stable hadrons in studies of the mass and quark-content dependence of mechanisms that influence the shapes of particle p_{T} spectra. The ALICE Collaboration has measured a centrality-dependent suppression of the ρ^0/π and $\mathrm{K}^{*0}/\mathrm{K}$ ratios in Pb–Pb collisions which can be described by EPOS calculations with UrQMD. Multiplicity-dependent suppression of the $\mathrm{K}^{*0}/\mathrm{K}$ ratio is also observed in pp and p–Pb collisions. In contrast, the ϕ/K ratio does not exhibit a centrality- or multiplicity-dependent suppression. This behavior may be due to the loss of the ρ^0 and K^{*0} signals caused by re-scattering of their decay products in the hadronic phase, while the ϕ decays mostly after the hadronic phase and is not affected. The $\Sigma(1385)^{\pm}/\Lambda$ and $\Xi(1530)^0/\Xi^-$ ratios are not observed to depend on system size or activity in pp and p–Pb collisions. Comparisons of resonance and proton $\langle p_{\mathrm{T}} \rangle$ values suggest that hydrodynamics can describe the p_{T} spectra of those particle in central Pb–Pb collisions, while other effects may be necessary to describe the p_{T} spectra in smaller collision systems. The ALICE Collaboration is currently extending these measurements to higher energies, other resonances ($\Lambda(1520)$ and $K^*(892)^{\pm}$), and other collision systems (*e.g.*, baryonic resonances in Pb–Pb collisions).

References

[1] Bleicher M and Stöcker H 2004 *J. Phys.* G **30** S111–8
[2] Markert C *et al* 2002 *AIP Conf. Proc.* **631** 533–52
[3] Vogel S and Bleicher M 2005 *Ricerca Scientifica ed Educazione Permanente* Supplemento N. 124 ed I Iori and A Bortolotti (Milan: Università degli Studi di Milano) pp 116–9 (*Preprint* nucl-th/0505027)
[4] Torrieri G and Rafelski J 2001 *Phys. Lett.* B **509** 239–45
[5] Rafelski J *et al* 2001 *Phys. Rev.* C **64** 054907
[6] Abelev B *et al* (ALICE Collaboration) 2015 *Phys. Rev.* C **91** 024609
[7] Adam J *et al* (ALICE Collaboration) 2016 *Eur. Phys. J.* C **76** 245
[8] Abelev B *et al* (ALICE Collaboration) 2012 *Eur. Phys. J.* C **72** 2183
[9] Wheaton S, Cleymans J and Hauer M 2009 *Comput. Phys. Commun.* **180** 84-106
[10] Stachel J, Andronic A, Braun-Munzinger P and Redlich K 2014 *J. Phys.: Conf. Ser.* **509** 012019
[11] Knospe A G *et al* 2016 *Phys. Rev.* C **93** 014911
[12] Beddall A, Beddall A and Bingül A 2008 *Acta Phys. Polon.* B **39** 173-80
[13] Abelev B *et al* (ALICE Collaboration) 2015 *Eur. Phys. J.* C **75** 1
[14] Adam J *et al* (ALICE Collaboration) 2016 *Phys. Lett.* B **758** 389-401
[15] Abelev B *et al* (ALICE Collaboration) 2013 *Phys. Rev.* C **88** 044910
[16] Abelev B *et al* (ALICE Collaboration) 2013 *Phys. Lett.* B **728** 25-38
[17] Abelev B *et al* (ALICE Collaboration) 2013 *Phys. Rev. Lett.* **111** 222301
[18] Fries R *et al* 2003 *Phys. Rev. Lett.* **90** 202303
[19] Fries R *et al* 2008 *Annu. Rev. Nucl. Part. Sci* **58** 177–205
[20] Qui Z *et al* 2012 *Phys. Lett.* B **707** 151–5
[21] Shen C *et al* 2011 *Phys. Rev.* C **84** 044903
[22] Bożek P and Wyskiel-Piekarska I 2012 *Phys. Rev.* C **85** 064915
[23] Minissale V *et al* 2015 *Phys. Rev.* C **92** 054904

Higher Moments of Net-Kaon Multiplicity Distributions at STAR

Ji Xu (for the STAR Collaboration)

Key Laboratory of Quark & Lepton Physics (MOE) and Institute of Particle Physics, Central China Normal University, Wuhan 430079, China
Lawrence Berkeley National Laboratory, Berkeley, CA 94720, US

E-mail: `jixu@lbl.gov`

Abstract. Fluctuations of conserved quantities such as baryon number (B), electric charge number (Q), and strangeness number (S), are sensitive to the correlation length and can be used to probe non-gaussian fluctuations near the critical point. Experimentally, higher moments of the multiplicity distributions have been used to search for the QCD critical point in heavy-ion collisions. In this paper, we report the efficiency-corrected cumulants and their ratios of mid-rapidity ($|y| < 0.5$) net-kaon multiplicity distributions in Au+Au collisions at $\sqrt{s_{\rm NN}} = 7.7$, 11.5, 14.5, 19.6, 27, 39, 62.4, and 200 GeV collected in 2010, 2011, and 2014 with STAR at RHIC. The centrality and energy dependence of the cumulants and their ratios, are presented. Furthermore, the comparisons with baseline calculations (Poisson) and non-critical-point models (UrQMD) are also discussed.

1. Introduction

The main goals of the high energy nuclear collisions are to explore the phase structure of strongly interacting hot and dense nuclear matter and map the quantum chromodynamics (QCD) phase diagram which can be displayed in the two dimensions (temperature (T) vs. baryon chemical potential (μ_B)). Fluctuations of conserved quantities, such as net-baryon (ΔN_B), net-charge (ΔN_Q) and net-strangeness (ΔN_S), have been predicted to be sensitive to the QCD phase transition and the QCD critical point [1–3]. The cumulants of the event-by-event distributions of these conserved quantities are directly connected to the thermodynamic susceptibilities computed with Lattice QCD [4–6] and in the Hadron Resonance Gas (HRG) model [7–10]. Thus, those cumulant ratios are equivalent to the ratios of various order susceptibilities as $C_4/C_2 = \chi_i^{(4)}/\chi_i^{(2)}$ and $C_3/C_2 = \chi_i^{(3)}/\chi_i^{(2)}$, where i indicates the conserved quantity.

Experimentally, we use net-kaon (ΔN_K) as proxy for net-strangeness. The fluctuations of conserved quantities have been widely studied experimentally and theoretically [1–3, 7, 11–14]. In this work, we report recent efficiency-corrected cumulants and cumulant ratios of the net-kaon (ΔN_K) multiplicity distributions measured in Au+Au collisions at $\sqrt{s_{\rm NN}} = 7.7$, 11.5, 14.5, 19.6, 27, 39, 62.4, and 200 GeV collected in 2010, 2011, and 2014 by STAR at RHIC.

2. Analysis Details

The STAR (Solenoidal Tracker At RHIC) detector at BNL has a large uniform acceptance at mid-rapidity and excellent particle identification capabilities. Energy loss (dE/dx) in the

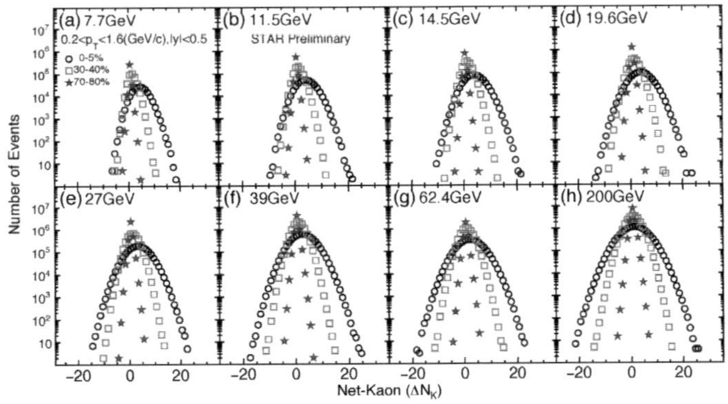

Figure 1. Uncorrected raw event-by-event net-kaon (ΔN_K) multiplicity distributions for Au+Au collisions at $\sqrt{s_{\rm NN}} = 7.7$, 11.5, 14.5, 19.6, 27, 39, 62.4, and 200GeV for 0-5% top central (black circles), 30-40% central (red squares), and 70-80% peripheral collisions (blue stars).

Time Projection Chamber and mass-squared (m^2) from the Time-Of-Flight detector are used to identify kaons. The K^+ (K^-) for net-kaon analysis are measured at mid-rapidity ($|y|$<0.5) within the transverse momentum region 0.2<p_T<1.6 GeV/c.

For the centrality selection, the uncorrected multiplicity distribution of primary charged particles, excluding kaons/anti-kaons within $|\eta|$<1.0 has been used for ΔN_K to avoid auto-correlations. The results are corrected for a finite centrality bin width [15].

Figure 1 shows the uncorrected event-by-event net-kaon multiplicity distributions for Au+Au collisions at $\sqrt{s_{\rm NN}} = 7.7$, 11.5, 14.5, 19.6, 27, 39, 62.4, and 200 GeV for ΔN_K in three centrality intervals. We observe that most central collisions have a wider distribution compared with peripheral collision. The peak of the net-kaon distributions shift slightly towards to the positive direction with the energy decrease. Detailed discussions about the efficiency correction and error estimation can be found in [15, 16].

3. Results and Discussion

All results have been corrected for finite detector reconstruction efficiencies, with the statistical uncertainties calculated based on the Delta Theorem [16]. Figure 2 shows the centrality dependence of cumulants (C_1, C_2, C_3, and C_4) of net-kaon multiplicity distributions in Au+Au collisions at $\sqrt{s_{\rm NN}}$ =7.7∼200 GeV. The red dashed lines represent the Poisson expectation. In general, those cumulants show a linear variation with the averaged number of participant nucleons and most C_3 and C_4 values are consistent with Poisson expectation within uncertainties.

Figure 3 shows the centrality dependence of $S\sigma$/Skellam and $\kappa\sigma_2$ of net-kaon multiplicity distributions for Au+Au collisions at $\sqrt{s_{\rm NN}}$ = 7.7 to 200GeV. The error bars are statistical errors and the caps represent systematic errors. For $S\sigma$/Skellam, the values are consistent with Poisson expectation for most central collisions within uncertainties. Most of the values of $\kappa\sigma^2$ are consistent with Poisson baseline, which is unity, within uncertainties.

The energy dependence of the volume-independent cumulant ratios (σ^2/M, $S\sigma$/Skellam, $\kappa\sigma^2$) for net-kaon multiplicity distributions in Au+Au collisions are presented in Figure 4. The blue bands give the predictions from the UrQMD calculations which do not include critical physics processes. One can see that the values of σ^2/M increase as the energy increases. The values of $S\sigma$/Skellam and $\kappa\sigma^2$ for net-kaon multiplicity distributions are consistent with both poisson and negative binomial distribution baseline within errors. Moments results from UrQMD (no

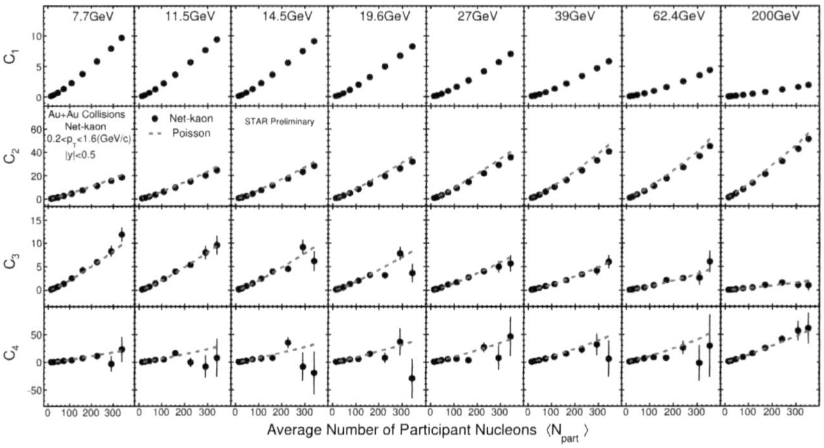

Figure 2. Centrality dependence of cumulants (C_1, C_2, C_3, and C_4) of net-kaon multiplicity distributions for Au+Au collisions at $\sqrt{s_{NN}}$ = 7.7, 11.5, 14.5, 19.6, 27, 39, 62.4, and 200GeV. The Poisson expectations are denoted as dotted lines. The error bars are statistical errors.

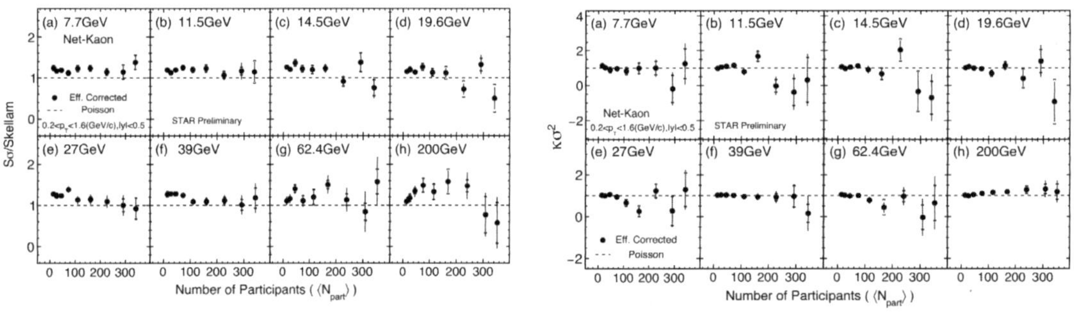

Figure 3. Centrality dependence of $S\sigma$/Skellam (Left) and $\kappa\sigma_2$ (Right) of net-kaon multiplicity distributions for Au+Au collisions at $\sqrt{s_{NN}}$ = 7.7, 11.5, 14.5, 19.6, 27, 39, 62.4 and 200GeV. The Poisson expectations are denoted as dotted lines. The error bars are statistical errors and the caps represent systematic errors.

Critical Point) show no energy dependence for $S\sigma$/Skellam and $\kappa\sigma^2$.

The measurements presented here will be further improved in the upcoming RHIC BES II in 2019-2020 which include an STAR detector upgrade. An inner TPC and Endcap TOF upgrade will enlarge the phase-space up to $|\eta|<1.5$ and down to p_T = 60 MeV/c. The Event Plane Detector at forward rapidities will allow for a better centrality estimation, suppressing auto-correlations.

4. Summary

In this paper, centrality and energy dependence of cumulants (up to forth order) and their ratios from $\sqrt{s_{NN}}$ = 7.7 to 200 GeV in Au+Au collisions for net-kaon multiplicity distributions are presented. Within uncertainties (statistical and systematic), $S\sigma$/Skellam and $\kappa\sigma^2$ values are consistent with the Poisson and negative binomial distribution baseline within errors at most central collisions. The results from UrQMD model calculations, where no-criticality is intended, show no energy dependence. The RHIC BES II will bring a larger event sample and a wider

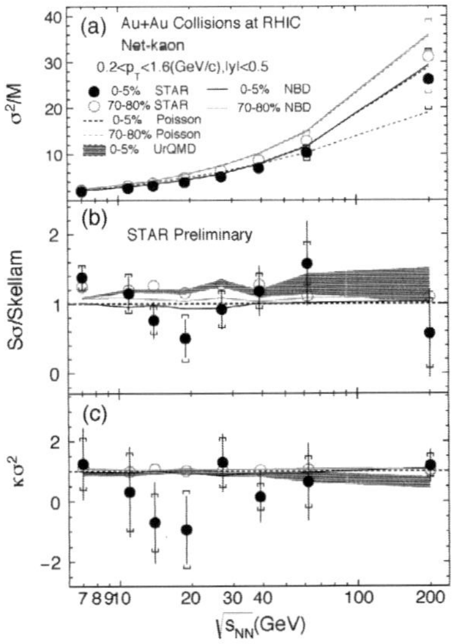

Figure 4. Energy dependence of cumulant ratios (σ^2/M, $S\sigma$/Skellam, $\kappa\sigma^2$) of net-kaon multiplicity distributions for top 0-5% central and 70-80% peripheral collisions at Au+Au collisions at $\sqrt{s_{\rm NN}}$ = 7.7, 11.5, 14.5, 19.6, 27, 39, 62.4, and 200GeV. The Poisson and NBD expectations are denoted as dotted and solid lines, UrQMD calculations are shown as blue bands. The error bars are statistical errors and the caps represent systematic errors.

phase-space to enhance the search for the QCD critical point in the high net-baryon density region $420 < \mu_B < 250$ MeV ($7.7 < \sqrt{s_{\rm NN}} < 19.6$ GeV) .

Acknowledgments

The work was supported in part by the MoST of China 973-Project No.2015CB856901, NSFC under grant No. 11575069, 11221504.

References

[1] M. A. Stephanov, Phys. Rev. Lett. 102, 032301 (2009).
[2] M. Asakawa *et al.*, Phys. Rev. Lett. 103 262301 (2009).
[3] M. A. Stephanov, Phys. Rev. Lett. 107 052301 (2011).
[4] R. V. Gavai and S. Gupta, Phys. Rev. D 78, 114503 (2008).
[5] M. Cheng *et al.*, Phys. Rev. D 79, 074505 (2009); A. Bazavov *et al.*, arXiv:1203.0784.
[6] S. Borsanyi *et al.*, JHEP01, 138 (2012).
[7] F. Karsch and K. Redlich, Phys. Lett. B 695, 136 (2011).
[8] P. Garg *et al.*, Phys. Lett. B 726, 691 (2013).
[9] Marlene Nahrgang *et al.*, arXiv: 1402.1238.
[10] J. Fu, Phys. Lett. B 722, 144 (2013).
[11] S. Gupta, X. Luo, B. Mohanty, H. G. Ritter and N. Xu, Science 332, 1525 (2011).
[12] Aggarwal M M *et al.* (STAR Collaboration) 2010 Phys. Rev. Lett. 105 022302
[13] X. Luo (STAR Collaboration), Nucl. Phys. A 904 911c (2013).
[14] X. Luo, arXiv:1512.09215
[15] X. Luo, J. Phys. G: Nucl. Part. Phys. 39 025008 (2012).
[16] X. Luo, Phys. Rev. C 91, 034907 (2015).
[17] L. Adamczyk *et al.*, (STAR Collaboration), Phys. Rev. Lett.113, 092301 (2014).

Behavior of universal critical parameters in the QCD phase diagram

Marcus Bluhm[1], **Marlene Nahrgang**[2,3], **Steffen A. Bass**[3] **and Thomas Schäfer**[1]

[1] Department of Physics, North Carolina State University, Raleigh, NC 27695, USA
[2] SUBATECH, UMR 6457, Université de Nantes, École des Mines de Nantes, IN2P3/CNRS. 4 rue Alfred Kastler, 44307 Nantes cedex 3, France
[3] Department of Physics, Duke University, Durham, NC 27708, USA

E-mail: mbluhm@ncsu.edu

Abstract. We determine the dependence of important parameters for critical fluctuations on temperature and baryon chemical potential in the QCD phase diagram. The analysis is based on an identification of the fluctuations of the order parameter obtained from the Ising model equation of state and the Ginzburg-Landau effective potential approach. The impact of the mapping from Ising model variables to QCD thermodynamics is discussed.

1. Introduction

Effective models for QCD thermodynamics [1, 2, 3] predict that the chiral phase transition in QCD is of first order at large baryon chemical potential μ_B. At $\mu_B = 0$, lattice QCD simulations established that the transformation between the denser deconfined quark matter phase at high temperature T and the low-T hadronic matter phase is an analytic crossover [4]. As a consequence, the first-order phase transition line should terminate in a second-order chiral critical point at some non-zero μ_B. A critical point is characterized through the growth of the fluctuations of the order parameter σ in the scaling region around it. In the thermodynamic limit, these fluctuations diverge at the critical point with a given exponent of the diverging correlation length ξ.

Experimentally, the phases of QCD matter are studied in heavy-ion collision experiments by varying the beam energy and the system size. The quest for finding the conjectured critical point in the QCD phase diagram has triggered an extensive effort culminating in the recently completed beam energy scan phase I at RHIC [5, 6, 7] and being continued with the NA61 experiment at CERN-SPS and the phase II of RHIC's beam energy scan program in the future. Signaling the presence of a critical point, large event-by-event fluctuations [8, 9] of conserved quantities such as electric charge or baryon number have been predicted. In reality, the actual growth of the correlation length is limited by the finite size and, more importantly, lifetime of the created system. Thus, higher-order non-Gaussian cumulants of the event-by-event distributions, which depend more strongly on the growth of ξ [10], are particularly interesting. These exhibit certain patterns in the QCD phase diagram [11, 12] which are governed by the qualitative behavior of universal critical parameters.

IOP Conf. Series: Journal of Physics: Conf. Series **779** (2017) 012074 doi:10.1088/1742-6596/779/1/012074

Assuming that QCD belongs to the same universality class as the three-dimensional Ising model [2, 3] allows one to identify the expectation value of the order parameter of the chiral phase transition with the order parameter in the spin model, the magnetization M. Fluctuations of the order parameter in the scaling region can then be determined from the critical equation of state for M. This enables us to study the behavior of the universal critical parameters in this work. A similar approach was previously used in [13, 14] to construct an equation of state with critical point for QCD matter.

In the following, we employ the linear parametric representation [15] of the critical equation of state for the magnetization with critical exponents $\beta = 1/3$ and $\delta = 5$

$$M = M_0 R^{\beta} \theta \,, \tag{1}$$

$$r = R(1-\theta^2) \,, \tag{2}$$

$$h = R^{\beta\delta} \tilde{h}(\theta) \,. \tag{3}$$

In this representation $M(r, h)$, which is a function of reduced temperature $r = (T - T_c)/T_c$ with spin model critical temperature T_c and of reduced external magnetic field $h = H/H_0$, is expressed in terms of the auxiliary variables $R \geq 0$ and θ as $M(R, \theta)$. The parametrization is uniquely defined within $-\theta_0 \leq \theta \leq \theta_0$, where for $\tilde{h}(\theta) = 3\theta(1 - 2\theta^2/3)$, which is an odd function of θ, one finds $\theta_0 = \sqrt{3/2}$ as the relevant non-trivial root. M_0 and H_0 are normalization constants of mass dimension one and three, respectively.

In spin model coordinates, the critical point is located at $r = h = 0$ ($R = 0$). At $h = 0$, one finds a first-order phase transition for $r < 0$ ($R > 0$ and $\theta = \pm\theta_0$), while for $r > 0$ there is a crossover ($R > 0$ and $\theta = 0$). The behavior of the order parameter in the scaling region is such that $M(r = 0, h) \sim |h|^{1/\delta} sgn(h)$ and $M(r, h \to 0^+) \sim |r|^{\beta}$ for $r < 0$, which can be assured by imposing conditions on M_0 and H_0 such that M is positive for $h \geq 0^+$ [13].

2. Order parameter fluctuations

The equilibrium cumulants quantifying the fluctuations of the order parameter in the scaling region are determined from derivatives

$$\langle (\delta\sigma)^n \rangle_c = \left(\frac{T}{V H_0} \right)^{n-1} \left. \frac{\partial^{n-1} M}{\partial h^{n-1}} \right|_r \tag{4}$$

of M with respect to the external magnetic field at fixed r. Explicit expressions in terms of R and θ based on the linear parametric representation can be found in [16]. For given r and h, these cumulants can then be determined by making use of Eqs. (2) and (3).

The cumulants of the fluctuations of the order parameter can, likewise, be determined from the corresponding probability distribution which depends on an effective action for the order parameter of Ginzburg-Landau type [16, 10]. Including up to quartic interaction terms one finds

$$\langle (\delta\sigma)^2 \rangle = \frac{T}{V} \xi^2 \,, \tag{5}$$

$$\langle (\delta\sigma)^3 \rangle = -2\lambda_3 \frac{T^2}{V^2} \xi^6 \,, \tag{6}$$

$$\langle (\delta\sigma)^4 \rangle_c = 6 \left(2(\lambda_3 \xi)^2 - \lambda_4 \right) \frac{T^3}{V^3} \xi^8 \,. \tag{7}$$

In the scaling region, the cubic and quartic interaction strengths depend on the correlation length as $\lambda_3 = \tilde{\lambda}_3 T (T\xi)^{-3/2}$ and $\lambda_4 = \tilde{\lambda}_4 (T\xi)^{-1}$, where we neglected small anomalous scaling dimension corrections in the cumulant expressions.

Strictly speaking, the expressions in Eqs. (5) - (7) are only valid in the scaling region as long as the correlation length is small compared to the macroscopic length scale of the considered system. Keeping this limitation in mind we can, nonetheless, identify the expressions for the cumulants following from Eq. (4) with those in Eqs. (5) - (7). This yields parametric expressions for ξ, $\tilde{\lambda}_3$ and $\tilde{\lambda}_4$ in terms of the auxiliary spin model variables reading

$$\xi = \sqrt{\frac{M_0}{H_0}} \frac{1}{R^{2/3}(3+2\theta^2)^{1/2}} , \tag{8}$$

$$\tilde{\lambda}_3 = \frac{2\theta(9+\theta^2)}{(3-\theta^2)(3+2\theta^2)^{3/4}} C , \tag{9}$$

$$\tilde{\lambda}_4 = 2\tilde{\lambda}_3^2 + 2\frac{(81-783\theta^2+105\theta^4-5\theta^6+2\theta^8)}{(3-\theta^2)^3(3+2\theta^2)^{3/2}} C^2 , \tag{10}$$

where $C = (T^2 H_0/M_0^5)^{1/4}$. The parameters $\tilde{\lambda}_3$ and $\tilde{\lambda}_4$ are found to be dimensionless and independent of R which implies that they remain finite in the entire domain of the parametric representation, while ξ is of inverse mass dimension and diverges as $R \to 0$.

According to Eq. (9), $\tilde{\lambda}_3$ is an odd function of h which is negative for $h < 0$. At $h = 0$, one finds $\tilde{\lambda}_3 = 0$ for $r > 0$, while for $r < 0$ the dimensionless parameter approaches $\tilde{\lambda}_3 \to \pm 7\, C/6^{1/4}$ as $h \to 0^{\pm}$. The qualitative behavior of the fourth-order cumulant $\langle (\delta\sigma)^4 \rangle_c$ is determined by the behavior of the parameter difference $2\tilde{\lambda}_3^2 - \tilde{\lambda}_4$. As evident from Eq. (10), this difference is an even function of h, which is positive for all $r < 0$ with maximum value $(2\tilde{\lambda}_3^2 - \tilde{\lambda}_4) = 128\, C^2/27$ at $h = 0$. For $r > 0$, instead, one finds an interval $-h_0 \leq h \leq h_0$ in which the parameter difference becomes negative with minimum value $(2\tilde{\lambda}_3^2 - \tilde{\lambda}_4) = -4\, C^2/9$ at $h = 0$. The size of this interval depends on the value of r and shrinks to zero, $h_0 \to 0$, as $r \to 0^+$.

3. Parameter behavior in the QCD phase diagram

The parameters $\tilde{\lambda}_3$ and $\tilde{\lambda}_4$ in Eqs. (9) and (10) are universal functions of the spin model variables in the scaling region. However, the mapping from r and h to μ_B and T in QCD thermodynamics is not universal but strongly model-dependent. By relating the density difference from the critical density in QCD to the magnetization M, this mapping has to satisfy, nonetheless, certain constraints near the critical point based on universality class arguments. The QCD critical point with baryon chemical potential μ_B^{cp} and temperature T^{cp} must be located at $r = h = 0$, positive r-values must correspond to the QCD crossover regime and positive h-values have to be realized in the denser phase.

In practice, these conditions can be assured by a simple linear mapping employing as auxiliary variables

$$\tilde{r} = \frac{(\mu_B - \mu_B^{\text{cp}})}{\Delta\mu_B^{\text{cp}}} \quad , \quad \tilde{h} = \frac{(T - T^{\text{cp}})}{\Delta T^{\text{cp}}} . \tag{11}$$

The parameters ΔT^{cp} and $\Delta\mu_B^{\text{cp}}$ relate scales in the spin model coordinate system with the unknown size of the critical region in QCD. As by definition $\tilde{r} > 0$ in the QCD first-order phase transition regime, one has to rotate $\tilde{r}$ to obtain r satisfying the above condition. Since the first-order phase transition line is expected to be bent, it is intuitive to perform this rotation such that the r-axis lies tangentially to the transition line in the QCD critical point. The exact orientation of the h-axis is, in contrast, less constrained. In line with [17], we opt for defining h parallel to the T-axis in QCD.

The corresponding behavior of the scaled parameter $\tilde{\lambda}_3/C$ and the scaled parameter difference $(2\tilde{\lambda}_3^2 - \tilde{\lambda}_4)/C^2$ in the QCD phase diagram is shown in Figures 1 and 2, respectively. Qualitatively, the observed patterns follow the behavior with r and h described in Section 2, but appear tilted to some extent given the particular mapping to QCD thermodynamics employed in [17].

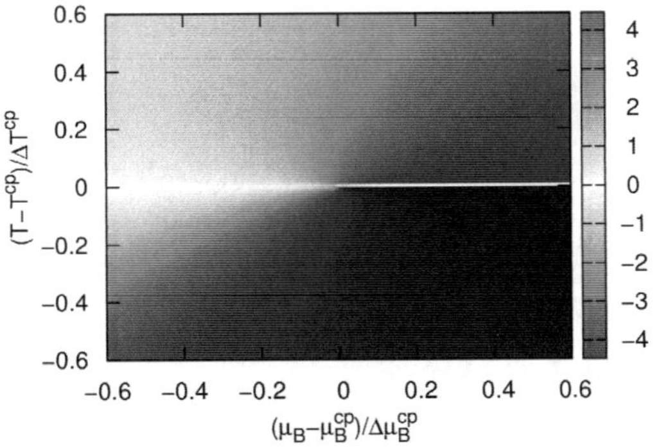

Figure 1. Scaled dimensionless parameter $\tilde{\lambda}_3/C$ from Eq. (9) as a function of $\tilde{r}$ and $\tilde{h}$, cf. text for details.

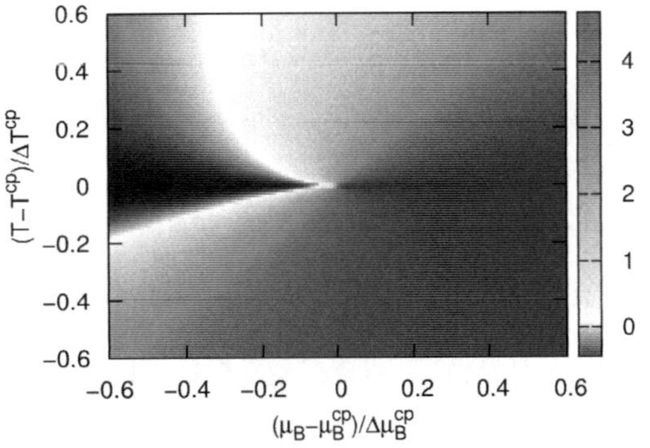

Figure 2. As in Figure 1 but for the scaled dimensionless parameter difference in Eq. (10), $(2\tilde{\lambda}_3^2 - \tilde{\lambda}_4)/C^2$.

4. Conclusions

We discussed the behavior of important parameters of the effective action near the QCD critical point in line with the three-dimensional Ising model universality class. The shown results are based on the mapping used in [17]. In future work, these parameters will serve as input for dynamical studies of critical fluctuations in heavy-ion collisions, similar to [18, 19].

Acknowledgments

The authors thank J.-W. Chen, S. Mukherjee and M. Stephanov for helpful discussions. M.N. acknowledges support from a fellowship within the Postdoc-Program of the German Academic Exchange Service (DAAD). This work was supported in parts by the U.S. Department of Energy under grants DE-FG02-03ER41260 and DE-FG02-05ER41367, and within the framework of the Beam Energy Scan Theory (BEST) Topical Collaboration.

References

[1] Asakawa M and Yazaki K 1989 *Nucl. Phys.* A **504** 668
[2] Berges J and Rajagopal K 1999 *Nucl. Phys.* B **538** 215 (*Preprint* hep-ph/9804233)
[3] Halasz A M, Jackson A D, Shrock R E, Stephanov M A and Verbaarschot J J M 1998 *Phys. Rev.* D **58** 096007 (*Preprint* hep-ph/9804290)
[4] Aoki Y, Endrodi G, Fodor Z, Katz S D and Szabo K K 2006 *Nature* **443** 675 (*Preprint* hep-lat/0611014)
[5] Adamczyk L *et al.* (STAR) 2014 *Phys. Rev. Lett.* **112** 032302 (*Preprint* arXiv:1309.5681 [nucl-ex])
[6] Adamczyk L *et al.* (STAR) 2014 *Phys. Rev. Lett.* **113** 092301 (*Preprint* arXiv:1402.1558 [nucl-ex])
[7] Adare A *et al.* (PHENIX) 2016 *Phys. Rev.* C **93** 011901 (*Preprint* arXiv:1506.07834 [nucl-ex])
[8] Stephanov M A, Rajagopal K and Shuryak E V 1998 *Phys. Rev. Lett.* **81** 4816 (*Preprint* hep-ph/9806219)
[9] Stephanov M A, Rajagopal K and Shuryak E V 1999 *Phys. Rev.* D **60** 114028 (*Preprint* hep-ph/9903292)
[10] Stephanov M A 2009 *Phys. Rev. Lett.* **102** 032301 (*Preprint* arXiv:0809.3450 [hep-ph])
[11] Asakawa M, Ejiri S and Kitazawa M 2009 *Phys. Rev. Lett.* **103** 262301 (*Preprint* arXiv:0904.2089 [nucl-th])
[12] Stephanov M A 2011 *Phys. Rev. Lett.* **107** 052301 (*Preprint* arXiv:1104.1627 [hep-ph])
[13] Nonaka C and Asakawa M 2005 *Phys. Rev.* C **71** 044904 (*Preprint* nucl-th/0410078)
[14] Bluhm M and Kämpfer B 2006 *Proc. of Science* CPOD(2006)004 (*Preprint* hep-ph/0611083)
[15] Schofield P 1969 *Phys. Rev. Lett.* **22** 606; Josephson B D 1969 *J. Phys. C: Solid State Phys.* **2** 1113
[16] Mukherjee S, Venugopalan R and Yin Y 2015 *Phys. Rev.* C **92** 034912 (*Preprint* arXiv:1506.00645 [hep-ph])
[17] Bluhm M, Nahrgang M, Bass S A and Schäfer T 2016 *Preprint* arXiv:1612.XXXXX [nucl-th]
[18] Nahrgang M, Leupold S, Herold C and Bleicher M 2011 *Phys. Rev.* C **84** 024912 (*Preprint* arXiv:1105.0622 [nucl-th])
[19] Herold C, Nahrgang M, Yan Y and Kobdaj C 2016 *Phys. Rev.* C **93** 021902 (*Preprint* arXiv:1601.04839 [hep-ph])

IOP Conf. Series: Journal of Physics: Conf. Series **779** (2017) 012075 doi:10.1088/1742-6596/779/1/012075

A particlization procedure for the NχFD model and the evolution of the net-proton kurtosis

C Herold[1], M Nahrgang[2,3], A Limphirat[1], C Kobdaj[1], Y Yan[1]

[1] School of Physics, Suranaree University of Technology, 111 University Avenue, Nakhon Ratchasima 30000, Thailand

[2] Department of Physics, Duke University, Durham, North Carolina 27708-0305, USA

[3] SUBATECH, UMR 6457, Universit?e de Nantes, Ecole des Mines de Nantes, IN2P3/CRNS. 4 rue Alfred Kastler, 44307 Nantes cedex 3, France

E-mail: herold@g.sut.ac.th

Abstract. We present an analysis of the net-proton kurtosis on the crossover side of the critical point within the model of nonequilibrium chiral fluid dynamics (NχFD). The chiral order parameter is propagated explicitly and coupled to an expanding fluid of quarks and gluons to describe the dynamical situation in a heavy-ion collision. After implementing a particlization routine into this model, we study the behavior of the net-proton kurtosis as driven by the characteristic structure of the net-baryon number susceptibility in the critical region.

1. Introduction

Extreme temperatures and densities may create an extreme state of matter, the so-called quark-gluon plasma (QGP), where quarks and gluons are deconfined and chiral symmetry is restored. The transition from hadron gas to QGP is of crossover type for zero baryochemical potential, but expected to turn into a critical point and first-order phase transition for larger net-baryon densities. Main support for this assumption is obtained from effective model studies [1] or Dyson-Schwinger equations [2]. In the vicinity of a conjectured critical point, fluctuations of the order parameter or the net-quark number as calculated from thermodynamic susceptibilities exhibit characteristic peak structures [3, 4]. Experiments such as NA49 at CERN or STAR at RHIC have been trying to measure such fluctuations on an event-by-event basis. Although recent STAR measurements have reported a nonmonotonic behavior of the net-proton kurtosis as function of the beam energy [5], a thorough understanding of the dynamical processes leading to the obtained signal is still due. We address this issue using the NχFD model [6, 7, 8, 9, 10] by implementing a particlization procedure. This means that by using the Cooper-Frye formula, we find particle distributions on hypersurfaces of constant energy-density from the fluid dynamical model as described in [11]. We are then in a position to study the behavior of the experimentally accessible net-proton kurtosis during an evolution in the crossover region near a critical point [12].

IOP Conf. Series: Journal of Physics: Conf. Series **779** (2017) 012075 doi:10.1088/1742-6596/779/1/012075

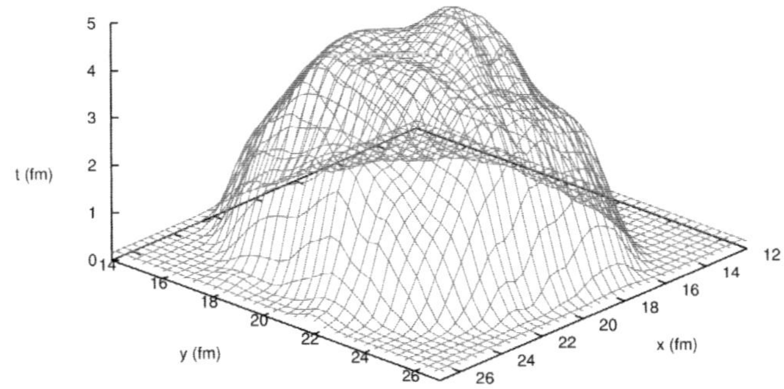

Figure 1. Freeze-out surface of constant energy density $e = 2e_0$ where the particlization procedure is performed after the crossover transition.

2. Particlization in the NχFD model

Starting point for our study is a quark-meson model with chiral condensate σ and dilaton field χ,

$$\mathcal{L} = \overline{q}\left(i\gamma^\mu\partial_\mu - g_{\mathrm{q}}\sigma\right)q + \frac{1}{2}\left(\partial_\mu\sigma\right)^2 + \frac{1}{2}\left(\partial_\mu\chi\right)^2 + \mathcal{L}_A - U_\sigma - U_\chi\ , \tag{1}$$

with the constituent gluon Lagrangian $\mathcal{L}_A$, the potentials U_σ and U_χ for the respective fields and the quark-meson coupling $g = 3.3$, determined from the constituent quark mass in vacuum. The nonequilibrium evolution during expansion and cooling after a heavy-ion collision is driven by the Langevin dynamics of the sigma field, derived from the two-particle irreducible effective action,

$$\partial_\mu\partial^\mu\sigma + \eta_\sigma\partial_t\sigma + \frac{\delta V_{\mathrm{eff}}}{\delta\sigma} = \xi\ , \tag{2}$$

with the mean-field potential V_{eff} obtained from integrating out the quark degrees of freedom. The damping coefficient η depends on T and μ as a result of the $\sigma \leftrightarrow q\bar{q}$ pair production process. It is related to the stochastic noise field ξ via

$$\langle\xi(t,\vec{x})\xi(t',\vec{x}')\rangle_\xi = \delta(\vec{x}-\vec{x}')\delta(t-t')m_\sigma\eta_\sigma\coth\left(\frac{m_\sigma}{2T}\right)\ . \tag{3}$$

To make qualitative comparisons to experimental observables, we implemented a Cooper-Frye freeze-out [13, 14], producing all non-strange particles from the UrQMD model [15, 16] along hypersurfaces of constant energy density. An example of such a hypersurface is shown in Fig. 1 for $e = 2e_0$ and the case of a crossover transition on which this study focuses on. We clearly see a smooth structure as opposed to inhomogeneities that may arise in the presence of spinodal decomposition at a first-order phase transition. Particles are produced until the total integrated energy of quark-gluon fluid and fields σ, χ is obtained:

$$e = e_{\mathrm{fluid}} + \frac{1}{2}\left(\frac{\partial\sigma}{\partial t}\right)^2 + \frac{1}{2}\left(\nabla\sigma\right)^2 + U_\sigma + \frac{1}{2}\left(\frac{\partial\chi}{\partial t}\right)^2 + \frac{1}{2}\left(\nabla\chi\right)^2 + U_\chi\ . \tag{4}$$

We furthermore ensure that net-momentum, net-charge and net-baryon number are exactly conserved in each event.

3. Net-proton kurtosis and net-baryon susceptibility

We focus on a crossover evolution near the critical point of the model given by eq. (1), corresponding to a trajectory shown in Fig. 2. Here the values of T and μ are calculated

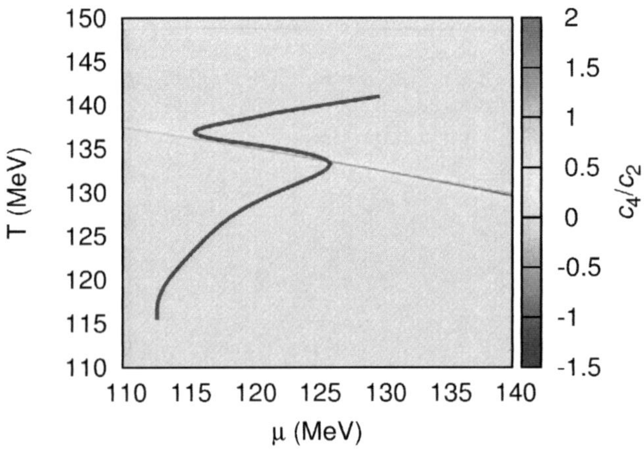

Figure 2. Event-averaged trajectory near the critical point. Although the region of enhanced susceptibility is narrow, the evolution remains there for an extended amount of time.

as volume-averages in a central region of the fireball. Initial conditions are obtained from the UrQMD transport model. Fig. 2 also shows the net-baryon susceptibility ratio c_4/c_2 which is related to the net-baryon kurtosis. In Fig. 3 we show this quantity along the trajectory, calculated for the corresponding values of T, μ and give a comparison to the net-proton kurtosis, determined from the particlization at certain points of constant energy density along the evolution. We see that the location of the minima in both curves are in clear correspondence. The two smaller maxima in c_4/c_2 are not visible in the kurtosis, though. This might on the one hand be due to a lack in resolution, on the other hand we may expect that in an inhomogeneous medium with varying values of T and μ for each hypersurface, the dominant contribution is usually provided by regions with the minimum peak in the susceptibilities so that effects of the small peaks, which are visible in the equilibrium calculation, become washed out in a dynamic inhomogeneous setup.

4. Summary and Outlook

In the work presented here, we have extended the NχFD model with a particlization procedure, an important step towards making quantitative predictions for future experiments and understanding previous results from the beam-energy scan program. With this we have come to understand the relation between susceptibilities and dynamically generated fluctuations in particle numbers as they are measured to determine the location of a possible QCD critical point. We saw that the dominant minimum in the kurtosis-related c_4/c_2 is reflected in a minimum around the same energy density in the net-proton kurtosis obtained from our nonequilibrium model.

In the future, we are going to include effects of sigma field fluctuations which we expect to couple to particle production, in particular (anti-)protons and pions. This could be included into the model via an interaction term $g\bar{p}\sigma p$ [17, 18, 19]. We are furthermore going to consider a subsequent hadronic afterburner.

Acknowledgments

This work is funded by Suranaree University of Technology (SUT) and by the Office of the Higher Education Commission under NRU project of Thailand. CH and CK acknowledge support from

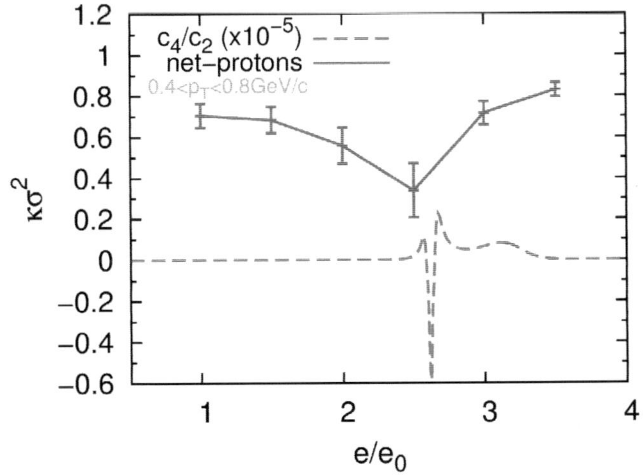

Figure 3. Net-proton kurtosis as function of freeze-out energy for a nonequilibrium evolution compared with the generalized susceptibilities c_4/c_2. Figure from [12].

SUT-CHE-NRU (FtR.15/2559) project. The computing resources have been provided by the National e-Science Infrastructure Consortium of Thailand, the Center for Computer Services at SUT and the Frankfurt Center for Scientific Computing. The authors thank Dirk Rischke for providing the SHASTA code that was used for the fluid dynamical simulation. MN acknowledges support from the U.S. Department of Energy under grant DE-FG02-05ER41367 and a fellowship within the Postdoc-Program of the German Academic Exchange Service (DAAD).

References

[1] Herbst T K, Pawlowski J M and Schaefer B J 2011 *Phys.Lett.* **B696** 58–67 (*Preprint* 1008.0081)
[2] Eichmann G, Fischer C S and Welzbacher C A 2015 (*Preprint* 1509.02082)
[3] Sasaki C, Friman B and Redlich K 2007 *Phys.Rev.* **D75** 074013 (*Preprint* hep-ph/0611147)
[4] Skokov V, Friman B and Redlich K 2011 *Phys.Rev.* **C83** 054904 (*Preprint* 1008.4570)
[5] Adamczyk L *et al.* (STAR Collaboration) 2014 *Phys.Rev.Lett.* **112** 032302 (*Preprint* 1309.5681)
[6] Nahrgang M, Leupold S, Herold C and Bleicher M 2011 *Phys.Rev.* **C84** 024912 (*Preprint* 1105.0622)
[7] Nahrgang M, Leupold S and Bleicher M 2012 *Phys.Lett.* **B711** 109–116 (*Preprint* 1105.1396)
[8] Herold C, Nahrgang M, Mishustin I N and Bleicher M 2013 *Phys.Rev.* **C87** 014907 (*Preprint* 1301.1214)
[9] Herold C, Nahrgang M, Yan Y and Kobdaj C 2014 *J.Phys.* **G41** 115106 (*Preprint* 1407.8277)
[10] Herold C, Nahrgang M, Mishustin I and Bleicher M 2014 *Nuclear Physics A* **925** 14 – 24 ISSN 0375-9474
[11] Huovinen P and Petersen H 2012 *Eur.Phys.J.* **A48** 171 (*Preprint* 1206.3371)
[12] Herold C, Nahrgang M, Yan Y and Kobdaj C 2016 *Phys. Rev.* **C93** 021902 (*Preprint* 1601.04839)
[13] Cooper F, Frye G and Schonberg E 1975 *Phys.Rev.* **D11** 192
[14] Cooper F and Frye G 1974 *Phys.Rev.* **D10** 186
[15] Bass S A, Belkacem M, Bleicher M, Brandstetter M, Bravina L *et al.* 1998 *Prog.Part.Nucl.Phys.* **41** 255–369 (*Preprint* nucl-th/9803035)
[16] Bleicher M, Zabrodin E, Spieles C, Bass S A, Ernst C *et al.* 1999 *J.Phys.* **G25** 1859–1896 (*Preprint* hep-ph/9909407)
[17] Stephanov M A, Rajagopal K and Shuryak E V 1999 *Phys.Rev.* **D60** 114028 (*Preprint* hep-ph/9903292)
[18] Stephanov M A 2009 *Phys.Rev.Lett.* **102** 032301 (*Preprint* 0809.3450)
[19] Athanasiou C, Rajagopal K and Stephanov M 2010 *Phys. Rev.* **D82** 074008 (*Preprint* 1006.4636)

IOP Conf. Series: Journal of Physics: Conf. Series **779** (2017) 012076 doi:10.1088/1742-6596/779/1/012076

Fast Dynamical Evolution of Hadron Resonance Gas via Hagedorn States

M Beitel, K Gallmeister and C Greiner

Institut für Theoretische Physik, Goethe-Universität Frankfurt am Main, Max-von-Laue-Str. 1, 60438 Frankfurt am Main, Germany

E-mail: gallmeister@th.physik.uni-frankfurt.de

Abstract. Hagedorn states (HS) are a tool to model the hadronization process which occurs in the phase transition region between the quark gluon plasma (QGP) and the hadron resonance gas (HRG). These states are believed to appear near the Hagedorn temperature T_H which in our understanding equals the critical temperature T_c. A covariantly formulated bootstrap equation is solved to generate the zoo of these particles characterized baryon number B, strangeness S and electric charge Q. These hadron-like resonances are characterized by being very massive and by not being limited to quantum numbers of known hadrons. All hadronic properties like masses, spectral functions etc. are taken from the hadronic transport model Ultra Relativistic Quantum Molecular Dynamics (UrQMD). Decay chains of single Hagedorn states provide a well description of experimentally observed multiplicity ratios of strange and multi-strange particles as the Ξ^0- and the Ω^--baryon. In addition, the final energy spectra of resulting hadrons show a thermal-like distribution with the characteristic Hagedorn temperature T_H. Box calculations including these Hagedorn states are performed. Indeed, the time scales leading to equilibration of the system are drastically reduced down to $2\dots5\,\mathrm{fm}/c$.

1. Introduction

Back in the 60's of the last century and before the advent of quantum chromodynamics (QCD) as the theory of strong interactions, R. Hagedorn [1] proposed the existence of a whole zoo of massive, unobserved hadronic resonance states. The spectrum of these particles, known as Hagedorn spectrum, exhibits the specific feature of being exponential in the infinite mass limit with the slope given by the so called Hagedorn temperature T_H. This temperature denotes the limiting temperature for hadronic matter since any partition function of a HRG with Hagedorn-like mass spectrum diverges as long as $T > T_H$. Above the Hagedorn temperature a new state of matter, namely the QGP, shall be realized. Thus, Hagedorn states provide a tool to understand the phase transition from HRG to QGP and back. The Hagedorn states are created in multi-particle collisions most abundantly near T_H which in our understanding equals to the critical temperature T_c. Hagedorn states are color neutral objects which are allowed to have any quantum numbers as long as they are compatible with its mass. The appearance of Hagedorn states in multi-particle collisions and their role was already discussed in [2, 3, 4, 5, 6]. Their appearance near T_c can explain, as shown in [4, 5, 6], the fast chemical equilibration of (multi-) strange baryons B and their anti-particles $\bar{B}$ at Relativistic Heavy Ion Collider (RHIC) energies. The inclusion of Hagedorn states in a hadron resonance gas model provides also a lowering of the speed of sound, c_s and of the shear viscosity over entropy density ratio η/s at the phase transition and being in good agreement with lattice calculations [7, 8, 9, 10].

2. Model and Results

The starting point of all calculations provided here is the postulate of the statistical bootstrap model (SBM) stating that fireballs consist of fireballs which in turn consist of fireballs etc. . As shown in [11] (cf. also [12]) the mathematical formulation of this postulate in the formulation of [13, 14] leads to the evolving bootstrap equation for the density of states,

$$\tau_{\vec{C}}(m) = \frac{R^3}{3\pi\, m} \sum_{\vec{C}_1,\vec{C}_2} \int_{m_1^0}^{m} \mathrm{d}m_1 m_1 \tau_{\vec{C}_1}(m_1) \int_{m_2^0}^{m-m_1} \mathrm{d}m_2\, m_2 \tau_{\vec{C}_2}(m_2) \times\ p_{\mathrm{cm}}(m, m_1, m_2)\, \delta^{(3)}(\vec{C} - \vec{C}_1 - \vec{C}_2)\ , \tag{1}$$

where the conservation of the quantum numbers baryon number, strangeness and electrical charge $\vec{C} = (B, S, Q)$ is guaranteed. A major aspect was to restrict on Hagedorn states, which just consist of two constituents in order to allow for generation processes as $2 \to 1$ and decay as $1 \to 2$. This is some necessary constraint in order to allow for a microscopical calculation in a transport model simulation based on cross sections. The above set of integral equations of Volterra type is solved numerically by disretizing the mass range and consecutively stepping to higher mass bins. The zeroth order input are the known hadrons according their spectral functions as implemented in the transport model UrQMD [15].

The radius R is the only free parameter of the model. Choosing reasonable values as $R = 0.8\,\mathrm{fm}$ ($1.0\,\mathrm{fm}$) yield Hagedorn temperatures, i.e. the slopes of the exponentially rising Hagedorn spectra $\tau_{\vec{C}}(m) \sim m^a \exp(m/T_H)$ as $T_H = 162\,\mathrm{MeV}$ ($145\,\mathrm{MeV}$) nearly independent of the quantum numbers, as indicated in fig. 1.

In order to treat the decay and the production of Hagedorn states, the decay width Γ and the production cross section $\sigma = \pi R^2$ are connected via detailed balance as

$$\Gamma_{\vec{C}}(m) = \frac{\sigma}{2\pi^2\, \tau_{\vec{C}}(m)} \sum_{\vec{C}_1,\vec{C}_2} \int_{m_1^0}^{m} \mathrm{d}m_1 \tau_{\vec{C}_1}(m_1) \int_{m_2^0}^{m-m_1} \mathrm{d}m_2\, \tau_{\vec{C}_2}(m_2) \times\ p_{\mathrm{cm}}^2(m, m_1, m_2)\, \delta^{(3)}(\vec{C} - \vec{C}_1 - \vec{C}_2)\ , \tag{2}$$

which represents the second major equation of the model. Corresponding Hagedorn spectra and their decay width are shown in fig. 1. One observes, that the slope decreases with increasing volume of the Hagedorn state. Correspondingly also the width decreases. Nevertheless, for larger masses, the width is nearly constant in the range of some hundreds of MeV, and also stays finite in the infinite mass limit.

Having the total decay width and the corresponding composition, one is able to calculate the branching ratios for the decay into light Hagedorn states and/or hadrons. Performing such cascading decay simulations, the resulting hadron multiplicity coincide very well with experimentally observed value[11]. This is visualized in fig. 2, where the multiplicity ratios for strange and multi-strange particles are compared with experimental values measured at LHC. The most striking result is the fact, that the energy spectra of these decay hadrons *look* thermal, i.e. the functional behavior is exponential, as shown in fig. 2. The slope parameter coincides with the underlying Hagedorn temperature.

All these ingredients were implemented into UrQMD [15], which can handle the creation and the decay of Hagedorn states and their component dynamically. This enables us to perform box calculations and study the time evolution and the equilibration of the system. The model is able to handle different initialization scenarios, as e.g. the initialization with only nucleons (or

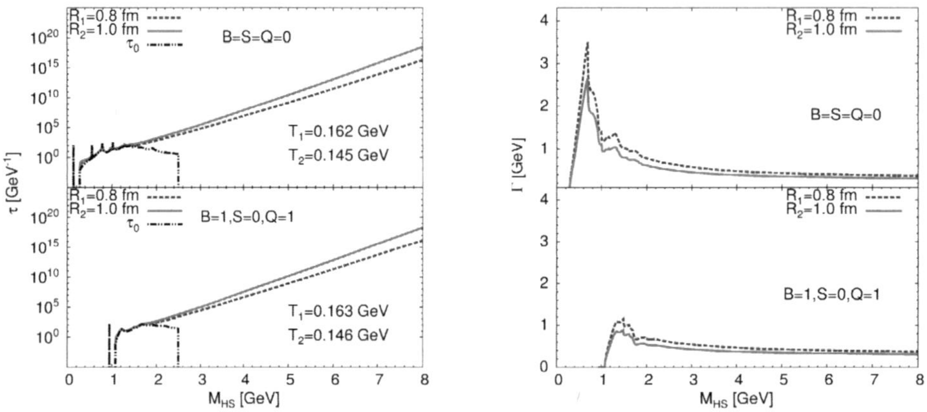

Figure 1. Hagedorn spectrum (left) and total decay width (right) for meson-like ($B = S = Q = 0$) (upper panel) and baryon-like ($B = 1, S = 0, Q = 1$) (lower panel) Hagedorn states. τ_0 denotes the sum of spectral functions of hadrons with the same quantum numbers.

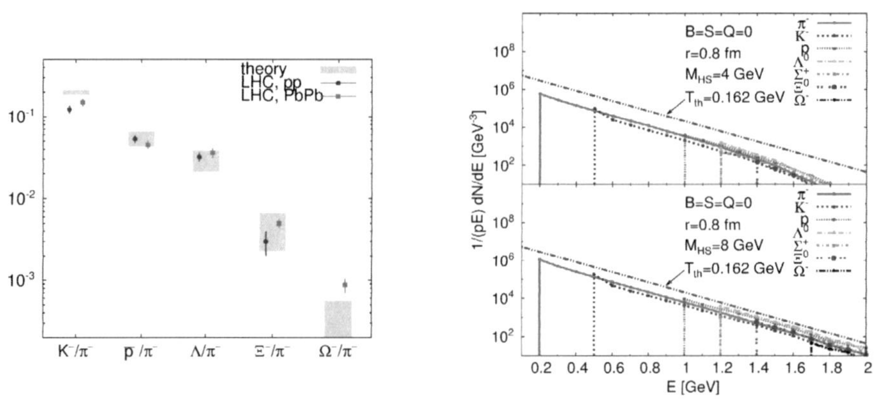

Figure 2. Left: Comparison of particle multiplicity ratios from theory vs. p-p at $\sqrt{s_{NN}} = 0.9\,\text{TeV}$ [16] and Pb-Pb at $\sqrt{s_{NN}} = 2.76\,\text{TeV}$ [17, 18, 19], both from ALICE at LHC. The bands show the range of calculations from cascade decay of a single charge neutral Hagedorn state with radius $R = 0.8\,\text{fm}$ and initial mass $M_{HS} = 4\,\text{GeV}$ and $M_{HS} = 8\,\text{GeV}$. Right: Energy spectra of hadrons stemming from cascade decay of charge neutral Hagedorn state with radius $R = 0.8\,\text{fm}$ and initial mass $M_{HS} = 4\,\text{GeV}$ and $M_{HS} = 8\,\text{GeV}$.

pions) with a given particle and energy density. While this may stand for some 'bottom-up' scenario, alternatively one may start with a given Hagedorn state distribution and look into the resulting hadron distributions. This indicates a 'top-down' scenario, where the Hagedorn states may stem from some preceding deconfined phase.

Fig. 3 shows the time dependence of Omega baryons for box calculations, where only Hagedorn states according to different energy densities were initialized. The given multiplcities at some time are the resulting muliplicities, if all particles would decay down to the stable remnants. One indeed observes very short time scales for the equilibration: For $t \leq 1 - 2\,\text{fm}/c$ the most hadrons have kinetically and chemically equilibrated. Even due to bad statistics for the multi-strange baryon Ω^- it is apparent that the chemical equilibration is established for $t = 5 - 10\,\text{fm}/c$. In addition we find, that the energy spectra of the hadrons again

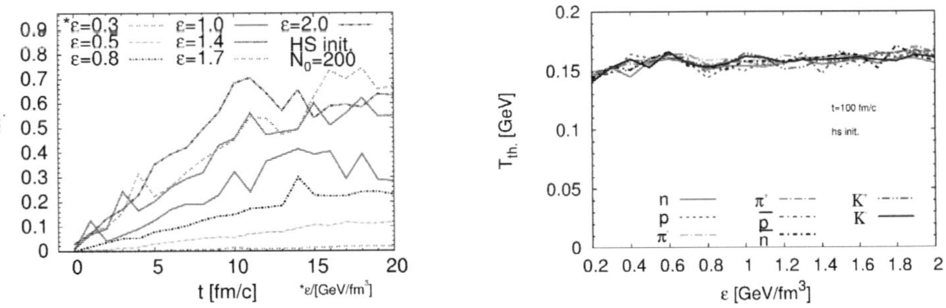

Figure 3. On the left the time dependence of Ω^- baryons in a box calculation is shown, initialized with Hagedorn states, for different initial energy densities as indicated in the plot. On the right the dependence of particle's Boltzmann slopes (temperatures) as function of energy density is depicted.

follow an exponential shape, with slope parameters very close to the Hagedorn temperature $T_H = 162\,\mathrm{MeV}$, see fig. 3.

3. Conclusion

We conclude, that Hagedorn states are a valuable tool for the understanding of the phase transition between a deconfined and a hadronic phase. While already the decay particles of one single Hagedorn state *look* thermal with the Hagedorn temperature as the only scale, a collective system of Hagedorn states leads to small equilibration times of hadrons in the order of few fm/c. Even exotic multi-strange baryons as the Ω^- seem to equilibrate on such short time scales. These remarkable features may also open the door for alternative descriptions of heavy ion collisions, where directly from a pure gluonic phase without quarks a transition to the confined phase is possible [20].

[1] Hagedorn R 1965 *Nuovo Cim. Suppl.* **3** 147–186
[2] Greiner C and Leupold S 2001 *J. Phys. G* **27** L95–L102
[3] Greiner C, Koch-Steinheimer P, Liu F M, Shovkovy I A and Stoecker H 2005 *J. Phys. G* **31** S725–S732
[4] Noronha-Hostler J, Greiner C and Shovkovy I 2008 *Phys. Rev. Lett.* **100** 252301
[5] Noronha-Hostler J, Greiner C and Shovkovy I 2010 *J. Phys. G* **37** 094017
[6] Noronha-Hostler J, Beitel M, Greiner C and Shovkovy I 2010 *Phys. Rev. C* **81** 054909
[7] Noronha-Hostler J, Noronha J and Greiner C 2009 *Phys. Rev. Lett.* **103** 172302
[8] Majumder A and Muller B 2010 *Phys. Rev. Lett.* **105** 252002
[9] Noronha-Hostler J, Noronha J and Greiner C 2012 *Phys. Rev. C* **86** 024913
[10] Jakovac A 2013 *Phys. Rev. D* **88** 065012
[11] Beitel M, Gallmeister K and Greiner C 2014 *Phys. Rev. C* **90** 045203
[12] Yellin J 1973 *Nucl. Phys. B* **52** 583–594
[13] Frautschi S C 1971 *Phys. Rev. D* **3** 2821–2834
[14] Hamer C and Frautschi S C 1971 *Phys. Rev. D* **4** 2125–2137
[15] Bass S A *et al.* 1998 *Prog. Part. Nucl. Phys.* **41** 255–369
[16] Aamodt K *et al.* (ALICE) 2011 *Eur. Phys. J. C* **71** 1594
[17] Abelev B *et al.* (ALICE Collaboration) 2013 *Phys. Rev. C* **88** 044910
[18] Abelev B B *et al.* (ALICE Collaboration) 2013 *Phys. Rev. Lett.* **111** 222301
[19] Abelev B B *et al.* (ALICE Collaboration) 2014 *Phys. Lett. B* **728** 216–227
[20] Stoecker H *et al.* 2016 *J. Phys.* **G43** 015105

IOP Conf. Series: Journal of Physics: Conf. Series **779** (2017) 012077 doi:10.1088/1742-6596/779/1/012077

High baryon densities achievable in the fragmentation regions at RHIC and LHC

Joseph Kapusta and Ming Li

School of Physics and Astronomy, University of Minnesota, Minneapolis, MN 55455 USA

E-mail: kapusta@physics.umn.edu, ml@physics.umn.edu

Abstract. We use the McLerran-Venugopalan model of the glasma energy-momentum tensor to compute the rapidity loss and excitation energy of the colliding nuclei in the fragmentation regions followed by a space-time picture to obtain their energy and baryon densities. At the top RHIC energy we find baryon densities up to 3 baryons/fm^3, which is 20 times that of atomic nuclei. Assuming the formation of quark-gluon plasma, we find initial temperatures of 200 to 300 MeV and baryon chemical potentials of order 1 GeV. Assuming a roughly adiabatic expansion it would imply trajectories in the $T-\mu$ plane which would straddle a possible critical point.

For more than three decades the high energy heavy ion community has been focussed on the central rapidity region following the seminal work of Bjorken [1]. The main reasons are (i) the energy density is expected to be higher there, (ii) the matter is nearly baryon-free making it relevant to the type of matter that existed in the early universe, and (iii) detectors in a collider can more easily measure particle production and correlations in a few units of rapidity around the center-of-momentum. The earlier work of Anishetty, Koehler and McLerran [2], which found that nuclei were compressed by a factor of 3.5 and excited to an energy density of about 2 GeV/fm^3 when they collide at extreme relativistic energies, was pursued only sporadically over the years [3, 4, 5, 6, 7]. Here we briefly report on our recent and ongoing work on this subject. We focus on the baryonic fireballs which emerge at large rapidity in the center-of-momentum frame. Our calculations are not relevant or applicable to the lower energy beam scans at RHIC, nor to future experiments at NICA and FAIR [8].

Consider central collisions of equal mass nuclei. We neglect transverse motion, which should not be important during the fraction of a fm/c time interval of relevance. Then the collision can be thought of as a sum of independent slab collisions each taking place at a particular value of the transverse coordinate $\mathbf{r}_\perp$ with the beam along the z-axis. The projectile slab has a 4-momentum per unit area in the center-of-momentum frame denoted by $\mathcal{P}^\mu_{\rm P} = (\mathcal{E}_{\rm P}, 0, 0, \mathcal{P}_{\rm P})$. Each slab loses energy and momentum to the classical color electric and magnetic fields produced in the region between the two receding nuclei, referred to as the glasma. This loss is determined by $d\mathcal{P}^\mu_{\rm P} = -T^{\mu\nu}_{\rm glasma} d\Sigma_\nu$ where $d\Sigma_\nu = (dz, 0, 0, -dt)$ is the infinitesimal four-vector perpendicular to the hypersurface spanned by dt, dz, and unit transverse area. The energy-momentum of the glasma has the form [9, 10]

$$T^{\mu\nu}_{\text{glasma}} = \begin{pmatrix} \mathcal{A} + \mathcal{B}\cosh 2\eta & 0 & 0 & \mathcal{B}\sinh 2\eta \\ 0 & \mathcal{A} & 0 & 0 \\ 0 & 0 & \mathcal{A} & 0 \\ \mathcal{B}\sinh 2\eta & 0 & 0 & -\mathcal{A} + \mathcal{B}\cosh 2\eta \end{pmatrix}. \tag{1}$$

The $\mathcal{A}$ and $\mathcal{B}$ are known analytical [10] and numerical (for SU(2)) [11] functions of proper time τ, while the dependence on space-time rapidity η follows from the fact that $T^{\mu\nu}_{\text{glasma}}$ is a second-rank tensor in a boost-invariant setting. The longitudinal position z_{P} at each $\mathbf{r}_\perp$ is a function of time. The $z_{\text{P}}(t)$ is related to the time t via the velocity $v_{\text{P}} = dz_{\text{P}}/dt = \tanh y_{\text{P}}$, where y_{P} is the momentum-space rapidity of the slab. The Lorentz invariant effective mass per unit area $\mathcal{M}_{\text{P}}$ is defined via the relations $\mathcal{E}_{\text{P}} = \mathcal{M}_{\text{P}} \cosh y_{\text{P}}$ and $\mathcal{P}_{\text{P}} = \mathcal{M}_{\text{P}} \sinh y_{\text{P}}$. The pair of differential equations describe both the loss of kinetic energy of the projectile nucleus and the internal excitation energy imparted to it during the collision. Thus $\mathcal{M}_{\text{P}}$ is not constant but increases with time, unlike the case of the string model [5].

Initial conditions are needed to solve the equations of motion. For that we turn to hydrodynamical descriptions of collisions at the top RHIC energy of $\sqrt{s_{NN}} = 200$ GeV. Reference [12] assumed that viscous hydrodynamics became applicable at $\tau = 0.6$ fm/c with $\varepsilon(r_\perp = 0, \tau = 0.6 \text{ fm/c}) = 30 \text{ GeV/fm}^3$. Extrapolating back to $\tau = 0$ using the results in [10] gives $\varepsilon_0 \equiv \varepsilon_0(r_\perp = 0, \tau = 0) = 123$, 142, and 158 GeV/fm^3 for UV cutoffs of $Q = 3$, 4 and 5 GeV, respectively. We use a Wood-Saxon nuclear density distribution ρ_{A} and assume as usual that the initial energy density is proportional to the square of the thickness function $\int_{-\infty}^{\infty} dz \rho_{\text{A}}(r_\perp, z)$

Figure 1 shows the momentum-space rapidity y_{P} of the central core of a gold nucleus as a function of proper time τ. The central core loses about 3 units of rapidity within the first 0.1 to 0.2 fm/c; this is a robust result, insensitive to the value of Q. When averaged over the whole nucleus the baryon rapidity loss is about 2.4. For 0-10% centrality BRAHMS [13, 14] found an average rapidity loss of about $2.05 + 0.4/-0.6$. This is consistent with our result, especially since we focus on 0% centrality for illustration. Figure 2 shows the excitation energy per baryon in units of the nucleon mass as a function of proper time. There is a slow but monotonic increase, unlike the rapidity loss whose asymptotic limit is reached within a few tenths of a fm/c.

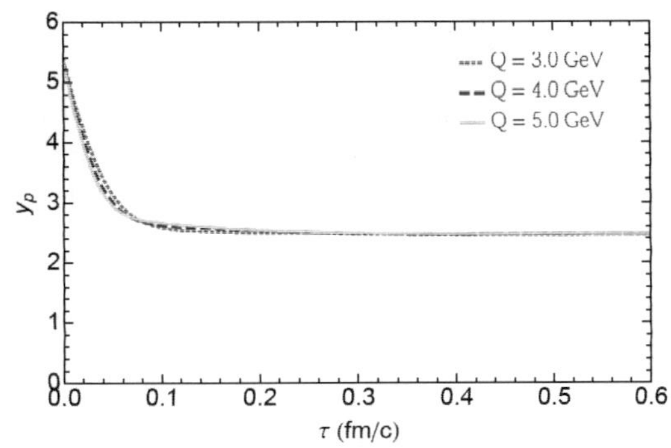

Figure 1. Rapidity of the central core of a Au projectile nucleus in the center-of-momentum frame for $\sqrt{s_{NN}} = 200$ GeV as a function of proper time. The result is insensitive to the choice of Q.

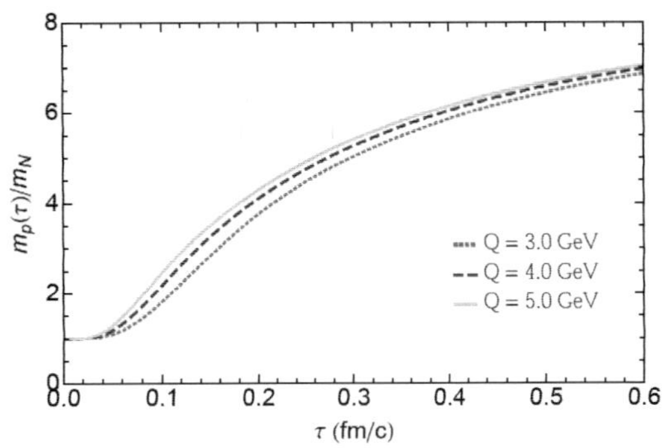

Figure 2. Excitation energy per baryon in the central core of a Au projectile nucleus in the center-of-momentum frame for $\sqrt{s_{NN}} = 200$ GeV as a function of proper time. The result is mildly sensitive to the choice of Q.

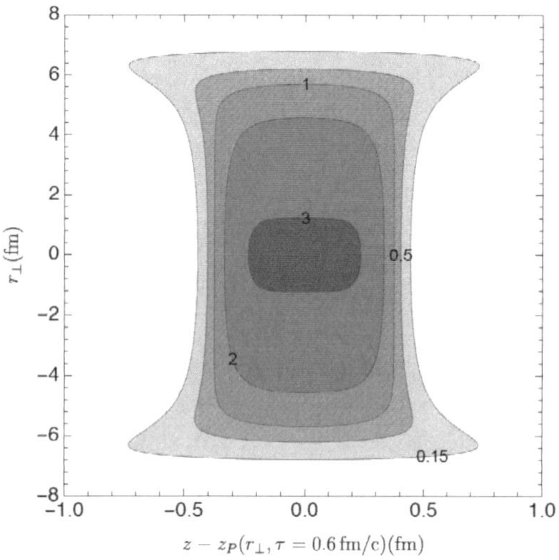

Figure 3. Contour plot of the proper baryon density for central collisions of Au nuclei at $\sqrt{s_{NN}} = 200$ GeV. The units are baryons per fm^3. The horizontal axis measures the distance in the local rest frame.

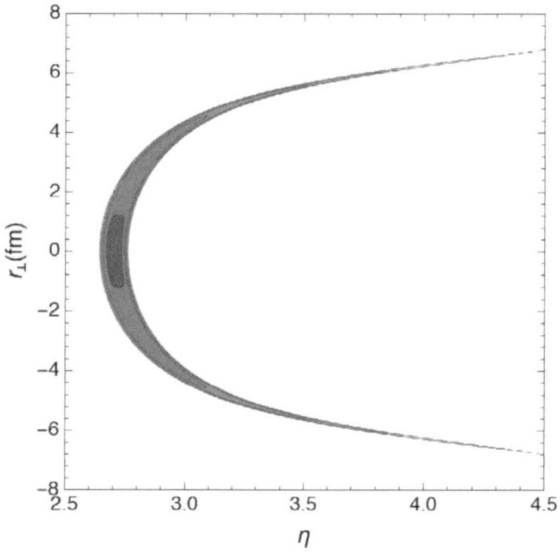

Figure 4. Contour plot of the proper baryon density at $\tau = 0.6$ fm/c for central collisions of Au nuclei at $\sqrt{s_{NN}} = 200$ GeV. The units are baryons per fm^3. The horizontal axis is the space-time rapidity.

The McLerran-Venugopalan model assumes that the nuclei can be treated as infinitesimally thin slabs. It does not address the space-time evolution of the individual nuclei. Anishetty *et al.* [2] gave a very simple and direct argument that the nuclear matter would be compressed by a factor of $\exp(\Delta y)$ where $\Delta y > 0$ is the rapidity loss (gain) of the projectile (target). It is a Lorentz invariant quantity which follows from the infinitely thin projectile sweeping through the target in the target rest frame. The argument was verified in a specific model in [4]. The local baryon density is $n_B(r_\perp, z') = \mathrm{e}^{\Delta y(r_\perp)} \rho_A \left(r_\perp, z' \mathrm{e}^{\Delta y(r_\perp)}\right)$ where $z' = z - z_P(r_\perp)$, all evaluated at $\tau = 0.6$ fm/c. Contours of the resulting baryon density are shown in fig. 3 when plotted in the z-direction and in fig. 4 when plotted against space-time rapidity.

The baryon rapidity distribution at $\tau = 0.6$ fm/c is shown in fig. 5. Also shown is the distribution smeared by a Gaussian with a width given by $\sqrt{T/m} = 0.16$.

Assume that the matter in the fireball does equilibrate on the time scale of 0.6 fm/c as argued in [2]. For simplicity consider a massless gas of noninteracting up, down, and strange quarks and gluons. The net strangeness in the fireball is zero. Let the up and down quark chemical potentials be equal to each other and to 1/3 of the baryon chemical potential μ_B. The equation of state is then given by the pressure as a function of temperature and baryon chemical potential by $P(T, \mu_B) = \frac{19\pi^2}{36} T^4 + \frac{1}{9} T^2 \mu_B^2 + \frac{1}{162\pi^2} \mu_B^4$. The energy density decreases with $r_\perp$ faster than the baryon density. Illustrative numbers are given in Table 1. It would be expected that the hydrodynamic expansion of the fireball would be approximately adiabatic, just as in the central rapidity region. If that is the case, then the values of the entropy per baryon estimated above would be in just the right range for the trajectories of the fluid elements in the $T - \mu_B$ plane to pass near or even through a possible critical point [16].

In conclusion, we have employed the McLerran-Venugopalan model to calculate the energy/rapidity loss of baryons in high energy heavy ion collisions. We found that the baryon densities in the fireballs outside the central rapidity region attain values an order of magnitude

Table 1. Illustrative results for central collisions of Au nuclei at $\sqrt{s_{NN}} = 200$ GeV. Uncertainties of 33% in s and n_B result in uncertainties in T and μ_B of 10%.

$r_\perp$(fm)	n_B(baryons/fm^3)	ε_P(GeV/fm^3)	T(MeV)	μ_B(MeV)	s/n_B
0	3.0	20.0	299	1061	26.2
5.25	1.5	5.5	205	1007	18.9

greater than normal nuclear matter. These findings suggest that further theoretical and experimental studies be done to probe the equation of state at the highest baryon densities achievable in a laboratory setting.

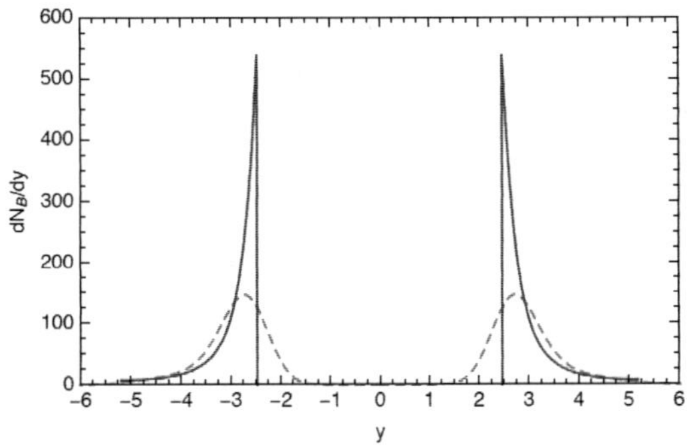

Figure 5. Baryon rapidity distribution at $\tau = 0.6$ fm/c with and without thermal smearing.

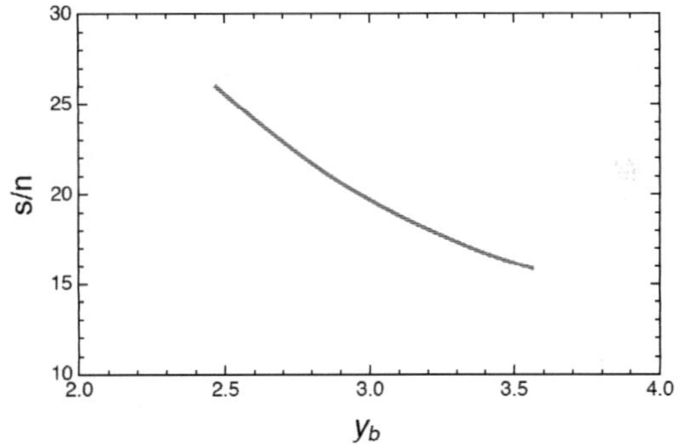

Figure 6. Entropy per baryon rapidity distribution at $\tau = 0.6$ fm/c. Beam energy or rapidity scan?

Acknowledgments

This work was supported by the U.S. Department of Energy grant DE-FG02-87ER40328.

References

[1] Bjorken J D 1983 *Phys. Rev.* D **27** 140
[2] Anishetty R, Koehler P and McLerran L 1980, *Phys. Rev.* D **22** 2793
[3] Csernai L P 1984 *Phys. Rev.* D **29** 1945
[4] Gyulassy M and Csernai L P 1986 *Nucl. Phys.* A **460** 723
[5] Mishustin I N and Kapusta J I 2002 *Phys. Rev. Lett.* **88** 112501
[6] Frankfurt L and Strikman M 2003 *Phys. Rev. Lett.* **91** 022301; Strikman M 2006 *AIP Conf. Proc.* **842**, 92
[7] Mishustin I N and Lyakhov K A 2007 *Phys. Rev.* C **76** 011603
[8] *The CBM Physics Book: Compressed Baryonic Matter in Laboratory Experiments* 2011 ed B Friman *et al* (Springer) Lect. Notes Phys. **814** pp 1-980
[9] Chen G, Fries R J, Kapusta J I and Li Y 2015 *Phys. Rev.* C **92**, 064912
[10] Li M and Kapusta J I 2016 *Phys. Rev.* C **94**, 024908
[11] Epelbaum T and Gelis F 2013 *Phys. Rev. Lett.* **111**, 232301
[12] Song H, Bass S A and Heinz U W 2011 *Phys. Rev.* C **83**, 024912
[13] Bearden I G *et al* [BRAHMS Collaboration] 2004 *Phys. Rev. Lett.* **93** 102301
[14] Arsene I C *et al* [BRAHMS Collaboration] 2009 *Phys. Lett.* B **677** 267
[15] Kurkela A and Vuorinen A 2016 *Phys. Rev. Lett.* **117**, 042501
[16] Asakawa M, Bass S A, Müller B and Nonaka C 2008 *Phys. Rev. Lett.* **101**, 122302

IOP Conf. Series: Journal of Physics: Conf. Series **779** (2017) 012078 doi:10.1088/1742-6596/779/1/012078

Analysis of hadron yield data within hadron resonance gas model with multi-component eigenvolume corrections

Volodymyr Vovchenko[1,2,3], **Horst Stoecker**[1,2,4]

[1]Frankfurt Institute for Advanced Studies, Goethe Universität Frankfurt, D-60438 Frankfurt am Main, Germany

[2]Institut für Theoretische Physik, Goethe Universität Frankfurt, D-60438 Frankfurt am Main, Germany

[3]Department of Physics, Taras Shevchenko National University of Kiev, 03022 Kiev, Ukraine

[4]GSI Helmholtzzentrum für Schwerionenforschung GmbH, D-64291 Darmstadt, Germany

E-mail: vovchenko@fias.uni-frankfurt.de

Abstract. We analyze the sensitivity of thermal fits to heavy-ion hadron yield data of ALICE and NA49 collaborations to the systematic uncertainties in the hadron resonance gas (HRG) model related to the modeling of the eigenvolume interactions. We find a surprisingly large sensitivity in extraction of chemical freeze-out parameters to the assumptions regarding eigenvolumes of different hadrons. We additionally study the effect of including yields of light nuclei into the thermal fits to LHC data and find even larger sensitivity to the modeling of their eigenvolumes. The inclusion of light nuclei yields, thus, may lead to further destabilization of thermal fits. Our results show that modeling of eigenvolume interactions plays a crucial role in thermodynamics of HRG and that conclusions based on a non-interacting HRG are inconclusive.

1. Introduction

Thermodynamic models have long been employed to estimate the temperatures reached in the relativistic heavy-ion collisions [1, 2, 3, 4]. Particularly the hadron-resonance gas (HRG) model has been quite successful in describing the hadron multiplicities data in relativistic nucleus-nucleus collisions [5, 6], possibly with additional introduction of parameters to regulate deviations from full chemical equilibrium [7, 8, 9].

In its simplest version the HRG is described as a gas of non-interacting hadrons and resonances. In a more realistic HRG model one has to take into account the attractive and repulsive interactions between hadrons. In this work we study the influence of short-range repulsive interactions between hadrons on thermal fits to hadron yield data.

2. Model description

The repulsive interactions between hadrons are modeled via eigenvolume (EV) correction of the van der Waals type [10, 11, 12, 13]. For a multi-component case there is more than a single way to model the eigevolume corrections. In our analysis we use two different formulations: the "Diagonal" eigenvolume HRG model from [14], and the "Crossterms" EV model introduced in [15] (see Ref. [16] for technical details). The "Diagonal" model is simpler but it is not consistent with the 2nd order virial expansion of the equation of state of hard spheres. The "Crossterms" model, on the other hand, while technically more involved, is consistent with that expansion.

In our analysis we include the established strange and non-strange hadrons listed in the Particle Data Tables [17], along with their decay branching ratios. This includes mesons up to $f_2(2340)$ and (anti)baryons up to $N(2600)$. The finite width of the resonances is taken into account by adding the additional integration over their Breit-Wigner shapes. The feed-down from strong and electromagnetic decays of the unstable resonances to the total hadron yields is included in the standard way. The full chemical equilibrium assumed in this work, i.e. we do not consider here possible under- or over-saturation of the light and/or strange quarks which improves the data description but also introduces additional parameters (see, e.g., Ref. [7]).

3. Mass-proportional eigenvolume at finite μ_B

First a bag-model inspired parametrization with the hadron eigenvolume proportional to its mass through a bag-like constant is employed, i.e.

$$v_i = m_i/\varepsilon_0. \tag{1}$$

Such eigenvolume parametrization had been obtained for the heavy Hagedorn resonances, and was used to describe their thermodynamics [10, 12], their effect on particle yield ratios [18], and compared to lattice QCD [19].

The EV-HRG model fits are using the hadron yield data of the NA49 collaboration, which include 4π yields of charged pions, charged kaons, Ξ^-, Ξ^+, Λ, ϕ, and, if available, Ω, $\bar{\Omega}$, measured in most central Pb+Pb collisions at $\sqrt{s_{\rm NN}} = 6.3, 7.6, 8.8, 12.3,$ and 17.3 GeV [20, 21, 22, 23, 24, 25, 26, 27]. Additionally, the data on the total number of participants, N_W, is identified with total net baryon number and is included in the fit.

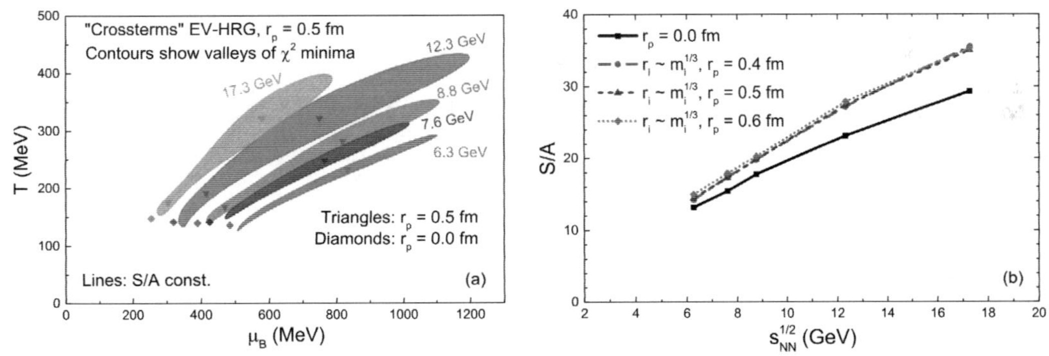

Figure 1. (a) Regions in the T-μ_B plane where the "Crossterms" eigenvolume HRG model with bag-like constant ε_0 fixed to reproduce the hard-core radius of $r_p = 0.5$ fm yields a better fit to the NA49 data as compared to the point-particle HRG model. The solid lines show the isentropic curves for the eigenvolume model, which go through the global χ^2 minima. (b) Collision energy dependence of entropy per baryon at global minima of thermal fits to NA49 data.

For illustration purposes we consider here only the "Crossterms" model and we fix the ε_0 to reproduce the hard-core proton radius of $r_p = 0.5$ fm. A much more detailed study regarding fits with mass-proportional eigenvolumes can be found in [16]. Figure 1a depicts the regions in the T-μ_B plane where the fit quality of NA49 data in the EV model is better than in the non-interacting HRG. The location of the fit minima within non-interacting HRG are shown by diamonds, and are consistent with the systematics established in numerous previous studies. At all five NA49 energies wide regions of improved χ^2 values are observed at high temperatures and chemical potentials. At given bombarding energy, for a given set of radii, the EV HRG model fits do not just yield a single $T - \mu_B$ pair, but a whole range of $T - \mu_B$ pairs, each with similarly good fit quality. These pairs form a valley in the $T - \mu_B$ plane along a line of nearly constant entropy per baryon, S/A. We also show the energy dependence of values of S/A (Fig. 1b). The S/A, extracted at different energies, is a robust observable: it is almost independent of the details of the modeling of the EV interactions and of the specific $T - \mu_B$ values obtained.

4. Two-component model and role of light nuclei

We also consider parameterization where all mesons are assumed to be point-like and where all baryons have a fixed hard-core radius $r_p = 0.3$ fm. As demonstrated in [28] this parameterization yields a good description of lattice data on the pressure at least up to $T = 190$ MeV. It was also shown in [29] that it leads to an irregular χ^2 profile of thermal fit to ALICE hadron yield data with wide double minimum structure in the 155-210 MeV temperature range. Here we investigate how inclusion of light nuclei influences the fit. For that we consider the 0-10% centrality ALICE data which has a rich amount of data regarding yields of light nuclei. The actual data used for fitting includes midrapidity yields of charged pions, charged kaons, and (anti)protons [30], (anti)Ξ^- and (anti)Ω [31], (anti)Λ and K^0_S [32, 33], ϕ [34], (anti)deuterons and (anti)^{3}He [35], and (anti)$^3_\Lambda$H [36]. The addition of light nuclei requires additional assumptions regarding their eigenvolumes. First we consider only addition of (anti)deuterons and compare two cases: (1) deuteron has same eigenvolume as baryons, i.e. $r_p = r_d = 0.3$ fm; (2) deuteron has a twice larger eigenvolume compared to baryons, i.e. $r_d \simeq 0.38$ fm. The results of the calculations of χ^2 temperature profile are depicted in Fig. 2a. Note that μ_B is also fitted at each T and found to be consistent with zero within fit errors.

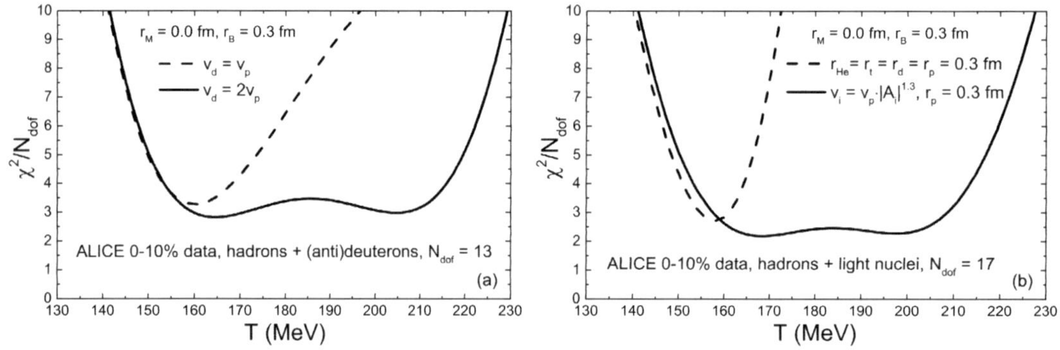

Figure 2. The temperature dependence of $\chi^2/N_{\rm dof}$ of fit to ALICE data on hadron + light nuclei yields in 0-10% most central Pb+Pb collisions at 2.76 TeV within eigenvolume HRG model with point-like mesons, baryons with hard-core radius of 0.3 fm, and different assumptions for eigenvolumes of light nuclei. In (a) only (anti)deuterons are added to fit with $r_d = r_p = 0.30$ fm (dashed line) and $r_d = 2^{1/3}\, r_p \simeq 0.38$ fm (solid line). In (b) light nuclei up to $^3_\Lambda$H are added with $r_{\rm He} = r_t = r_d = r_p = 0.3$ fm (dashed), and $v_i = v_p \cdot |A_i|^{1.3}$ (solid), where $|A_i|$ is mass number.

If deuterons are assumed to have same eigenvolume as protons then the χ^2 profile has a regular structure with a minimum at $T \simeq 161$ MeV. Thus, even though inclusion of light nuclei may be questionable in general, it would seem that even inclusion of only deuterons into thermal fit would stabilise them with regards to the modeling of eigenvolume corrections. This conclusion, however, is illusory. Changing the deuteron eigenvolume to a physically more motivated value equal to twice that of proton one gets a very different χ^2 profile: a two-minimum structure in a wide 155-210 MeV temperature range. This change is attributed to a larger suppression of (anti)deuteron yield at higher temperatures due to larger eigenvolume. It is remarkable that fit can be so sensitive to the properties of only single particle species. Similarly, the inclusion of yields of ^{3}He and $^3_\Lambda$H does not help to stabilise the fit, as seen in Fig. 2b. This is so despite the remarkably large number of degrees of freedom $N_{\rm dof} = 17$. We therefore conclude that introduction of light nuclei into thermal fits leads to a further destabilisation of the fit as it requires non-trivial assumptions regarding their eigenvolumes.

5. Conclusions

It is shown that thermal fits are very sensitive to the details of the modeling of the eigenvolume interactions. At given collision energy, the EV HRG model fits do not just yield a single $T - \mu_B$ pair, but a whole range of $T - \mu_B$ pairs, each with similarly good fit quality. The entropy per baryon, on the other hand, appears to be a robust observable: its extracted value depends weakly on the modeling of hadron eigenvolumes. Finally, we explore the effect of including light nuclei into thermal fits and, somewhat surprisingly, find that results are even more sensitive to the assumptions regarding their eigenvolume parameters.

References

[1] Stoecker H, Ogloblin A A and Greiner W 1981 *Z. Phys.* A **303** 259
[2] Molitoris J J, Hahn D and Stoecker H 1985 *Prog. Part. Nucl. Phys.* **15** 239
[3] Stoecker H and Greiner W 1986 *Phys. Rept.* **137** 277
[4] Hahn D and Stocker H 1987 *Phys. Rev.* C **35** 1311
[5] Cleymans J and Satz H 1993 *Z. Phys.* C **57** 135
[6] Andronic A, Braun-Munzinger P and Stachel J 2006 *Nucl. Phys.* A **772** 167
[7] Letessier J and Rafelski J 2008 *Eur. Phys. J.* A **35** 221
[8] Becattini F, Manninen J and Gazdzicki M 2006 *Phys. Rev.* C **73** 044905
[9] Petrán M, Letessier J, Petráĉek V and Rafelski J 2013 *Phys. Rev.* C **88** 034907
[10] Hagedorn R and Rafelski J 1980 *Phys. Lett.* B **97** 136
[11] Gorenstein M I, Petrov V K and Zinovev G M 1981 *Phys. Lett.* B **106** 327
[12] Kapusta J I and Olive K A 1983 *Nucl. Phys.* A **408** 478
[13] Rischke D H, Gorenstein M I, Stoecker H and Greiner W 1991 *Z. Phys.* C **51** 485
[14] Yen G D, Gorenstein M I, Greiner W and Yang S N 1997 *Phys. Rev.* C **56** 2210
[15] Gorenstein M I, Kostyuk A P and Krivenko Y D 1999 *J. Phys.* G **25** L75
[16] Vovchenko V and Stoecker H 2016 *Preprint* 1606.06218
[17] Olive K A *et al.* [Particle Data Group Collaboration] 2014 *Chin. Phys.* C **38** 090001
[18] Noronha-Hostler J, Noronha J and Greiner C 2010 *J. Phys.* G **37** 094062
[19] Albright M, Kapusta J and Young C 2014 *Phys. Rev.* C **90** 024915
[20] Afanasiev S V *et al.* [NA49 Collaboration] 2002 *Phys. Rev.* C **66** 054902
[21] Alt C *et al.* [NA49 Collaboration] 2006 *Phys. Rev.* C **73** 044910
[22] Alt C *et al.* [NA49 Collaboration] 2008 *Phys. Rev.* C **77** 024903
[23] Alt C *et al.* [NA49 Collaboration] 2008 *Phys. Rev.* C **78** 034918
[24] Alt C *et al.* [NA49 Collaboration] 2008 *Phys. Rev.* C **78** 044907
[25] Alt C *et al.* [NA49 Collaboration] 2005 *Phys. Rev. Lett.* **94** 192301
[26] Anticic T *et al.* [NA49 Collaboration] 2012 *Phys. Rev.* C **85** 044913
[27] Friese V [NA49 Collaboration] 2002 *Nucl. Phys.* A **698** 487
[28] Andronic A, Braun-Munzinger P, Stachel J and Winn M 2012 *Phys. Lett.* B **718** 80
[29] Vovchenko V and Stoecker H 2015 *Preprint* 1512.08046
[30] Abelev B *et al.* [ALICE Collaboration] 2013 *Phys. Rev.* C **88** 044910
[31] Abelev B B *et al.* [ALICE Collaboration] 2014 *Phys. Lett.* B **728** 216
[32] Abelev B B *et al.* [ALICE Collaboration] 2013 *Phys. Rev. Lett.* **111** 222301
[33] Schuchmann S 2015 *PhD Thesis* (University of Frankfurt)
[34] Abelev B B *et al.* [ALICE Collaboration] 2015 *Phys. Rev.* C **91** 024609
[35] Adam J *et al.* [ALICE Collaboration] 2016 *Phys. Rev.* C **93** 024917
[36] Adam J *et al.* [ALICE Collaboration] 2016 *Phys. Lett.* B **754** 360

On the Use of the Tsallis Distribution at LHC Energies.

J. Cleymans

UCT-CERN Research Centre and Physics Department, University of Cape Town, South Africa

E-mail: jean.cleymans@uct.ac.za

Abstract. Numerous papers have appeared recently showing fits to transverse momentum (p_T) spectra measured at the Large Hadron Collider (LHC) in proton - proton collisions.This talk focuses on the fits extending to very large values of the transverse momentum with p_T values up to 200 GeV/c as measured by the ATLAS and CMS collaborations at $\sqrt{s} = 0.9$ and 7 TeV. A thermodynamically consistent form of the Tsallis distribution is used for fitting the transverse momentum spectra at mid-rapidity. The fits based on the proposed distribution provide an excellent description over 14 orders of magnitude. Despite this success, an ambiguity is noted concerning the determination of the parameters in the Tsallis distribution. This prevents drawing firm conclusions as to the universality of the parameters appearing in the Tsallis distribution.

It was shown in numerous recent publications that fits based on the Tsallis distribution [1] give a good description of transverse momentum distributions measured at the LHC [2, 3, 4, 5, 6, 7, 8, 9, 10]. Some of these fits extend to values of p_T up to 200 GeV/c [11, 12, 13, 14, 15] and provide an excellent description over 14 orders of magnitude in the transverse momentum spectrum.

Of the many forms proposed, one in particular [16, 17] leads to a consistent version of thermodynamics for the particle number, energy density and pressure. In particular, the momentum distribution of particles is given by the following expression

$$E\frac{dN}{d^3p} = gVE\frac{1}{(2\pi)^3}\left[1 + (q-1)\frac{E-\mu}{T}\right]^{-\frac{q}{q-1}}. \tag{1}$$

where g is the degeneracy factor, E the energy, p the momentum, V the volume of the system and μ is the chemical potential. The advantages are that all thermodynamic consistency conditions are satisfied:

$$N = V\left.\frac{\partial P}{\partial \mu}\right|_{T,V} \qquad \text{etc...},$$

and the parameter T truly deserves its name since

$$T = \left.\frac{\partial E}{\partial S}\right|_{V,N}.$$

In terms of the rapidity and transverse mass variables, $E = m_T \cosh y$, Eq. (1) becomes (at mid-rapidity $y = 0$ and for $\mu = 0$)

$$\left.\frac{d^2N}{dp_T\,dy}\right|_{y=0} = gV\frac{p_T m_T}{(2\pi)^2}\left[1+(q-1)\frac{m_T}{T}\right]^{-\frac{q}{q-1}}. \tag{2}$$

Integrating the above expression over the rapidity variable leads to the equivalent form first derived in [18]

$$\begin{aligned}\left.\frac{d^2N}{dp_T\,dy}\right|_{y=0} &= \left.\frac{dN}{dy}\right|_{y=0}\frac{p_T m_T}{T}\left[1+(q-1)\frac{m_T}{T}\right]^{-q/(q-1)}\\ &\times\frac{(2-q)(3-2q)}{(2-q)m^2+2mT+2T^2}\\ &\times\left[1+(q-1)\frac{m}{T}\right]^{1/(q-1)}\end{aligned} \tag{3}$$

or, showing only the dependence on the transverse variables:

$$\begin{aligned}\left.\frac{d^2N}{dp_T\,dy}\right|_{y=0} &= \left.\frac{dN}{dy}\right|_{y=0} p_T\frac{m_T}{T}\left[1+(q-1)\frac{m_T}{T}\right]^{-q/(q-1)}\\ &\times(\text{factors independent of } p_T)\end{aligned} \tag{4}$$

At large transverse momenta the asymptotic behavior is

$$\lim_{p_T\to\infty}\left.\frac{d^2N}{dp_T\,dy}\right|_{y=0} \propto p_T\left[\frac{p_T}{T}\right]^{-1/(q-1)}$$

which shows that the scale is being set by the temperature T and the asymptotic behavior is set by q. It is highly sensitive to small deviations from 1 in this last variable. The upper limit for q is given by

$$q < \frac{4}{3}. \tag{5}$$

For larger values of q the integrals become divergent [19].

The above power law has been used to fit the p_T spectra of charged particles measured by the ATLAS [20] and CMS [21] collaborations in [14]. The ATLAS collaboration has reported the transverse momentum in an inclusive phase space region taking into account at least two charged particles in the kinematic range $|\eta| < 2.5$ and $p_T > 100$ MeV [20]. The CMS collaboration has presented the differential transverse momentum distribution covering a p_T range up to 200 GeV/c, the largest range ever measured in a colliding beam experiment [21]. The fits are presented in Figs. (1) and (2).

The results can be compared to those obtained in [7, 11, 12] where very good fits to transverse momentum distributions were presented. The quality of the fits is confirmed albeit with different values of the parameters since a different version is being used. The resulting parameters are listed in table 1. To show the quality of the fits we also present the ratio of the measured values over the fitted values in Fig. (3). It is quite remarkable that the transverse momentum distributions measured up to 200 GeV/c in p_T can be described consistently over 14 orders of magnitude by a straightforward Tsallis distribution.

Despite this success, an ambiguity has been noted concerning the determination of the parameters [23] in the Tsallis distribution. This is shown explicitly in Fig. (3) where fits are presented to the distributions of proton measured by the CMS collaboration [24]. In the figure,

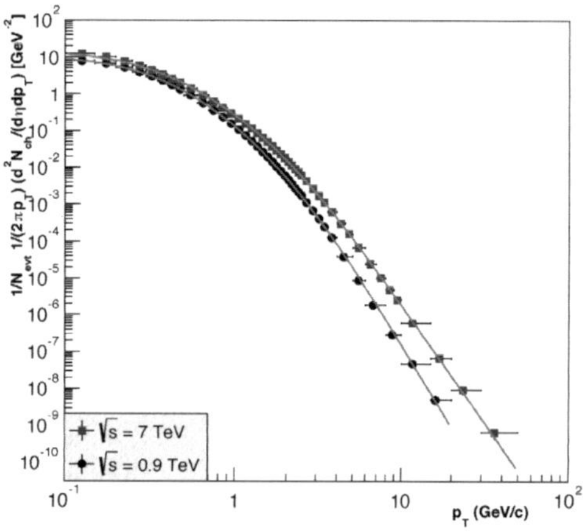

Figure 1. Charged particle multiplicities as a function of the transverse momentum measured by the ATLAS collaboration for events with $n_{ch} \geq 2$, p_T >100 MeV and $|\eta| < 2.5$ at $\sqrt{s} = 0.9$ and 7 TeV in proton - proton collisions [20] fitted with Tsallis distribution [14].

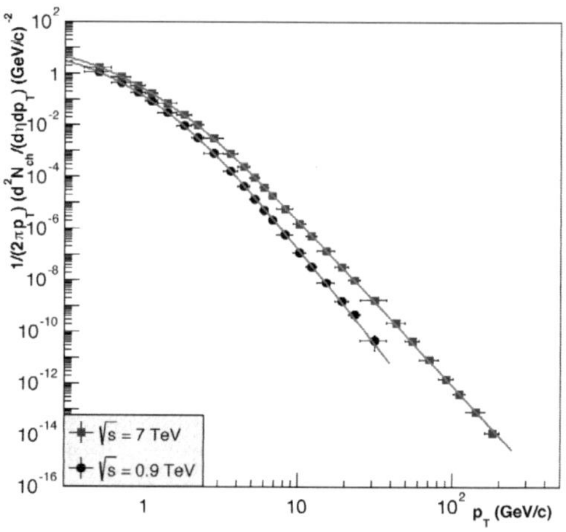

Figure 2. Charged particle differential transverse momentum yields measured within $|\eta| < 2.4$ by the CMS collaboration in proton - proton collisions at $\sqrt{s} = 0.9$ and 7 TeV [21] fitted with the Tsallis distribution [14].

Experiment	$\sqrt{s}$ **(TeV)**	q	T **(MeV)**	R **(fm)**	χ^2/NDF
ATLAS	0.9	1.129 ± 0.005	74.21 ± 3.55	4.62 ± 0.29	0.66/36
ATLAS	7	1.150 ± 0.002	75.00 ± 3.21	5.05 ± 0.07	4.35/41
CMS	0.9	1.129 ± 0.003	76.00 ± 0.17	4.32 ± 0.29	0.65/17
CMS	7	1.153 ± 0.002	73.00 ± 1.42	5.04 ± 0.27	0.52/24

Table 1. Values of the q, T and R parameters and χ^2/NDF obtained from fits to the p_T spectra measured by the ATLAS [20] and CMS [21] collaborations.

as an example, three different sets of variables are shown reproducing the measurements. This shows clearly that a larger interval in the transverse momentum is needed before the parameters can be determined accurately. The solid black line corresponds to a temperature $T = 29.6$ MeV, the dashed line has $T = 52.3$ MeV and the dash-dotted line has $T = 73.00$ MeV as determined from the fit to charged particles at the same energy. The first two choices for T and the corresponding changes in q and dN/dy give equally acceptable fits despite the large change especially in the temperature T. Thus no firm conclusion can be drawn at present as to the universality of the parameters appearing in the Tsallis distribution as given above.

References

[1] C. Tsallis, J. Stat. Phys. **52** (1988) 479-487.
[2] G. Biró, G.G. Barnaföldi, T.S. Biró, K.Ürmössy, arXiv:1608.0143.
[3] H. Zheng, L. Zhu, Adv. High Energy Physics, 2016 (2016) 9632126.
[4] Y-Q Gao, F-H Liu, Indian J. Phys. **90** (2016) 319.
[5] H. Zheng, L. Zhu, Adv. High Energy Physics, 2015 (2015).
[6] H. Zheng, L. Zhu, A. Bonasera, Phys. Rev. **D 92** (2015) 074009.
[7] G. Wilk, Z. Wlodarczyk, Entropy **17** (2015) 384.

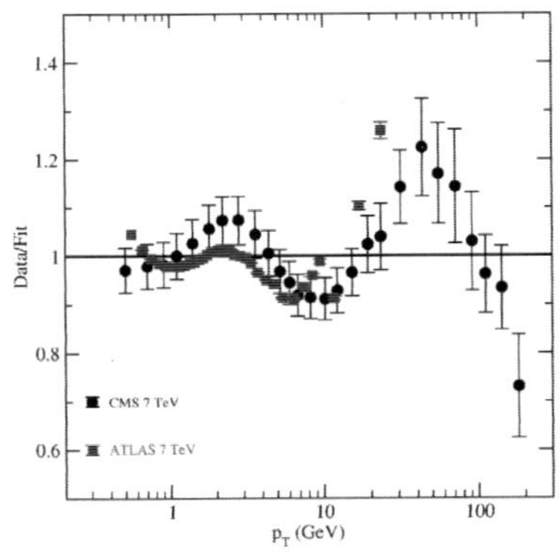

Figure 3. Ratio of the transverse momenta of charged particles as measured by the CMS [22] and ATLAS [20] collaborations in $p-p$ collisions at $\sqrt{s} = 7$ TeV and the fit values listed in Table 1.

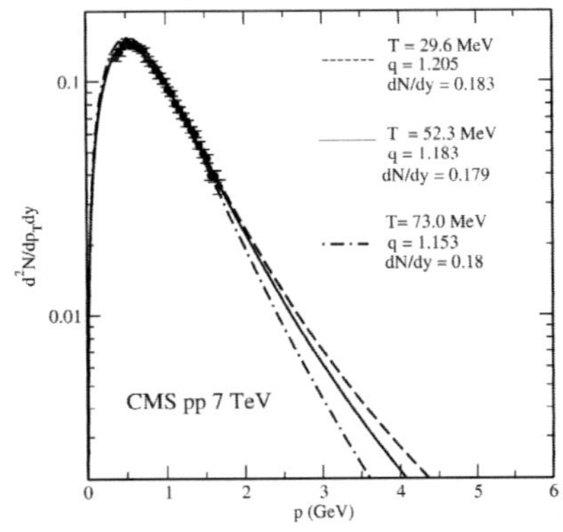

Figure 4. Distribution of protons measured by the CMS collaboration [24] in $p-p$ collisions at 7 TeV. Three different Tsallis fits are shown with the parameters indicated in the figure.

[8] L. Marques, J. Cleymans, A. Deppman, Phys. Rev. **D 91** (2015) 054025.
[9] K. Ürmössy, G. G. Barnaföldi, S. Harangozó, T. S. Biró and Z. Xu, arXiv:1501.02352.
[10] J. Cleymans, G. I. Lykasov, A. S. Parvan, A. S. Sorin, O. V. Teryaev, D. Worku, Phys. Lett. **B 723** 351, 2013.
[11] C.Y. Wong, G. Wilk Phys. Rev. **D 87** (2013) 114007.
[12] C.Y. Wong, G. Wilk, Acta Phys. Polonica **B 43** (2012) 2047.
[13] C. Y. Wong, G. Wilk, L. J. L. Cirto, C. Tsallis, Phys. Rev. **D 91** (2015) 114027.
[14] M.D. Azmi and J. Cleymans, Eur. Phys. J. **C 75** (2015) 430.
[15] M. D. Azmi, J. Cleymans, J. Phys. **G 41**, 065001, 2014.
[16] J. Cleymans and D. Worku, J. Phys. **G 39**, 025006 (2012).
[17] J. Cleymans and D. Worku, Eur. Phys. J. **A 48**, 160 (2012).
[18] M. Rybczynski, Z. Wlodarczyk, Eur. Phys. J. **C 74**, 2785 (2014).
[19] T. Bhattacharyya, J. Cleymans, S. Mogliacci, arXiv:1608.08965.
[20] ATLAS Collaboration, G.Aad et al. New J. Phys. **13** (2011) 053033.
[21] CMS collaboration, S. Chatrchyan, et al. , JHEP **1108**, 086 (2011).
[22] CMS Collaboration, V.Khachatryan et al. JHEP **02**, 041 (2010) & JHEP 05, 064 (2011); Eur. Phys. J. **C** 72, 2164, 2012.
[23] L. Barnby, private communication.
[24] CMS collaboration, Eur. Phys. J. **C 72** (2012) 2164.

IOP Conf. Series: Journal of Physics: Conf. Series **779** (2017) 012080 doi:10.1088/1742-6596/779/1/012080

Updates to the p+p and A+A chemical freeze-out lines from the new experimental data

V. V. Begun[1], V. Vovchenko[2,3,4], M. I. Gorenstein[2,5]

[1] Institute of Physics, Jan Kochanowski University, PL-25406 Kielce, Poland
[2] Frankfurt Institute for Advanced Studies, Goethe Universität Frankfurt, D-60438 Frankfurt am Main, Germany
[3] Institut für Theoretische Physik, Goethe Universität Frankfurt, D-60438 Frankfurt am Main, Germany
[4] Department of Physics, Taras Shevchenko National University of Kiev, 03022 Kiev, Ukraine
[5] Bogolyubov Institute for Theoretical Physics, 03680 Kiev, Ukraine

E-mail: viktor.begun@gmail.com

Abstract. We show that the new data on mean multiplicities measured in p+p and A+A collisions together with the updated list of resonances lead to the significant changes of the obtained freeze-out lines. The new A+A line gives much smaller temperatures at high collision energies and agrees with the values obtained at the LHC. The newly obtained p+p line is much closer to the A+A line than previously expected, and even touches it in the region where the K^+/π^+ horn appears in the data. It indicates that the temperatures that will be obtained in the beam energy and system size scan by the NA61/SHINE Collaboration might be very close. However, our analysis shows that the chemical potentials could be very different for the same energies in A+A and p+p. It adds more puzzles to the set of surprising coincidences at the energies close to the possible onset of deconfinement.

The main motivation of these studies is the new p+p data from the NA61/SHINE and HADES Collaborations at $\sqrt{s_{NN}} = 3.2 - 17.3$ GeV [1, 2, 3, 4, 5], as well as the new A+A data from HADES, and the updated A+A data from the NA49 Collaborations at $\sqrt{s_{NN}} = 2.2 - 17.3$ GeV, see [6, 7, 8, 9, 10] and references therein. A detailed comparison between the description of the p+p and A+A data at the discussed energies is important, since the ratio of positively charged kaons to protons, K^+/π^+, has a sharp maximum at $\sqrt{s_{NN}} = 7.6$ GeV, which was predicted as one of the signals of the onset of deconfinement in Ref. [11], see also the summary of other deconfinement signals and the latest experimental outcome in [12].

The updated NA49 A+A data contain more particles, some of them are with different error bars, while others are excluded. The chemical freeze-out analysis is performed in the framework of the hadron resonance gas (HRG) model. We use the latest set of resonances from THERMUS package [13] with masses $M \leq 2.4$ GeV, while previous analyzes included only the resonances with $M \leq 1.7$ GeV, see, e.g., [14, 15]. We also exclude the σ and κ resonances from the particle list due to the reasons described in [16]. The combination of these factors alters the freeze-out line obtained within the HRG see [10] and Fig. 1. Our new line in Fig. 1 is the result of the fit to the temperatures T and baryon chemical potentials μ_B obtained by us at particular collision energies. The fit functions for the $T_{\mathrm{A+A}}(\mu_B)$ and $\mu_B(\sqrt{s_{NN}})$ are the same as in [15], however, the obtained parameters are rather different:

$$T_{\mathrm{A+A}}(\mu_B) \;=\; a \;-\; b\mu_B^2 \;-\; c\mu_B^4\,, \qquad\qquad \mu_B(\sqrt{s_{NN}}) \;=\; \frac{d}{1\;+\;e\sqrt{s_{NN}}}\,, \tag{1}$$

$$a = 0.157\,\mathrm{GeV},\; b = 0.087\,\mathrm{GeV}^{-1},\; c = 0.092\,\mathrm{GeV}^{-3},\; d = 1.477\,\mathrm{GeV},\; e = 0.343\,\mathrm{GeV}^{-1}. \tag{2}$$

They suggest that the freeze-out line goes much lower than the previous estimates from [15] at small μ_B, i.e., at large collision energies. For $\mu_B = 0$ Eqs. (1) and (2) give the temperature $T \simeq 157$ MeV, which was surprisingly obtained in HRG at the LHC, see discussion in [17]. The difference in the obtained freeze-out parameters is important for the analysis of heavy nuclei production. Our results agree well with the LHC [18], as well as with the NA49 data [19].

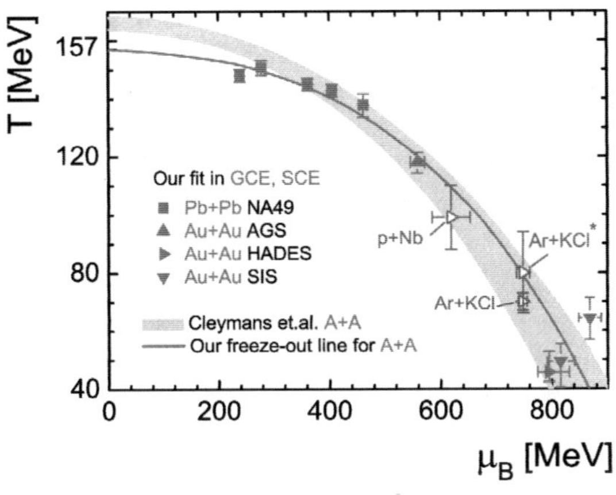

Figure 1. The freeze-out line in central A+A collisions. The grey band is the previous parametrization from [15]. The solid line is our new fit from [10], see Eqs. (1), (2). The points correspond to different collision energies. The p+Nb and Ar+KCl points are from the independent analysis [20], and were not included in the fit. The calculations are mostly done in the grand canonical ensemble (GCE). For the small energies of the old SIS and new HADES data the exact conservation of strangeness was taken into account within the strangeness canonical ensemble (SCE) [10].

The new p+p data are much more precise than the previous world data in that region. It is a difficult test for transport models, see, e.g., [10, 21]. The previous calculations of the chemical freeze-out parameters in p+p within HRG [22] were performed for larger collision energies $\sqrt{s_{NN}} \geq 17.3$ GeV. The obtained temperature had a large uncertainty due to the uncertainty in the existing data at that time. The p+p chemical freeze-out parameters for $\sqrt{s_{NN}} < 17.3$ GeV are calculated for the first time in [10].

We performed the fit of the p+p data in the HRG with the latest table of resonances and excluded σ, as discussed above. The analysis is done within the full canonical ensemble (CE) with exact conservation of electric charge, baryon number, and strangeness. Our $T_{\mathrm{p+p}}(\mu_B)$ line is obtained in two steps. The $T_{\mathrm{p+p}}(\sqrt{s_{NN}})$ dependance is the straight line fit to the points obtained within our CE HRG analysis of the new p+p data [10]. The corresponding $\mu_B(\sqrt{s_{NN}})$ are calculated from the primordial CE HRG multiplicities of neutrons and anti-neutrons using the relation between the average baryon number in GCE and the exact baryon number in CE[1], $\langle B\rangle_{\mathrm{GCE}} = B_{\mathrm{CE}}$, see Eqs. (7-11) in [24]. The neutrons and anti-neutrons are chosen, because they carry only one charge – the baryon number, and one can use the analytic formulas for the CE with one charge conservation. The combined result for the $T(\mu_B)$ in A+A and in p+p is shown in the right panel of Fig. 2.

[1] An alternative way is to fit the p+p data within the GCE, requiring that the obtained temperature is equal to that in the CE, and, additionally, demand that the average baryon number, electric charge, and strangeness, equal to the corresponding exact CE values in p+p, $\langle B\rangle_{\mathrm{GCE}} = B_{\mathrm{CE}} = 2$, $\langle Q\rangle_{\mathrm{GCE}} = Q_{\mathrm{CE}} = 2$, and $\langle S\rangle_{\mathrm{GCE}} = S_{\mathrm{CE}} = 0$. This method is applicable only in thermodynamic limit, i.e. for large enough systems, but gives practically the same $T_{\mathrm{p+p}}(\mu_B)$ line as in the exact case.

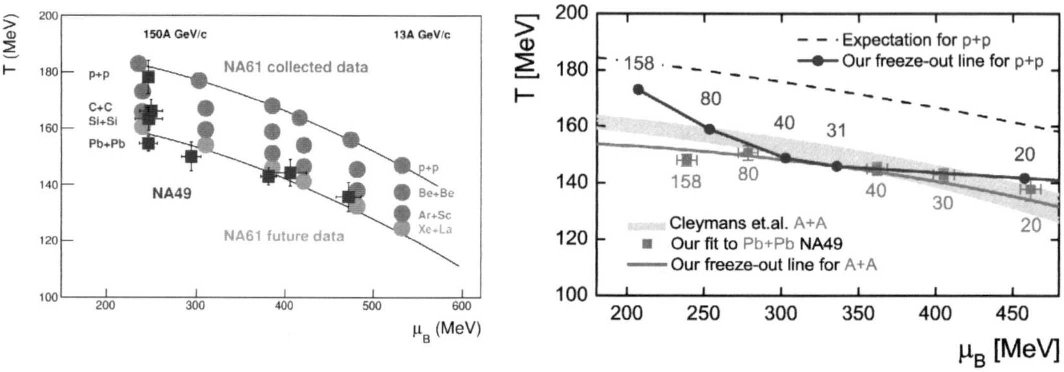

Figure 2. Left: The expectation of the NA61/SHINE Collaboration (full circles), and the $T(\mu_B)$ values obtained within the HRG with lighter resonances $M_{\rm res} \leq 1.7$ GeV (full squares with errors) from Ref. [23]. Right: The $T(\mu_B)$ values obtained for p+p and A+A in our analysis (dots and solid lines), compared to the expectations in A+A (grey band) and in p+p (dashed line). The numbers indicate the positions of the fit results for the corresponding collision energies in the lab frame E_{lab} in the A GeV units.

One can see that the obtained $T_{\rm p+p}(\mu_B)$ line behaves very differently, compared to the expectation of the NA61/SHINE in left panel of Fig. 2. The expected positions of the $T(\mu_B)$ points for the intermediate systems should be shifted vertically. They are also much closer to the A+A line than expected. A similar conclusion can be done looking at the p+Nb and Ar+KCl points in Fig. 1. It means that the freeze-out temperatures obtained in the energy and system size scan at the SPS [25] can be very similar[2]. The p+p and A+A lines even touch in the most interesting region $E_{lab} = 30 - 40A$ GeV ($\sqrt{s_{NN}} = 7 - 9$ GeV). However, the chemical potential is larger in A+A for 70 MeV and for 60 MeV at $E_{lab} = 30A$ and $40A$ GeV, correspondingly, which can be summarized as follows:

$$\text{Expectation} \qquad T_{\rm p+p} \gg T_{\rm A+A}\,, \qquad \mu_B^{\rm p+p} \simeq \mu_B^{\rm A+A}\,, \tag{3}$$

$$\text{Our result} \qquad T_{\rm p+p} \simeq T_{\rm A+A}\,, \qquad \mu_B^{\rm p+p} \ll \mu_B^{\rm A+A}\,. \tag{4}$$

The error bars for $T_{\rm p+p}$ are still quite large due to the small number of measured multiplicities, see [10]. Therefore, more data are needed to make a firm conclusion. We found that the minimal set should include particles possessing all three conserved charges B, S, Q, for both p+p and A+A, for example, $\pi^{\pm}$, $K^{\pm}$ and p, $\bar{p}$ particles. If the picture seen in right panel of Fig. 2 will preserve after the new measurements, then it will add more puzzles to the strange coincidences happening at these energies.

Acknowledgments

V.V.B. was supported by Polish National Science Center grant No. DEC-2012/06/A/ST2/00390. M.I.G. was supported by the Program of Fundamental Research of the National Academy of Sciences of Ukraine.

References

[1] Abgrall N *et al.* (NA61/SHINE) 2014 *Eur. Phys. J.* **C74** 2794 (*Preprint* 1310.2417)
[2] Agakishiev G *et al.* (HADES) 2014 *Phys. Rev.* **C90** 015202 (*Preprint* 1403.6662)

[2] One can still see a clear differences looking at the obtained radii of the systems [10].

[3] Pulawski S (NA61) 2015 *PoS* **CPOD2014** 010 (*Preprint* 1502.07916)
[4] Agakishiev G *et al.* (HADES) 2015 *Phys. Rev.* **C92** 024903 (*Preprint* 1505.06184)
[5] Aduszkiewicz A (NA61/SHINE Collaboration) 2015 Report from the NA61/SHINE experiment Tech. Rep. CERN-SPSC-2015-036. SPSC-SR-171 CERN Geneva URL http://cds.cern.ch/record/2059310
[6] Anticic T *et al.* (NA49) 2012 *Phys. Rev.* **C85** 044913 (*Preprint* 1111.2588)
[7] Anticic T *et al.* (NA49) 2011 *Phys. Rev.* **C84** 064909 (*Preprint* 1105.3109)
[8] Blume C, Gazdzicki M, Lungwitz B, Mitrovski M, Seyboth P and Strobele H (NA49 Collaboration) 2013 Compilation of NA49 numerical results Tech. Rep. 1075059 v.4 CERN Geneva URL https://edms.cern.ch/document/1075059/4
[9] Lorenz M (HADES) 2014 *Nucl. Phys.* **A931** 785–789
[10] Vovchenko V, Begun V V and Gorenstein M I 2016 *Phys. Rev.* **C93** 064906 (*Preprint* 1512.08025)
[11] Gazdzicki M and Gorenstein M I 1999 *Acta Phys. Polon.* **B30** 2705 (*Preprint* hep-ph/9803462)
[12] Wilczek A (NA61/SHINE) 2015 *7th International Conference on Physics and Astrophysics of Quark Gluon Plasma (ICPAQGP 2015) Kolkata, West Bengal, India, February 2-6, 2015* (*Preprint* 1510.08239) URL https://inspirehep.net/record/1401218/files/arXiv:1510.08239.pdf
[13] Wheaton S and Cleymans J 2009 *Comput. Phys. Commun.* **180** 84–106 (*Preprint* hep-ph/0407174)
[14] Becattini F, Manninen J and Gazdzicki M 2006 *Phys. Rev.* **C73** 044905 (*Preprint* hep-ph/0511092)
[15] Cleymans J, Oeschler H, Redlich K and Wheaton S 2006 *Phys. Rev.* **C73** 034905 (*Preprint* hep-ph/0511094)
[16] Broniowski W, Giacosa F and Begun V 2015 *Phys. Rev.* **C92** 034905 (*Preprint* 1506.01260)
[17] Floris M 2014 *Nucl. Phys.* **A931** 103–112 (*Preprint* 1408.6403)
[18] Stachel J, Andronic A, Braun-Munzinger P and Redlich K 2014 *J. Phys. Conf. Ser.* **509** 012019 (*Preprint* 1311.4662)
[19] Anticic T *et al.* (NA49) 2016 (*Preprint* 1606.04234)
[20] Agakishiev G *et al.* (HADES) 2016 *Eur. Phys. J.* **A52** 178 (*Preprint* 1512.07070)
[21] Vovchenko V Yu, Anchishkin D V and Gorenstein M I 2015 *Nucl. Phys.* **A936** 1–5 (*Preprint* 1408.5493)
[22] Becattini F and Heinz U W 1997 *Z. Phys.* **C76** 269–286 [Erratum: Z. Phys.C76,578(1997)] (*Preprint* hep-ph/9702274)
[23] Gazdzicki M and Seyboth P 2016 *Acta Phys. Polon.* **B47** 1201 (*Preprint* 1506.08141)
[24] Begun V V, Gorenstein M I and Zozulya O S (Bogolyubov Institute for Theoretical Physics, Kiev) 2005 *Phys. Rev.* **C72** 014902 (*Preprint* nucl-th/0411003)
[25] Gazdzicki M (NA61/SHINE) 2009 *J. Phys.* **G36** 064039 (*Preprint* 0812.4415)

IOP Conf. Series: Journal of Physics: Conf. Series **779** (2017) 012081 doi:10.1088/1742-6596/779/1/012081

Near and Far from Equilibrium Power-Law Statistics

T.S. Biró[1], G.G. Barnaföldi[1], G. Bíró[1] and K.M. Shen[2]
[1] Wigner Research Centre for Physics, Budapest, Hungary
[2] Central China Normal University, Wuhan, China

E-mail: biro.tamas@wigner.mta.hu

Abstract. We analyze the connection between p_T and multiplicity distributions in a statistical framework. We connect the Tsallis parameters, T and q, to physical properties like average energy per particle and the second scaled factorial moment, $F_2 = \langle n(n-1)\rangle / \langle n\rangle^2$, measured in multiplicity distributions. Near and far from equilibrium scenarios with master equations for the probability of having n particles, P_n, are reviewed based on hadronization transition rates, μ_n, from n to $n+1$ particles.

In this paper we approach the hadronization problem in high energy physics from statistical viewpoint. In a balanced version of decay and growth processes a simple master equation arrives at a final state including the Poisson, Bernoulli, negative binomial and Pólya distribution [1]. Such decay and growth rates incorporate a symmetry between the observed subsystem and the rest of a total system as a rule. For particle physics problems P_n is the probability of having n particles (or other quanta). In an avalanche type process the simplest assumptions about elementary rates in the master equation result in the exponential distribution with constant rates and in the power-law tailed Waring [2] distribution with linear preference rates. In this short paper we review relevant random filling patterns in phase space and treat thermal parameters as averages. We present master equations classified for describing dynamical stochastic processes near and far from equilibrium, and in particular analyze stationary distributions for large n. In this limit a set of coupled ordinary differential equations is replaced by a partial differential equation.

Following Einstein, we assume that the available phase-space volume at a given total energy, E, is filled evenly in statistical equilibrium or in a stationary state. Then the blind chance to find a part of the phase space, a.o. a single particle with energy $\omega \approx m_T - m$, will be given by $\Omega(\omega)\Omega(E-\omega)/\Omega(E)$. Beyond the one-particle phase space factor, $\Omega(\omega)$, the rest occurs as an environmental weight factor

$$p(\omega) = \frac{\Omega(E-\omega)}{\Omega(E)}. \tag{1}$$

In the simplest, idealized case the phase space volume is just a hypersphere with radius E, in dimension n: $\Omega(E) \propto E^n$. The above weight factor for picking out a single particle with energy ω becomes:

$$p(\omega) = \left(1 - \frac{\omega}{E}\right)^n. \tag{2}$$

The often cited thermodynamical limit then leads to a Boltzmann–Gibbs factor

$$p(\omega) \sim \lim_{\substack{E\to\infty, n\to\infty \\ E/n=T}} \left(1 - \frac{\omega}{E}\right)^n = e^{-\omega/T}, \tag{3}$$

interpreting the kinetic temperature as $T = E/n$.

However, experimentally studied physical systems are often far from the thermodynamical limit. In *small systems* fluctuations can be large, and an averaging over several millions of events is done for the histogram bins in obtaining p_T spectra. In thermal models ω is a monotonic rising function of p_T. Such an average of the above ratio,

$$\langle p(\omega) \rangle = \sum_{n=0}^{\infty} P_n \left(1 - \frac{\omega}{E}\right)^n, \tag{4}$$

is interesting to be inspected for famous n-distributions. For the Poisson we obtain

$$\langle p(\omega) \rangle^{\text{POISSON}} = \mathrm{e}^{-\langle n \rangle \, \omega / E}, \tag{5}$$

and for the negative binomial distribution (NBD)

$$\langle p(\omega) \rangle^{\text{NBD}} = \left(1 + \frac{\langle n \rangle}{k+1} \frac{\omega}{E}\right)^{-k-1}. \tag{6}$$

For a general event distribution we expand eq.(4) for $\omega \ll E$ and compare the result with the Tsallis–Pareto distribution

$$p(\omega) = \left(1 + (q-1)\frac{\omega}{T}\right)^{-\frac{1}{q-1}}. \tag{7}$$

Expanding both expressions (4,7) up to terms quadratic in ω we obtain

$$T = \frac{E}{\langle n \rangle}, \qquad \text{and} \qquad q = \frac{\langle n(n-1) \rangle}{\langle n \rangle^2}. \tag{8}$$

The parameter T is the event-averaged kinetic temperature, while $(q-1)$ is a measure of the non-Poissonity in the multiplicity distribution. These findings can easily be generalized by comparing the environmental factor (1) with the Tsallis–Pareto distribution (7):

$$\langle p(\omega) \rangle = \left\langle \mathrm{e}^{S(E-\omega)-S(E)} \right\rangle \approx \left(1 + (q-1)\frac{\omega}{T}\right)^{-\frac{1}{q-1}}. \tag{9}$$

Expanding in $\omega \ll E$ both sides up to quadratic terms we obtain

$$\frac{1}{T} = \langle S'(E) \rangle = \langle \beta \rangle, \qquad q = 1 - \frac{1}{C} + \frac{\Delta\beta^2}{\langle \beta \rangle^2}. \tag{10}$$

This interprets the parameter T as the inverse of the average $\beta = S'(E)$ and $(q-1)$ as the non-Gaussianity in β fluctuation. For $q = 1$ the variance $\Delta\beta$ would be $1/T\sqrt{C}$ [3]. Here $\langle S''(E) \rangle = -1/CT^2$.

We discuss the following schemes of master equations. The diffusion class describes near equilibrium stochastic evolution of the probability, P_n, with the growth rate form n to $n+1$, μ_n and the decay rate to $n-1$, λ_n:

$$\dot{P}_n = [(\lambda P)_{n+1} - (\lambda P)_n] - [(\mu P)_n - (\mu P)_{n-1}]. \tag{11}$$

The avalanche class describes a contesting race eventually achieving stationary branching ratios:

$$\begin{aligned} \dot{P}_0 &= \langle \gamma \rangle - (\gamma_0 + \mu_0) P_0, \\ \dot{P}_n &= \mu_{n-1} P_{n-1} - (\mu_n + \gamma_n)\, P_n \qquad \text{for } n \geq 1. \end{aligned} \tag{12}$$

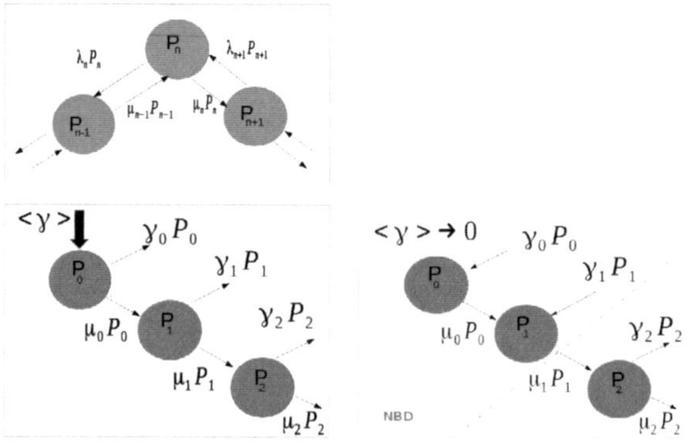

Figure 1. Comparison of near equilibrium, diffusion like and far equilibrium, avalanche like models. The graphs on the left side demonstrate the respective structures of the equations, on the right side the particular one suggested to generate the NBD distribution in Ref.[4].

Here the decay rate, γ_n, describes exit from the chain, a reduction in n is not assumed. Fig.1 depicts the difference between these classes. By studying the stationary branching in the avalanche type dynamics, we are especially interested in the mean aging model, where for $\forall n : \gamma_n = \gamma$ and therefore $\langle\gamma\rangle = \gamma$. In the stationary limit: $P_n(t) \to Q_n$, and from $\dot{Q}_n = 0$ one obtains

$$Q_n = \frac{\mu_{n-1}}{\mu_n + \gamma} Q_{n-1} = \cdots = \frac{\gamma}{\mu_n} \mathrm{e}^{-\sum\limits_{j=0}^{n} \ln(1+\gamma/\mu_j)}. \tag{13}$$

Now, constant growth rates, $\mu_j = \sigma$ lead to the exponential distribution

$$Q_n = \frac{1}{1+\sigma/\gamma} \mathrm{e}^{-n \cdot \ln(1+\gamma/\sigma)}. \tag{14}$$

Linear preference growth rates $\mu_j = \sigma(j+b)$ $(b > 0)$ lead to the Waring distribution [2; 5; 6],

$$Q_n = \frac{\gamma}{\gamma + b\sigma} \frac{\Gamma(n+b)\,\Gamma(b+1+\gamma/\sigma)}{\Gamma(b)\,\Gamma(n+b+1+\gamma/\sigma)}. \tag{15}$$

This distribution has a power-law tail for large n as $\sim n^{-1-\gamma/\sigma}$. For negligible decay compared to the growth, $\gamma \ll \sigma$, one arrives at the Zipf distribution. It is interesting to note that it has been suggested by Osada et.al. [4] that the special dynamics with $\gamma_n = \sigma(n - kf)$ and $\mu_n = \sigma f(n+k)$ leads exactly to the NBD as stationary distribution,

$$Q_n = \binom{n+k-1}{n} f^n (1+f)^{-n-k}. \tag{16}$$

We note that for large n the avalanche dynamics effectively uses the variable $x = n\Delta x$ and solves $\frac{\partial}{\partial t}P(x,t) = -\frac{\partial}{\partial x}(\mu(x)\,P(x,t)) - \gamma(x)P(x,t)$. The stationary distribution is

$$Q(x) = \frac{K}{\mu(x)} \mathrm{e}^{-\int\limits_0^x \frac{\gamma(u)}{\mu(u)}\,du}. \tag{17}$$

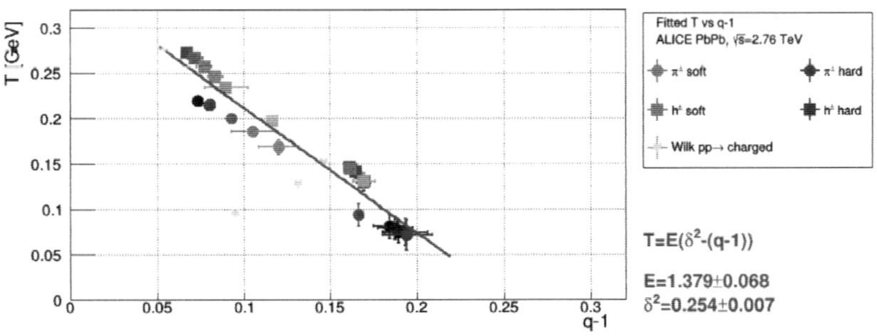

Figure 2. The Tsallis parameters T vs $q-1$ from fits to the 2.76 TeV ALICE PbPb data.

In this framework the constant rate $\mu(x) = \sigma$ leads to the exponential, and the linear rate, $\mu(x) = \sigma(x+b)$ to the Tsallis–Pareto stationary distribution [7]

$$Q(x) = \frac{\gamma}{\sigma b}\left(1+\frac{x}{b}\right)^{-1-\gamma/\sigma}. \tag{18}$$

A connection between p_T and n distributions is hinted at in finding different q values for different participant numbers in experiments [8–10]. Our model form eqs.(8,10) predicts $T = E(\delta^2-(q-1))$ with $E = CT \approx 1.4$ GeV and $\delta = \Delta\beta/\langle\beta\rangle \approx 0.5$. Fig.2 shows our fit results to the LHC ALICE PbPb data at 2.76 TeV. Darker points belong to more central collisions. We also distinguish between soft ($p_T < 5$ GeV, red data points) and hard ($p_T > 3$ GeV, blue data points) spectral parts. For fit parameters and further details see legend. The green stars are data from pp collision [11], lying on a $T = E(q-1)$ line with $E \approx 1$ GeV. The AA points seem to favor $\delta^2 = 0.25$ independent of $\langle n\rangle$, while the pp points $\delta^2 \propto 1/\langle n\rangle$, meaning a constant $f = \langle n\rangle/k$ value in the NBD distribution.

Acknowledgments: Discussions with Zoltán Néda and András Telcs are gratefully acknowledged. This work has been supported by NKFIH (OTKA K 104260, 120660 projects) and by a bilateral Chinese–Hungarian governmental project, TeT 12CN-1-2012-0016. GGB thanks the János Bolyai scholarship of the Hungarian Academy of Science.

References

[1] T. S. Biró, Z. Néda, arxiv: 1606.05737.
[2] J. O. Irwin: J. Roy. Stat. Soc. A **126** (1963) 1
[3] T. S. Biró, P. Ván, G. G. Barnaföldi, K. Ürmössy, Entropy **16** (2014) 6497
[4] T. Osada, N. Nakajima, M. Biyajima, N. Suzuki, Prog. Theor. Phys. **98** (1997) 1289
[5] P. L. Krapivsky, G. J. Rodgers, S. Redner: Phys. Rev. Lett. **86** (2001) 5401-5404
[6] A. Schubert, W. Glänzel: Scientometrics **6** (1984) 149
[7] S. Thurner, F. Kyriakopoulos, C. Tsallis, Phys. Rev. E **76** (2007) 036111
[8] ALICE collaboration, Phys. Lett. B **720** (2013) 52
[9] PHENIX collaboration, Phys. Rev. Lett. **101** (2008) 232301
[10] J. T. Mitchell and the PHENIX Collaboration, Nukleonika **51** (2006) S89
[11] G. Wilk, Z. Wlodarczyk, AIP Conf.Proc. **1558** (2013) 893

Prototype sector production for the STAR inner TPC upgrade

Chi Yang (for the STAR Collaboration)
University of Science and Technology of China, 230026, Hefei, Anhui, China
E-mail: chiyang@ustc.edu.cn

Abstract. The STAR experiment at the Relativistic Heavy Ion Collider (RHIC) is upgrading the Inner TPC sectors (iTPC). By increasing the number of inner pad rows from 13 to 40 and renewing the inner sector wires, this major detector upgrade will improve the rapidity coverage from $|\eta| < 1$ to $|\eta| < 1.5$, provide better momentum resolution, and better energy loss (dE/dx) resolution. The iTPC upgrade is crucial to STAR Beam Energy Scan Phase II (BES-II) program, which will provide in-depth understanding on QCD phase diagram and in-medium modification. In this paper we report on progress on the iTPC sector construction. The iTPC module fabrication techniques and testing results from the first full size prototype are presented.

1. Introduction

The STAR experiment [1] is dedicated to study the fundamental properties of the deconfined QCD medium (Quark-Gluon Plasma, QGP) via ultra-relativistic heavy-ion collisions at the RHIC [2]. To improve many results by an order of magnitude for the measurements done with STAR at lower collisions energies, where a critical point may be observed, the BES-II program is scheduled from year 2019 to year 2020 with collision energies from 7.7 GeV to 19.6 GeV. This important program calls for detector upgrades including iTPC, eTOF (endcap Time of Flight) and EPD (Event Plane Detector). The TPC [3] is the main detector in STAR for tracking and PID in the central region, which provides precise charged-particle tracking, momentum measurement, and particle identification in high multiplicity heavy-ion collisions [4, 5]. The major goals of the iTPC upgrade are as follows: wider rapidity coverage, better momentum resolution, and better dE/dx resolution [6]. The current inner TPC pad row geometry does not have continuous coverage at all radii. Distances between nearby pad rows are greater than 50mm while the pad rows are 11.5mm tall. Therefore, 80% of the inner pad plane area has no readout. In the iTPC upgrade, we will get continuous coverage on η, doubling the number of pads in the inner sector pad plane and increasing the pad size. This will increase the percentage of sampled track path length from 20% to 95%, which means we will design "continuous" pad rows on pad plane. The electronics will be upgraded accordingly due to the doubled number of readout pads. In addition, the multi-wire proportional chambers (MWPCs) in inner sectors will be renewed to reset the time for wire aging on anode wires due to the increasing integrated and instantaeous luminosity delivered by RHIC. In general, the goal for the projects is to replace all 24 existing inner sectors in the STAR TPC with new, fully instrumented, sectors.

2. Physics impact of the iTPC upgrade

The enhanced measurement capabilities of STAR after the iTPC upgrade are crucial for BES-II and provide a major benefit for many analyses, especially fluctuations (Kurtosis) and dielectron. The search for a possible critical point [7, 8] in the QCD phase diagram is one of the most interesting and important topics in heavy ion physics. The critical point, if it exists and if it can be identified, would provide a landmark in the phase diagram of nuclear matter and guide further experimental and theoretical studies of QCD under a wide range of conditions. RHIC has completed Phase-I of the beam energy scan program (BES-I) with center-of-mass beam energies per nucleon in Au+Au collisions of 39, 27, 19.6, 14.5, 11.5 and 7.7 GeV. The proposed BES-II will focus on an in-depth study of energies below 20 GeV with typically 20 times the statistics as in the same energy region exploited in BES-I. The iTPC upgrade extends the rapidity coverage by 50%. Wider rapidity coverage is important for this correlation study. For dielectron measurements [9], the iTPC upgrade will reduce hadron contamination from a dominant source of uncertainty to an expected statistical uncertainty of only 10%. With good statistics and better uncertainty control, in-medium modification behavior can be well studied in BES-II via dielectron measurements. Models comparison with different in-medium ρ broadening scenarios may give more clear physics messages.

3. The first prototype production

The first prototype of iTPC sector was fabricated in Shandong University. Design parameters compared with the old design can be found in Tab. 1. In the following we described the three steps of the MWPC production.

Table 1. Parameters for the original and new STAR TPC pad planes.

Item	Inner	Outer	iTPC	Comments
Pad Pitch (center to center)	3.35×12	6.70×20	5.00×16	mm
Isolation gap between pads	0.5	0.5	0.5	mm
Pad Size	2.85×11.5	6.20×19.5	4.5×15.5	mm^2
Number of pads	1750	3940	3440	
Anode to pad plane spacing	2	4	2	mm
Anode voltage	1170V	1390V	~1120V	20:1 S/N
Anode Gas Gain	3770	1230	~2000	nominal
Anode Wire diameter	20μm	20μm	20μm	Au plated W
Anode Witch Pitch	4	4	4	mm

3.1. Wire winding

A wire winding machine is used to wind the wires on an Aluminum frame. Wire tension is kept via a sensor and wire pitch is controlled by stepper motor when winding. Before mounting the wires on TPC sector, they are checked when they are on wire frames. A laser system is designed to check the wire tension and pitch. Figure 1 shows a layout of this laser system. Synchronized with gas jet, the laser is used to scan each wire. Fundamental oscillator frequency can be derived from the voltage fluctuation transformed of laser absorption via a photodiode. Then wire tension can be calculated with known wire parameters. Wire pitch can be obtained simultaneously. Figure 2 shows the distribution of measured anode wire tensions. The desired mean value is 0.5N. All wire tensions are qualified (within 0.5±0.05N). In the wire winding step, the precision of wire pitch is about 50μm, which will be further improved to less than 10μm by an wire comb in wire mounting step.

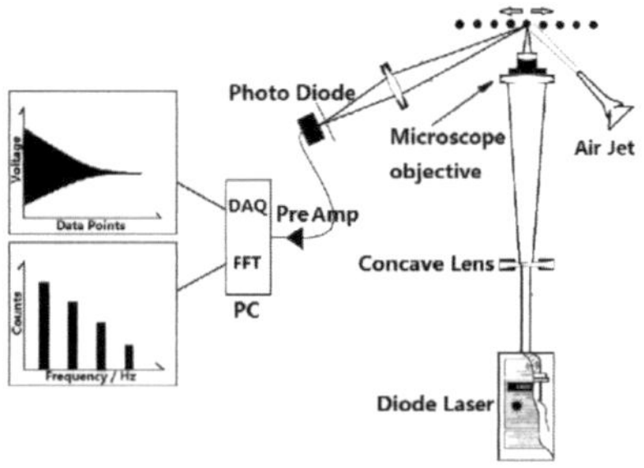

Figure 1. Layout of the laser system for wire tension measurement. Layouts of photodiode transition and Fast Fourier Transform are shown on left.

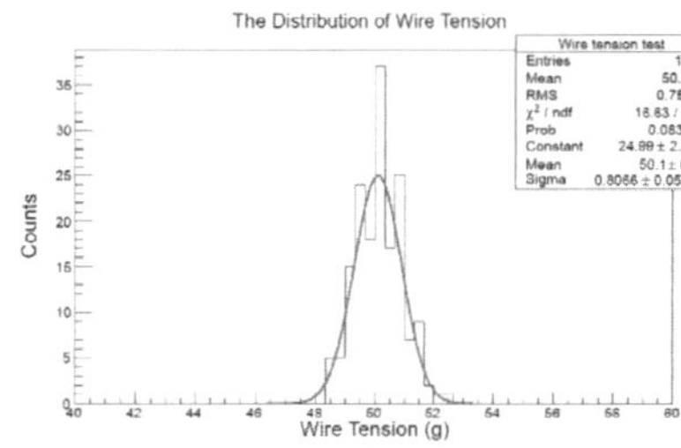

Figure 2. Wire tension distribution of anode wires. The desired mean value is set to 0.5N.

3.2. Wire mounting

The Aluminium supporting strong backer made by University of Texas, Austin with the glued pad plane is seen in Fig. 3. The pad plane and side wire mounts are reproduced with original drawings in China. The tolerance of height from pad plane to the bottom of strong back is required as 10μm. Same tolerance standard is required for three wire planes. The qualified wire frames should be further mounted on the wire mount PCB which are installed on strong backer. Fig. 4 shows a photo in this wire mounting process. The wire tension and pitch are firstly guaranteed by wire winding machine. And then, wire pitch and height are modified by wire combs. These wire combs consist two parts. One structure like comb is to fix wire pitch while the other structure with straight edge is to fix wire height. After using wire comb, the height and pitch tolerance of wires are controlled to within 10μm. After gluing and soldering, the installed wires need double check on wire tension and pitch. This double check shows good consistency with the test results before wire mounting.

Figure 3. Pad plane glued on strong backer. The Aluminum strong backer shown in bottom is to provide mechanical support for all three wire frames.

Figure 4. Wire frame laying on a sector in wire mounting process. Two protection covers are used to prevent unnecessary touching on wire comb.

3.3. Sector testing based on cosmic-ray

A testing system using a cosmic-ray trigger has been built in Shandong University. The trigger system contains two layers of scintillator readouts by PMTs. When a cosmic-ray muon passes across, the coincidence of two PMT signals provides the trigger signal. The first prototype of iTPC sector is located in Plexiglass chamber with an Aluminum supporting base. The size of this test chamber is $92 \times 76 \times 76.5$cm. With one layer of cathode and one layer of ground, combined with a simplified field cage, the electric field uniformity is guaranteed. The operation gas P10 (90% Ar + 10% CH_4) is flowed into the testing chamber with good air tightness. With a DAQ similar to what STAR uses now, cosmic-ray signal can be collected. Figure 5 shows the ADC versus time bin (represent the ionized electron drift time, ~100ns/bin) distribution of 30k events recorded. The cutoff at about time bin 170 is due to the drift length of the testing chamber. Since we observe a decrease of the signal with drift length there is likely an effect of absorbtion due to gas purity. This is under investigation. Some cosmic-ray tracks can be "seen" by pushing back the readout signals into 3 dimension (pad position X vs row position Y vs relative drift time Z). Figure 6 shows an example of the reconstructed cosmic-ray "track" left in the testing chamber. Cluster finder algorithm will be applied in our local track reconstruction soon.

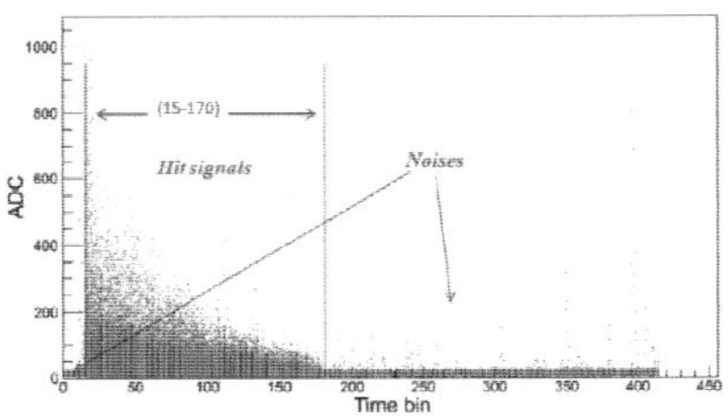

Figure 5. ADC versus time bin distribution of 30k recorded events in cosmic-ray test.

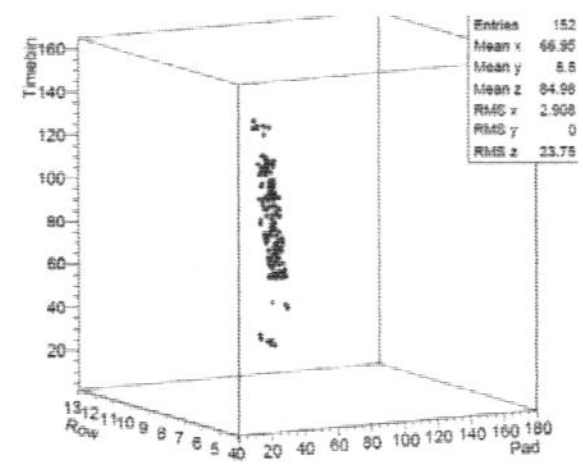

Figure 6. Observed ionization signal positions left in drift area for a cosmic-ray track.

4. Summary and Outlook

The STAR inner TPC upgrade is underway. This upgrade will improve many STAR physics results in statistics, acceptance coverage and systematic uncertainties. A complete construction plan is ready, aiming the BES-II program in year 2019 and 2020. First inner sector prototype has been produced with qualified wire tension, pitch and height. Testing system based on cosmic-ray has been built and operated while the data is taken smoothly. Additional testing for this first prototype is ongoing.

References

[1] K. H. Ackermann *et al.* 2003 *Nucl. Instr. Meth. A* **499**, 624
[2] M. Harrison, T. Ludlam, and S. Ozaki 2003 *Nucl. Instr. Meth. A* **499**, 235
[3] M. Anderson *et al.* 2003 *Nucl. Instr. Meth. A* **499**, 659
[4] H. Bichsel 2006 *Nucl. Instr. Meth. A* **562**, 154
[5] Y. Xu *et al.* 2010 *Nucl. Instr. Meth. A* **614**, 28
[6] "A Proposal for STAR Inner TPC Sector Upgrade (iTPC)" 2015 `https://drupal.star.bnl.gov/STAR/files/Summary_September14th_iTPC_final.pdf`
[7] Y. Aoki, G. Endrodi, Z. Fodor, S. D. Katz and K. K. Szabo 2006 *Nature* **443**, 675
[8] M. A. Stephanov 2009 *Phys. Rev. Lett.* **102**, 032301
[9] L. Adamczyk *et al.* [STAR Collaboration] 2014 *Phys. Rev. Lett.* **113**, 022301
L. Adamczyk *et al.* [STAR Collaboration] 2015 *Phys. Rev. C* **92**, 024912.

A fixed-target programme at the LHC (AFTER@LHC)

Barbara Trzeciak, on behalf of the AFTER@LHC study group [1]
Utrecht University, Princetonplein 1, 3584 CC Utrecht, The Netherlands

E-mail: b.a.trzeciak@uu.nl

Abstract. We report on the perspectives for hadron, heavy-ion and spin physics with a multi-purpose fixed-target programme using the LHC multi-TeV proton and heavy-ion beams (AFTER@LHC). This would be the most energetic fixed-target experiment opening new domains of particle and nuclear physics and complementing current and future collider programmes. Thanks to the large boost, one can fully access –with conventional detectors– the backward hemisphere in the center-of-mass system which allows for studies of the largely uncharted high-x region ($x_F \to$ -1).

1. Introduction

AFTER@LHC is a proposal for a multi-purpose fixed-target experiment using the multi-TeV proton and heavy-ion beams of the LHC [1, 2, 3]. The physics cases of such an experiment covers 3 main fields : (i) the understanding of the high-x gluon, antiquark and heavy-quark content in the nucleon and nucleus; (ii) the dynamics and spin of gluons inside (un)polarised nucleons; (iii) heavy-ion collisions from mid to large rapidities between SPS and RHIC energies.

The most relevant advantages of a fixed-target mode compared to a collider mode are: (1) an easy access to high Feynman $|x_F|$ domain by studying the far backward region in the center-of-mass system (c.m.s); (2) the versatility of the target species; (3) the possibility to polarise a target; (4) the very high luminosities with either dense targets or high intensity beams. All this can be realized in a setup not affecting the LHC performance itself.

With the 7 TeV proton beam colliding on a fixed target, the c.m.s energy per NN collisions ($\sqrt{s_{NN}}$) is 115 GeV, with a rapidity shift $\Delta y^{\rm lab}_{\rm c.m.s} \approx 4.8$ ($y_{\rm c.m.s} = 0 \to y_{\rm lab} = 4.8$). The 2.76 TeV Pb beam provides $\sqrt{s_{NN}} \approx 72$ GeV and the rapidity shift of 4.3. This gives the same c.m.s energy for $p+p$, $p+d$, $p+A$ collisions with the LHC proton beam, and for Pb+p, Pb+A collisions with the LHC lead beam. Due to the large boost, the entire forward hemisphere, $y_{\rm c.m.s} > 0$, falls within an angle of $0^\circ < \theta_{\rm lab} < \mathcal{O}(1^\circ)$ w.r.t to the beam axis in the laboratory frame. Due to the large occupancy, the forward region is challenging to access. On the contrary, the backward hemisphere, $y_{\rm c.m.s} < 0$, corresponds to larger angles in the laboratory frame and is easily accessible using standard detection techniques without the need to cope with constraints from the beam pipe. This provides an access to the largely uncharted backward physics with partons having a momentum fraction $x_2 \to 1$ in the target ($x_F \to$ -1), for different systems.

Several technological options allowing one to perform fixed-target experiments with the LHC beams are currently under investigation: (1) installing an internal gas target [4, 5, 6]; (2) placing a wire target in the beam halo [7]; (3) deviating a part of the beam halo with a bent-crystal onto an external target [8, 9]. The advantages of the first option is the possibility of combining it with an existing detector, such as LHCb or ALICE. Furthermore, by installing a HERMES-like Storage Cell [10] as a target, polarised gases can be used such that an ambitious spin-physics

[1] M. Anselmino, R. Arnaldi, S.J. Brodsky, V. Chambert, C. Da Silva, J.P. Didelez, M.G. Echevarria, E.G. Ferreiro, F. Fleuret, Y. Gao, B. Genolini, C. Hadjidakis, I. Hřivnáčová, D. Kikola, A. Klein, A. Kurepin, A. Kusina, J.P. Lansberg, P. Lenisa, F. Lyonnet, L. Massacrier, C. Lorcé, A. Nass, C. Quintans, F. Rathmann, P. Robbe, I. Schienbein, E. Scomparin, J. Seixas, H.S. Shao, A. Signori, E. Steffens, N. Topilskaya, U.I. Uggerhøj, R. Ulrich, Z. Yang.

program can be conducted. The bent-crystal option can be used in two ways: (i) with a dedicated beamline which is probably the most costly option; (ii) keeping the particles deviated by the bent crystal inside the beampipe as a split beam which would then interact with a target adapted to the beampipe. Further investigations are needed to tell whether this beam splitted option is compatible with a polarised target like that of [11] and the use of LHCb or ALICE.

2. Physics highlights and projection studies

2.1. Set-up for projection studies

The simulations presented here focus on quarkonium and Drell-Yan (DY) production by exploring di-muon ($\mu^+\mu^-$) pair production from different sources. The signal (quarkonia and DY) and different di-muon background sources are separately studied in order to keep a good control over the input distributions and the normalization of these different sources. The quarkonium signal, DY and correlated open-charm and beauty backgrounds ($c\bar{c}, b\bar{b}$) were simulated with HELAC-ONIA [12, 13]. The events were processed with PYTHIA 8 [14] to perform the hadronisation, to account for initial/final-state radiations and for the decay of resonances. The uncorrelated background (mostly μ coming from pion and kaon decays) was obtained from minimum bias $p+p$ collisions generated with PYTHIA 8. Afterwards, a fast simulation was performed to account for realistic detector-resolution and particle-identification performances. The detector response –momentum resolution, μ identification efficiency and π/K misidentification probability with μ– was simulated with a detector setup similar to the LHCb detector [15] with a pseudorapidity coverage of $2 < \eta_{\rm lab} < 5$. More details on the applied procedure can be found in Ref. [16]. First projections for $p+A$ and $A+A$ collisions are done by applying a scaling factor to $p+p$ simulations, with no nuclear effects considered (not even isospin corrections). For the quarkonium and DY simulations, where the full background is considered, the uncorrelated one is subtracted only assuming the like-sign technique (that is, the worst case scenario in terms of the expected statistical uncertainties). Integrated luminosities taken for the simulations and expected to be achieved within one year are: $p+p$: 10 fb^{-1}, $p+A$: 100 pb^{-1} and $A+A$: 7 nb^{-1}.

2.2. QCD and nuclear PDFs (nPDFs) studies at high x with a proton beam

Among the goals of AFTER@LHC are high-x studies for particle and astroparticle physics. The most relevant features of AFTER@LHC in this case are the very high luminosities, the access towards very low p_T and backward rapidities; it would indeed be a multi-purpose experiment providing detailed studies of heavy-flavour, quarkonia and prompt photon production. In order to solve long standing debates in quarkonium hadroproduction (see [17]), precise measurements of cross-sections, correlations and polarisations of *a large number of* quarkonium states are essential. Such measurements can be conducted at AFTER@LHC thanks to the large expected quarkonium yields – typically 10^9 for charmonia and 10^6 for bottomonia over a year accounting for the branchings. AFTER@LHC will also give a unique opportunity to access C-even quarkonia ($\chi_{c,b}, \eta_c$) and associated-production channels [18] in a new energy and rapidity domain. In $p+p$ and $p+A$ collisions, quarkonium studies can be performed over wide transverse momenta and rapidities beyond that of RHIC experiments [16].

Thanks to the access to the target-rapidity region, the gluon, antiquark and heavy-quark distributions in the proton, neutron and nuclei can then be extracted at mid- and high-momentum fractions x. The wide rapidity coverage will help to study nuclear effects in the antishadowing, EMC and Fermi motion regions (essentially $x_2 > 0.05$). The search of a possible *gluon* EMC effect could be crucial to understand the origin of the quark EMC effect and its connections with short-range correlations in nuclei. Furthermore, a precise knowledge of nPDFs is important for studies of heavy-ion collisions as fundamental inputs for the initial states of the collisions. In [16, 21], the impact of the gluon densities on the nuclear modification factors of prompt and non-prompt J/ψ and Υ in p+Pb and Pb+Pb was evaluated using the EPS09 parametrization. The effect of coherent energy loss was also studied in [22]. An example of the

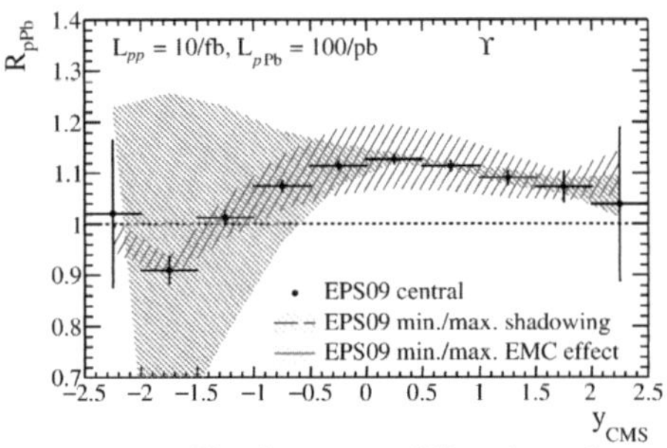

Figure 1. Nuclear modification factor for Υ vs. $y_{\mathrm{c.m.s.}}$ in p+Pb collisions at $\sqrt{s_{NN}} = 115$ GeV (from [16]). Points represent EPS09 [19] central prediction with statistical uncertainties based on expected yields at AFTER@LHC. Uncertainties of the nPDF in EPS09 assuming min./max. shadowing and min./max. EMC effect are shown as the blue and red band, respectively.

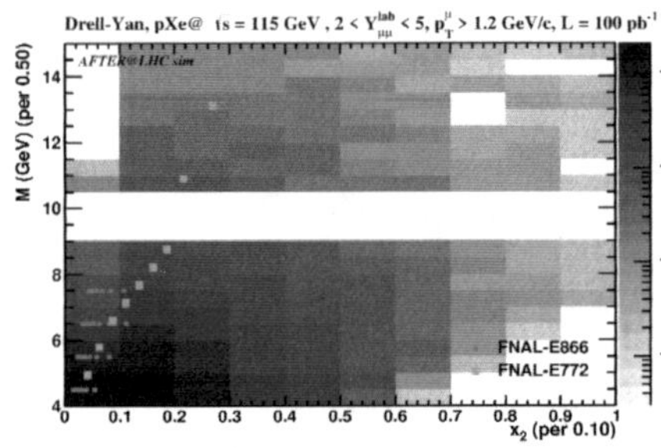

Figure 2. DY yields in the di-muon invariant mass (excluding Υ states region) and x_2 bins in p+Xe collisions with existing DY $p + A$ data used for nPDF fits [20] – p+Xe at $\sqrt{s_{NN}} = 115$ GeV.

nPDF uncertainties propagated to Υ R_{pA} is given on Figure 1 and is compared to projected experimental uncertainties, which are much smaller in the ALICE and LHCb y acceptance.

nPDFs can be further constrained by DY measurements. Figure 2 shows the huge expected DY yields as a function of the di-muon invariant mass and x_2 in p+Xe collisions – similar yields are expected for $p + p$ collisions. The existing DY $p + A$ data currently used for global nPDF fits are also shown in red and green (see [20]). It is obvious that AFTER@LHC offers a unique coverage towards the unexplored high x region in $p + A$ with extremely high yields (in $p + p$, $p + A$). The expected precision on a nuclear modification factor (R_{pXe}) measurement is presented in Figure 3 for 3 rapidity bins corresponding to different x_F ranges. The statistical uncertainties are dominated by uncertainties from the uncorrelated background subtraction with like-sign techniques. These could probably be reduced using event mixing down to a level where systematical uncertainties would then become dominant. We stress that going to the very backward rapidities ($2 < y_{\mathrm{lab}} < 3$) –which probably is the most interesting one physics wise, renders the DY signal even more clean since the quark-induced processes are favoured and the background is reduced accordingly.

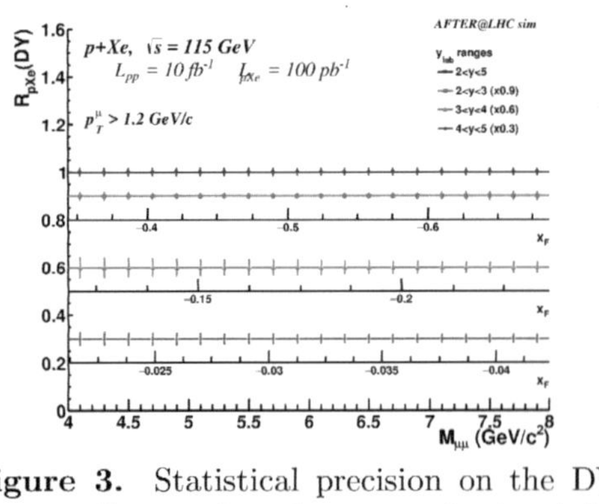

Figure 3. Statistical precision on the DY R_{pXe} *vs* di-muon invariant mass (and x_F) for 3 rapidity ranges – p+Xe at $\sqrt{s_{NN}} = 115$ GeV.

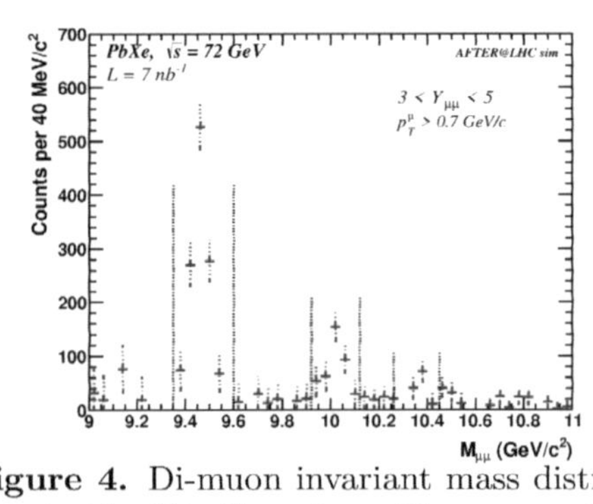

Figure 4. Di-muon invariant mass distribution in the $\Upsilon(nS)$ region. The uncorrelated background was subtracted through the like-sign technique – Pb+Xe at $\sqrt{s_{NN}} = 72$ GeV.

Giving the large expected yields, one can also measure the single spin asymmetries (A_N) for all quarkonia states and DY with unheard of precision and quantify the gluon and quark Sivers effects. For more insights into the spin physics opportunities, see [23, 24, 25, 26, 27].

IOP Conf. Series: Journal of Physics: Conf. Series **779** (2017) 012083 doi:10.1088/1742-6596/779/1/012083

2.3. Physics highlights for heavy-ion collisions

One of the prime objectives of heavy-ion physics at high-energy facilities is the search and characterisation of the Quark-Gluon Plasma (QGP). AFTER@LHC with $\sqrt{s_{NN}} = 72$ GeV provides a complementary coverage to the RHIC and SPS experiments in a temperature and baryon chemical potential region where the QGP is expected to be formed. Thanks to the aforementioned assets of the fixed-target mode, the longitudinal expansion of the hot medium can be studied using the wide available rapidity coverage with colliding systems of different sizes with the LHC Pb beam. With the rapidity scan one can probe the medium at different average temperatures and energy densities and infer the temperature dependence of the shear viscosity of the QGP liquid [28]. In addition, one can perform precise measurements of the QGP properties with hard probes. Figure 4 shows expected $\Upsilon(nS)$ yields in Pb+Xe collisions with a excellent separation of the 3 Υ states. What is also important for hot-medium investigations is the target versatility since it offers a unique opportunity to study the cold nuclear matter effects and their factorization. The ideal probe for such checks is DY as it is immune from the final state QCD matter effects in AB collisions. AFTER@LHC is probably the only facility where this can be done. Finally, precise charm and beauty measurements will address the long standing question of heavy-quark energy loss in the medium [29] - collisional *vs* radiative energy loss.

3. Conclusions

We have reported on the proposal of performing fixed-target experiments with the multi-TeV LHC proton and lead beams. We have presented a selection of physics topics with projection studies. Such a new experimental endeavour opens the way to a number of unique and precise measurements, many of which can not be performed elsewhere, significantly advancing our knowledge in the fields of heavy-ion, hadron, spin and astroparticle physics.

References

[1] S. J. Brodsky, F. Fleuret, C. Hadjidakis, J. P. Lansberg, *Phys. Rept.* **522**, 239 (2013).
[2] J. P. Lansberg, S. J. Brodsky, F. Fleuret, C. Hadjidakis, *Few Body Syst.* **53**, 11 (2012).
[3] J. P. Lansberg, *et al.*, *PoS* **QNP2012**, 049 (2012).
[4] C. Barschel, Ph.D. thesis, RWTH Aachen U. CERN-THESIS-2013-301 (2014).
[5] M. Ferro-Luzzi, *Nucl. Instrum. Meth.* **A553**, 388 (2005).
[6] R. Aaij, *et al.*, *JINST* **9**, P12005 (2014).
[7] A. B. Kurepin, N. S. Topilskaya, *Adv. High Energy Phys.* **2015**, 760840 (2015).
[8] W. Scandale, *ECFA LHC Workshop, Aachen, Germany, October 4-9, 1990* (1990), pp. 760–764.
[9] E. Uggerhoej, U. I. Uggerhoej, *Nucl. Instrum. Meth.* **B234**, 31 (2005).
[10] C. Barschel, P. Lenisa, A. Nass, E. Steffens, *Adv. High Energy Phys.* **2015**, 463141 (2015).
[11] C. Brown, *et al.*, Letter of Intent for a Drell-Yan Experiment with a Polarized Proton Target (2014).
[12] H.-S. Shao, *Comput. Phys. Commun.* **184**, 2562 (2013).
[13] H.-S. Shao, *Comput. Phys. Commun.* **198**, 238 (2016).
[14] T. Sjostrand, S. Mrenna, P. Z. Skands, *Comput. Phys. Commun.* **178**, 852 (2008).
[15] A. A. Alves, Jr., *et al.*, *JINST* **3**, S08005 (2008).
[16] L. Massacrier, *et al.*, *Adv. High Energy Phys.* **2015**, 986348 (2015).
[17] A. Andronic, *et al.*, *Eur. Phys. J.* **C76**, 107 (2016).
[18] J.-P. Lansberg, H.-S. Shao, *Nucl. Phys.* **B900**, 273 (2015).
[19] K. J. Eskola, H. Paukkunen, C. A. Salgado, *JHEP* **04**, 065 (2009).
[20] K. Kovarik, *et al.*, *Phys. Rev.* **D93**, 085037 (2016).
[21] R. Vogt, *Adv. High Energy Phys.* **2015**, 492302 (2015).
[22] F. Arleo, S. Peigne, *Adv. High Energy Phys.* **2015**, 961951 (2015).
[23] M. Anselmino, U. D'Alesio, S. Melis, *Adv. High Energy Phys.* **2015**, 475040 (2015).
[24] K. Kanazawa, Y. Koike, A. Metz, D. Pitonyak, *Adv. High Energy Phys.* **2015**, 257934 (2015).
[25] T. Liu, B.-Q. Ma, *Eur. Phys. J.* **C72**, 2037 (2012).
[26] L. Massacrier, *et al.*, *Int. J. Mod. Phys. Conf. Ser.* **40**, 1660107 (2016).
[27] J. P. Lansberg, *et al.*, *EPJ Web Conf.* **85**, 02038 (2015).
[28] G. Denicol, A. Monnai, B. Schenke, *Phys. Rev. Lett.* **116**, 212301 (2016).
[29] D. Kikola, *Adv. High Energy Phys.* **2015**, 783134 (2015).